# Cuban Confederate Colonel

# Cuban Confederate Colonel

*The Life of Ambrosio José Gonzales*

## ANTONIO RAFAEL DE LA COVA

The University of South Carolina Press

© 2003 University of South Carolina

Cloth edition published by the University of South Carolina Press, 2003
Paperback edition published in Columbia, South Carolina,
by the University of South Carolina Press, 2008

www.sc.edu/uscpress

Manufactured in the United States of America

17 16 15 14 13 12 11 10 09 08   10 9 8 7 6 5 4 3 2 1

The Library of Congress has cataloged the cloth edition as follows:

De la Cova, Antonio Rafael, 1950–
   Cuban Confederate colonel : the life of Ambrosio José Gonzales / Antonio Rafael de la Cova.
   p. cm.
   Includes bibliographical references and index.
   ISBN 1-57003-496-6 (cloth : alk. paper)
   1. Gonzales, Ambrosio José. 2. Confederate States of America. Army—Officers—Biography. 3. Soldiers—Confederate States of America—Biography. 4. Cubans—Confederate States of America—Biography. 5. United States—History—Civil War, 1861–1865—Participation, Cuban. 6. United States—History—Civil War, 1861–1865—Campaigns. 7. Plantation owners—South Carolina—Biography. 8. Revolutionaries—Cuba—Biography. 9. Cuba—History—Insurrection, 1849–1851—Biography. I. Title.
   E467.1.G59D4 2003
   973.7'82—dc21
   [B]                                                                     2003048359

ISBN: 978-1-57003-844-0 (pbk)

*For Dr. Ambrose Gonzales Hampton (1926–2001)*

# Contents

List of Illustrations / ix
List of Maps / xi
Preface / xiii
Introduction / xix
Abbreviations / xxvii

- **One**     Cuban Conspirator / 1
- **Two**     Freedom Fighter / 35
- **Three**     Filibuster Fiascos / 71
- **Four**     Patronage Pursuer / 101
- **Five**     Davis Dispute / 141
- **Six**     Confederate Colonel / 172
- **Seven**     Honey Hill / 209
- **Eight**     Reconstruction Retailer / 251
- **Nine**     Family Feud / 286
- **Ten**     Paralytic Patriot / 324

Epilogue / 365

*Appendix A*   167 Participants of the 1850 López
              Expedition / 369

*Appendix B*   338 Participants of the 1851 López
              Expedition / 375

Notes / 381
Bibliography / 465
Index / 493

# Illustrations

*following page 250*

Ambrosio José Gonzales's Havana University diploma, May 2, 1839
Cuban Independence general Ambrosio José Gonzales
Narciso López
John Anthony Quitman
Filibuster monument, Louisville, Ky.
Harriett Rutledge Elliott Gonzales
The oak allée of Oak Lawn Plantation, Colleton County, S.C.
The Anchorage, Beaufort, S.C.
A Maynard rifle, with its accoutrements and case
Confederate colonel Ambrosio José Gonzales
Confederate general Pierre Gustave Toutant Beauregard
Ambrosio José Gonzales's oath of allegiance, July 28, 1865
Charleston race riot, June 24, 1866
Advertisement for George Page & Company sawmills and steam engines
Baltimore residence of Ambrosio José Gonzales in 1872
Photograph of elderly Ambrosio José Gonzales surrounded by spirit faces
Narciso Gener Gonzales wearing his Cuban Liberation Army uniform
William Elliott Gonzales
Ambrose E. Gonzales
Gertrude Gonzales
Ana Rosa Gonzales
Home for Incurables (ward), Bronx, N.Y.
Monument of Ambrosio José Gonzales, Woodlawn Cemetery, Bronx, N.Y.

# Maps

Cuba and the Caribbean Sea / 40
City of Cárdenas, Cuba / 46
The Georgia Coast, 1850 / 73
The South Carolina Coast, 1861 / 157
Charleston Defenses, 1861–65 / 170
Plan of the Battle of Honey Hill, November 30, 1864 / 230
Social Hall Plantation, 1872 / 262

# Preface

While reading a book on the American Civil War in the summer of 1991, I learned that a Cuban compatriot, Ambrosio José Gonzales, had been a Confederate colonel and chief of artillery of the Department of South Carolina, Georgia, and Florida. It made me remember a popular Cuban saying: "Cubans are like white rice: they mix with everything." Cuban history books shed only a few brief references about Gonzales, including that he had been a Cuban revolutionary general and the first Cuban to shed his blood for the independence of his homeland. This further aroused my curiosity, so I consulted my friend in Miami, Florida, the prolific ninety-year-old Cuban historian Herminio Portell Vilá, who directed me to his three-volume biography of Gen. Narciso López. Portell Vilá mentioned that the sons of Gonzales had founded the *State* newspaper in Columbia, South Carolina. Unfortunately, Portell Vilá passed away in January 1992, and I never had the chance to discuss with him the discrepancies I found in his work.

When I telephoned the *State*, I was referred to the charming Carol Moore, secretary to Exec. Vice Pres. Page Morris. After explaining my research project, she mailed me a copy of S. L. Latimer Jr.'s *The Story of The State, 1891–1969, and the Gonzales Brothers* (1970), which contained a biographical chapter on Gen. Ambrosio José Gonzales. Moore also mentioned that her boss was related to Gonzales and had named his son Ambrose Gonzales Morris, and that descendant Mary Hampton had inherited the family's antebellum Oak Lawn plantation. Mary Hampton, the widow of former Columbia mayor Kirkman Finlay, gave me permission to visit Oak Lawn and stated that Ann Fripp Hampton was the family genealogist and that the general's great-great-grandson, Dr. Ambrose Gonzales Hampton, lived in town. Ann Hampton sent me a good amount of biographical data on Gonzales, his sons, and their father-in-law, William

Elliott. During a lengthy telephone conversation with the amiable Dr. Hampton, I learned that he possessed an oil painting of Gonzales in Confederate uniform and that he had been the physician of his great-aunt, Harriett R. E. Gonzales, the general's youngest daughter, who resided at Oak Lawn until her death in 1957. He also mentioned that a biography of Narciso Gener Gonzales, the general's son, titled *Stormy Petrel*, had been written by Lewis Pinckney Jones.

In fall 1991, after reading *Stormy Petrel*, I telephoned Professor Jones to inquire if he knew the location of Social Hall plantation, mentioned in his book as the Gonzales residence during Reconstruction. None of the Gonzales descendants had heard of it. Jones sent me a road map of Colleton County, South Carolina, indicating that Social Hall had been somewhere south of Green Pond, between the Combahee and the Ashepoo Rivers. A number of plantations appeared on the map in that area but the closest one to Social Hall Creek was Airy Hall plantation. H. L. "Buck" Limehouse, owner of Airy Hall, informed me on the telephone that Ambrose Gonzales was on the title chain of his property, and after a long conversation about the Civil War, he invited me to stay a weekend at the plantation. A call to the South Caroliniana Library at the University of South Carolina in Columbia revealed that their manuscript collection contained some Gonzales letters.

I decided to spend a three-week Christmas vacation in South Carolina doing research on Gonzales. After meeting Dr. Hampton, I went to the South Caroliniana Library, where a staff member indicated that five boxes of uncataloged Gonzales family documents had been left on deposit there twenty-five years earlier by descendant Cecilia McMaster. I made copies of numerous letters, land deeds, the general's Cuban passport, and a map he had drawn of the boundaries of Social Hall plantation. Dr. Hampton was thrilled with this discovery and lamented how the general had been neglected and misrepresented in historical accounts. He encouraged my idea of writing a biography and offered accommodations in his lakeside mansion and in his home on Tradd Street in Charleston, where I spent months doing research. My research trips expanded to the National Archives in Washington, D.C., to New York, and to New Orleans. In Charleston, Dr. Hampton introduced me to Peter Manigault, chairman of the board of the *News and Courier*, whose great-granduncle, Confederate major Edward Manigault, had served under Colonel Gonzales during the defense of Charleston. Although the Edward Manigault diary has been published, I am grateful to Peter for having let me work with the original.

I then headed for the South Carolina lowcountry determined to find Social Hall plantation. Buck Limehouse and his son Chip studied the 1872 sketch drawn by Gonzales and then laid out on a table a large Colleton County map that hung on their wall. My nineteenth-century map perfectly matched the modern property boundaries, showing that Social Hall plantation was the neighboring Ashepoo plantation, owned by the late Gaylord Donnelley. An examination of the Airy Hall plantation title chain revealed that the general's son, Ambrose Gonzales, had bought the property in 1911 and sold it five years later. Chip Limehouse provided a tour of Social Hall, where we discovered in the high pine area, thirty-eight feet above sea level, the brick foundation remnants of the Gonzales cabins.

Gregory Lane, a Civil War buff and metal detector hobbyist from Yemassee, provided a tour of the Honey Hill battlefield. Gonzales had commanded the artillery during that battle, defeating a Union force four times superior in number. Lane showed me the Union landing site, the trenches where the first skirmish occurred, and the Honey Hill earthwork fortification in a desolate wooded area. In nearby Beaufort, South Carolina, I met Prof. Stephen R. Wise, a Civil War author and director of the Parris Island Museum, who treated me to lunch at the Anchorage House, the former home of William Elliott. Wise introduced me to Willis "Skipper" Keith, an authority on the history of James Island, South Carolina. Skipper drove me around what remains of the James Island Civil War trenches and earthworks that Gonzales had been instrumental in fortifying with artillery. He served as a guide through the former Fort Johnson, Fort Pemberton, Fort Lamar, and the headquarters that Gonzales occupied at McLeod's House and Royall's House.

Because Gonzales was such a prolific writer, I found many of his letters in depositories throughout the United States. My investigations have left me indebted in gratitude to a number of people, especially the staffs of the following institutions: the National Archives in Washington, D.C.; Fort Sumter National Monument; the South Carolina Historical Society in Charleston; the Caroliniana Library and the South Carolina Department of Archives and History in Columbia; the Special Collections

Department of the Duke University Library; and the Southern Historical Collection at the University of North Carolina, Chapel Hill. Individuals who offered special assistance include Arthur Bergeron, Civil War Historian and Director of Pamplin Park, Virginia; Lynda L. Crist, Editor of the Jefferson Davis Papers at Rice University; Nathaniel C. Hughes, author of five Civil War books; Prof. Robert E. May of Purdue University, the dean of filibuster era historians; Prof. John Super of West Virginia University; Prof. John McKivigan of Indiana University–Purdue University Indianapolis; Tim Rues, Director of Constitution Hall, Lecompton, Kansas; Past Grand Master J. Roy Crowther, Grand Historian of the Grand Lodge of Free and Accepted Masons of Florida; Paul Davies, a Falls Church, Virginia, antique gun collector, owner of a Le Mat revolver and various Maynard rifles, of which Gonzales was a sales agent; Louise Bailey, a local historian in Flat Rock, North Carolina, where Gonzales spent the summers at the home of his in-laws; Librarian John H. Christian, Bryan-Lang Historical Library in Woodbine, Georgia; Allen Drury, of Folkston, Georgia; Ralph and Maria Galliano, of Tyson's Corner, Virginia; Matthew D. Fike, of Auburn, Indiana; and Civil War reenactors Gary and Sherry Belk of Louisville, Kentucky. I would also like to thank Dr. Hampton's widow, Joanne, and Danish graduate student Ulf Christensen, who provided copies of documents from the Cuban National Archive. Arthur Bergeron, Skipper Keith, and Stephen Wise read the Civil War chapters and provided valuable suggestions. Special mention goes to my wife, Carlina, who throughout this work provided love and support.

After I amassed a large amount of Gonzales material, the last piece to fit into the puzzle was the location of his grave. Newspaper accounts indicated that he was buried in Woodlawn Cemetery in the Bronx. The name was familiar: my great-aunt Margot González-Abreu had been interred there a decade earlier. The cemetery office pointed to the location of the Gonzales plot on a hillside parallel to 233rd Street. On my way there, I wondered if his tombstone contained information I did not have. To my surprise, Gonzales lay in an unmarked grave. To mark the spot, I planted chrysanthemums and displayed three small flags: Cuban, American, and Confederate. Standing there, I thought of the many Cuban patriots who fought for Cuban liberty and freedom, some of whom are also in unmarked graves. When I informed Dr. Hampton of my finding, he ordered a monument and requested that I write the epitaph. I chose two remembrances. The first was pronounced over a century earlier by Cuban independence apostle José Martí: "the first Cuban wounded in combat for the liberty of Cuba"; the second, by Jefferson Davis, praised Gonzales as "a soldier under two flags but one cause; that of community independence."

# Introduction

As the American Civil War entered its second year, the *Charleston Courier* announced: "Lieutenant-Colonel A. J. Gonzales.—No citizen, native or adopted, has labored more zealously, efficiently, and disinterestedly for South Carolina since the opening of the war, than General A. J. Gonzales, as he is known to his friends." Thirty years later, Cuban independence leader José Martí wrote in his New York émigré revolutionary newspaper *Patria*: "General Ambrosio José Gonzales was the first Cuban wounded in combat for the liberty of Cuba." Baptized a Roman Catholic in 1818, Gonzales after migrating to the United States converted to the Episcopalian Church, making him the earliest documented Cuban to embrace the Protestant faith. Moreover, Gonzales was one of the most important Cuban filibuster leaders of the 1850s and a significant Confederate artillery officer. In spite of these acclamations, no one has yet done a complete biographical interpretation of the life and times of Gonzales. He has often been omitted from accounts of the attack on Fort Sumter, the siege of Charleston, and the Battle of Honey Hill even though he played a prominent role in those events. This historical slight is consistent with Gonzales's life. His career was blighted by bad luck and tragedy, and his finances were never stable or predictable. His endless optimism led his friends to compare him to Micawber, a character in Charles Dickens's *David Copperfield*.

The son of a Cuban planter and educator, Gonzales from an early age attended school in New York City, where he was greatly influenced by Democratic politics and other American institutions. A voracious reader, he became multilingual and multicultural before returning to Cuba. Gonzales was a talented musician and singer, an expert marksman, and a superb billiard player. Although he earned a law degree from the University of Havana, the corrupt Spanish colonial judiciary system, which discriminated against native-born Creoles, forced him to seek employment as a college teacher. His disgust at what he perceived as abuse and injustice inspired him to fight for lost causes. At a time when Freemasons were persecuted by the Spanish Catholic Crown, he joined the fraternity. Freemasons would figure prominently throughout his life and provide aid and support. In 1848, his patriotic anxieties and Masonic connections brought Gonzales into a conspiracy, patterned on the Texas model, to annex Cuba to the United States. Despite his lack of military experience, Gonzales became adjutant general to revolutionary Gen. Narciso López and served as his principal translator. The following year they established headquarters in Washington, D.C., and sought assistance from American politicians, Mexican War veterans, and Freemasons for two filibuster expeditions that invaded Cuba and for two other failed attempts. Gonzales differentiated between these activities and the later incursions of William Walker into Mexico and Nicaragua, calling them a "foraging" band.

The United States' isolationist policy toward Cuba, starting in the early nineteenth century, did not actively pursue the island's annexation, despite massive American westward expansion. Repeated annexation requests by Cuban revolutionaries and their allies were rejected by all administrations, whether Whig or Democrat. The Whigs shunned Cuban expansionism because it would infuse a large slave population into the United States. The Democrats, on the other hand, generally favored annexation, but officially pursued it only through purchase proposals. The Polk administration, which made the first offer, failed to acquire the island because they did not take into account Spanish pride and recalcitrance in selling one of its last colonies and its most economically productive. President Pierce, in spite of favoring expansionism in his inaugural address, applied the same nonintervention policy toward Cuba as his Whig predecessors.

This work identifies for the first time the origin of the term *filibuster* in relation to military invasions. The initial enterprise, the frustrated Round Island expedition of 1849, includes a previously unknown link between Col. Robert E. Lee and three attempted filibuster departures from New York. The second expedition of some six hundred American volunteers landed in the city of Cárdenas on May 19, 1850. Gonzales, described as the "soul" of the movement, was wounded during the initial skirmish, an event that gave him immortality as the first Cuban to shed his blood for his country's independence. After taking Cárdenas, the filibusters were quickly routed back to the United States, where Gonzales and other conspirators were arrested by the federal government for violation of the 1818 Neutrality Act. Three trials in New Orleans ended in hung juries after Gonzales matched his wits on the witness stand against special prosecutor Judah Benjamin, later renowned as the "brains of the Confederacy."[1]

Before the trials even concluded, Gonzales became a leading organizer in Georgia for the frustrated April 1851 *Cleopatra* filibuster expedition. This endeavor had collected at least two thousand men, firearms for each individual, ten artillery pieces, and three steamers that were to sail from New York, Georgia, Florida, and Louisiana to invade Cuba. The federal authorities disbanded the enterprise with the help of a paid informant of the Spanish consul in New York. Gonzales raised thousands of dollars and hundreds of volunteers, and he got Georgia governor George W. Towns to donate state militia artillery, firearms, and accoutrements. Historians have generally overlooked his activities and focused instead on John O'Sullivan and the New York participants of the *Cleopatra* affair who were detained rather than Gonzales, who went underground for six months to evade an arrest warrant issued by Pres. Millard Fillmore. Gonzales eventually surrendered in Savannah, but the charges of violating the Neutrality Act were later dismissed. The following year, he wrote a widely disseminated political pamphlet promoting Cuban annexation to the United States. These intrigues brought Gonzales into close contact with influential Southern politicians, including Jefferson Davis, John A. Quitman, Mirabeau Buonaparte Lamar, Thomas J. Rusk, Pierre Soulé, John Slidell, Louis T. Wigfall, Stephen Mallory, Robert Toombs, John Henderson, and Carolinians John C. Calhoun, Francis W. Pickens, James Henry Hammond, and James Chesnut Jr. These men and other solons would also provide Gonzales with recommendations during his lifelong quest for political patronage.

Gonzales, a Union Democrat, never owned a slave, yet he regarded the institution as essential for Cuba's sugar monoculture. He denounced the slave trade as "horrific," believing in gradual emancipation through government remuneration. Gonzales feared that immediate abolition in Cuba would produce racial massacres like in Haiti. The Cuban annexation movement played an important role in the causation of the American Civil War. Since Cuba was an agricultural slave society, its annexation was spurned by the North and coveted by the South. Northerners considered the filibusters part of a slave power conspiracy to expand their dominions into the Caribbean basin, and Southerners responded to this opposition in Congress with increasing demands for secession.

Gonzales has been described as "perhaps the earliest volunteer" to the Confederate cause and "among the last in service—three weeks after Appomattox."[2] In 1856 he married Harriett Rutledge Elliott, daughter of William Elliott, a planter, statesman, and writer from Beaufort, South Carolina. Gonzales's sense of duty and obligation to his in-laws and his adopted state provided an overt motivation for his Confederate service. He also longed to achieve on the battlefield the rank of general, since his title of Cuban revolutionary general derived from an adjutant general commission. Gonzales served as chief of artillery of the Department of South Carolina, Georgia, and Florida (1862–65) under Generals John Pemberton, Pierre Gustave Toutant Beauregard, Samuel Jones, William Hardee, and Joseph E. Johnston. This work describes his bureaucratic activities, which included a concurrent period as department chief of ordnance. Gonzales's repeated attempts to attain the rank of general were always endorsed by his superior commanding officers, but Confederate president Jefferson Davis misinterpreted a precipitous 1861 letter requesting promotion. No matter how much he later tried to appease Jefferson Davis, Gonzales could not sway the susceptible Davis, who held in contempt those who disagreed with him. I have substantiated this flaw in the Confederate president's character by citing the historiography dealing with this defect. Gonzales is probably the only Confederate officer whose recommendation for promotion to brigadier general was denied six times.

Early in the war, Gonzales invented a "siege train" flying artillery to quickly mobilize heavy guns to enemy disembarkment positions. He zealously contributed to the defense of Charleston against repeated Union encroachments. As chief of artillery, he transferred his headquarters during six months to the forward positions on James Island and devised planting land mines in the Battery Wagner gorge, which helped impede a now-famous Union attack there, spearheaded by the African American 54th Massachusetts Regiment. After the initial charge was repulsed with hand-to-hand combat, Gonzales recommended that sixty double-barreled shotguns be sent to Battery Wagner for its defense. These weapons would have a lethal effect at close combat range.

Historians have repeatedly failed to give proper credit to Gonzales for strengthening Battery Wagner and his other military contributions. His exclusion from historical narratives is puzzling. For example, during the battle of Honey Hill, Gonzales commanded the artillery that enabled twelve hundred Confederates to defeat five thousand Union soldiers. It was his finest moment. Yet he is omitted from recent accounts of that engagement and the biography of Gen. Gustavus Woodson Smith, even though Gonzales appears in the Confederate battle reports cited by the author. A 1994 South Carolina Civil War iconography has also overlooked Gonzales.

After the war, tragedy, bad luck, and sorrow shadowed Gonzales for the rest of his life. During Reconstruction, he was a Charleston merchant for six months, until his business was destroyed by a race riot. He then became a planter and sawmill operator in the lowcountry until a general economic crisis and thievery drove him into bankruptcy. The Cuban was innovative as the first person to promote Chinese and European indentured farmers in South Carolina, a policy later embraced by state politicians. The tragic death of his wife shortly afterward prompted a tempestuous relationship with the Elliotts that lasted a lifetime. His sister-in-law Emily, frustrated in her hopes of marrying the bereaved widower, attacked his weakest point, scornfully dedicating the rest of her life to turning his six children against him. Gonzales never married again. A religious man who sang in church choirs most of his life, he turned to spiritualism in his late years. He worked as a teacher and translator until his death in 1893 due to progressive spinal myelitis.

To complete this biography, I have analyzed filibuster movement aspects previously overlooked or slightly investigated by historians. Filibuster conspirators left few records of their activities or movements because they were engaged in an enterprise that violated the United States Neutrality Act and was punishable with death in Cuba. This work cites newspaper hotel registry and ship passenger lists to provide accurate data on individuals and to reveal the role of some activists, such as dentist George A. Gardiner, who were overlooked by previous historians. I also analyze the social background of the filibusters and have compiled the most complete list available of 167 of the six hundred participants in the Cárdenas expedition and 338 members of the final invasion.

The role of Freemasonry in the filibuster movement, previously indicated by scant reference, is explored here in detail. The fraternity produced the initial introduction of participants in the movement, stimulated the entry of many in the rank and file, and provided protection from federal authorities trying to suppress their activities. Proceedings from Grand Lodges with membership rosters are used to link the fraternity to these activities. Freemasonry bonded Northern and Southern politicians and generals on behalf of Cuban annexation. Northern supporters included Caleb Cushing, Stephen A. Douglas, Daniel S. Dickinson, Mansfield Lovell, and William J. Worth. Southern Masonic backers included John A. Quitman, Mirabeau Lamar, Thomas J. Rusk, Pierre Soulé, Robert Toombs, Gustavus W. Smith, and John Henderson. The filibuster movement faced its stiffest opposition from avowed anti-Masons, including Presidents Zachary Taylor and Millard Fillmore, Secretary of State Daniel Webster, and *New York Tribune* editor Horace Greeley.

Above all, the following work presents, for the first time, a thorough account of the life of Ambrosio José Gonzales. Little has been available about Gonzales, besides brief biographical sketches. The first two sketches appeared in 1850 with the publication of Richardson Hardy's *The History and Adventures of the Cuban Expedition* and O.D.D.O.'s *History of the Late Expedition to Cuba*. In December 1892, when Gonzales had only seven months left to live, José Martí's *Patria* printed a laudatory, two-page biography by Gonzalo de Quesada, erroneously portraying Gonzales as an "Abolitionist, like Lincoln." When the general passed away, the *State* newspaper in Columbia, South Carolina, edited by his son Narciso Gener Gonzales, published a lengthy obituary. Yet only one paragraph was dedicated to Gonzales's Confederate service, and the rest of the article dealt mostly with his struggle for Cuban independence, which was finally achieved a decade later. In 1911, former Confederate cavalryman Ulysses R. Brooks devoted to Gonzales a sixteen-page chapter in his book *Stories of the Confederacy* that included eleven articles reprinted from Charleston newspapers mentioning the Cuban's Confederate role.[3]

Cuban historian Emeterio S. Santovenia wrote a four-page biography of Gonzales in 1928 in *Huellas de Gloria*, taken mostly from the *Patria* article. In 1944, Herminio Portell Vilá dedicated a fourteen-page chapter to Gonzales in his book *Vidas de la Unidad Americana*. It included some errors such as the general's birth date and his mother's maiden name (which Hispanics use as their second last name) and claimed that during the Civil War "Gonzalez [sic] fought from Virginia to Florida and also in Tennessee and Mississippi," when, in fact, Gonzales hardly left South Carolina and did not visit the two latter states during the conflict.[4]

The next Gonzales biography to appear was a ten-page article in the *South Carolina Historical Magazine* in 1955 by Prof. Lewis Pinckney Jones. He wrote that his article was a spinoff from his 1952 doctoral dissertation, *Carolinians and Cubans: The Elliotts and Gonzales, Their Work and Their Writings*, which was later published as *Stormy Petrel: N. G. Gonzales and His State*. Jones largely omitted Gonzales's Confederate career. The *State* printed a two-column biography of General Gonzales in 1966 titled "A Fighter for Freedom." The following year, a brief rehash of Cuban historical accounts about Gonzales was published by Fermín Peraza Sarausa in his *Diccionario Biográfico Cubano*, a monumental task that he began in Cuba in 1951. He wrote this last volume while exiled in Florida. In 1970, S. L. Latimer Jr. published *The Story of The State, 1891–1969, and the Gonzales Brothers*, reprinting the 1893 obituary dedicated to General Gonzales. His last biographical note appeared in 1995 in Bruce S. Allardice's *More Generals in Gray*.[5]

While Gonzales was a voluminous writer, as evidenced by his Confederate correspondence and his many letters scattered in over twenty manuscript collections throughout the United States and Cuba, he did not leave an autobiography. His only works in print are a sixteen-page pamphlet, *Manifesto on Cuban Affairs Addressed to the People of the United States* (1853); a series of ten articles about Cuba that appeared in the *Detroit Free Press* five years later; two lengthy 1884 articles in the *New Orleans Times Democrat* describing his participation in the Cuban independence movement from 1848 to 1851; and the sixty-eight-page booklet, *Heaven Revealed: A Series of Authentic Spirit-Messages, from a Wife to her Husband, Proving the Sublime Nature of True Spiritualism* (1889), published twenty years after the tragic death of his wife.[6]

I have relied on a wide variety of primary and secondary sources to bring Ambrosio José Gonzales into the historical record. They show the influence Gonzales had on historical events during the Cuban filibuster movement, the American Civil War, and Reconstruction in South Carolina, as well as the vicissitudes of a Cuban exile in the United States during the last half of the nineteenth century.

# Abbreviations

ACB     Angel Calderón de la Barca
AEA     Ministerio de Asuntos Exteriores, Archivo General, Madrid
AEG     Ambrose Elliott Gonzales
AGP     Ambrosio Gonzales Papers, South Caroliniana Library, University of South Carolina
AHN     Archivo Histórico Nacional, Madrid
AIG     Letters Received by the Confederate Adjutant and Inspector General, U.S. National Archives
AJG     Ambrosio José Gonzales
AN     Archivo Nacional, Havana
ANE     Annie Elliott
BLH     Baker Library, Harvard University
CCP     Caleb Cushing Papers, Library of Congress, Washington, D.C.
CMR     Compiled Service Records of Confederate General and Staff Officers, and Nonregimental Enlisted Men, Reel 108, Ambrosio Jose Gonzales
CSW     Letters Received by the Confederate Secretary of War, 1861–65, RG 109, U.S. National Archives
DCH     Dispatches from United States Consuls in Havana, U.S. National Archives
DHU     Dearborn Collection, Confederate Officers, Houghton Library, Harvard University
DUL     Duke University Library
EGP     Elliott-Gonzales Family Papers, University of North Carolina
EEL     Emily Elliott
GFP     Gonzales Family Papers, South Caroliniana Library, University of South Carolina

| | |
|---|---|
| HRG | Harriett Rutledge Elliott Gonzales |
| JD | Jefferson Davis |
| JDD | Jefferson Davis Papers, Duke University Library |
| JDP | Jefferson Davis Papers, Tulane University Library |
| JMC | John Middleton Clayton Papers, Library of Congress |
| JPP | James K. Polk Papers, Library of Congress |
| JQP | John A. Quitman Papers, Department of Archives and History, State of Mississippi |
| LC | Library of Congress |
| MAC | Maynard Arms Company Collection, Museum of American History, Smithsonian Institution |
| MAE | Mrs. Ann Elliott |
| MBJ | Mary Barnwell Johnstone |
| MLS | Miscellaneous Letters of the Department of State, National Archives |
| NA | United States National Archives |
| NGG | Narciso Gener Gonzales |
| OR | Official Records of the Union and Confederate Armies |
| ORN | Official Records of the Union and Confederate Navies |
| PBP | The Papers of P. G. T. Beauregard, Library of Congress |
| PGTB | Pierre Gustave Toutant Beauregard |
| REE | Ralph Emms Elliott |
| RG | Record Group |
| SCA | South Carolina Department of Archives and History |
| SCL | South Caroliniana Library, University of South Carolina |
| SHC | Southern Historical Collection, University of North Carolina |
| TRE | Thomas Rhett Smith Elliott Papers, Duke University Library |
| TSE | Thomas Rhett Smith Elliott |
| WE | William Elliott |
| WEJ | William Elliott Jr. |

## Cuban Confederate Colonel

# One

## CUBAN CONSPIRATOR

The Latin baptismal prayers resounded in the San Carlos Cathedral in Matanzas, Cuba, on Thursday afternoon, October 15, 1818. A Roman Catholic priest ritually dribbled water from a decanter onto a baby's forehead. The infant, born twelve days earlier, was christened Ambrosio José Cándido González Rufín. The extended name is a Hispanic tradition that adds the mother's maiden name to the surname. The child was cradled in his godmother's arms, while his maternal grandfather served as the godfather. The ceremony concluded with the peal of church bells, summoning outside a swarm of street urchins. They followed the celebrants, exiting the cathedral in pairs, to the child's home. The parents, who traditionally did not attend the baptism, anxiously waited at their residence. Upon receiving the arriving baptismal party, they began the traditional celebration with all the pomp and grandeur of contemporary aristocratic social events.[1]

The proud father, thirty-year-old Ambrosio José González Perdomo, was from the village of Managua, in Havana province. His father, José Antonio González, a native of Santa Cruz de Tenerife, Canary Islands, had emigrated to Cuba, where he married Rosalia Perdomo, from Santa María del Rosario, Havana province. González Perdomo had wed Crescencia Josefa Gertrudis Rufín de Torres on January 1, 1818. She was the daughter of Bernardino Rufín, from Havana, and Josefa María del Rosario de Torres, of a prominent Matanzas family. The Rufín name was the

Spanish derivative of her Ruffini grandparents, who had migrated to Cuba in the mid-eighteenth century from Genoa, Italy. They were descendants of Paul Ruffini, the renowned mathematician, author, physician, professor, and president of the University of Modena and of the Italian Institute of Sciences.[2]

After the baptismal celebrants entered the González home, the godfather appeared on a balcony with a servant holding a kettle containing gold coins of various denominations, from the glittering ounce to the small escudo. There were also souvenir gold medallions with the baby's name and birthdate. The ragamuffins who had followed the procession scurried for advantageous positions among the growing public on the street. The godfather, or *compadre*, who shared the responsibility for the child's upbringing, manifested his wealth by flinging fistfuls of gold coins at the crowd. This custom was later prohibited by the governor as the gamins increasingly brawled over the alms. The festivity then began with a seven-course dinner and exquisite wine. It ended with an animated dance, accompanied by singing and the tunes of the family piano. After Ambrosio's birth, his parents had three daughters: Gertrudis Dominga, in 1819, Isabel María, in 1820, and Brígida de los Dolores, three years later.[3]

González Perdomo had been raised on a farm and lacked an education until adulthood. He received a college degree in Havana and went to Matanzas in 1810 to teach mathematics. Five years later, the Matanzas City Council established a primary public school for one hundred children in a City Hall assembly room. They selected González Perdomo in March 1816 to head the institution. The new principal was "a modest teacher, but a neophyte; in Matanzas he introduced the teaching of the political constitution and the Copernican astronomical geography." The curriculum included "Christian doctrine, Reading, Writing, Arithmetic, Grammar, Spanish Orthography, Urbanity, and Geography," taught by González Perdomo, aided by an assistant. School hours were from 8:00 A.M. to 11:00 A.M. and from 2:00 P.M. to 5:00 P.M.[4]

The "highly esteemed" schoolmaster, who was also an engineer and a sugar planter, resigned on January 24, 1817, because his "short salary" of $750 yearly was insufficient. González Perdomo was rehired the following year at $1,400 annually, and his friend Tomás Gener Buigas was appointed the school inspector. The thirty-year-old Gener had received a law degree from Columbia College in New York City. The work of the public school and its headmaster was evaluated in 1818 during a two-day official visit by two royal inspectors and Matanzas governor Juan Tirry Lacy. Eighteen outstanding students demonstrated their knowledge of "Christian principles, a very moving lecture on prose and verse, presented beautiful map drawings, grammatical analysis, definitions on the blackboard and the globe, and geographic map descriptions." González Perdomo was praised in the report to the Royal Intendant for the considerable learning and manners imparted to his disciples.[5]

In 1821, after Governor Tirry was superseded and Gener had gone to Madrid as elected parliamentarian to the Spanish Cortes, González Perdomo took his family to Havana, where he established a new institute. During the next two years, the Matanzas public school had two ineffectual principals. The city council then induced González Perdomo in May 1823 to return to his former post with a three-year contract and a $2,000 annual salary. The following year, young Ambrosio began his primary education in his father's institution. His classmates included Pedro José Guiteras Font, a lifetime friend, and José Jacinto Milanés, a future national poet. Ambrosio's childhood innocence was marred by the tragic death of his mother, after which his father sent him to a private school in the United States. The father and son left Havana in the brig *Burdett* with thirteen other passengers, including three children, arriving in New York City on April 20, 1828. Nine-year-old Ambrosio enrolled in the French Institute on Bank Street, a semimilitary academy managed by the brothers Louis and Hyacinth Peuquet, two former officers in Napoleon Bonaparte's army. The elder, Louis, a former artillery captain, had in his torso a bullet from the battle of Waterloo. Hyacinth, a former cavalry captain, was an excellent mathematician who thoroughly drilled his students in the study of fractions.[6]

Gonzales quickly made friends with Louisiana classmate Pierre Gustave Toutant Beauregard, the future Confederate general, who was similar in age and social background. Among their other Cuban aristocratic classmates, Beauregard would later recall those of the Pizarro, Yznaga, and León families of Havana and the children of the Gener and Ruiz families from Matanzas. The Peuquet brothers inflamed their pupils' minds with glorious epics of courage, valor, and honor. The youths were fascinated by stories about the cavalry charges that Hyacinth successfully led at Austerlitz. Gonzales was especially impressed with Louis's account of how his army had crossed the Great St. Bernard Pass between Switzerland and Italy, a previously unaccomplished feat. The French army had struggled to get its cannons through the ten-foot-deep Alpine snow until it fashioned large sleds from felled trees. This innovative feat in military history would later inspire Gonzales to develop new tactics as a Confederate artillery officer.[7]

Meanwhile, in Matanzas, González Perdomo continued teaching until his renewed contract expired in June 1829. Due to an economic crisis that year, the Cuban public schools suffered a "temporary decadence," so González Perdomo acquired a two-year contract as a Matanzas coastal inspector. With José Antonio Velasco he also established a daily newspaper, *Diario de Matanzas*, which they published from December 1, 1829, to February 20, 1831. At the same time, the new publisher was appointed the official printer of the Matanzas government. During a trip to New York City to visit his son, González Perdomo wed twenty-three-year-old Emilia Gauffreau Berault on May 3, 1830. The marriage was ratified in Matanzas the following month. This union produced three sons: Próspero, Ignacio, and Emilio. In January 1835, González Perdomo and his wife founded Our Lady of Guadalupe School, where she served as schoolmistress and both shared the teaching load.[8]

Ambrosio Junior had four years of schooling in New York City, becoming fluent in Spanish, English, French, and Italian before returning to Cuba with the highest class honors, "thoroughly Americanized," and imbued in the democratic ideology of the Jacksonian era. He enrolled in San Cristóbal de Carraguao High School in Havana, a preparatory institute whose literary director was the renowned educator and abolitionist José de la Luz Caballero. The teenager and his 180 classmates were taught under the newly introduced explanatory method with much cultural emphasis. Other boarding students from Matanzas included José Luis Alfonso García de Medina, Miguel Aldama Alfonso, Pedro José Guiteras Font, José Manuel Ximeno, and Teodoro Dalcour, whose friendship lasted a lifetime. After graduation, Gonzales entered the University of Havana, receiving a law degree on May 2, 1839, at the age of twenty. He was soon disgusted with the corruption of the Spanish colonial judiciary system, which discriminated against native-born Creoles by circumscribing their practice to mere notarial duties. Ambrosio became a professor of mathematics, geography, Latin, and modern languages in San

Cristóbal de Carraguao School. He was socially active among the Havana aristocracy. A Matanzas socialite later recalled his reputation as a "courteous ladies' man." His circle of friends included the American consul in Havana, thirty-nine-year-old Nicholas Philip Trist, former private secretary to president Andrew Jackson. Trist was married to the granddaughter of Thomas Jefferson and had studied law under the former president. Trist owned the Flor de Cuba sugar plantation and had invested in the Cuba Mining Company.[9]

The sudden death of Ambrosio's fifty-seven-year-old father on May 2, 1845, left him inconsolably distressed. The patriarch was interred in the Matanzas General Cemetery the following day. He had recently suffered "various adverse vicissitudes in his material interests," prompting his former students to purchase his gravestone, which they inscribed: "To the memory of the virtuous Ambrosio José González, from his various thankful disciples." Ambrosio mitigated his sorrow by spending the next two years traveling in Spain and the United States. Upon returning to Havana, he resumed teaching at San Cristóbal de Carraguao and maintained close relations with Matanzas intellectuals who notably figured with him in the movement to annex Cuba to the United States.[10]

The annexation of Texas in March 1845 and the nascent Manifest Destiny expansionist movement initiated a popular campaign for acquiring Cuba, with greater momentum in the South. Florida senator David Levy Yulee soon presented a Senate resolution for President James Knox Polk to negotiate with Spain to obtain the island but withdrew it before it was debated. That year, Vice Pres. George Dallas toasted the annexation of Cuba at a Fourth of July banquet and in a letter two years later favored appropriating the island. The issue was postponed in Congress due to the Mexican War and the Wilmot Proviso banning slavery in newly acquired territories.[11]

In the spring of 1848, the Havana Club was organized secretly to initiate Cuba's entry into the Union. It was headed by thirty-eight-year-old attorney, planter, poet, and novelist José Luis Alfonso García de Medina, vice president of the Havana Railway Company. Other members included his cousin and brother-in-law, twenty-eight-year-old magnate and planter Miguel Aldama Alfonso, who owned five sugar mills with fifteen hundred slaves; Carlos Núñez del Castillo, director of the Savings Bank of Havana; forty-three-year-old merchant Domingo de Goicouría;

thirty-three-year-old attorney, poet, and novelist José Antonio Echeverría; attorney Manuel Rodríguez Mena; thirty-six-year-old attorney, novelist, and educator Ramón de Palma Romay, and twenty-six-year-old educator and poet Rafael María de Mendive Daumy. These Creoles were friends of Ambrosio José Gonzales, who joined the Havana Club at the age of thirty. Others who joined with Gonzales were thirty-five-year-old attorney Cirilo Villaverde, author of the acclaimed antislavery novel *Cecilia Valdés*; nineteen-year-old journalist José Agustín Quintero; thirty-eight-year-old agronomist Francisco de Frías Jacott (the count of Pozos Dulces); forty-two-year-old attorney Anacleto Bermúdez; twenty-seven-year-old journalist Pedro Angel Castellón; and Venezuelan consul Manuel Muñoz Castro.[12]

The Havana Club met in the Aldama mansion using the code of secrecy practiced in clandestine Masonic lodges. Each member adopted a covert code name. Gonzales was known as *Germán*. The conspirators included aristocrats, intellectuals, Freemasons, abolitionists, and slave-owning sugar planters. They frequently traveled to the United States and Europe, and they abhorred colonial government abuses, corruption, economic stagnation, and high taxes. Gonzales, Quintero, Alfonso, Frías, and Goicouría had been educated in the United States. They desired to install American democratic institutions in their country and to make its three provinces individual Southern states. Some plotters feared that Spain, pressured by the abolitionist policies of England and France, would end slavery and ruin the island's growing sugar economy. The 1848 French revolution that emancipated the slaves in the French Caribbean colonies by decree on April 27 and the threat of a Republican regime in Spain that would abolish slavery in Cuba gave a sense of urgency to the annexationist plan. These insurgents favored annexation because it would guarantee their chattel property. Those opposed to slavery, like Villaverde and Goicouría, believed that joining the Union would increase white emigration and foment industrialization, which would then induce gradual emancipation. Gonzales, who was not a slave owner, regarded slavery as an economic necessity for Cuba. He opposed the "horrid traffic" of slaves, believing in gradual emancipation through government remuneration.[13]

A derivative of the Havana Club, the clandestine Cuban Council, was created in New York City by Gaspar Betancourt Cisneros,[14] Cristóbal F. Mádan Mádan,[15] Miguel Teurbe Tolón,[16] José Aniceto Yznaga Borrell, and

Pedro de Agüero Sánchez. They were in contact with revolutionary delegations in Havana, Matanzas, Trinidad, Puerto Príncipe, and Santiago de Cuba. They published *La Verdad,* a bilingual revolutionary newspaper begun in January 1848 using the *New York Sun*'s press. The newspaper ran articles from secret correspondents in Cuba and Puerto Rico. It was edited by forty-one-year-old Jane McManus Storm,[17] under the pseudonym Cora Montgomery, and advocated "the independence of Cuba, Canada, and all colonies." *La Verdad* proposed for an independent Cuba gradual slave emancipation "without ruin to all classes, and to the industry and productiveness of the island. They have a lesson of warning in the condition of Jamaica, Hayti and St. Domingo."[18]

In May 1848, the Havana Club unanimously agreed that since Cubans lacked military experience, an invasion to overthrow the colonial regime was only feasible by contracting an American general with five thousand Mexican War veterans. These volunteers would provide a quick victory to avoid a prolonged civil war that might obliterate their wealth and provoke a slave insurrection similar to Haiti's. The Havana Club sent gymnast Rafael de Castro to locate fifty-four-year-old Maj. Gen. William Jenkins Worth in occupied Mexico and offer him $3 million to lead the expedition. The Cubans were aware of the military exploits of this veteran of the War of 1812 and the Seminole War. During the Mexican War, Worth led the attack and seized the bishop's palace at Monterrey, assumed to be impregnable. He participated in the capture of Veracruz and the battles of Churubusco and Molino del Rey, and he headed the assault upon Mexico City through the gate of San Cosme. Worth had also advocated the cause of Manifest Destiny in a letter he sent to Secretary of War William Learned Marcy in October 1847, which prompted both the *New York Sun* and the *New York Herald* to call for his nomination as Democratic presidential candidate. The *New Orleans Picayune* reported three months later that Worth's missive favored the United States' annexation of Mexico, Central America, and, "in due course of time, the island of Cuba." Worth's division evacuated Mexico City on June 17, 1848, arrived in Jalapa five days later, and did not leave Mexico until July 15. His brother-in-law, Maj. John T. Sprague, wrote that Worth was first approached with the Cuba offer by a "committee of gentlemen" in Puebla while departing Mexico. According to Gonzales, after Worth spoke with the emissaries at Jalapa, he accepted their offer, contingent upon resigning his military commission and the disbanding of his volunteer troops in Mexico.[19]

Another separatist conspiracy in Cuba, led by Gen. Narciso López, had been brewing in the central Trinidad region for over a year. López was born in Caracas, Venezuela, on October 29, 1797, to a wealthy merchant family and joined the Spanish army at the start of the Latin American wars of independence. When the royal troops evacuated Venezuela, the twenty-one-year-old colonel followed them to Cuba, where he married Dolores de Frías Jacott, a member of a rich Creole planter family. According to his biographer, López maintained the bachelor vices common to contemporary Spanish army officers: he led a licentious life, mutually wooing aristocratic ladies and lower-class women, producing an unaccounted number of out-of-wedlock offspring, and gambling his pay on playing cards, cockfights, or any game of chance. His enemies would later say that he had squandered his wife's fortune at the gambling table, causing their separation, and that he was reduced to borrowing money from acquaintances to keep betting.[20]

In 1827, López was transferred to Spain, participating in the Carlist War and the Portugal campaign in 1834 as aide-de-camp to Gen. Gerónimo Valdés Sierra. Two years later, he was promoted to brigadier general and governor of Valencia, and ascended to field marshal in 1839. The following year, the Spanish Crown faced a revolt from enlightened army officers who wanted to install a republican form of government. López, who had been initiated into the Masonic fraternity in Madrid, sided with the liberals, many of whom were Freemasons. These officers dethroned the regency of María Cristina, described by historian Emilio Roig de Leuchsenring as "a frivolous and dissolute woman" with a "necessity to expand her sensual temperament." Army commander in chief general Baldomero Espartero became regent from 1840 to 1843. General Valdés was appointed captain general of Cuba in May 1841, and he offered López, recently named senator from Seville, the post of lieutenant governor of Matanzas. Two years later, López was named lieutenant governor of Trinidad and president of the Executive and Permanent Military Commission, which prosecuted political cases.[21]

When the Spanish liberal officers were deposed in September 1843, Capt. General Valdés was replaced by Leopoldo O'Donnell Jorris, whose brother, Carlos, had been defeated by López during the Carlist War.

O'Donnell quickly demoted López to garrison duty and began a reign of persecution against liberals and Freemasons. López, with economic assistance from his brother-in-law, Francisco de Frías Jacott, tried making a living in the sugar industry in Cienfuegos, coffee growing in Vuelta Abajo, and copper mining, without favorable results. Extremely bitter toward the Spanish Catholic monarchy, which had stripped him of his former wealth and power, López started plotting against the government in 1847, "to revolutionize Cuba and Porto Rico, and separate them from Spain." As worshipful master of the irregular Masonic Táyaba Lodge in Trinidad, he recruited some brethren, including forty-year-old Isidoro Armenteros Muñoz, owner of the San Luis and Laberinto sugar mills and a lieutenant colonel in the cavalry militia.[22]

The conspiracy, encompassing the nearby cities of Cienfuegos, San Fernando de Camarones, Manicaragua, and Trinidad, was supported by some Spanish officers and López's friends and former subordinates among the rank and file, along with planters and peasants. Another secret cell, established in Matanzas, met at the drugstore of Francisco Javier de la Cruz and included Blas Ximeno, Dr. Santiago de la Huerta, Dr. Antonio Escoto, attorney Félix Govín, and others. The leader of the movement in Cienfuegos was attorney and councilman José G. Díaz de Villegas. López also recruited thirty-four-year-old José María Sánchez Yznaga, whom he had first met four years earlier at a Trinidad cockfight ring. Sánchez Yznaga and his brother Saturnino owned the Santa Bárbara Sugar Mill and its slave force in Cienfuegos.[23]

Cuban historian Vidal Morales believed that López desired only to separate Cuba from Spain by whatever means available, without concern for the future. Herminio Portell Vilá, who based his seminal three-volume, sixteen-hundred-page biography of López principally on the unpublished Cirilo Villaverde diary, expounded a subsequent Villaverde revisionist thesis that López was a nationalist who manipulated his followers to favor his secret agenda of establishing a permanent Cuban Republic. Villaverde claimed in 1892: "I was, am, and never will be anything else than for independence, and could swear that Gaspar Betancourt Cisneros and Narciso López were also. My state of health does not allow me to argue on this matter." This contrasted with an anonymous López biography published by Villaverde in 1849, indicating that his plan had always been "Independence and Annexation to the American

Union." López was ideologically an authoritarian *caudillo* pragmatic opportunist, garnishing political support by telling each faction what they wanted to hear, very similar to contemporary Mexican general Antonio López de Santa Anna.[24]

In May, López informed American consul Robert Blair Campbell in Havana of his planned revolt. The native South Carolinian, a bilingual sexagenarian, had been a state senator and had served as a nullifier representative to Congress from 1834 to 1837. Campbell was commissioned general of South Carolina troops during the nullification movement and was appointed U.S. consul at Havana in 1842. Campbell's defiance of the captain general's capricious treatment of the 1,256 Americans living in Cuba was admired by the separatist Creoles. López gullibly assumed that the consul would assist or provide American recognition to his movement, but Campbell warned that another plot was being fomented by the Havana Club. This was verified to López by his friend José Antonio Echeverría, the Havana Club secretary, who advised him to postpone his uprising, scheduled for June 24, 1848, until Worth's invasion with American volunteers. López also learned of the republican uprising in Madrid that had been crushed on March 26. Upon returning to Cienfuegos, he told José Sánchez Yznaga that a republic had been proclaimed in France, that the same would be done in Spain, and that Cuba should follow the peninsular movement, under his leadership. If Spain followed the French example and abolished slavery, he argued, Cuba would need assistance from the United States. Because it "lacked the precise and necessary force to contain the blacks, it was more convenient to ask for annexation to the United States from where immediately would arrive the necessary forces for the indicated object." The revolt in France created an economic depression in Havana, paralyzing its commerce and forcing some of its most important merchant firms into bankruptcy. As a result, a growing number in the business community regarded annexation to the United States as their only salvation.[25]

While López awaited the Worth invasion, he informed sixty-four-year-old Pedro Gabriel Sánchez, the father of José Sánchez Yznaga, of his insurrectional plans. López told him that Spain would eventually lose the island, that the plotters included prominent people, and that the steamer *Guadalquivir* would transport five thousand Americans to assist in the annexation of the island. After the invasion, plans were delayed. On July

4, 1848, the elderly Sánchez denounced the conspiracy to the local authorities, motivated by "serving his country and freeing his family from a catastrophe, particularly his son José Sánchez Yznaga, whom General López had seduced." He said that an American war schooner had been scheduled to arrive in Cienfuegos with "one thousand muskets, gunpowder and bullets" on June 24. López had chosen that day because it was the city's feast day. The insurgents, dressed in carnival costumes for the occasion, would attack and take over the garrison, while the bulk of the troops patrolled the streets during the event. The general told Saturnino Sánchez Yznaga, José's brother, that once he triumphed, "an independent republican government would be installed, and that afterwards if it was convenient they would annex themselves to the United States of America or continue independent." José Sánchez Yznaga and José G. Díaz de Villegas were quickly detained and sent to El Príncipe prison in Havana. Alejo and Pedro José Yznaga Hernández and Francisco and Lucas de Castro were arrested with the conspirators' weapons and ammunition. The Spanish authorities seized in Trinidad letters implicating López in the insurrection. Two days later, the governor of Cienfuegos summoned López on an urgent matter. The general, aware of the arrest of his followers, departed for Matanzas. The next day, he boarded the brig *Neptune*, arriving in Bristol, Rhode Island, on July 23 "with his son and servant," the youthful mulatto Pedro Velazco, and proceeded to New York the next day.[26]

The day López fled, Consul Campbell wrote to Commodore M. C. Perry of the USS *Cumberland* in Havana Harbor that there was much revolutionary fervor on the island. Prudent annexationists were trying to hinder it, but a general that he knew—i.e., López—was very anxious to make an immediate *pronunciamiento*. Attempts were being made to contain López, and some of his friends, disenchanted with his impetuousness, were using all their energies to impede him from taking action. Campbell added that many people doubted the general's capacity and prudence.[27]

The Havana Club then commissioned Ambrosio José Gonzales to travel to the United States and offer $3 million to General Worth for an armed invasion of Cuba. On August 5, Gonzales left Havana on the steamship *Crescent City* for New Orleans, which had served as a transit point for the American troops returning home from Mexico. But it would take him several days to catch up to the general. Worth and his staff, which included Col. Henry Bohlen, had taken quarters in the luxurious St. Charles Hotel on July 20 before leaving for Washington the next morning. He was scheduled to appear on August 2 before a court of inquiry to respond to Gen. Winfield Scott's charge that he had written illicit private letters detailing active operations during the Mexican campaign.[28]

Gonzales arrived in New Orleans on August 10, having stowed away in a stateroom of the steamship *Crescent City* because he had problems obtaining a passport. The day before landing, Gonzales signed a card along with thirty-five other passengers, expressing their satisfaction with the ship's captain and the "extremely comfortable and agreeable voyage." Gonzales tried to be anonymous by signing "Ambrosio G. Rufín," but after disembarking, the newspaper list of passengers identified him as "A. Gonzalez." After arriving, our subject began signing his surname with its last letter as an "s." Two other Cuban separatists on the voyage, Antonio Yznaga del Valle and Juan Armenteros, were affiliated with the Cuban Council. Yznaga owned sugar plantations in Louisiana and Trinidad, Cuba. Gonzales had given his credentials to an American passenger in Havana and asked that they be deposited at the New Orleans post office under a fictitious name. Not finding the documents there, he later surmised that they had somehow been intercepted by Secretary of State James Buchanan.[29]

The *New Orleans Picayune* announced on August 10 that General Worth had reached Washington on July 29. Two days earlier, the same newspaper had heralded the "Arrival of an Insurrectionary Fugitive"—that is, General López—in New York. Gonzales, conscious of these events, immediately departed New Orleans on the same route taken by Worth, starting with an eighteen-hour steamer ride to Mobile, Alabama. He then had a forty-hour voyage on another steamer up the Alabama River to Montgomery. From there, the Havana Club emissary went by railroad to Opelika, Alabama, and rode ninety-three miles on the four-horse mail coach to the train depot at Griffin, Georgia. Gonzales continued on the Georgia Railroad, which advertised "state rooms and berths for night travel," via Atlanta and Augusta, arriving on August 18 at the Charleston Hotel, in Charleston, South Carolina. The passenger fare thus far was $39.50. He quickly departed on the night mail boat to Wilmington,

North Carolina, continuing for three days by railroad to Washington, D.C., via Petersburg and Richmond, Virginia. Meanwhile, the charges against Worth had been dismissed in the capital. He went to New York City on August 13 and proceeded to the Ocean House resort in Newport, Rhode Island, where the Cuban emissary finally caught up with him.[30]

Gonzales used ritualistic signs, code words, and a secret-grip handshake to identify himself as a brother Freemason to Worth before extending the Havana Club offer. He claimed that Worth gave him "perfect credence at the outset" and accepted his proposition. The Cuban returned with Worth to New York City, where the general registered on August 27 at the Astor House. Gonzales introduced him to brother Freemasons General López and Gaspar Betancourt, who reiterated the Havana Club financial offer and outlined the projected invasion plans. Afterward, Worth invited Gonzales on a trip to his native city of Hudson, New York.[31]

The next day, Gonzales left with Worth and his coterie on the steamer *Hendrick Hudson* up the Hudson River. They stopped at the United States Military Academy at West Point, where the Cuban was introduced by Worth to Commandant of Cadets Charles F. Smith, Prof. Gustavus Woodson Smith, a Freemason, and other army officers. Upon arriving at the general's birthplace, a reception committee escorted Worth and his friends to several carriages for the cavalcade to Hudson House. Gonzales attended the ceremony on the afternoon of the 29th at the courthouse, where Worth was presented with a sword from the citizenry. The general and his companions then returned to the Hudson House for a complimentary dinner. Gonzales sat at the head table with Worth, Col. Henry Bohlen, Capt. John T. Sprague, and other dignitaries.[32]

Worth was highly tempted by the Cuban financial offer because his home in Watervliet, New York, had been sold for tax arrears during his long absence. Gonzales suggested that Worth send Colonel Bohlen to Cuba to make a financial agreement with the Havana Club. Bohlen was a thirty-eight-year-old, German-born, wealthy Philadelphia liquor merchant and Dutch consul. López wrote from New York to Havana Club secretary José Antonio Echeverría on September 6 that Bohlen was en route to Cuba representing Worth, after having reached an accord. Three days later, Bohlen sailed from New York City to Havana on the steamship *Falcon*. He returned with a financial pledge from the Havana Club leaders, who provided the plans of major cities and fortifications requested by Worth. Betancourt Cisneros wrote to a friend on October 19 that "an envoy had returned from the island with good news that I cannot confide to the pen." According to Gonzales, another Worth confidant, Lt. Col. James J. Duncan, was also part of the plot.[33]

Gonzales later traveled to Washington, D.C., where General Worth introduced him to several government officials including Treasury secretary Robert J. Walker and navy secretary John Y. Mason, who was also a Freemason. The Cuban returned to New York after the U.S. War Department assigned Worth to command the Eighth and Ninth Military Departments in the West on November 7. Worth was ordered to report in person to Maj. Gen. Zachary Taylor in Louisiana. Taylor had just won the presidential election that day and agreed to resign his military commission in three months. Worth left Washington, D.C., on December 5, and after a public tribute in Pittsburgh, Pennsylvania, headed south on a Mississippi River steamer to assume his new position.[34]

A few days prior to Worth's departure, Gonzales wrote to Nicholas P. Trist, the former Havana consul now practicing law in West Chester, Pennsylvania, asking for letters of introduction to people in the capital, where he would soon reside. Trist, who that summer had negotiated the Treaty of Guadalupe Hidalgo ending the Mexican War, hastily wrote to South Carolina Democratic representative Isaac Edward Holmes recommending Gonzales as "a gentleman of high intelligence" in whom he had great confidence because of the Cuban's "honourable principles & generous sentiments." Holmes was a member of Masonic Union Kilwinning Lodge, No. 4, in Charleston, South Carolina. The former consul was apparently aware of the Cuban revolutionary plans, because he warned Holmes that Gonzales was involved in a business of a "perilous nature to himself" and warned of using "the utmost caution." Trist then requested that his name be mentioned by the Carolinian to people who would be conducive to helping Gonzales achieve his "object."[35]

The coming of a Whig administration to power in March gave Gonzales only three months in which to attempt to influence the outgoing Democrats to promote Cuban annexation, just as Texas had achieved statehood only two days before Polk's inauguration. The Whigs were "opposed to the extension of the national boundaries" while the Democrats favored the "extension of the area of freedom." Taylor, a Louisiana slave owner, had won the Whig nomination supported by Southern delegates after he renounced opposition to the expansion of slave territory. By the time he entered the White House, however, Taylor had been influenced by William H. Seward, whose antislavery territorial expansion goals pointed toward Canada and Alaska rather than the Caribbean Basin. Taylor also opposed Freemasonry and soon after assuming office declared that he was "definitely not a Mason." Taylor's vice president shared his views on Freemasonry: during the anti-Masonic period of the 1830s, Millard Fillmore had been "one of the most bitter critics of the fraternity which he characterized as 'organized treason.'" A number of prominent Whig politicians, including William Seward, Thaddeus Stevens, Charles Sumner, and Taylor's secretary of state Daniel Webster had either belonged to the Anti-Masonic Party in the 1830s or denounced the brotherhood.[36]

Gonzales established residency in the fashionable Irving Hotel in the capital, claiming that the revolutionary mission that brought him to the United States precluded his return home. In pursuit of Cuban annexation he made powerful political connections, including Massachusetts lawyer and legislator Caleb Cushing, who introduced Gonzales to President Polk in December. Freemasonry also was a common bond among Gonzales, Cushing, Polk, and Vice Pres. George Dallas. The president had commissioned Cushing a brigadier general during the Mexican War after he raised a volunteer regiment. A former Whig representative, Cushing favored states' rights and opposed abolition. He went to Washington, D.C., in December when Polk planned to name him commissioner in the Mexican boundary dispute, but soon returned to Massachusetts because of his father's terminal illness. Gonzales afterward began visiting Polk's niece at the White House.[37]

While in the capital, Gonzales received various letters of introduction, including one from *La Verdad* editor Jane McManus Storm, dated January 4, 1849, for New York's antiabolitionist Democratic senator Daniel S. Dickinson, who was a Freemason and favored the annexation of Cuba and Canada. She highly recommended Gonzales, describing him as "a republican patriot in theory as well as action" whose statements and opinions were those of a "true representative of true men." Storm asked Dickinson to introduce the Cuban to liberal Northern Democrats. During the 1844 presidential campaign, these Democrats had promoted the annexation of Texas and Oregon. They tried to reconcile the free-soil ideology of the North and the proslavery sentiment of the South by calling for the annexation of Cuba and Canada. The Taylor administration was chastised for adhering to the Whig policy opposing the extension of the national boundaries and the "area of freedom."[38]

Gonzales was joined in Washington on February 4, 1849, by López, José Sánchez Yznaga, and John Carlos Gardiner, who had arrived from New York via Baltimore. Gardiner, who also used the names Carlos Gardiner and Charles Gardiner, roomed with his brother, dentist George A. Gardiner, while López and Sánchez Yznaga checked into the elegant Willard's Hotel, on Pennsylvania Avenue near the White House. Gonzales, his future wife, and his in-laws would frequent the hotel during the next decade. The Willard, still fashionable today, reveled in its velvet-tufted carpets, "in the finely arranged apartments; in the fare and its admirable cooking; in the satin like cleanliness of the bed rooms; in the beverages; in the never tiring personal attentions of Willard, and his immediate associates." George A. Gardiner had appeared before the Spanish consul in Charleston, South Carolina, on December 30, 1848, and received a visa to travel to Havana two days later on the steamer *Isabel*. The consul, answering an inquiry from his superior, stated that he believed him to be the "Carlos S. Gardner [sic], suspected of complicity in the criminal attempts of former General Narciso López." The consul reported that no one he questioned in Charleston seemed to know Gardiner, including Pedro de León and Guillermo Tirry Loynaz, the Marqués de la Cañada de Tirry, who traveled from Washington with him. He also wrote that Gardiner had not been to Havana previously.[39]

Two days after arriving in Washington, D.C., the Cubans moved to the Irving Hotel. Gonzales, frequently traveling in and out of the capital, registered at the same hotel three weeks later. Sánchez Yznaga, who had jumped bail and fled Cuba after being charged in the López conspiracy of 1848, was sentenced in absentia the following month to eight years in exile, while López was expelled from the army and prohibited from ever returning to Spanish territory. López wrote on February 23 to conspirator Juan Manuel Macías Sardina, who was leaving New York for Cuba. The twenty-three-year-old Macías, a Matanzas native with a "fine personal appearance," had resided in the United States since July 1846. The general asked Macías to raise money for an April expedition from former mayor and attorney José Elías Hernández, twenty-nine-year-old poet Benigno Gener Junco, and others in Matanzas. Macías was then to rush the money to Alonso Betancourt in Philadelphia.⁴⁰

The first politician to call on General López was Sen. John C. Calhoun of South Carolina. In a meeting with López, Gonzales, and Sánchez Yznaga, the Carolinian supported Cuban annexation. He told López, while Gonzales translated, that "the South ought to flock down there in 'open boats,' the moment they hear the tocsin." Shortly thereafter, Gonzales was invited by a prominent Southern senator to the Senate recess room to confer with Calhoun and four other senators, including Mississippian Jefferson Davis and Texan Thomas Jefferson Rusk, interested in the Carolinian's views on Cuban annexation. Calhoun spoke favorably of the Cuban cause, the right of Americans to assist in case of insurrection, and his contempt of European intervention. Everyone present apparently concurred with him. However, North Carolina Democratic representative Abraham Watkins Venable, an antiannexationist, told the House after Calhoun passed away that he had attended the meeting with the Cubans and that Calhoun disapproved of their plans and opposed annexation.⁴¹

Calhoun's views changed a few months later as he intensified his opposition to the Wilmot Proviso, which prohibited slavery in the territories. In a Senate speech on the proposed occupation of Yucatán, he favored Cuba remaining in the hands of Spain "as it has been the policy of all administrations ever since I have been connected with the Government." If Spain relinquished its hold on Cuba, however, "it shall not be into any other hands but ours." Calhoun, an Anglophobe, believed that British possession of the island would be a threat to United States national security. In a private interview with Gonzales, Calhoun acknowledged that the populations of Cuba and Virginia had similar proportions of whites and blacks, and that the island could add at least two states to the South. Yet Calhoun procrastinated on immediate action by telling Gonzales, "You have my best wishes, but whatever the result, as the pear, when ripe, falls by the law of gravity into the lap of the husbandman, so will Cuba eventually drop into the lap of the Union."⁴²

Gonzales translated for López in March when they conferred with Sen. Jefferson Davis, chairman of the Committee on Military Affairs, in his Washington, D.C., boardinghouse. They proposed that Davis lead a liberating expedition to Cuba in exchange for one hundred thousand dollars, to be immediately deposited in his wife's name, and after victory, an additional equal amount or a very fine coffee plantation. The senator turned down the offer but recommended Maj. Robert E. Lee as the only man in whom he had implicit confidence. Davis later recalled that the Cuban leaders met with Lee when he was captain of engineers in Baltimore and made a generous offer for his military services. After conferring with the Cubans, Lee went to the capital and discussed the matter with Davis, who twenty years later stated that Lee had declined "a proposition for foreign service against a government with which the United States was at peace." Davis and Lee, who were not Freemasons, did not get involved in the affair as deeply as some Southern members of the fraternity. Gonzales then offered the leading role of the expedition to Senator Rusk, a Freemason, who replied that "whenever the proper time for such action should arrive," he would raise five hundred Texans and lead them himself.[43]

While lobbying on behalf of Cuban annexation on Capitol Hill, Gonzales, probably encouraged by Caleb Cushing and George A. Gardiner, applied for the secretaryship of the Commission for the Adjustment of Mexican Claims. The board was established by an act of Congress on March 3, 1849, providing for three commissioners and a secretary to be appointed by the president of the United States. Gonzales wrote to the outgoing President Polk the next day, just prior to Zachary Taylor's inauguration, asking to be recommended to the new president or his secretary of state, John Middleton Clayton, for the Mexican Claims Commission position. This was the first of numerous attempts he made to acquire political patronage jobs during the next forty years. To augment his chances for government employment, Gonzales became an American citizen in the Circuit Court of Washington County on March 26, 1849. The Naturalization Act granted citizenship to any free white who lived in the United States for at least three years before the age of twenty-one and who (for three years before taking the oath) professed his *bona fide* intention to become an American citizen. During the oath-taking ceremony, in which he renounced all foreign allegiance and fidelity, "particularly to the Queen of Spain," Gonzales was accompanied by Dr. Gardiner, who served as a character witness.[44]

Gardiner, the son of a physician, was born in New York. He completed his medical studies in South Carolina and also practiced dentistry. In the spring of 1840, Gardiner went to Mexico City, became fluent in Spanish and developed an interest in mineralogy. Four years later, he moved to San Luis Potosí, staking a claim on some abandoned silver mines. The dentist allegedly invested sixty thousand dollars in mining machinery and implements before being expelled from his property in October 1846 by Gen. Antonio López de Santa Anna's army during the Mexican War. Gardiner went to the U.S. Army camp at Tampico the following year, where he was contracted as assistant surgeon in the military hospital. He returned to the United States in July 1848 with the last of the evacuated troops. Gardiner then traveled from New Orleans to Washington, D.C., with Jefferson Davis. In late 1849, the dentist filed a $1,650,000 petition with the Mexican Claims Commission for the loss of his silver mine, and was subsequently awarded $450,000.[45]

Before Gonzales was sworn as an American citizen, President Taylor appointed William Carey Jones as secretary to the Board of Commissioners of the Mexican Claims Commission. Three days after his naturalization oath, Gonzales wrote to Caleb Cushing to request a recommendation for a Department of State position, indicating that he was "equally versed in the English, French, Spanish and Italian languages." He explained that he intended to stay in the United States until Cuba was also "*a land of the free*" and gave his return address care of Dr. George A. Gardiner, Washington, D.C. But Gonzales never obtained employment from the new Whig administration.[46]

The following week, López traveled to Philadelphia and checked into Jons Hotel on April 6 to await a "young friend" sent by John C. Gardiner who would have an offer for them. Four days later, the general wrote to Gardiner that the person had not yet appeared, and asked him to give Gonzales a letter of introduction for their contact and instructions to locate him in Philadelphia. López stated that their plans had "a very good aspect," but needed to be accelerated in case of possible variations in the expected offer. Meanwhile, in Cuba, some of the independence leaders were tried in absentia by a military court in Havana for their conspiratorial activities. López was sentenced to death by firing squad; Miguel Teurbe Tolón received a ten-year ultramarine prison term; Sebastian

Morales, eight years incarceration and deportation to the Philippines; and Cirilo Villaverde and José Sánchez Yznaga, six years confinement and perpetual banishment from Cuba. By late April, the movement leadership was back in Washington, D.C., where they were joined by John L. O'Sullivan and Lt. Col. William H. Bell, a Seminole War veteran and "worthy Democrat" appointed by President Taylor a few months later to the post of Indian subagent for the Missouri frontier. Bell described himself as "poor, without a name," after losing his left arm in the Battle of Monterrey with the First Regiment of Mississippi Rifles.[47]

The revolutionary plans took a downturn when Worth suddenly died of cholera on May 7, 1849, and Lt. Col. James J. Duncan succumbed to fever in Mobile two months later. Gonzales later blamed the transfer of Worth to Texas by a "powerful and rival influence" and his sudden demise for frustrating their goals. However, another conspirator suggests that Worth's death might have been a moot point by May: four years after Worth's death, Sánchez Yznaga wrote that the Havana Club had withdrawn its $3 million offer in early 1849, "declaring that it was impossible to gather that sum." Worth's biographer was unable to clarify this matter.[48]

Gonzales, still using Washington as a base of operations, returned to the Irving Hotel in mid-May. The following month, he traveled with López to New York City. The conspirators held a meeting in the boardinghouse of Mrs. Clara Levis, at 39 Howard Street, where they conceived the Cuban flag. Masonic emblems and the red, white, and blue tricolor of liberty went into the design. Master Mason Miguel Teurbe Tolón drew three oblong horizontal blue stripes, separated by two white stripes, to represent the three regions into which Spain divided Cuba. López superimposed on the banner's left an equilateral triangle, resembling a Master Mason's apron, "for besides its Masonic significance it is also a striking geometrical figure." He rejected placing the Masonic All-Seeing Eye in the center of the triangle, as it was difficult to embroider. Instead, they used "the Five-pointed Star of the Texas flag because it also carries a symbolic meaning," representing the Masonic five points of fellowship.[49]

During the first week of July 1849, Gonzales and other revolutionary leaders recruited for the Cuban expedition in New York, Boston, Philadelphia, Baltimore, New Orleans, Mobile, and Pascagoula, Mississippi. In New Orleans, enlistment was conducted by Mexican War veterans

George William White, Walter F. Biscoe, and Charles C. Campbell. The twenty-eight-year-old White had migrated from Ireland to New York City, working as a clerk in a Bowery store before moving to the Crescent City. He became a shopman in a linen draper's store and was later a clerk in a Magazine Street dry goods establishment. A balance sheet later given to Mississippi governor John Anthony Quitman indicates that Gonzales and the Cuban conspirators raised $368,267.01 and acquired the steamers *Sea Gull* and *Fanny* for the venture. A government informant correctly indicated that eight hundred recruits would take the steamer *Fanny* at Cat Island, ten miles south of Gulfport, Mississippi, in late August for an invasion of southern Cuba. The *Fanny*'s first officer was Callender Irvine Fayssoux, a five-foot-four, wiry, bearded, twenty-eight-year-old native of St. Louis, Missouri. Fayssoux joined the Texas Republic Navy in 1841 and the U.S. Navy during the Mexican War. His explosive temperament prompted three duels and got him involved in filibuster movements for over a decade. According to John L. O'Sullivan, the New Orleans expedition and another from New York would rendezvous off Cape Catoche, Yucatán, and thence "proceed together to the South Side of the Island."[50]

Colonel White traveled to New York City in mid-July to meet with General López and Gonzales and to coordinate expedition preparations. They were later joined by Mexican War veterans Lewis Carr of Philadelphia and Kentuckians John T. Pickett and Theodore O'Hara. The latter was a twenty-eight-year-old lawyer and Democratic newspaper editor. Pickett was a twenty-five-year-old Kentucky attorney who had been U.S. consul at Turks Island, a post he resigned to join the expedition. As a West Point cadet in 1841, he had too "wild and erratic a disposition to remain long enough to graduate." Conspirator George A. Gardiner arrived a few weeks later. Another Kentuckian, Col. E. B. Gaither, was authorized to raise five hundred recruits in his state. A lieutenant colonel from Ohio mustered in Cincinnati "quite a number" of expeditionaries. On July 25, John L. O'Sullivan wrote from New York to Thomas Ritchie, editor of the *Washington Union*, the Democratic Party organ, that regarding Cuban independence, "events are not only shaping themselves but *fast moving*."[51]

The same optimism was shared by Gonzales, who in late July went to the summer resort of Fauquier White Sulphur Springs, generally known as Warrenton Springs, six miles south of Warrenton, Virginia, at the present-day junction of State Roads 802 and 687. Visitors arrived from Washington, D.C., on a one-day coach trip over winding roads through wooded hills. The aristocracy traditionally spent the summer months at the Virginia mountain-spring resorts. Fauquier White Sulphur Springs, located on the slope of a breezy knoll, was the largest and most elegant of the vacation spots. The four-story Pavilion Hotel on the summit was flanked by a semicircle of eight washed-brick cottages with slate roofs. The sulphur and iron waters at the spring house, ritually consumed by the guests in abundant daily doses, were erroneously believed to have medicinal benefits. From the hotel entrance, peering over the hillside treetops, was a majestic view of the Blue Ridge Mountains. Gonzales hobnobbed among the six hundred visitors, including president Zachary Taylor's son, many uniformed Mexican War veterans, and "the Secretaries of the Spanish and the Russian Legations." He most likely occupied the bachelors' quarters, the four-story, seventy-room "Rowdy Hall" opposite the hotel. It housed billiard tables, bowling alleys, a bar room, and a pistol gallery where Gonzales spent many hours improving his shooting skills. Other attractions included a hurdle race and tournament, a fancy ball, a masquerade, and brilliant fireworks.[52]

The Virginia General Assembly was in session at the springs after fleeing a cholera epidemic in Richmond. In the evenings, after the daily ball, Gonzales entertained the guests with classical songs. His voice had extraordinary clarity and tone, as well as powerful ranges of pitch. On Friday, July 27, the Cuban sang with great spirit *Suona La Tromba,* from Vincenzo Bellini's 1835 opera *I Puritani.* Three days later, there was a banquet for Ellwood Fisher, a forty-year-old native Virginian residing in Cincinnati, who had championed Southern rights in his pamphlet *Lecture on the North and South.* More than one hundred Virginia legislators were seated at two parallel tables in the large drawing room. Among the guests was Mississippi senator Henry S. Foote, a Union Democrat, who addressed the gathering. After the speeches were pronounced and the customary honors were presented, Col. Joseph W. Spaulding of the Virginia press introduced Gonzales as a Cuban patriot dedicated to "the advancement of human liberty" and gave a toast to his health. Gonzales "returned the thanks of himself and his countrymen, and trusted that he would be permitted to intrude the name of a country, which, though actually out of the pale of the American Union, he felt happy to say, was familiar to the ear, and akin to the heart of the American people." Gonzales then proposed a toast: "To the coming sister of the South—to the future gem State of the Union—the Island of Cuba. She will prove a bulwark against external foes, an inexhaustible source of internal prosperity. May her disenthralment form the complement of American Liberty."[53]

In a fiery speech with "wild tones of defiance," Gonzales denounced the atrocities committed by the Spaniards in his homeland. His comments were published in various Southern newspapers. The response of the Whig *Louisville Journal* was reprinted in the *National Intelligencer*: "Whatever may be said of the modesty of a Cuban in toasting his island as the future 'gem State' of the Union, long before the public of this country has expressed a wish to have such a 'gem' in our national casket, or national nose, or in our national ear, there can be no question that when Mr. Gonzalez ranked Cuba as more valuable than any of the States of our Confederacy, he offered very little flattery to our present sisterhood of Commonwealths, the poorest of which would feel shocked at the impudence of Cuba if Cuba were to assume to be her equal. We propose that Mr. 'Ambrosio Gonzalez,' the gentleman who is all the way from Cuba, have leave to withdraw his toast."[54]

That same morning, July 30, Col. George White engaged three vessels in transferring some four hundred recruits from New Orleans and others previously encamped on Cat Island to Round Island, three miles southwest of Pascagoula, Mississippi. Secretary of State John Clayton, warned by the U.S. attorneys in New Orleans and Mobile, quickly informed President Taylor and wired the Spanish envoy extraordinary and minister plenipotentiary, Angel Calderón de la Barca, who was vacationing in Glen Cove, Long Island, to return to Washington for consultation. Born in Buenos Aires to a Spanish colonial bureaucrat, the fifty-nine-year-old diplomat was a veteran of the Napoleonic Wars. He had been at his U.S. post since 1835, excluding a three-year stint in Mexico, and had married the Scottish immigrant Frances "Fanny" Erskine Inglis in New York City. After being summoned to the capital by Clayton, Calderón de la Barca quickly activated the Spanish "system of espionage in New York, Philadelphia, and other cities on the Atlantic coast," which since 1830 had monitored the political activities of Cuban exiles. President Taylor ordered Navy secretary William Ballard Preston to send the Home Squadron in Pensacola, Florida, to Cat Island to suppress the movement.[55]

Clayton then drafted a presidential proclamation on August 10, warning that Americans intent on the "criminal" invasion of Cuba would be subject to "heavy penalties" and would forfeit any U.S. protection. The document instructed all law enforcement officers to arrest the offenders "for trial and punishment." Clayton sent a copy to Taylor in Harrisburg, Pennsylvania, who returned it the next day with his "entire approbation." The presidential proclamation appeared in newspapers nationwide. In New York, the Spanish royalist newspaper *La Crónica* spearheaded a derisive campaign against Cuban independence, threatening that "Spain would arm the slaves and lay the island in ashes and blood, sooner than permit the loss of its dominion by the revolution of the people." The Whig press rallied behind the president in denouncing the expeditionaries. The *Commercial Advertiser* stated that "their character and conduct would be no other than that of filibusters and pirates." Contrary to prior historical assertions, this is the first time that the term *filibuster* was applied to the López movement. The Spanish royalist periodicals credited the New York Whig press with having coined the term. The *Gaceta de La Habana,* quoting Havana's *Diario de la Marina,* indicated that "pirates and filibusters, is how the expeditionaries are called in the New York newspapers." *La Crónica* continued referring to the Round Island event as a "filibuster expedition." The U.S. district attorneys in New York, Philadelphia, Baltimore, and Boston were notified by Clayton that recruitment efforts in their cities were underway to invade Cuba.[56]

The day after the presidential proclamation was issued, a Cuban conspirator in Washington, D.C., wrote a cryptic letter to Gonzales at Fauquier White Sulphur Springs, signed "Your Friend." He warned that "the *uncle* cannot delay any further to take means. I am making my efforts to retard it," and that Colonel White, with seven hundred men, "was ready to direct to the peninsula [Yucatán], and not the island [Cuba]." A final enigmatic paragraph concluded, "May God grant you the intelligence to comprehend this letter." Gonzales had already left the resort after reading the presidential proclamation, and the message was forwarded to him in Washington, D.C.[57]

Expedition preparations in New York City were well underway by mid-August. Newspaper notices of hotel arrivals and letters at the post office indicate that López, Gonzales, O'Sullivan, Col. James W. Breedlove,

Capt. James C. Marriott, Pedro José Yznaga Hernández, and other leading conspirators, were arriving in and departing from Manhattan frequently. They were joined by Kentuckian Burwell B. Sayre and Cirilo Villaverde, recently escaped from a Cuban jail, charged with being a *La Verdad* correspondent. Col. Lewis Carr held a large Cuba meeting in Lafayette Hall, in which "several bodies of men, numbering over a hundred each, have organized themselves into regular military divisions." Col. Robert E. Lee, who earlier had been offered command of the expedition, registered at the Irving House one week prior to the scheduled departure. When it was postponed because of the late arrival of a steamer, Lee departed on August 15 for the Ocean House summer resort in Newport, Rhode Island, but then returned to New York in a fortnight, three days before the next sailing date. After the expedition was again delayed, Lee temporarily sojourned to West Point.[58]

The volunteers on Round Island were subjected to a U.S. Navy blockade on August 28 and ordered to disperse. The following day, Gonzales and López, unaware of these events, had breakfast at the Washington, D.C., home of State Department official translator Dr. Robert Greenhow and his wife, Rose O'Neal Greenhow. She, described as "tall, handsome, and graceful," later gained notoriety as a Confederate spy and became a Southern martyr after drowning during the shipwreck of a blockade runner. The Greenhows were ardent supporters of Sen. John C. Calhoun and sympathized with Cuban annexation. After their guests left, Mrs. Greenhow promptly wrote to Calhoun that the expedition would leave New York in three days aboard a steamer with one thousand men. An additional fifteen hundred recruits would sail in another vessel from New Orleans. Mrs. Greenhow erroneously concluded that "the Government here are in *the secret* and have done no more in the matter of the *Proclamation* than *regard for appearances demanded.*"[59]

Calhoun had already learned of the expedition from John L. O'Sullivan, who wrote to him on August 24 from Atlanta, while on his way to New Orleans. O'Sullivan had asked the Carolinian to assist the Cuban independence cause by calling out for Southern volunteers. Calhoun, who had favored Cuban annexation, was embroiled in his Senate fight against the Wilmot Proviso prohibiting slavery in the territories. Battling tuberculosis, he died seven months later without publicly endorsing the annexationists, believing that the Cuba question would turn his "threshold" issue from an internal to an external contest, and might provoke its demise.⁶⁰

Two weeks later, on September 6, New York U.S. attorney J. Prescott Hall received orders from Secretary of State Clayton to seize the expeditionary vessels. The steamer *Sea Gull* was confiscated at Staten Island with 203 boxes containing muskets, rifles, pistols, swords, uniforms, cartridges, gunpowder, and one howitzer. Robert A. Parrish Jr. and some forty Spaniards and Cubans found on board were released. The steamer *New Orleans*, with a capacity for nine hundred men, was seized at the Grand Street wharf with 377 tons of coal, 120,000 rations in 200 barrels, camp equipment, water and medical stores. Lewis Carr was arrested and released on five-thousand-dollars bail, and warrants were issued for Edward Wier, James C. Marriott, William W. McFarlan, Pigot, and Clark. That same day, Col. Robert E. Lee, who had spent the previous week at West Point, arrived in Manhattan for the third time in three weeks. Lee's biographers have never clarified his apparent connection with these events. President Taylor disbanded the expedition to "preserve the neutrality of the country," but "it was not his wish that those concerned in it, thus far, should be proceeded against criminally." The steamer *Fanny* was seized by the U.S. marshal in New Orleans and the last of the volunteers on Round Island departed on October 11. In Cuba, Capt. Gen. Federico Roncali, the count of Alcoy, requested that the monarchy send muskets and warships to Cuba and create a local militia. At the end of October, conspirators James W. Breedlove and Dr. George A. Gardiner left New York for New Orleans. After their arrival, Col. George White departed the Crescent City for Manhattan on the steamship *Ohio*.⁶¹

The failure of the enterprise further widened a growing schism between General López and Cristóbal Mádan, the president of the Cuban Council in New York City. Mádan decided that another military leader was needed, so he negotiated in Kentucky a proposition with thirty-one-year-old Col. John Stuart Williams, an officer of the Fourth Regiment Kentucky during the Mexican War, where his dashing conduct won him the sobriquet "Cerro Gordo." Mádan returned to Manhattan on November 13, and the next day created the Council of Superior Government, to supercede the Cuban Council. Its members signed an agreement giving Williams the rank of major general of the expeditionary force, under the orders of the general in chief and of the council. Initial attempts by López to invoke unity failed due to his rigid *caudillo* mentality. On November 17, he sent Ambrosio Gonzales to offer Mádan a new conciliatory proposition. Gonzales had recently returned to New York, after stopping at the Irving Hotel in Washington, D.C., two weeks earlier. Gonzales managed to negotiate a satisfactory agreement, but the next day the impetuous López wrote Mádan that to avoid "frequent conflicts" harmful to the liberation struggle, he would seek better fortune elsewhere.[62]

The leadership feud polarized the small Cuban exile community. Mádan and his followers gathered on November 22 at the home of Gaspar Betancourt Cisneros and drafted the charter for the clandestine Cuban Council of Organization and Government. This faction, superseding the two previous Mádan councils, was supported by the Havana Club. López, on the other hand, blamed the Havana Club for the Round Island expedition failure and for the delay of his June 1848 insurrectional plans. He maintained that it was time for "military action and not a civil Government." The general invited his opponents to attend a public meeting to settle their dispute and establish an energetic cooperation that would assure "adding the star of Cuba to those that shine on the glorious flag of the American Union."[63]

The Cuban Council of Organization and Government was covertly established on December 3, 1849, with Mádan as president, signing all documents under the pseudonym *León Fragua del Calvo,* Gaspar Betancourt Cisneros as vice president/treasurer, and Miguel Teurbe Tolón as secretary. López responded two days later by publicly creating his own Junta for Promoting the Political Interests of Cuba. They issued a proclamation, apparently written by Gonzales, whose signature headed those of José Sánchez Yznaga, Cirilo Villaverde and Juan Manuel Macías. Copies were mailed to sixteen Whig and Democratic newspapers. The document left a blank space for the signature of Pedro de Agüero, who was in New Orleans liquidating the Round Island expedition debts. However, upon his return to New York, de Agüero instead joined Mádan's organization.[64]

The growing hostility of Mádan's Council forced López, Gonzales, and their group to move their headquarters to Washington, D.C. They began lobbying among the politicians returning to the capital for the congressional session starting on December 3. Gonzales arrived on the sixteenth at the Irving Hotel, his former residence the previous winter, where his friend Sen. Daniel S. Dickinson was now a guest. Two days later, Mádan checked into the United States Hotel. He met with Secretary of State John Clayton and denounced the Cuban Junta and their plans. When Gonzales heard of the meeting, he quickly informed López.[65]

On the evening of December 21, Gonzales attended a White House reception with three friends of the First Family, prompting dour stares from the Spanish diplomats present. The Cuban was introduced to fifty-three-year-old John Henderson, who was a Freemason, a Mississippi state militia brigadier general, and a former Whig senator. After exchanging the secret Masonic greeting, Henderson arranged to meet Gonzales at the White House two days later for further consultation. Hours before seeing Henderson, Gonzales met with Jane Storm, who introduced him to sympathetic legislators. Gonzales later presented Henderson with a "precise" account of his revolutionary plans, and the Mississippian indicated that the Junta could rely on his help in New Orleans. Henderson was a political ally of Gen. John Anthony Quitman, elected governor of Mississippi the previous month, and an avid supporter of Calhoun's doctrines favoring slavery and states' rights.[66]

A few days after the Henderson meeting, Gonzales joined Round Island conspirators Col. Theodore O'Hara and Lt. Col. John T. Pickett, who introduced him to thirty-year-old Maj. Thomas Theodore Hawkins, a Freemason from a prominent family in Newport, Kentucky. Hawkins had "a slight and delicate frame," and "a dark complexion, with [a] black moustache." During the Mexican War, Hawkins had been an adjutant of the First Regiment of Kentucky, was promoted in 1847 to second lieutenant of Company K, 16th Regiment, U.S. Infantry, and fought gallantly at Buena Vista. The trio offered to raise and pay for a regiment of Kentuckians, agreeing to later meet with Gonzales and López in Louisville. At Gonzales's request, General López joined him at the Irving Hotel on December 29, arriving with Juan Macías, John Carlos Gardiner, and Col. George White. On New Year's Day, John O'Sullivan and Kentucky filibuster lieutenant R. S. Triplett reached Washington. Villaverde and Doctor George A. Gardiner rendezvoused there within two days.[67]

Four Cubans arrived at Willard's Hotel on January 8, 1850, where O'Sullivan had returned three days earlier after a trip to New York. They were José A. Portocarrero, G. J. Martínez, and Round Island fund raisers

Antonio Yznaga del Valle and Pedro José Yznaga Hernández. One of them met with Sen. Sam Houston of Texas and stated that the Cuban people were "very impatient to start the revolution," and that the landing of a popular leader with just five hundred men would be catalyst enough to initiate it. Houston forwarded this information to the López Junta. A week later, seven exiled Hungarian revolutionaries, led by Col. László Ujházy, who had arrived in the United States the previous month, registered at the Irving Hotel. Among them was Col. Janos Pragay, who had been an adjutant general in the army of Louis Kossuth and would figure prominently in the last López expedition to Cuba. This convergence of filibuster activists indicates that the Junta's plans were in the final stages.[68]

These movements did not go unnoticed by the adroit Spanish minister in Washington who, through his espionage network, knew more about filibuster plans than the American government. On January 19, Calderón de la Barca complained to Secretary Clayton that the Junta leaders were smuggling insurrectional proclamations into Cuba, selling bonds to pay for recruits, training with weapons, and holding meetings in New York, New Orleans, and other places, assisted by American citizens. The López proclamations were addressed, "To the Lovers of Liberty in Cuba," "To the Spanish Army in Cuba," and to the Cubans, Spaniards, and men of all nations on the island, outlining his ideals and promising to liberate them in March. The minister's dispatch concluded that Spanish-American relations would be strained unless a respected voice of authority warned those Americans who were being dangerously misled. President Taylor instructed Clayton to inform the federal attorneys in New York, Washington, and New Orleans "to keep a vigilant watch" on such activities. This was a moot issue with Mádan's Cuban Council, which had decided the previous month not to prepare another expedition until Taylor's term expired in 1853. In January, Mádan told Colonel Williams in the capital that the council had annulled his contract. He also spoke with annexation supporter Sen. Daniel S. Dickinson at the Irving Hotel "in terms of decided disapprobation of the anticipated López or other expedition against that Island."[69]

López departed New York to join Gonzales in Washington, D.C. He arrived on February 5 at the Irving Hotel, where conspirator George A. Gardiner had registered five days earlier. John O'Sullivan and Frederick

Henry Quitman, son of the Mississippi governor, had checked into Willard's Hotel twelve days later. O'Sullivan brought copies of the *United States Magazine and Democratic Review*, which contained his laudatory sixteen-page biography of General López, indicating that "his plan for Cuba has always been Independence and Annexation to the American Union." Gonzales, López and, O'Sullivan met in the Capitol's Hall of Columns with a group of senators that included Mississippi Democrats Jefferson Davis and Henry S. Foote. When Foote repeated intelligence supposedly known only by Mádan, O'Sullivan felt betrayed by his brother-in-law and "cried like a child." López suddenly arose and told Gonzales to inform the senators that they were obviously under the influence of a "malignant person," and that he hoped that in time they would realize who Mádan really was. The revolutionary leaders soon departed for Baltimore, where on February 20 López wrote to Macías and Villaverde in New York that O'Sullivan was on his way to tell them what had transpired. The general requested at least three hundred dollars for personal expenses, forwarded to New Orleans under his pseudonym "N. de Oriola." Soon after López and Gonzales left Washington, the Spanish minister was informed by his espionage network that they were headed south. On February 24, Calderón de la Barca relayed the news to the captain general of Cuba and requested his consul in Charleston to watch for López.[70]

Gonzales and López left Baltimore on the thirty-two-hour Monongahela railroad and stagecoach route to Pittsburgh. From there, an Ohio River steamer took them on a two-day trip to Cincinnati, where they checked into the Broadway Hotel on February 26. The filibuster leaders departed the next day on the steamer *Telegraph No. 1* for Louisville, Kentucky, where López signed the Louisville Hotel registry as "N. Oriola," and Gonzales penned his own name. When they met with O'Hara, Pickett, and Hawkins the following day, López "exhibited correspondence with some of the leading citizens of Cuba, urging him to come to their assistance as soon as possible—*alone,* if need be." Gonzales indicated that the landing would occur "where a large number of the people were already organized and armed" in readiness to join them. This would be "the signal for a general rising of the people." The Kentuckians agreed to raise a skeleton regiment that would swell with local volunteers after disembarkment. The Cubans then sailed on the Mississippi River, stopping at Vicksburg, Mississippi, where López remained while Gonzales proceeded to New Orleans to find John Henderson.[71]

The fifty-three-year-old Henderson lived in a ten-thousand-dollar estate at Pass Christian, Mississippi, with his family. He commuted by steamer to his law office at 16 St. Charles Street in New Orleans, which he shared with his thirty-two-year-old son, attorney John Henderson Jr. Henderson introduced Gonzales to prominent annexation sympathizers in New Orleans, including thirty-four-year-old state senator and attorney Laurent J. Sigur, proprietor of the *New Orleans Delta*. Sigur, a cigar smoker, was described by an R. G. Dun and Company credit agent as "a man of ability and of undoubted integrity. His family is rich, his wife [a widow] has means of her own and there is no doubt that he disposes of all the requisites to make the paper prosper."[72]

Meanwhile, on March 12, Colonel O'Hara wrote to militia captain William Hardy, a twenty-five-year-old "tall and athletic" Democrat activist in Covington, Kentucky, to "recruit a number of men to aid in revolutionizing the Island of Cuba." Hardy had been a sergeant in Company B, Second Regiment Kentucky during the Mexican War. He agreed to raise a five-hundred-man contingent in northern Kentucky and southwestern Ohio and to procure one hundred thousand dollars, a thousand muskets, one hundred kegs of gunpowder, one hundred swords, and five hundred uniforms. Hardy then advertised in the newspapers and stated at a public meeting in Covington that he was organizing a company to work in the California gold mines.[73]

Three days later, Henderson accompanied Gonzales to meet López in Vicksburg and provided three hundred dollars for their expenses. The trio went on March 17 on the Vicksburg and Jackson Railroad to the state capital, where they met Gov. John Anthony Quitman in the Executive Mansion. The fifty-year-old Quitman, son of a Lutheran minister, had graduated from Hartwick Seminary in his native New York before studying law in Ohio. In 1821, he moved to Natchez, Mississippi, became a planter and married the daughter of a prominent citizen. Six years later, he was elected as a Democrat to the state legislature and in 1835 presided over the state senate. Quitman, who owned several plantations and many slaves, politically identified with the states' rights nullifiers. In 1836 he organized a company to fight for Texas independence, and was later appointed major general of the Mississippi Militia. When the Mexican

War began, Quitman was commissioned brigadier general by President Polk. He participated in the Battle of Monterrey and led the bloody assaults on Chapultepec Castle and Belen Gate, which prompted the fall of Mexico City. He later gave Polk a detailed plan for the permanent occupation of Mexico. Congress awarded Quitman the brevet of major general along with a sword, and Mississippi granted him the governorship, which he assumed on January 10, 1850. As the "Father of Mississippi Masonry," Quitman identified with Gonzales and López, whom he greeted with the Masonic handshake.[74]

Also present at the meeting with the Cubans were the three justices of the Mississippi High Court of Errors and Appeals: Chief Justice William L. Sharkey and Associate Judges Cotesworth Pinckney Smith and Samuel S. Boyd. The first two were Freemasons. Sharkey, a fifty-three-year-old veteran of the battle of New Orleans, became a Master Mason in 1825 in Washington Lodge No. 3, where Henderson was affiliated. An Old Line Whig state legislator, he was later elected to the bench on four successive terms. The previous year, Sharkey had been chosen to preside over the convention of Southern states, scheduled for June in Nashville, to discuss the issue of slavery in the territories. The fifty-four-year-old Smith, like Henderson, had started his law practice in Wilkinson County. He was later elected Whig state representative and senator before serving as a judge. Boyd had arrived from Maine twenty years earlier and had gained prominence as a lawyer.[75]

The Cubans offered Quitman in writing "the office and powers of general-in-chief of the organization, movement, and operations of all the military and naval force which shall or may be employed in behalf of the contemplated revolution." In the document, written by Gonzales, López consented to act as second in military command to Quitman, who would be remunerated for his services, along with all other filibusters. The letter also stated that "to give better assurance to these propositions, it is intended that General López shall repair to Cuba with all dispatch, and at once raise the standard of Cuban independence, and will, from the field of revolution, furnish General Quitman [proof] that the people of Cuba approve these suggestions, and will welcome his presence to aid their cause as herein indicated." The three Mississippi Supreme Court judges provided legal opinions on ways to circumvent the Neutrality Act. A map of Cuba was spread on a conference table. López pointed to the positions of the twelve Spanish infantry regiments, each containing twelve hundred men, while Quitman gave tactical military advice. Despite the urging of his friends to lead the enterprise, Quitman concluded that he "could not engage until the people of Cuba, by their own free act, should first erect the standard of Independence," as occurred in Texas.[76]

The next day, the governor responded that his "devotion to the cause of civil liberty, and to the extension of the glorious republican principles of our government to the adjoining states of America" strongly urged him to agree to their offers, but that his official duties temporarily restricted him. He then added, "It is possible, however, that, after a short period, these obligations, which my sense of duty now imposes on me, will cease to exist. In that event, should circumstances be favorable, I should be disposed to accept your proposals." Quitman then orally agreed to meet the Cubans in New Orleans in fifteen days. On March 20, Gonzales wrote the governor from Natchez that he and López consented to his terms of landing in Cuba after the flag of independence had been raised. Gonzales added that López would remain in Natchez for a few days, that Henderson was departing that evening for neighboring Wilkinson County, and that he was sailing to New Orleans where Quitman could safely write to him by addressing the initials A.J.G., to Post Office Box S 154.[77]

Gonzales had managed, in less than two years, to rise from an obscure conspirator to become a leading figure in the Cuban annexation movement. López trusted him more than anyone else. Gonzales used his persuasive skills initially to mediate the López-Mádan feud. His captivating personality, charm, refinement, ideology, and Masonic affiliation had gained him in a short time the trust and friendship of nationally renowned generals, politicians, and government officials, and obtained entry into the White House. He had become a citizen of the United States, and his earlier studies in New York and travels throughout the country had thoroughly Americanized him. Gonzales strongly believed that his homeland, under the oppression of Spanish colonial despotism, could best benefit by enjoying the freedoms guaranteed under the American Constitution. Statehood would also bring economic prosperity and insure the institution of slavery, on which the Cuban sugar monoculture was built. These goals were not easy to achieve, especially when the expansion of slave territory had become such a contested issue in Congress. All legal avenues of self determination were denied to the Cuban people, so Gonzales chose the road of revolutionary change, even though he lacked the military experience for such an endeavor. The hour was fast approaching when he would have to abandon the speaker's podium and take up the gun.

## Two

# FREEDOM FIGHTER

Two days after Gonzales reached New Orleans on March 21, 1850, he informed Governor Quitman that there was a three-week departure delay, making the governor's presence there of "incalculable value." He told of receiving "cheering news" from California, which he was forwarding to López, and would tell Quitman upon his arrival in New Orleans. Gonzales asked the governor to appeal to his personal friends in raising a Mississippi volunteer reserve in case of a last moment shortage of recruits. He enclosed articles from New York newspapers indicating that the expedition would rendezvous in Santo Domingo, regarding it as a favorable error that would divert the Spanish navy.[1]

López believed that the Cuban Council of Organization and Government was leaking expedition information to the New York press. On April 2, he wrote from New Orleans to José Aniceto Yznaga that Mádan, Betancourt, and de Agüero had betrayed him. Two days later, John Henderson arrived at the St. Charles Hotel and gave Gonzales a letter from Quitman, with questions regarding their preparations. López soon departed on a Mississippi River steamer to an undisclosed destination for nine days. That evening, Gonzales was introduced to Peter Smith, son of Judge Cotesworth Pinckney Smith, and M. J. Bunch, who had arrived at the Veranda Hotel from Memphis, Tennessee, the previous day. Bunch had served two years in the Mexican War and fought at the battle of Buena Vista. The two men gave Gonzales letters of introduction from

Quitman and received authorization to raise a skeleton regiment in Mississippi. Bunch was promised the rank of lieutenant colonel, and Smith that of major. Capt. A. Mizell, a thirty-two-year-old Hinds County constable, was allowed to recruit an independent cavalry company attached to the Mississippi regiment.[2]

Gonzales replied to Quitman's letter on April 5, indicating that the Kentucky "emigrants" would not arrive until the day of sailing to insure greater secrecy. They were led by efficient Mexican War veterans, some with West Point military training, and all belonged "to the best families in the State." Gonzales wrote that a meeting was to be held that night to negotiate the price of a Panama line steamer capable of transporting six hundred men. They expected to sail on May 1, with a force ranging from two hundred to six hundred men. López would choose the landing site at the last moment, basing his decision on the number of men available and the news from Cuba. Gonzales told Quitman that Hungarian revolutionaries in New York had offered their services to the Cuban cause and were willing to sell them sixteen twelve-pounder cannons. Also, Junta member Cirilo Villaverde had recently received a letter from his brother, who claimed insurrectional leaders in western Pinar del Río province had "several thousand men already armed and munitioned" who would rise once López disembarked, and "seize the forts of Cabañas, Mariel and Bahía Honda, and the towns of Guanajay, Mantua and Pinar del Río," the latter the provincial capital. The Gonzales letter mentioned that in Cuba "all are expecting us. Some with dread, some with hope and anxiety."[3]

During the first week of April, twenty-four-year-old Chatham Roberdeau Wheat introduced himself to Gonzales, "begging to be allowed to go" with them. The Alexandria, Virginia, native had graduated in 1845 from the University of Nashville, and was a brevet captain during the Mexican War. Wheat settled in New Orleans on July 7, 1848, after returning with the troops. He joined the Louisiana bar and became active in Whig politics. His father, Episcopalian clergyman John Thomas Wheat, later wrote that his son's reasons for joining the Cuban expedition were "not only from universal feelings of philanthropy, but for the patriotic purpose of aggrandizing the South. This latter consideration was pressed by several prominent Southern statesmen, his Mason personal friends, who anticipating the coming strife with the North, saw in the acquisition of Cuba as a new Slave State, a vast resource on the event of revolution." Wheat was more precise when he told attorney Thomas R. Wolfe that he wanted to be a major general before the age of twenty-four. Gonzales accepted Wheat when he identified himself as a Freemason and proposed to raise and equip a skeleton regiment and to charter a transport vessel. The Junta commissioned Wheat as a colonel with the brevet rank of brigadier general and gave him permission to organize one of their three regiments.[4]

Filibuster volunteers were streaming into New Orleans by the hundreds. Gonzales and the officers established their headquarters in the luxurious St. Charles Hotel. López resided two blocks away, in the twenty-two-thousand-dollar home of Laurent Sigur, at 96 Custom House Street (today Iberville Street) in the French Quarter. On April 10, Dr. George A. Gardiner, the character witness for Gonzales during his naturalization ceremony, registered at Hewlett's Hotel. Twenty-four hours later, the steamer *Martha Washington* docked in nearby Freeport at 3:00 A.M., where the weary Ohio and Kentucky volunteers sought lodging. William Hardy and Henry H. Robinson, leaders of the Cincinnati group, proceeded to Hewlett's Hotel in New Orleans and located John Henderson, who arranged an interview with Gonzales and Lt. Col. John T. Pickett. Checking into the St. Charles Hotel that day were filibusters John McFarland McCann, a twenty-one-year-old Episcopalian minister from Paris, Kentucky, who had edited the *Cincinnati Nonpareil* and tutored at St. John's College during the previous two years, and Henry Theodore Titus, a twenty-eight-year-old New Jersey–born former Philadelphia postal clerk with "dark brown eyes and hair; standing well over six feet in height and weighing 250 pounds." He was described by one expedition chronicler as "a fine officer."[5]

Junta founder Juan Manuel Macías checked into the St. Charles Hotel on April 17 under the alias "M. Sardina," his mother's maiden name. He was accompanied by P. Duval of Buenos Aires, Argentina, a veteran fighter of the revolutions in his homeland, in Italy and Hungary. José Sánchez Yznaga arrived two days later from New York, registered at the Veranda Hotel, and later moved to the St. Charles. Two other Cubans joining the expedition in New Orleans were Round Island conspirator José Manuel Hernández Canalejos and forty-six-year-old journalist and educator Francisco Javier de la Cruz Rivero, who had fled Matanzas the previous month under the accusation of being a member of the López Junta. De la Cruz, a Bayamo native, had moved to Matanzas in 1830, establishing an apothecary store. Since 1845 he had published two geography books, a history of Cuba, and a novel. Hernández, a native of Matanzas, had served under Simón Bolívar in South America. He was the son of Dr. Juan José Hernández Cano, "a distinguished patriot of Cuba, who was poisoned in his dungeon by the Spanish authorities." The two newcomers assisted Gonzales with covert expedition preparations.[6]

Gonzales was instrumental in acquiring the steamer *Creole* on April 20. The paddle-wheeled vessel, built in New York City in 1841, measured 165 feet in length and twenty-two feet in breadth, weighed 306 tons, had one deck, no mast nor galleries, a round tuck, and a billet head. The *Creole,* a passenger and mail transport between New Orleans and Pensacola, Florida, had been refitted and painted nine months earlier. It was three-fourths owned by J. L. Day, Abraham Wolff, and the commercial firm J. & R. Geddes, and one-fourth owned by Capt. R. A. Hiern. The *Creole*'s agent was forty-year-old Ohio-born Robert Geddes, who, with his older brother and business partner, John, resided in the St. Charles Hotel. Geddes visited John Henderson at his law office to discuss payment of the sixteen-thousand-dollar sale price. He refused to accept a larger amount in Cuban bonds signed by López, Gonzales, and Sánchez Yznaga. Henderson gave Geddes ten thousand dollars in cash and a six-month note for two thousand dollars secured with a property deed at Pass Christian. The balance went to Captain Hiern. Henderson accompanied Geddes to the customhouse, where the *Creole* bill of sale was consigned to William H. White of New Orleans, who had been present at both interviews.[7]

Gonzales then met with forty-seven-year-old Brig. Gen. Jean Baptiste Donatien Augustin, commander of the Louisiana Legion, a state militia unit, and member of Masonic Perfect Union Lodge No. 1, in New Orleans, who procured the weapons for the expedition. Augustin obtained from the Louisiana state arsenal 398 muskets, 46 percussion pistols, 16 flint pistols and 60 cavalry sabers. The day the arms were secured, Louisiana Militia colonel James W. Breedlove, a Round Island conspirator, chartered the barque *Georgiana* with Capt. Rufus Benson, his mate Joseph A. Graffan, and three sailors "to go to Chagres [Panama] with passengers, for $500 per month." Breedlove was a sixty-year-old native Tennessean, a former treasurer of the Second Municipality, and the New Orleans federal collector in the 1830s. He was a partner in the small commission merchant and cotton firm Boyers, Breedlove and Company. The *Georgiana*, owned by Benson in partnership with eight other men, had been built a few months earlier in Lincolnville, Maine. The vessel was described as having "one deck and three masts," adding that "her length is one hundred and two feet; her breadth is twenty-five feet nine and 1/2 inches; her depth is ten feet four and 1/2 inches; and that she measures two hundred forty-three 68/95 tons; that she is a bark, has a square stern; no galleries, and a billet head." The one thousand coal sacks the *Georgiana* carried were for fueling the steamer *Creole*.[8]

The *Georgiana* cleared port at 9:00 P.M. on April 25 with the Kentucky Regiment, except Lieutenant Colonel Pickett, who remained with López. As the barque sailed downstream in the darkness, Lt. Richardson Hardy, the expedition chronicler, looked back and saw that "three men stood upon the pier, waving adieus long as their eyes could discern the bark. They seemed to be overjoyed at our safe departure, and filled with admiration for all on board. Those men were—Narciso López, Ambrosio J. Gonzales, and Gen. John Henderson." The *Georgiana* stopped the next afternoon six miles from the mouth of the Mississippi River. A fishing smack appeared alongside, piloted by Laurent Sigur, with Maj. Thomas Theodore Hawkins and nineteen-year-old Lt. Albert W. Johnson of the Kentucky Regiment. The officers boarded the *Georgiana* along with ten boxes containing 250 Louisiana Arsenal brown muskets with bayonets, and some 10,000 ball cartridges. During the transaction, Adjutant Henry Titus and Richardson Hardy made a muster roll of the passengers, which Sigur returned to López in New Orleans, along with farewell letters to mail.[9]

The day the *Georgiana* left, John Henderson hired Hugh M. Sherman as the *Creole*'s chief engineer. The following day, Governor Quitman arrived at the St. Charles Hotel and quietly used his political contacts and Masonic brethren to assist the revolutionaries. A Mississippi state officer provided fifty "Mississippi" rifles, the 1842 Yauger model weapon that got its nickname after a local regiment used it with great skill during the Mexican War. The armament was completed with the arrival of two hundred Jennings patent rifles from the North. Quitman agreed to lead a reinforcement expedition a few weeks after the López landing prompted

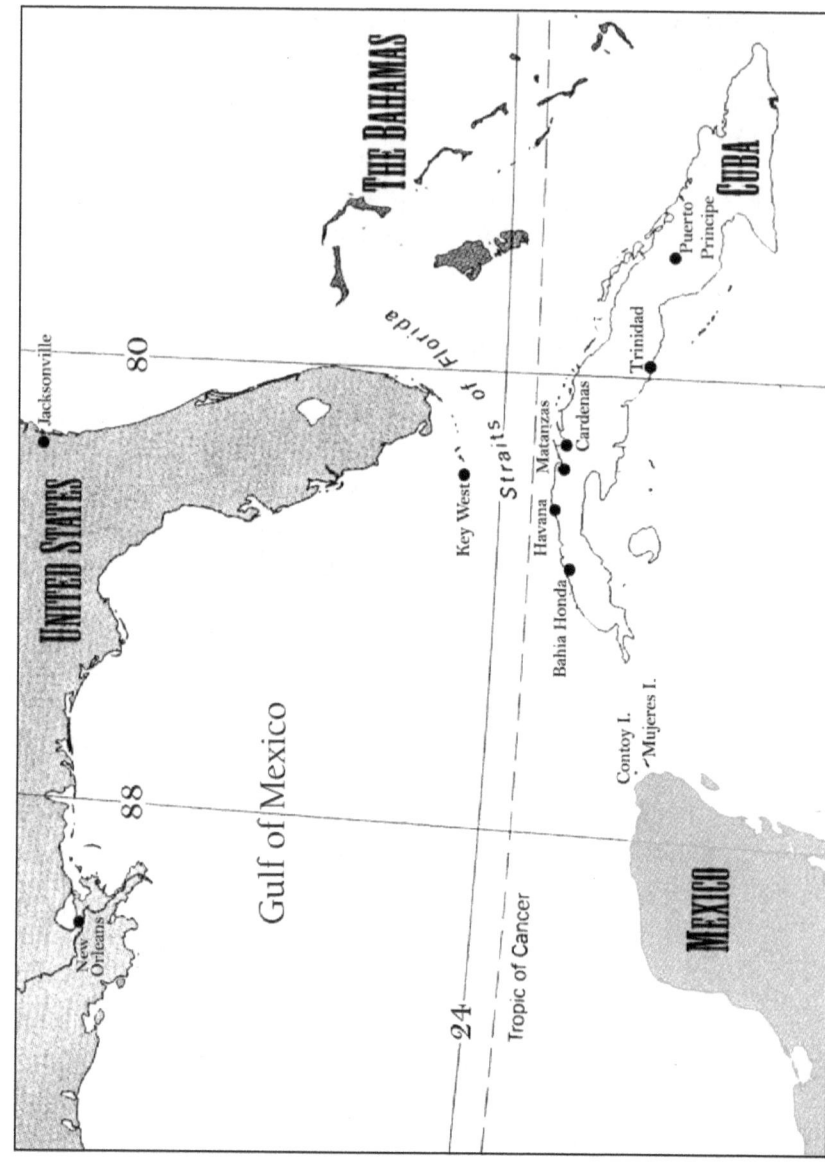

*Cuba and the Caribbean Sea*

a popular uprising. Gonzales introduced Chatham Roberdeau Wheat, leader of the Louisiana regiment, to Quitman. According to Wheat, the second contingent led by Quitman would depart "between the 1st and 15th June" and his friend Gen. Robert Armstrong, was pledged to it. Henderson stated that Quitman, after receiving the "call," would resign the governorship and head for Cuba. The correspondent of the *New York Herald* wrote a week later that López, Gonzales, Henderson, and Sigur were in the Crescent City, and that the first expeditionary regiments had already sailed for the rendezvous point. The article claimed that the next group would depart in a few days, and that "a distinguished officer of the late war, and at present the Executive of an adjoining State, will follow with the *corps de reserve,* and take command of the entire forces of the new Republic."[10]

The New Orleans Spanish consul, Juan Ygnacio Laborde Rueda, reported to Minister Angel Calderón de la Barca on April 26 that fifteen hundred men were to be transported to Cuba via Chagres, Panama, on four ships. He ascertained that "another vessel is freighted by Mr. Breedlove . . . to take out more of the men" and that the leaders were López, Gonzales, and the other Junta members. That same day, the *New Orleans Picayune* published a note from the *New York Herald* stating that "Gen. Lopez has sailed for St. Domingo with the view of assisting the Spanish inhabitants against the [Haitian] blacks, and then making the island a base of operations against Cuba."[11]

Four days later, on April 30, Cotesworth Pinckney Smith, associate judge of the Mississippi High Court of Errors and Appeals, arrived at the St. Charles Hotel. He quickly endorsed a $20,000 revolutionary bond also signed by Gonzales, López, and José Sánchez Yznaga, bearing the Cuban coat of arms and the seal of the Provisional Government. Total expedition expenses were estimated by Gonzales to be $40,000 to $50,000. Henderson calculated some $50,000 since March 20, including $19,000 invested in the *Creole.* Sánchez Yznaga wrote the total at $37,500, all of it donated by Americans, including one person who sold his only slave to meet an urgent expense. The cash from the bond was used on May 1, when James W. Breedlove, as agent for Chatham Wheat, chartered the brigantine *Susan Loud* with her captain, Simeon Pendleton, his mate Thomas G. Hale, four sailors, and a cook for $600 monthly to take passengers to Chagres.[12]

The next day, about 150 Louisiana Regiment volunteers were ready to sail. J. C. Davis carried his carpetbag aboard the *Susan Loud,* docked in the Second Municipality. He later received the rank of captain and command of Company B of the Louisiana Regiment and penned his memoirs of the expedition under the pseudonym O.D.D.O. Richardson Hardy wrote that the Louisiana Regiment privates were "mostly men of degraded character." Kentucky Regiment commander Theodore O'Hara later complained to Quitman that "the riffraff which Col. Wheat picked up promiscuously in New Orleans were but a burden to the expedition." Gonzales accompanied López and others to the wharf, where a paddle towboat pulled the *Susan Loud* to the mouth of the Mississippi River.[13]

On May 4, the *New York Herald* correspondent wrote that the expedition, headed by López, Gonzales, Henderson, and Sigur, would "sail in a few days." He noticed that the Spanish consul "has his spies, thick as blackberries, throughout the city." One of the secret agents approached Gonzales, claiming to be a dealer of horses and mules in Cuba, and wanted to help López by providing transport at the landing point. Gonzales communicated his suspicions to the general, who approved a ruse to misdirect the Spanish authorities. Gonzales later told the informant that the expedition was going to the southeastern coast of Santa Cruz, distant from their objective on the northwestern shore.[14]

Two days later, thirty-six-year-old banker James Robb wrote to President Taylor that a "formidable" military organization had been raised and was departing for Cuba. He was unable to provide the names of those involved, but he suggested that the government should send a strong naval force to the Gulf "to act as circumstances may require." The banker stated that since the captain general of Cuba had the authority to emancipate the slaves in case of a civil war, Taylor should warn him that such action would precipitate American intervention. Robb had also asked his friend, Gen. James Hamilton in South Carolina, to see the president immediately on this matter. Another informant, William L. Hodge, on May 7 wrote from New Orleans to Taylor: "The last of the Cubans leave this evening, accompanied by Generals Lopez, Gonsalves [sic], etc. They expect to land on the island within a week from this time. Their rendezvous is not at Chagres, but much nearer the island." Hodge wrongly estimated there were between six thousand and eight thousand expeditionaries and that to protect American interests from a general emancipation unleashed against them "a large naval force should be present" in Cuba.[15]

The *Creole* was piloted by Capt. Armstrong Irvine Lewis, a thirty-two-year-old Virginian who resided in Orleans Parish for more than twelve years with his wife, their two children, and his mother-in-law. He was slender, middle height, with "even features, dark complexion and hair, and of quick, active appearing habits." The steamer departed with General López, his servant Pedro Velazco, nephew Pedro Manuel López, some 130 men of the Mississippi Regiment, about twenty stragglers from Louisiana and Kentucky, Argentinian P. Duval, and five Cubans: Gonzales, José Sánchez Yznaga, Juan Manuel Macías, José Manuel Hernández, and Francisco de la Cruz Rivero, who were aristocratic intellectuals without military experience. Mádan's divisive campaign against López affected the expedition recruitment within the small Cuban exile community.[16]

After the *Creole* left New Orleans, it stopped seven miles south, at the river depot of the Mexican Gulf railroad, to receive crated weapons. At the plantation landing of Antoine Landier in St. Bernard parish, Henderson appeared in a steam ferry boat from Mississippi with Associate Judge Cotesworth Pinckney Smith and a General Cooper, carrying a large supply of boots and compounds. The material was moved to the *Creole*, which proceeded to sea at 4:00 A.M. on May 9. Upon reaching international waters, the regimental colors and the *Creole*'s Cuban flag were unfurled. Gonzales helped distribute a broadside titled "Soldiers of the Liberating Expedition of Cuba!" calling the volunteers to free Cuba from Spain's tyranny and the perpetual threat of converting the island into something worse than Haiti. The future Republic would "add another glorious Star to the banner" of the United States. López promised the men that "within a few weeks" after victory, they would each receive four thousand dollars and be allowed to return home.[17]

The *Creole* caught up with the *Susan Loud* at 1:00 P.M. on Friday, May 10, at the rendezvous point of latitude 26° north, longitude 87° west. The Cuban flag fluttered from the mastheads of both vessels. Gonzales boarded the brig to inquire about the health and spirits of the men after eight days at sea. He then gave "a glowing description" of the bearing of the Mississippi Regiment and of all the officers and men. Gonzales provided instructions on the method of nighttime procedures before returning to the steamer. The next day was spent transferring the passengers from the *Susan Loud* to the *Creole* in boats. Captain Pendleton, summoned by López to the *Creole*, later stated that when he boarded the steamer, "Gen. Gonzales and Captain [Thomas L.] Fisher took me by the arms and led me to López, saying they must detain me." The skipper was compelled to pilot the steamer to Cárdenas. The *Creole*, followed by the *Susan Loud*, then sailed to find the *Georgiana*.[18]

The following morning, Sunday, May 12, some officers worried that the Spaniards might intercept them on the high seas, and requested to arm the men for such an emergency. The "ever able and efficient" Gonzales "caused several long and short boxes to be opened," and distributed the weapons. He made a lasting impression on Richardson Hardy, who described him as a "polished man of letters" and "a deep powerful thinker," with "the extreme benevolence and kindness of heart apparent in every feature." After chatting with Gonzales for five minutes, Hardy recalled, "He will display to you the most erudite knowledge of character and the general world. I shall say that deep policy and mental activity were his distinguishing characteristics." Gonzales was depicted as having "a heavy, and somewhat sluggish, yet strongly marked countenance," a "strong, and full, and restless dark eye," and displaying occasionally "powers of oratory of no ordinary character."[19]

The *Georgiana* was sighted at 8:00 A.M. on May 14, anchored under the lee of Contoy Island, five miles northeast of the Yucatan peninsula. The *Creole* left the *Georgiana* at Contoy and traveled to Mujeres Island. The volunteers camped and drilled on the beach for the next two days, while the "ever energetic" Gonzales engaged most of the local natives in hauling water and supplies aboard the steamer. The next day, López assembled the regimental officers on board the *Creole*, who appointed all officers below them, with the general's approval. Newton C. Breckenridge, a twenty-two-year-old Kentucky farmer and member of Masonic Clarke Lodge No. 51, in Louisville, received the rank of captain and command of Company G, Louisiana Regiment. López named Gonzales his adjutant general, José Sánchez Yznaga as his first aide-de-camp, and Juan Manuel Macías and Captain Murry as his second aides-de-camp. Murry, an Irishman, was a veteran of warfare in Spain, Lombardy, and Hungary. The general assigned to his staff Col. G. N. L. Smith, a member of Masonic Yazoo Lodge No. 42, in Yazoo City, Mississippi, and Capt. Beverly Mathews.

On the morning of the sixteenth, before the *Creole* returned to Contoy, the officer of the day reported the desertion of thirteen men of the Louisiana Regiment.[20]

Meanwhile, back at the *Georgiana*, seventeen dissenters demanded their return to the United States. When the *Creole* arrived, coal, provisions, and troops were transferred from the brig. All the Kentuckians, except two who remained on the barque with the protesters, boarded the *Creole*. A filibuster recalled that "General López assembled the men and addressed them through General Gonzales, telling them that such of them as did not wish to go to Cuba could now have permission to return to the United States in the *Georgiana*." Seventeen men on the *Susan Loud* and eight from the *Creole* defected to the *Georgiana*. Col. M. J. Bunch, leader of the Mississippi Regiment, received military command of the *Creole*, which departed for Cuba at 1:00 A.M. on the seventeenth. A report indicated that the Liberation Army consisted of 610 men, including 240 in the Kentucky Regiment, whose "long beards and dusky countenances gave them a fierce, uncivilized aspect."[21]

After breakfast at dawn, the "ever indefatigable and industrious" Gonzales distributed the weapons and ammunition. The fifty Mississippi rifles were assigned to the Kentucky Regiment, which had the best leadership and organization. The old flint muskets went to the Louisianians and the Jennings patent rifles to the Mississippians. The expeditionaries were issued red flannel shirts and "a black cloth cap, with a Lone Star cockade." The captains wore white pants, the lieutenants black, and the troops of various shades and stripes. Almost every man carried a Bowie knife or a revolver.[22] Capt. J. C. Davis, who regarded Gonzales as an "extraordinary man," later wrote:

> But see Gonzales upon our merry little craft. He is the embodiment and incarnation of ubiquity—here, there, and every where—now ready to intervene, and palliate, and remove entirely, any acerbity of feeling that may spring up among the officers; and again, pausing for hours by the side of unclean privates, to afford every explanation and dissipate every doubt and fear. Without ostentation or intrusion, he enters into the most minute domestic arrangement of the boat. I have seen him, wearied with the cares and exertions of a hot tropical day, surrender his berth to some gaping officer, and with joy on his face

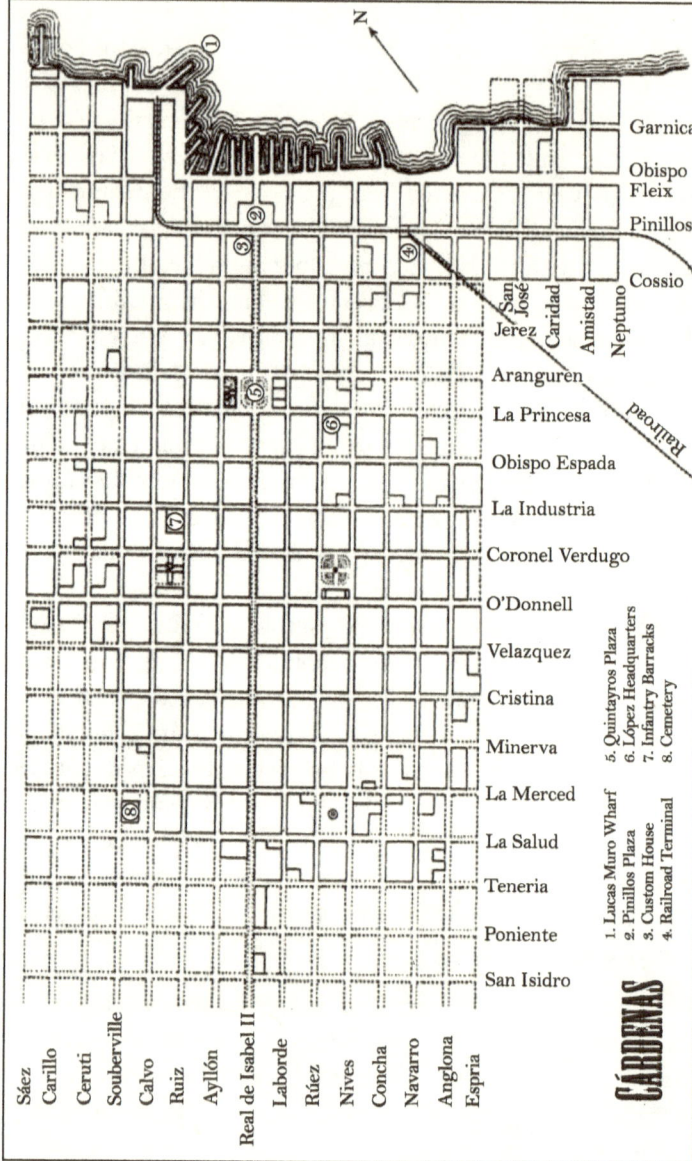

The City of Cárdenas, Cuba

betake himself to slumber upon the hard deck, without even the consolation of a knapsack for a pillow.

His labors are the more arduous, as he is the only medium through which the commander in chief diffuses his wishes to the force under his command.[23]

That evening, López revealed the operational plan to the field officers. The landing had been projected for the southwestern coast of Pinar del Río, in the Vuelta Abajo region, where Gonzales claimed they "had 4,000 troops in commission." The general now decided on a nighttime surprise of the unfortified northern coastal town of Cárdenas. They would then continue by railroad thirty miles west to Matanzas, guarded by fifteen hundred troops, attacking at 7:00 A.M. the rear of the fortress facing the sea. After its capitulation, one hundred filibusters would go by train to within nine miles of Havana, destroy the bridge and tracks to impede reinforcements from leaving the capital, and return to Matanzas. López believed that by then his three skeleton regiments would be bulging with volunteers and three new ones raised, totaling some five thousand mounted troops. When the Quitman reinforcements were due a week later, López expected to have thirty thousand men encamped outside the walls of Havana.[24]

In the capital, Capt. Gen. Federico Roncali, the count of Alcoy, received information about the expedition from dispatches sent by the New Orleans consul on two schooners and from Cuban fishermen who had seen the expedition vessels at Contoy. The *Official Gazette* published a citation on the eighteenth for Gonzales and other Junta leaders to appear within nine days in court, or be found guilty by default, of conspiracy to revolutionize Cuba and Puerto Rico to obtain their independence. Roncali organized in Havana the Regiment of Noble Neighbors militia, with three thousand mostly Spanish volunteers commanded by wealthy merchant officers, and ordered nineteen armed vessels under Adm. Francisco Armero Peñaranda, commandant general of the Spanish naval forces in the Antilles, to cruise the coast. Armero's squadron had been reinforced that week with the arrival from Spain of Rafael de Aristegui Vélez, the count of Mirasol, in the frigate *Esperanza* with forty-two guns and a crew of 366, and the war steamers *Blasco de Garay* and *Pizarro* with troops, weapons, artillery officers, and engineers. The *Pizarro,* built in

England the previous year, was 220 feet long, with 350-horse-power engines and a top speed of 15 knots. It had 122 crew members, four 32-pounder guns, and two 68-pounder Paixan guns. Armero personally commanded the *Pizarro,* accompanied by the frigate *Habanero,* to intercept the López expedition.[25]

The war steamers arrived at Contoy on the eighteenth at 4:00 P.M. The *Georgiana* and the *Susan Loud,* anchored a mile apart and displaying no flag, were boarded without resistance. Admiral Armero arrested Capt. Rufus Benson, his pilot Joseph A. Graffan, the mate of the *Susan Loud,* Thomas G. Hale, the seven crew members of both vessels, and the forty-two deserters. Attorney Archibald B. Moore, a *Creole* deserter, "alone, confessed openly his knowledge of, and participation in, the enterprise." The Spaniards used death threats and put noosed ropes around the necks of some manacled prisoners to elicit admissions, before taking them and the ships back to Cuba.[26]

After sundown on the eighteenth, the expeditionaries were assembled on the *Creole* deck for a grand review. Gonzales accompanied the general and his staff, as López deeply fixed "his keen black eyes upon every man." The general harangued the filibusters in Spanish while Gonzales translated. Gonzales afterward distributed sixty rounds of cartridges per man. Upon opening the ammunition boxes on deck, López "called out loudly to Gonzales" that glowing sparks from the chimneys were falling around. As the *Creole* passed Varadero Beach, rounding Hicacos Peninsula to penetrate Cárdenas Bay, Gonzales contemplated the shores of his family's estate. Lt. Richardson Hardy observed, "Never shall I forget the exulting looks and tones of Gen. Gonzales, as he stood gazing upon his native land,—the proscribed and condemned exile—the wanderer—returning upon his heartless oppressors in the power of patriotism and steel. 'Ah!' he exclaimed between shut teeth, 'we'll soon sway it over them!'" The Piedra Key lighthouse, at the harbor entrance, fifteen miles from the town wharves, was sighted at 10:00 P.M. Captain Pendleton, steering at half speed, tried to recall the location of the shipping channel he previously used, to avoid the numerous islets, sand banks, and coral reefs in the enormous harbor. *Creole* fireman William Wilcox, a twenty-two-year-old New Yorker, fearing what was about to unfold, jumped overboard and swam to a nearby cay, where he hid for two days before surrendering. Three hours later, after the moon disappeared, the motionless and silent liberators saw the faint lights of Cárdenas in the distance.[27]

Cárdenas, founded in 1828 by Bernabé Martínez de Pinillos, the count of Villanueva, was a prosperous commercial town with paved streets. The rectangular city grid, flanked by impassable swamps, contained nineteen thoroughfares north to south and fifteen east to west. There were four parks, the largest being Quintayros Plaza, four blocks south of the docks. The royal census of 1851 indicated a population of 4,524, including 885 slaves, comprising all classes and social conditions, living in seventy-three rubble masonry homes, 232 dwellings of wood and shingles and five palm-frond shacks. Its establishments included two drug stores, two taverns, five clothing stores, twenty-five dry goods stores, eight bakeries, eight inns, seven café/billiard halls, two barber shops, ten cobblers, eight attorneys, five physicians, one veterinarian, four blacksmiths, two cooperages, five carpenters, four tailors, two silversmiths, a horologist, two saddleries, eight tobacco shops, two hat stores, one furrier, a coppersmith, a butcher, one tin shop, a foundry, five limestone ovens, twenty charcoal stores, a fire station, and a cockfight ring. Nine months before the expeditionaries landed, Cárdenas had "a considerable [number] of American merchants and American planters, having estates in its vicinity." The waterfront, with wooden warehouses and sixteen wharves spread over a mile, had opened to international commerce on January 1, 1850. The military forces assigned to Cárdenas and its periphery were the infantry regiments of León and Nápoles and a squadron of mounted Royal Lancers, a local peasant outfit. The total number of troops in town on May 19 has been estimated at between sixty and one hundred men.[28]

The *Creole,* instead of docking at the steamer pier, headed for the longest wharf, more than one hundred yards in length, belonging to the Lucas Muro sugar warehouse, with an adjacent railroad track. The steamer abruptly grounded about 3:30 A.M. during low tide, a few yards from the wooden dock. The first mate, Round Island veteran Callender Irvine Fayssoux, jumped overboard clenching a rope with his teeth. He climbed the wharf and used the cord to pull a long plank from the gunwale, sitting on its end to stabilize it. López was impatient to land, but Gonzales suggested that since he weighed less, he should test the firmness of the board. The adjutant general walked to the middle of the gang walk, jumped a few times, and beckoned López to proceed. The general immediately encountered Port Capt. Patricio Montojo, an old acquaintance, who awaited their arrival assuming that it was a Spanish war steamer.[29]

Montojo was sent with four escorts to tell Lt. Gov. Florencio Ceruti to present himself, but after quickly turning a corner, he fled to the railroad yard outside of town. Montojo rode an express locomotive to Matanzas. After a mile and a half, he had the railroad workers lift a long section of track to contain the invaders. Montojo gave details of the events to the Matanzas governor, who sent him to Havana by steamer, notifying Captain General Roncali by 5:00 P.M. The count of Alcoy quickly issued an edict declaring the island in a state of siege, proclaiming martial law, and ordering that all invaders or anyone assisting them "be immediately shot." While Montojo was fleeing Cárdenas, watchman Juan García, who observed the landing, raced through town blowing a whistle and shouting "los americanos." Jail warden Nicolás Cárcamo, awakened by the noise, immediately notified Ceruti. The lieutenant governor sent him to retrieve the León Infantry Regiment from their barracks five blocks away. When they arrived, Ceruti left two sentinels on Real de Isabel II Street and ordered Capt. Andrés Segura, Second Lieutenant Aguado and some sixty soldiers to his office on the upper level of the Capitular House. Segura, an experienced infantry commander, considered this a detrimental strategy.[30]

The 610 expeditionaries were delayed over an hour while disembarking in single file over the plank. The six companies of Colonel O'Hara's Kentucky Regiment landed first and silently assumed formation. Their color bearer, William Redding, "carried the only flag which the invaders had at Cárdenas," donated by New Orleans ladies. O'Hara quickly detached sixty men under Lt. Col. John T. Pickett, to occupy the railroad yard, a mile and a half from the steamer and three-quarters of a mile southwest of the outskirts of town. This group consisted of Capt. John Allen's company and part of Capt. Albert W. Johnson's company, accompanied by translator Francisco de la Cruz Rivero, and included train engineers to conduct them to Matanzas. Pickett's unit impressed a watchman and an old fisherman on the street, who swiftly led them to their objective. O'Hara was waiting for interpreter Juan Manuel Macías when López ordered him to immediately march into town and capture the infantry garrison. The Kentuckians forced a mulatto passerby to take them to the barracks which, unknown to the expeditionaries, was not at the plaza but five blocks further away on the corner of Ruiz and Industria streets. O'Hara later erroneously claimed that the frightened man led them in the wrong direction, by taking them south on Ruiz Street, one block west of the plaza.[31]

The Muro warehouse owner, Joaquín Queipo, and his bookkeeper, Antonio Ojeda, were also impressed into service as guides. Wheat's Louisiana Regiment proceeded in the darkness to the town square, on a street parallel to the right flank of O'Hara. They were followed by Bunch's Mississippi Regiment, headed in the same direction, one block to the left of the Kentucky column. Capt. Achilles L. Kewen, with a detachment of nineteen Mississippians from Company A, occupied the railroad terminal on Pinillos Street, to prevent the departure of the 6:00 A.M. train, capturing twenty-four employees and a dozen armed men. When O'Hara assumed that he was being led astray, he countermarched his regiment north on Ruiz Street, encountering López, Gonzales and the general staff. Queipo, assigned to O'Hara as a guide, led the Kentuckians to Quintayros Plaza.[32]

The plaza was a rectangular open area with tropical shade trees, two blocks wide, and divided by the main artery of Real de Isabel II. On the northeast corner of Aranguren Street stood the massive stone jailhouse with its large doors and "immense iron-barred windows," while on the southeast corner of La Princesa and Laborde streets was the two-story rubble masonry Capitular House. It was owned by Alejandro Rodríguez-Capote y de la Cruz, who resided with his fourth wife Ursula and seventeen children on the upper floor. The ground level contained a tailor shop, and the mezzanine served as the lieutenant governor's residence and the Municipal Council office. On the west side of the plaza, on Ayllón Street, was the Immaculate Conception church, a Greco-Roman-style structure with twin eighty-foot bell towers. Above its entrance was a gas-illuminated clock encased in glass, six feet in diameter, with Roman numerals.[33]

As the Kentuckians approached the plaza, O'Hara thought that his guide, "stupefied with fear," gave confusing directions on the location of the infantry barracks. Queipo indicated that the jail was not the barracks, but his captors assumed otherwise. The Spaniard was goaded toward the prison gate, where the guard gave out the customary challenge, "Who lives?" Queipo correctly replied, "Spain," but when further asked, "What people?" he answered, "Citizen—but behind are the Americans," and bolted. The sentinel pronounced various rapid challenges before discharging his musket, and the Kentuckians responded with a deadly barrage. The fifteen soldiers inside the jail fired a volley at the advancing column, disabling a few filibusters, including O'Hara, who was shot in the thigh. Command of the Kentucky Regiment passed to Maj. Thomas Theodore Hawkins, instead of disgruntled Maj. William Hardy. The role of Hardy during the battle does not appear in his brother's book nor in the three official reports by the Kentucky Regiment officers. Hardy later filed his own report, stating the actions of the regiment, but omitting details of his own participation.[34]

The Mississippi Regiment arrived on the scene, and Lt. Col. M. J. Bunch ordered Captains Keating and Hawkins, of Companies B and C, respectively, to have their men fire at the prison. López advanced within a few feet of the jail windows, "careless of the tremendous firing on both sides," demanding their surrender, but only gained entrance after the Mississippians battered down the heavy doors. Gonzales rushed inside with the vanguard, while the defenders escaped through the back door, leaving behind a lone sentry in charge of the detainees. The guard fearlessly stood his ground, thrusting his bayoneted musket at the intruders. As the attackers were about to shoot him, López shouted against harming such a brave man. Thirty-three male and female inmates, only four of whom were white, were freed. All the prisoners fled except one who joined Ceruti and three incarcerated soldiers who joined the filibusters. An enthusiastic expeditionary wrote in the jail registry book: "This ends Spanish tyranny in the paradise of the World."[35]

Meanwhile, Captain Hale with Company D of the Mississippi Regiment had gained access to the corner house opposite the jail and across from the Capitular House, from where they responded to the gunfire of Ceruti and his men. When Colonel Wheat heard the barrage from a few blocks away, he credulously assumed it was a salute honoring López, called three cheers for "López and Liberty," and rushed toward the supposed jubilee. Sergeant Francis Boggess heard the cry, "Run boys, run, or you will lose all the fun." Wheat led the Louisianians into the plaza from the opposite side, near the Capitular House. A detachment of Spanish soldiers retreating from the jail into their path, fired on the Louisianians from within a few yards. Wheat, falling with a slight shoulder wound, shouted: "Louisianans! Your colonel is killed! Go and avenge his death!" In the ensuing skirmish, a few Spaniards were hit, while others fled eight miles to the town of Guamacaro, or sought refuge in the Capitular House. Wheat was taken to the boardinghouse of Mrs. Wergener Woodbury, on the corner of Real de Isabel II and Aranguren Street, to the left of the jail, which became a provisional filibuster hospital. The landlady was the sister-in-law of Levi Woodbury, an Associate Justice on the U.S. Supreme Court. Wheat passed command of the regiment to the one-armed Lt. Col. William H. Bell.[36]

López attached to the Louisiana Regiment the Mississippians of Captain Mizell's Independent Company and the Kentuckians in Capt. Henry H. Robinson's Company D, and ordered them to attack the Capitular House. He sent the rest of the Kentucky Regiment under Major Hawkins four blocks down Laborde Street to the O'Donnell Street intersection, connecting the eastern and western entrances into town. The Kentuckians formed a defensive line across O'Donnell Street, where a small body of Spanish horsemen had been reportedly assembling. At Quintayros Plaza, Lieutenant Colonel Bell directed Sergeant Boggess' company to storm the Capitular House. The men picked up "large stones, four men to each, with the Stevedore yell, they began battering the doors, and it did not take long to burst those strong oak doors to splinters." But the tailor shop on the ground floor had no access to the upper level. López ordered the attackers to leave the building after Ceruti sent an envoy with a surrender proposal. The offer was a diversionary tactic for some twenty soldiers who were seen escaping through the rear of the Capitular House. Bell sent Maj. George B. Hayden and his Louisianians to the area, where a fleeing officer dropped his sword, one soldier surrendered, and another, ignoring a warning to halt, was killed by Capt. Henry C. Foster, a second lieutenant of the Louisiana Volunteers during the Mexican War.[37]

This left thirty-five soldiers defending Ceruti's headquarters. Lieutenant Colonel Bunch noticed "López, without a body guard, fearlessly exposing himself to the fire" from the Capitular House. He assigned Mizell's Independent Company to provide cover. López summoned Bell's command, "with instructions not to fire on the building while marching to the Plaza." The Louisianians were met by a fusillade from the upper windows, hitting Lt. E. L. Jones and mortally wounding in the abdomen Lieutenant Sexias, of Mizell's company. Captain Robinson's company occupied a residence opposite the Capitular House, from where they fired on the building. Gonzales, always close to López during the battle, "exposed himself in the thickest of the fight and danger, passing from place to place communicating the orders of the commander in chief."[38]

The Louisiana Regiment "surrounded the square containing the Governor's palace, returning the fire that constantly proceeded from it." One filibuster recalled being shot at from the second-story windows "and the parapet of the roof, and we could only see them from the flash of their pieces. We got in some of the houses opposite, sometimes returning their fire, till daybreak . . . though they kept picking us off all the time." A bullet snipped Capt. A. Mizell's goatee. When the Jennings patent rifles assigned to his outfit failed to discharge properly, Mizell cursed at his weapon, broke it over a stone, and sent his men back to the *Creole* for muskets. The shouts of the invaders, a vestige of the Mexican War, were described as "most appalling."[39]

When dawn broke at 6:00 A.M., López had Gonzales accompany him to demand the surrender of the Capitular House. As they crossed the plaza, the Spaniards noticed the adjutant general's flamboyant uniform, mistaking him for the leader. Two shots fired from the upper floor hit Gonzales in the left thigh, knocking him down. López carried him to safety, probed the wound, and exclaimed that it was not fatal, due to the lack of dark arterial blood. In agony, the adjutant general was taken to the makeshift hospital. This inopportune event propelled Gonzales into immortality as the first native to shed his blood in combat for Cuban independence. Capt. J. C. Davis, who described Gonzales as "the life and soul of the Expedition," stated that the wounds kept the adjutant general laid up for the next three weeks. Davis believed that the battle-hardened López came out unscathed due to "the celerity of his movements; it was impossible to take a correct aim at him, unless the Spaniards could shoot on the wing, which I am told they cannot do." Lt. Richardson Hardy was amazed by the physique of the fifty-two-year-old Venezuelan: "Of medium size, a body of more compactness and strength, and agility, could hardly be imagined."[40]

An enraged López ordered torching the Capitular House at about 7:00 A.M. Bolts of cloth taken from the tailor's shop were kindled and thrown through the upper floor windows. Staff Captains Elliott and Hall, "standing out in the middle of the street exposed to a continuous fire," each shot a man upon the parapets. Ceruti and his soldiers put up a fierce resistance before the flames and smoke forced some of them to jump to the lower roof of an adjacent residence, which collapsed under them. Ceruti and his men, almost out of ammunition, opted to surrender. Lacking a white flag, Miss Regla Rodríguez-Capote, a Capitular House resident, offered her white petticoat, which was waved from a window on a bayonet. Ceruti emerged at 8:00 A.M., his head bandaged, accompanied by Captain Segura, Second Lieutenant Aguado and Alejandro Rodríguez-Capote, who begged López to extinguish the fire consuming his home.[41]

Ceruti announced the capitulation of Cárdenas and relinquished his sword to Theodore P. Byrd, adjutant of the Louisiana Regiment. The joyous "loud and long shouts of the men" were heard by Private Marion C. Taylor of Lieutenant Colonel Pickett's detachment at the railroad depot a mile away. As Captain Segura gave up his sword, he pointed to the lieutenant governor, saying, "Because of his fault," since he had initially opposed entrenching his troops in the building. Lieutenant Hardy regarded Ceruti as "a very fine looking, dignified and soldierly man." Sergeant Boggess felt that the Spaniards "had fought like devils." A group of soldiers who remained in the Capitular House and refused to surrender, killed and wounded several expeditionaries. This prompted a filibuster squad to enter the rear of the building, "where they found some Spaniards whom they instantly bayoneted and cut to pieces with sabers." Ceruti and some "forty men" were marched to the jailhouse by Capt. Abner C. Steed's Company A of the Louisiana Regiment, with Juan Macías acting as interpreter. The U.S. consul in Havana reported a few days later to his government that López had captured "about seventy soldiers."[42]

Capt. J. C. Davis, of the Louisiana Regiment, ascertained that the invaders had "three killed and nine wounded," that morning. Kentuckian Richardson Hardy placed the toll at "some six or eight killed, and twelve or fifteen wounded; the Spanish loss was probably about the same, notwithstanding that they had fought most of the time behind impenetrable walls." Major Hawkins reported that the Kentucky Regiment "lost eight, killed and wounded," while Lieutenant Colonel Bell indicated "some twenty" Louisianian casualties. John Sidney Thrasher, an anonymous correspondent in Havana for various American newspapers, went to Cárdenas five days later. He reported in the *New Orleans Picayune* under the pen name "Peregrine" that "eighteen were found dead and between thirty and forty wounded" that morning. The injured from both sides were attended by filibuster physicians John C. Bates, Thomas J. Kennedy, and surgeon Stull with his assistant A. A. Josephs, along with the local surgeon, Dr. Padrines. The wounded were sent to the *Creole* on two-wheeled *volantes*, while Gonzales rode the carriage of assessor Blas Dubouchet.[43]

Upon the evacuation of the Capitular House, the cathedral bell tolled the incendiary alarm. When the fire brigade arrived with two water engines, the filibusters astonished the townspeople by helping extinguish the blaze. The Kentucky Regiment returned from O'Donnell Street, the troops pitched a field tent in the plaza and stacked their arms, pickets were stationed at a jewelry store and on the outskirts, while a squad buried the dead in the municipal cemetery. The expeditionaries, who had not eaten or slept in twenty-four hours, searched for food and drink or took a nap. López, initially designating the jail as his headquarters, went with Adjutant Henry Titus and two others to the Municipal Council office, and expropriated $5,132.75. They proceeded to the Custom House on Real de Isabel II Street, where Titus demanded the "availables" from chief agent Pedro Nadal. Customs Administrator Tomás Fernández de Cossío, had just retrieved $15,952 for safekeeping and fled with his family to a French brigantine in the bay. Nadal offered López the gold-rimmed spectacles that his boss had forgotten. Titus seized the administrator's horse, and a three-key iron box, which was hauled to the jail by two slaves. Lt. Richardson Hardy recalled the "amusing sight to see the gallant Adjutant guarding the safe several squares by himself, sword in hand, having thrown away his coat and hat in the heat of the engagement." The safe, requiring separate keys held by three administrators, was breached with difficulty, yielding $1,492. At the harbor, Fernández de Cossío encountered ten armed customs agents who avoided the fight by remaining at their posts. He ordered them to board the foreign vessels at the harbor anchorage, one mile from the *Creole,* where many women and children had sought refuge, fearing that it was a slave uprising or a jail breakout.[44]

López returned to Quintayros Plaza, ceremoniously raised the Cuban flag in front of the church, and called on the citizens to gather in the square. He issued an edict that anyone failing to relinquish their horses or firearms within one hour would be summarily executed. This netted "a few old shotguns, and forty or fifty fine horses." Customs officer José M. Navarro was given a stack of printed proclamations to distribute throughout Cárdenas. These broadsides, emblazoned with the revolutionary coat of arms, were addressed, "The Spanish Army in Cuba," "To the Lovers of Liberty in Cuba," "To the Spanish Peninsulars," and "Narciso Lopez, Chief of the Cuban Forces," to the citizens at large. About fifty people, motivated by concern and curiosity, responded by delivering "a number of old swords and fowling pieces," but none was willing to use them. The general delivered a patriotic speech denouncing tyranny, in which he announced the establishment of a provisional government until the royalists were expelled "and the will of the people could be freely expressed." Sergeant Boggess indicated that López "still believed that as soon as the news reached the inhabitants that he had landed, he would receive large reinforcements of men and cannon."[45]

The general invited the crowd to fight for their country, but none volunteered join the foreign invaders. José Rosario Pineda, a mulatto bricklayer and fireman, cried out, "Viva la Libertad!" His coworker, the free black Anacleto González, shouted, "Long live the English and death to the Spaniards: we are free!" Pharmacist Félix Llanes carried independence proclamations to Lagunillas, and mulatto carpenter Pedro Ordaz assisted rounding up the few available horses for the expeditionaries. Canary Islander Bernardino Hernández, a thirty-year-old innkeeper, provided his best horse for the general. Felipe N. Gotay, a "tall, handsome" native of Peñuelas, Puerto Rico, also joined the filibusters that day. He was not a Cárdenas resident, spoke English, and had military experience, but nothing else is known about him. Gotay may have been a former Spanish soldier, a traveler, or an employee of the railroad or a nearby sugar mill. Sergeant Boggess recalled how the blacks "had a holiday and they were singing and dancing, thinking they would be released from slavery."[46]

Capt. J. C. Davis claimed that the inhabitants did not revolt because they were mostly Castilian Spaniards unsympathetic to the Creoles and loyal to the Crown. Lieutenant Hardy noticed that most of the citizenry "merely walked about, bowing and scraping to the red shirts." His brother William attributed "indifference and abject timidity" to their lack of response. Seven months earlier, a concerned writer had warned in the *New Orleans Picayune* that any attempt to form a Cuban republic would have "disastrous results," due to the prevalent "low state of education" of the masses and the despotic principles "imbued in their ideas of rule and sovereignty." López had expected his former subordinates, the populace, and most of the American expatriates residing in Cárdenas to join him, but Captain Davis recalled that "*the emigrants from the United States at Cardenas were as hostile to us as the Cubans.*" Edmund T. Doyle, a New York City merchant living there with his sister, visited Lieutenant Governor Ceruti in prison and "offered him his wardrobe and his services." In spite of this generosity, Doyle's sister complained to Secretary Clayton six weeks later about government reprisals against her brother and the other Americans, who were "treated like dogs, no respect paid them."[47]

One of the López proclamations called for a conscript army, but the general did not enforce this unpopular measure. He angrily expressed his frustration after the populace failed to support him. López returned to the jail at 10:00 A.M. and harangued the defeated Spanish troops, appealing more to his former subordinates than to the citizenry. Twenty-five soldiers and a sergeant of the Cárdenas Company of the León Infantry Regiment "responded with a loud shout, stripped off their uniform," donned filibuster red shirts, and swore loyalty to López. The general then rode on horseback to inspect the railroad depot outside of town and informed Lieutenant Colonel Pickett and his men that Gonzales, Wheat, O'Hara and others were wounded, and that some Kentuckians had been killed.[48]

The exemplary behavior of the invaders inspired the local merchants to open for business that Sunday morning. Lieutenant Colonel Bell indicated that "much apparent kindness was manifested on the part of the citizens toward the troops, who, with the exception of those detailed to duty, were passing from place to place through the town in pursuit of rest and refreshment." U.S. consul Robert Campbell informed his government three days later that "private property of all kinds was respected." Thrasher reported that after the battle, "the men dispersed themselves over the town. No one disturbed them, nor did they interfere with anyone. They ate and drank in the shops, paying for what they received, and the inhabitants say, were exceedingly polite to everyone." The expeditionaries celebrated their victory with rum and brandy and soon got *"pretty well corned."*[49]

Late that morning, Felipe Gauneaurd and Basilio N. Tosca, both independence sympathizers in their mid-twenties, arrived in Cárdenas from their nearby sugar mills. They met with their old friend Juan Manuel Macías, who was guarding the jailhouse, and inquired about their acquaintances exiled in the United States. When told that Gonzales was on the *Creole,* they immediately went to visit him. After reminiscing with Gonzales, Gauneaurd and Tosca returned at noon to have lunch with López and a group of officers at *El León de Oro.* They saw López stomp his foot on the ground and exclaim: "I made a mistake, we should have gone to Matanzas." Lacking sufficient horses to mount his entire command, the general decided to take them there by train.[50]

Spanish loyalist holdouts sporadically ambushed the invaders. At noon, an American officer scouting the town on horseback was shot dead by a Spanish innkeeper named Carricarte. Two filibuster deserters attempting to leave Cárdenas were gunned down, one on the corner of Laborde and O'Donnell streets and another on the road to Lagunillas. Five other deserters, Captain Dupeau, a Mexican War veteran; George Warner of Evansville, Indiana; nineteen-year-old William Kelly of Cincinnati; and two Scotchmen, Thomas Stevenson and Pvt. McGreggor, threw away their arms and red shirts. The three Americans appealed to U.S. vice consul George Bell for protection, asserting that "they had been first deceived and then forced to take arms," but he refused to intervene. Bell was a British subject employed as a clerk in a commercial house. Stevenson surrendered that afternoon in the home of a laundress to revenue guard Antonio Vázquez, manifesting that "he wanted to stay because he had been deceived."[51]

By 2:00 P.M., the tide had entered the bay, allowing the refloated *Creole* to dock properly. López ordered part of the Louisiana Regiment and two Mississippi companies to return to the steamer and transfer the ammunition and stores to a train at the railroad station for the trip to Matanzas. Quartermaster Lt. Thomas P. Hoy supervised the operation, using animal-drawn wagons and some thirty-five slaves. Lieutenant Colonel Bunch volunteered to assist on board, while Maj. Peter Smith with one company of the Mississippi Regiment and Captain Mizell's

Independent Company, were sent to the plaza. During the hour spent unloading the vessel, López rode on horseback around town, checking the position of his troops and the situation at the railroad station and the wharves. He wavered on his next move. His undisciplined troops were scattered, four of his principal officers—leaders O'Hara and Wheat, and staff members Gonzales and Captain Murry—lay wounded, and the populace had not rallied to his cause.[52]

After the *Creole* was unloaded by 3:00 P.M., López sent Mississippi Regiment Adjutant J. Collins with an order for Lieutenant Colonel Bunch at the steamer to forward a company to bolster the Kentuckians at the plaza. Bunch ignored the command, opting to remain safely on board with his men. An hour later, a horseman from Matanzas arrived "at full speed down the street, with his hat off and holding up a white handkerchief." He dismounted at the square calling for López, who greeted him with an embrace, and they began "talking excitedly in Spanish." He told López that all the Matanzas troops and artillery were heading for Cárdenas. Herminio Portell Vilá speculated that the newcomer might have been the enigmatic Puerto Rican Felipe N. Gotay. About 4:00 P.M., the general called Maj. Thomas Hawkins of the Kentucky Regiment and Lieutenant Colonel Bell of the Louisiana Regiment to the second floor of a residence on Rúez Street, between Princesa and Obispo streets, two blocks from the Quintayros Plaza, where he had moved his headquarters.[53]

López informed the officers that a massive enemy force would arrive by train from Matanzas in nine hours. Having failed to take Cárdenas by surprise, he decided to reboard the *Creole* and go to the Vuelta Abajo region of Pinar del Río, where "he would find a force organized and ready to support him." Months earlier, one of the Villaverde brothers had claimed that two thousand armed rebels in Vuelta Abajo were awaiting their arrival. Commissary Jonathan D. Rush McHenry was ordered to load on the steamer the armament transferred to the train. Lt. Col. John T. Pickett and the Kentuckians were recalled from their position at the railroad yard outside of town. Quartermaster L. C. Thomas employed a slave force to load the captured weapons and flags on the *Creole*. The officers commenced gathering their dispersed troops, which took some time, as some of the men were drunk and wandering about the streets.[54]

Lieutenant Colonel Bell led the rest of the Louisiana Regiment on Quintayros Plaza back to the steamer. They were followed by Captain

Steed's company, conveying Lieutenant Governor Ceruti, Captain Segura, commander of the León Infantry Regiment, and his second lieutenant Aguado, who were taken as hostages. Colonel O'Hara requested that Major Hawkins send Capt. Henry Robinson's company to the *Creole*, to facilitate reloading and supplying the vessel with water and the only available fuel, a low-grade coal dust. A slave gang used for this task tried to remain on board but they were "kept off the steamer by force." Hawkins moved his command to the plaza, consisting of "about 85 or 90 men" from Fielding C. Wilson's Company H, a portion of Albert W. Johnson's outfit, and the companies of Captains Knight and John A. Logan, the first led by Lt. Joseph C. Dear. When López perceived after 5:00 P.M. that his earlier command to Lieutenant Colonel Bunch had been ignored, he ordered Maj. Peter Smith and his two Mississippi companies and the Louisiana Regiment to board the *Creole*. The general relayed to Major Hawkins that an enemy force of infantry and lancers was approaching the town. Hawkins was ordered to form the Kentuckians on Quintayros Plaza, to protect the embarkation, and a bugler would signal their return to the steamer. The major remained on Real de Isabel II, assigning Captain Johnson's company to Ayllón Street, on his right flank, and sending Lieutenant Dear to Laborde Street, on his left.[55]

When Lieutenant Colonel Pickett and his command were one block from the *Creole,* they encountered López at Pinillos Plaza. Pickett inquired if he should assume command of the Kentucky Regiment, but the general declined, since they were about to depart, and ordered Pickett to hold that intersection until further notification. Pickett then boarded the *Creole* and conversed with Gonzales, O'Hara, and some wounded Kentuckians. Thereafter he went to Quintayros Plaza with a surgeon who served as interpreter to obtain necessities for the wounded. Upon returning to his post, the lieutenant colonel "heard the firing of volleys of musketry in the Plaza."[56]

The Louisianians and Mississippians had just neared the wharf when "a severe firing" was heard coming from Quintayros Plaza. Lieutenant Colonel Bell immediately ordered Captain Steed on board with the prisoners, and proceeded down Real de Isabel II. López ordered him to form his command at the foot of that thoroughfare near the *Creole,* "as he anticipated an attack from a cross street." Major Smith, standing on the pier, asked Bunch, who was on the steamer's hurricane deck, "to order out his command to the relief of the Kentuckians; this he declined to do." Bunch then attempted to sever the mooring cable with his sword but was stopped when Captain Lewis "drew his pistol and stood by the rope, declaring he would shoot the first man who attempted to cut it." Lieutenant Colonel Bunch later publicly denied this action, which subsequently led to three bloody duels, in one of which he wounded the skipper. When the commotion subsided, Smith, still on the dock, again urged Bunch to employ his force. The lieutenant colonel replied that "he would not take the responsibility of so exposing the men under his command." Smith, a subordinate in rank, could not usurp Bunch's command and announced that he would lead those wishing to assist the Kentuckians. He was "instantly joined by nearly every man on board; Col. Bunch then said that he too would go."[57]

The 120 Spaniards approaching the plaza included 50 infantry soldiers of the León and Nápoles infantry regiments, 40 peasant militia lancers, and 30 civilian riders, mostly Vizcaino Spaniards from the nearby sugar mills. They had been mustered in neighboring towns by León Martínez Fortún, commander of the Guamacaro post, and Lt. José N. Morales, in charge of a lancer patrol in El Recreo and Lagunillas. Before Hawkins could post the Kentuckians throughout the plaza, Spanish infantry stealthily gained cover in the houses on Laborde Street and opened fire. They mortally wounded the unsuspecting Chaplain John McCann, who was carried off by his companion, Major Dixon. Major Hawkins hastened the movement of Lieutenant Dear's company to that flank, "and they were just in time to effect their object." The lancers advancing north on Ayllón Street ran into Captain Johnson's company.[58]

Thirty-two-year-old Corp. Feliciano Carrasco turned on his saddle toward Second Lieutenant Morales and uttered, "With your permission, my lieutenant," before charging alone at the Kentuckians with his raised sword. He was emulating a heroic French officer at Waterloo, who lunged into his opponents trying to find an aperture by which to penetrate the line. A volley of gunfire brought down the galloping horse, injuring Carrasco. He recovered and attacked the expeditionaries, imparting two-handed blows with his sword. Juan Manuel Macías engaged the corporal in a spirited fencing duel. Carrasco, wounded fourteen times, fell dead, shouting, "Viva Isabel II." After witnessing this act of suicidal heroism, Morales remained behind and ordered his men to turn one block east to Real de Isabel II. They stopped three blocks south of the plaza, "going through a variety of brilliant movements preparatory to a charge."⁵⁹

From that distance, Hawkins estimated the cavalry column of platoons to have "sixty or seventy" men. A "considerable body of infantry" advancing on Laborde Street engaged Lieutenant Dear's company on the left flank of the plaza. According to Major Hawkins, Dear "found some difficulty in repulsing" the superior number of the enemy, but "by great exertion on his own part and the gallant assistance of Adjutant Titus and Sergeant Major McDonald, they were finally driven back with considerable loss." Lt. Richardson Hardy recalled Titus as "'Gallant Harry!' Jovial and laughing even in the midst of fight; and a perfect Ajax in courage and proportions." As the action unfolded, the rooftops around the plaza, the houses and the by-streets, "were full of citizens, gazing with intense anxiety upon the scene."⁶⁰

The Spanish infantry that occupied the houses on Laborde Street kept a "severe fire" on the two companies with Major Hawkins in the center of the plaza. This persisted for about half an hour until the Kentuckians heard the bugle call to retire to the steamer. Hawkins never received "the reinforcements promised" of Bunch's troops, but his regiment managed to drive back "with severe loss" the enemy on his front and flank. The major drew in the units on Ayllón and Laborde Streets, and moved the two center companies back one block to Jerez Street. They were joined there by López, accompanied by his servant Pedro Velazco, Juan Manuel Macías and other aides-de-camp. The lancers, two blocks south on Real de Isabel II and Industria Streets, were emboldened by their withdrawal, and went "thundering on in gallant style."⁶¹

Hawkins had his two center companies in the middle of the street, while those of Captain Johnson and Lieutenant Dear were on the sidewalks. When the first platoon of lancers bore down on the Kentuckians, those on the street scrambled up on the sidewalks with the others. As the riders rushed past "at full speed," the filibusters "poured in a raking fire which brought to the ground nearly the whole body, some eight or ten of them escaping by one of the cross streets." These cavalrymen, encountering the Louisianians at the foot of the street, were "despatched" with a volley of musketry. One of the riders killed was an Irish American emigrant named O'Reiley who that morning had given a sumptuous breakfast to the filibuster officers. No Louisianians were injured during this encounter. On board the *Creole,* the prostrate and agonizing Gonzales heard the firefight and shouting, but was helpless to assist.[62]

A second line of "thirty to forty" lancers then advanced down Real de Isabel II while Hawkins and his men retreated one block to Cossio Street. They were joined there by Lieutenant Colonel Pickett and Lt. Burwell B. Sayre, with fifteen men from Captain Johnson's company. Both Kentucky commands regrouped in time to receive the second cavalry charge "in the same manner." The lancers who "dashed by at headlong speed" past the gauntlet of musketry encountered Capt. John Allen's company at the next intersection on Pinillos Street. Hawkins and Pickett regrouped their units back at Captain Allen's corner, where Pickett assumed command of all the Kentuckians. López then ordered them to fall back toward the steamer. Pickett halted his column about 150 yards from the pier at Pinillos Plaza, where he "barricaded the street with a double row of empty sugar hogsheads, and posted Captain Johnson's company, in two equal detachments, in ambuscade." The Spaniards made no further move after 8:00 P.M. and the orders to embark were given an hour later.[63]

The filibuster officers' reports described a great number of Spaniards being "despatched" in the afternoon encounter. Lieutenant Hardy purported that "out of a hundred, seventy or eighty lay killed and wounded," but Consul Campbell informed the Department of State on May 22 that "the lancers lost eight men and ten horses," and that among the expeditionary wounded was "the Creole Gonzales." Colonial official Justo Zaragoza indicated that the royalist losses were only Carrasco and "a corporal and three lancers." Private Marion C. Taylor, of Captain Allen's company, claimed that his outfit killed seven lancers, and that the filibusters had forty casualties that afternoon. Pickett and Hawkins reported losses of three officers and five privates killed, and nineteen wounded. Among the dead was Lt. James J. Garnett, a Mexican War veteran shot "through the brain," and the wounded included a "terribly mangled" Capt. John A. Logan, who was carried back to the *Creole.* Two aides and an officer were accidentally wounded by the discharge of pistols thrown on deck by those clambering aboard. The final toll for the expeditionaries that day was twenty-six killed, some sixty wounded, and seven deserters. The ten Spaniards killed were interred in La Cabaña fortress in Havana under an obelisk bearing their names. Cárdenas physician Antonio García Ortega later reported having assisted fifteen wounded defenders. There is no information on the number of injured soldiers attended to by other local residents and physicians.⁶⁴

As the *Creole* departed Cárdenas at 9:00 P.M. on May 19, the Spaniards fired a parting volley from the wharf. The additional passengers on board included three hostages, the twenty-five soldiers and a sergeant who defected from the Cárdenas garrison, and seven stowaway male slaves. Gonzales, although prostrated in the cabin with the others wounded, was given command of the expedition by López. He appointed Lieutenant Colonel Pickett as field officer of the day. Five miles from Cárdenas, an erroneous "mark twain" call by an improvised leadsman grounded the *Creole* on a sandbar. The receding tide prompted "a scene of horror rarely equaled" for all the expeditionaries. In a desperate refloat effort, Gonzales shouted orders as three tons of provisions, weapons, and all the ammunition, except eight boxes, were thrown overboard. The Spaniards later retrieved the ammunition boxes and exhibited them as war trophies. Capt. Armstrong Lewis and Callender Fayssoux desperately worked the engine, but the *Creole* refused to budge. Ninety passengers were ferried on two yawls to an island a quarter mile away, when the steamer dislodged just before daylight.⁶⁵

After everyone had reembarked, Gonzales participated in a war council in which López expressed a desire to land at Mantua, in the Vuelta Abajo region and "send the *Creole* back for Gen. John Anthony Quitman and his troops and munitions." He was initially supported by Gonzales and the other Cubans, Colonel Wheat, Major Hardy, Capt. John Allen and Adjutant Henry Titus but "the greater part of the company officers and consequently nearly all the rank and file, would not assent." Colonel O'Hara "declared the proposition to be madness." The Hardy brothers later described the general's plan as "desperate and reckless." Only seven Louisianians and some fifteen or twenty Kentuckians expressed their willingness to follow López, along with the Spanish deserters. The dissidents cited "the scarcity of ammunition, the absence of artillery, the scant supply of coal for the vessel, the limited quantity of water, and the tardiness with which the Cubans at Cárdenas joined the liberating standard."⁶⁶

López resigned his command and requested to be landed at Mantua with those willing to follow him. Captain Lewis stated that rerouting was impossible, since they barely had enough fuel to reach Key West. A large amount of coal and water had been consumed while dislodging from the sandbar. The few officers who had been willing to follow López, now declined the attempt. The "completely demoralized" filibusters mutinied, forcing the helmsman to take them directly to Key West. A fishing smack was hailed near Piedra Key to release the lieutenant governor and his two companions, who promised no harm to the five deserters remaining in Cárdenas. Sergeant Boggess recalled how the prisoners embraced López and Gonzales, kissing them on the cheek in gratitude, before departing. Upon exiting, Ceruti "waved adieu with his handkerchief in handsome style."[67]

When Ceruti returned to his post, he ordered the arrest of filibuster collaborators Bernardino Hernández, José Rosario Pineda, Anacleto González, Félix Llanes, Pedro Ordaz, Felipe Gauneaurd, and Basilio N. Tosca. Hernández was sentenced to death, and the others received prison terms of up to two years. The five deserters who remained behind were rounded up and sent to Matanzas on May 24 for a summary court-martial. Thomas Stevenson, who had assisted a wounded Spanish lancer, got "ten years labor on the public works." The others were executed by firing squad and interred in the General Cemetery, where Gonzales had laid his father to rest five years earlier.[68]

Two mortally wounded officers expired aboard the *Creole* after leaving Cárdenas Bay. Capt. John A. Logan and Lieutenant Sexias, stitched up in blankets weighed with thirty pounds of lead, were ceremoniously buried at sea at 3:30 P.M., on May 20, at latitude 24.3° north, longitude 81.5° west in the Strait of Florida. Suddenly, the Spanish war steamer *Pizarro*, which had dropped off the Contoy prisoners at Havana and departed there at 11:00 A.M., appeared from the west, spewing a tall column of black smoke. Gonzales estimated that since the *Creole* had cut its engines and was not making smoke, it was overlooked by the warship, which proceeded toward Key West. Admiral Francisco Armero Peñaranda, accompanied by two hundred grenadiers, had told the Matanzas port captain that if he "found López and the Creole at Key West, 'he would capture them and settle the hash.'" The wind increased in the evening and by dark "it was blowing a gale and continued through the night." The *Creole* missed the northwest pass of the Florida Keys, anchoring on the south side of the reef, some forty miles from Key West, "where it was decided to remain until morning."[69]

As dawn broke on May 21, Capt. William Smart, a wrecker, approached in his schooner. Captain Lewis offered to hire him as a pilot, but when Smart demanded $1,500, Lewis "threatened to kick him overboard," before agreeing on a $150 fee. The wrecker guided the *Creole* to the northern inside channel of the reefs. Ten miles further, as the vessel approached the Sand Key lighthouse at 10:00 A.M., the lookout shouted "steamer ahead." López mounting the pilot house, trained his telescope on the vessel for some time, before growling, "*Carajo,* Pizarro." The wounded Gonzales craned his neck to peer from his berth through the cabin window toward the horizon. The war steamer, which had been at the northwest pass that morning looking for the expedition, was sailing fifteen miles away in the opposite direction. A three-hour chase commenced when Armero spotted the *Creole* raising the Stars and Stripes. The *Creole* began skirting the sandbank chain, whose shallow waters kept the heavier *Pizarro* at a distance. When the *Creole* was within eight miles of Key West, the *Pizarro* was two miles away and gaining. Pandemonium broke out as some filibusters darted about stripping off their red shirts and hiding their weapons in the hold. Gonzales ordered them to desist, to keep the vessel trim, while Captain Lewis told the engineer to stoke the boilers and "blow the steamer to hell or send her in port ahead of the *Pizarro*."[70]

The coal dust taken at Cárdenas had slowed the *Creole* to five miles an hour and the last shovelful was thrown on the fire as the Key West lighthouse appeared. Thirty barrels of rosin that Gonzales had ordered at New Orleans for such an emergency were fed into the furnace. All available kindling, including bunks and furniture, were used for fuel, along with the last rations of bacon and discarded red shirts. The *Creole* increased velocity to ten knots, but the *Pizarro,* with a maximum speed of fifteen knots, continued to gain on them. Captain Lewis, furiously pacing the deck, kept kicking the smokestack signaling the engineer to put on more steam. The *Pizarro* took on a local pilot to cross the reefs, but ran aground for twenty minutes and then had to arc around the island chain. The sight of the Key West wharf was greeted by the filibusters with a shout. Hundreds of spectators were on the dock and the house tops, "watching the chase with intense anxiety."[71]

Thirty-six hours after leaving Cárdenas, Captain Lewis entered Key West, six miles ahead of the *Pizarro,* with only two barrels of rosin left.

Port regulations forced him to await clearance at the quarantine grounds. A terrified Lieutenant Colonel Bunch jumped into one of the yawls hanging on the steamer port side "and stood, bowie-knife in hand, ready to cut the boat loose at the first fire, and make for the shore." As the *Creole* struck the pier, the *Pizarro* "passed within 100 yards" with its ports raised at battle stations, "without firing the customary salute, paying no attention to the Health officer." Gonzales recalled: "As I look[ed] around through the cabin window, I could distinctly see the faces of the admiral and his officers standing on the upper deck to have recognized them had I known them, and the gunners at their guns, with the old-fashioned port fires burning."[72]

Bunch headed the mad scramble for shore that nearly swamped the *Creole*. The U.S. Navy surveying cutter *Petrel*, commanded by Lieutenant Rodgers, raced in between both vessels, preventing the Spaniards from firing a broadside into the *Creole*. As the wounded were evacuated, Gonzales gave the Masonic Grand Hailing Sign of Distress to the crowd on shore. Thirty-seven-year-old attorney Stephen Russell Mallory, the future Confederate secretary of the navy, ordered four of his servants to carry Gonzales home on their shoulders. Mallory, who had been the collector of customs until his political dismissal the previous year, was the correspondent for the Whig *New York Tribune* although he was a Democrat. Gonzales received great care and had one of the musket balls extracted from his thigh. The filibusters were sheltered a mile from Key West, in the U.S. Army barracks, vacated by the soldiers who had gone to suppress a Seminole uprising.[73]

Customs collector Samuel James Douglass hoisted the confiscation flag on the *Creole* for breach of the revenue laws, and Lieutenant Rodgers retrieved the weapons left on board. The *Pizarro* dropped anchor at 1:30 P.M., and Admiral Armero's offer to protect the Key West citizenry from the "pirates" was rejected. A rumor circulated that Armero was offering fifty thousand dollars for the delivery of López, who "did not show himself in the street during the day." The general slept that night on the barracks porch, while his aides took turns patrolling the area. The following day, May 22, the seven stowaway slaves were handed over to Armero under a federal court order returning them to their owner. That evening, after a fishing vessel arrived from Havana with a message for the *Pizarro*, the steamer immediately sailed.[74]

The filibusters then began departing Key West. Private Marion C. Taylor sold his double-barreled shotgun to obtain passage on a schooner to New Orleans with about two hundred filibusters. Sergeant Boggess and seventeen others took a sloop to Charlotte Harbor, Florida, where he and a few others settled permanently. López gave fifteen hundred dollars taken at Cárdenas to his officers, who distributed eleven hundred dollars among the wounded and the Spanish deserters, returning the rest to the general. The mail steamship *Isabel* arrived at midnight, taking López, José María Sánchez Yznaga, and other officers and men to Savannah. When informing Washington of these events, Key West district attorney William R. Hackley proposed that the filibuster staff and officers be prosecuted for violating the Neutrality Act.[75]

The failure of the expedition, according to Kentucky Regiment commander Theodore O'Hara, was due to "the fatal consequence of an indiscriminate enlistment of men," especially the "riffraff" of the Louisiana Regiment. Lieutenant Hardy blamed it on "the fatal error of landing at Cárdenas, instead of going to Mantua in the first place" and described the entire campaign as "a harum-scarum business." A number of other factors were also responsible. The impetuous López had spent only six weeks hastily gathering weapons, transportation, and six hundred recruits. His military strategy was minimal and illusive, and at times he seemed to improvise, such as when he changed the place of landing at the last moment. He failed to send word to Vuelta Abajo, where it was alleged that two thousand rebels lay waiting; an uprising there could have diverted government troops to that end of the island. And López failed to anticipate the fear and apathy that greeted the foreign invaders at Cárdenas. The Texas model for independence collapsed when even the local American residents refused to join the insurrection. The capability to achieve further military victories was dampened by insubordination. Although the *Creole* had grounded approaching the Cárdenas dock, López failed to take a harbor pilot along with the hostages to secure his quick departure.

Gonzales had been a key organizer and moving force of the expedition. His role as López's translator, traveling companion, and adjutant general involved him in every facet of the conspiracy to liberate Cuba from Spain. Gonzales served as the main link between Governor Quitman and the Cubans. He had been instrumental in recruiting volunteers, procuring weapons, and signing the bonds to raise funds. His activities did not go unnoticed by informants who were reporting to President Taylor and the Spanish minister. Gonzales impressed his fellow expeditionaries as being erudite, benevolent, energetic, industrious, ubiquitous, and an extraordinary orator. He was not only the first Cuban to disembark but also the first native to shed his blood for Cuban liberty. Gonzales was not discouraged by his wounds or the filibuster failure, and he was determined to try again to free his homeland from Spanish colonialism and make it a state of the Union.[76]

## Three

# FILIBUSTER FIASCOS

After the failure of the Cárdenas expedition, Gonzales remained in Key West at the Stephen Mallory home, recovering from his wounds. López, Sánchez Yznaga, and several followers departed Thursday, May 23, 1850, on the steamer *Isabel* from Havana, arriving in Savannah at 4:00 A.M. the following day. They registered at the City Hotel on the river front, prompting popular excitement. John G. Doon, the U.S. deputy postmaster and Spanish vice consul, quickly filed a complaint against López with Secretary of State John Clayton, who on Saturday afternoon telegraphed an arrest order for violating the Neutrality Act. López, detained by the federal marshal at 8:30 P.M., demanded an instant hearing. U.S. District Court judge John C. Nicoll scheduled the session for 10:00 P.M. López was represented free of charge by forty-three-year-old Robert Milledge Charlton and Col. William B. Gaulden, solicitor general of Georgia, Liberty County planter, and member of Masonic Zerubbabel Lodge No. 15 in Savannah. The eloquent Charlton had been a state legislator, Savannah mayor, U.S. district attorney, and judge of the Eastern Judicial Circuit. U.S. district attorney Henry Williams requested a postponement until Monday morning but was overruled. He then presented four local witnesses López had recently spoken with, who gave sketchy second-hand accounts of the Cárdenas events, but none could affirm that the expedition had been organized in the United States. After a few minutes' deliberation, Nicoll released López at a quarter before midnight because "no sufficient evidence had been shown to justify his commitment."[1]

López and his coterie departed for New Orleans the following morning, Sunday, May 26, by way of the railroad-stagecoach-steamer route. Spanish minister Angel Calderón de la Barca sent a note to Secretary Clayton, demanding that López "and the rest of the pirates who are at Key West" be charged under the Neutrality Act. Clayton immediately instructed the federal prosecutors in Mobile and New Orleans to arrest López. Meanwhile, the filibusters in Key West departed within a week, except most of the wounded, including Gonzales.[2]

López surrendered to the federal marshal in New Orleans on June 7 and was taken before U.S. District Court judge Theodore McCaleb, accompanied by his counsels, John Henderson, Seargent Smith Prentiss, and Laurent J. Sigur. The *Picayune* court reporter described López as "about 45 years of age, middle size, rather stout, dark complexion, with very black eyes, and black eye brows, high forehead, hair slightly gray, and with gray whiskers under his throat." Prosecutor Logan Hunton told the court that López had been arrested on the affidavit of Spanish consul Juan Ygnacio Laborde Rueda. The following day, López was remanded for trial under a three-thousand-dollar bond provided by Democratic activist James H. Caldwell, recorder of the Second Municipality.[3]

The U.S. marshal in Key West received orders to detain filibuster leaders Gonzales, Theodore O'Hara, Chatham Wheat, John Pickett, William Bell, and Thomas Hawkins, along with *Creole* Capt. Armstrong Lewis, and to remit them to New Orleans for trial. Wheat and Lt. Col. M. J. Bunch slipped away. These two men were the only commanding officers of the Army of Liberation who neglected to issue a military report of operations to Adjutant General Gonzales. While recovering in the Mallory home, Gonzales met influential Key West merchants William H. Wall and Asa F. Tift, the former junior warden of Masonic Dade Lodge No. 14. They each reserved an escape vessel for him, while Mallory kept a fast rowboat ready to take Gonzales to a hiding place in Boca Chica Key. Instead of fleeing, Gonzales surrendered to the federal marshal, who escorted him to New Orleans on June 15 in a chartered sloop.[4]

Six days later, a New Orleans federal grand jury indicted Narciso López, Ambrosio Gonzales, John A. Quitman, John L. O'Sullivan, Theodore O'Hara, Laurent Sigur, John Henderson, Chatham Roberdeau Wheat, Cotesworth Pinckney Smith, Jean Baptiste Donatien Augustin, Thomas Theodore Hawkins, John T. Pickett, J. A. Hayden, M. J. Bunch, Peter

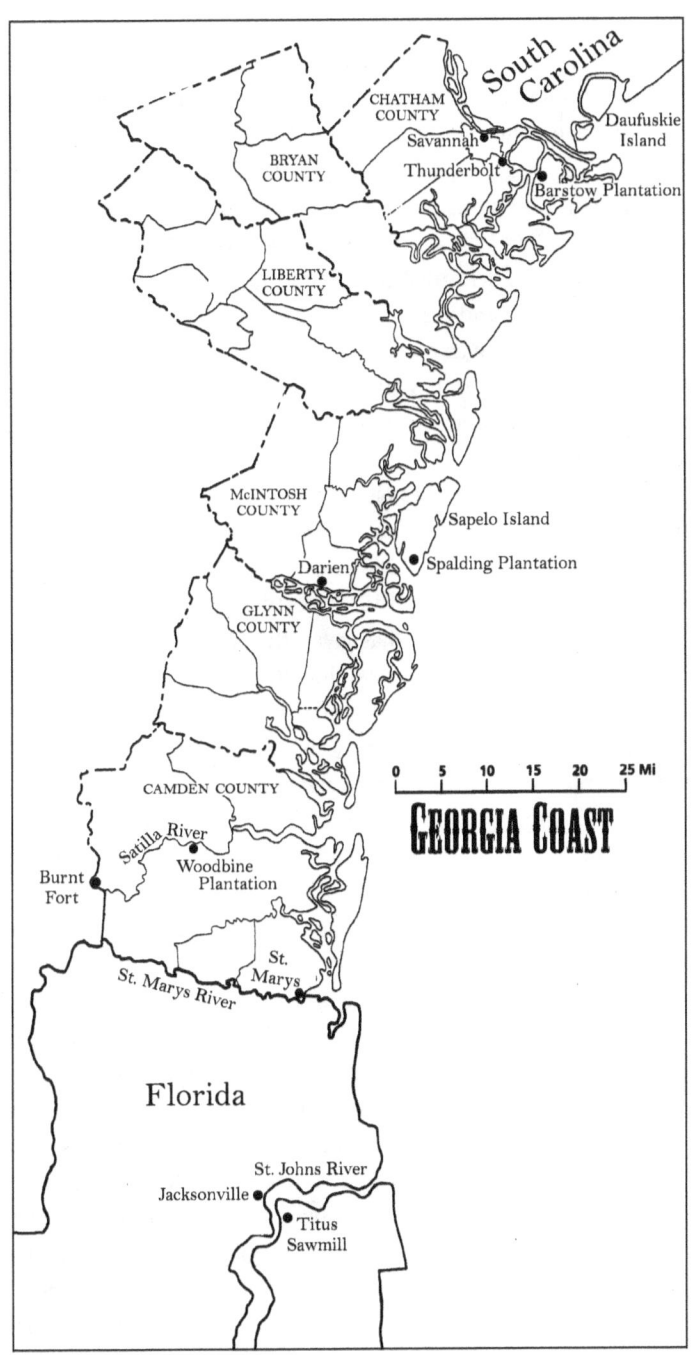

*The Georgia Coast, 1850*

Smith, and William H. Bell for violating the Neutrality Act. The accused "entered into Bonds for appearance at the next December term of Circuit Court." Two days after the indictment, Gonzales accompanied López, Sánchez Yznaga, and Felipe N. Gotay, a Puerto Rican who had joined them at Cárdenas, to a Masonic festivity in Gainsville, Mississippi. They arrived on June 23 and were received by their hosts, forty-eight-year-old Col. Thomas J. Ives, a former state senator, attorney A. B. Bacon, Dr. Gouldin, Ciro Butler, and others. Gonzales translated for López the acclamation rendered by Ives on behalf of the welcoming committee, proclaiming "a homely but hearty welcome to a soil consecrated to liberty, where we have homes and sanctuaries for our friends, arms and graves for our enemies."[5]

After the Cubans returned to New Orleans, Gonzales departed on July 2 with a letter from Henderson to John Quitman. They wanted the governor to lead a renewed effort to liberate Cuba. In spite of the Cárdenas failure, Henderson was optimistic that the action had "*planted the thought,* has embodied the *idea,* has given the fever to Cuban Revolution & independence that will never be obliterated." The letter stated that the South could provide more than one hundred thousand men if necessary, that the best officers of the U.S. Army "would contribute largely to your assistance," and that a couple of war steamers would be procured to sweep the Spanish navy off the coast of Cuba. Henderson indicated that Gonzales would personally give Quitman "all the Cuban items from which you will determine your course."[6]

For the new endeavor, Gonzales and López went on a recruitment drive, which was closely monitored by Spanish agents. Minister Calderón de la Barca informed the new secretary of state, Daniel Webster, that López had "left New Orleans in company with an obscure individual called Gonzales, wounded in Cárdenas, and who styles himself general. Both arrived at Mobile on the 11th of this month, and were to embark on the 13th in the direction of the north." The two "shameless villains" and "disturbers of the peace," stopped a few days later in Macon, Georgia, where they "found many sympathizers." They continued by train to Savannah, arriving at the Pulaski House on July 21. The Cubans left written petitions to join Solomon's Lodge No. 1, the oldest and most historic Masonic temple in Georgia. Two days later, while Gonzales remained in Georgia, López briefly stopped in Charleston, South Carolina, and continued to New York City, where he registered incognito at the Astor House on August 7, moving to Barnum's Hotel within a week.[7]

The expedition leaders, who had been indicted the previous December in Havana with some members of Mádan's Cuban Council, were tried in absentia by the Permanent Military Committee of the Council of War for the Island of Cuba. López was excluded since he was already under a death sentence by default for the 1848 conspiracy. The tribunal, by unanimous vote on August 19, sentenced to death by garrote: Ambrosio José Gonzales, José M. Sánchez Yznaga, Cirilo Villaverde, Juan Manuel Macías and Pedro de Agüero; to ten years overseas imprisonment and perpetual banishment from Cuba and Puerto Rico: Victoriano de Arrieta, Gaspar Betancourt Cisneros, and Cristobal Mádan, "with payment of costs, and also the damages sustained by individuals and the State from the invasion of Cárdenas." That month, the Military Tribunal also dismissed all charges against forty-two Contoy prisoners, seven crew members of the *Georgiana* and *Susan Loud,* confiscated both vessels, and sentenced Capt. Rufus Benson, his pilot Joseph A. Graffan, and Thomas G. Hale, the mate of the *Susan Loud,* to eight years at hard labor. The trio was pardoned by the Spanish queen three months later.[8]

When López returned to Savannah on September 8, he was pleased to see that during the previous two months Gonzales had organized in Georgia an extensive base of support. Its activists included Savannah mayor Dr. Richard Wayne, Dr. Richard Dennis Arnold, Savannah alderman Solomon Cohen, John E. Davis, John Lama, Alonzo B. Luce and cigar makers Antonio Ponce and Tomás M. Rosis, among others. Wayne, Arnold, Davis, Lama and Ponce were affiliated with Solomon's Lodge No. 1. Cohen, a wealthy attorney and former postmaster, owned twenty-three slaves, served as the local Democratic party leader, and was a prominent member of the Mikve Israel congregation. The thirty-four-year-old Lama was a liquor and grocery merchant who emigrated in 1842 from France. Ponce, a thirty-six-year-old Spaniard, had resided in Savannah for over a decade, had married a local lady, and owned four slaves who worked in his cigar factory. Rosis, a thirty-eight-year-old poet from Havana, had recently married a Savannah woman, owned a small "Segar, Snuff, and Tobacco Store" on Bay Street, and had a female child slave. An

R. G. Dun and Company credit agent described him as a "clever, industrious man, honest and attentive to business," with "good character." Luce, a thirty-five-year-old New Yorker, proprietor of Our House and Restaurant on Bay Street, was a member of Zerubbabel Lodge No. 15 and belonged to the Third Beat Company of the Georgia Militia. Ponce, Rosis, Davis, and Arnold were members of the Republican Blues infantry, First Regiment, Georgia Militia. Their company captain, John W. Anderson, also supported the filibusters.[9]

After Gonzales introduced López to the new adherents, the general returned on September 9 to Solomon's Lodge No. 1, where an "Extra Meeting" was held to give him the first two Masonic degrees of Entered Apprentice and Fellow Craft. Another function was held the following evening, where López received the Master Mason degree before leaving for New Orleans. He later recommended to O'Sullivan and Sánchez Yznaga in New York that they join the fraternity. Gonzales remained in Georgia, expanding the filibuster support network throughout the state. He returned on November 12 to Savannah on the steamer *Ivanhoe* from Burnt Fort, Georgia, where he established a rendevous point for the next expedition. Nine days later, he attended a meeting of Solomon's Lodge No. 1 and received the Entered Apprentice and Fellow Craft degrees, paying a fifteen-dollar fee. Gonzales soon left for Jacksonville, Florida, where he coordinated activities with Henry Theodore Titus, who two months earlier had established a sawmill there on the St. Johns River.[10]

Gonzales returned to Savannah on December 6, accompanied by filibuster activists Charles H. Hopkins, William Henry Mongin, and Randolph Spalding. Hopkins, a thirty-eight-year-old McIntosh County planter and militia colonel, later became Titus's brother-in-law. The thirty-four-year-old Mongin, a private in the Georgia Hussars cavalry of the First Regiment, Georgia Militia, was a planter on Daufuskie Island, South Carolina, and resided in Savannah with his wife Isabel R. Habersham, a cousin of Gonzales's future father-in-law, William Elliott. In coastal McIntosh County, the twenty-eight-year-old Spalding was a cotton and sugar planter on Sapelo Island, where his father, Rep. Thomas Spalding, owned many slaves and a grand colonial residence which today still exhibits Cuban rainwater *tinajones*. Four days later, Solomon's Lodge No. 1 held an extra meeting where Gonzales was "raised to the Sublime degree of Master Mason with the proper Masonic Instructions from the Worshipful

Master" Richard T. Turner. Those attending the ceremony included George Cuyler, who provided lodging for Gonzales, and filibuster activists John Lama and Antonio Ponce.[11]

Six days later, on December 16, the filibuster trial started in the U.S. Circuit Court of New Orleans, with Judge Theodore McCaleb. The court clerk called out the names of defendants Narciso López, Theodore O'Hara, John Pickett, Thomas Hawkins, William H. Bell, Chatham Roberdeau Wheat, John Henderson, Cotesworth Pinckney Smith, Armstrong Irvine Lewis, Laurent J. Sigur, Jean Baptiste Donatien Augustin, Ambrosio Gonzales, John Quitman, and John O'Sullivan, the last three not being present. Quitman had refused to answer the charges until the expiration of his term as governor. Gonzales would not arrive at the St. Charles Hotel until December 17, and O'Sullivan followed a week later.[12]

The trial resumed on January 2, 1851, before a "densely crowded" courtroom. U.S. district attorney Logan Hunton hired attorney Judah P. Benjamin as special prosecutor, due to the magnitude of the case. Benjamin would later become Whig U.S. Senator from Louisiana in 1853 and Confederate secretary of war and secretary of state. The jury was empaneled four days later, when Hunton and Henderson, representing himself, made their opening statements. Henderson challenged the indictment and argued for four hours on the vagueness and phrasing of the Neutrality Act, completing his argument the next day. Benjamin then substantiated the accusations. Henderson submitted a motion compelling the district attorney to produce three letters he exchanged with Secretary of State Clayton about the expedition. On Friday, January 10, the prosecution presented expeditionaries William Redding and John Reed, and former Contoy prisoners Archibald B. Moore and John H. Finch, who detailed their involvement. Benjamin then entered as evidence the *Creole*'s port clearance. The next morning, Robert Geddes took the stand and gave a full version of the *Creole* sale transaction made the previous April with Henderson. During his cross-examination, Henderson admitted to the jury that he had purchased the *Creole* for the expedition. The prosecution then entered as final evidence the port clearances of the *Georgiana* and *Susan Loud*, and rested its case.[13]

Gonzales and Laurent J. Sigur were the only witnesses called by the defense. The Cuban took the witness stand after midday on Saturday, keeping a ramrod-straight posture during the interrogation, conscious that all eyes in the courtroom gazed upon him. He gave an account of his arrival in the United States in 1848 as a political exile. Gonzales indicated that López had been sentenced to death by the Spanish Crown for his revolutionary activities in Cuba. He added that the Junta formed publicly in New York was headed by López, and that three of its four Cuban members disembarked in Cárdenas. The adjutant general stated that the Junta took advice from congressmen and "officers of high station" on how to accomplish their project "without violating the laws of the United States." Gonzales, whose vanity characteristically flared when he was under great stress, testified that he had been "perhaps more intimate than any other person with General López since his arrival in the United States. He neither speaks nor understands English, and I have acted as interpreter" in most of the conversations with Henderson.[14]

Gonzales said that he was introduced to Henderson at the White House, that López met him in early April, and that Henderson was not present when the expedition arrangements were made with the Kentuckians. He emphasized that "no act of military organization occurred in the United States," the Junta had merely arranged "an emigration," and that the *Susan Loud* departed from New Orleans without weapons. Special prosecutor Benjamin cautioned the witness against swearing to something he had not observed. This attempt to bait Gonzales was effective. The Cuban replied that "from his intimate connection with the expedition, he perhaps knew as much, if not more, than most who were engaged in it in relation to all the details." Under stress, his ego betrayed him; he admitted to an extensive knowledge of events, which proved a stumbling block during cross-examination. Gonzales, who had a law degree but had not practiced the profession, should have known better.[15]

Gonzales spoke the truth when stating that Henderson was not involved in the financial or military part of the *Susan Loud* emigration, but did not say that those arrangements were made by Chatham Roberdeau Wheat. He admitted departing in the *Creole,* which retrieved weapons in the lower Mississippi, but omitted mention of Henderson's participation. The witness indicated that "promises were made in the United States of commissions for offices under the republic of Cuba," but that, in his opinion, "these promises, whether verbal or written, were not commissions." The only commissions authorized by López and the Junta were those issued in writing on blank forms the day before landing, and it was not until they were in international waters that he was appointed adjutant general. Gonzales emphasized that "Americans had the right to emigrate with or without arms." He admitted choosing Mujeres Island as the rendezvous because it was outside of American jurisdiction. While there under the American flag, he added, they could not be seized by Spanish warships because the vessels' paper authorized them to go from New Orleans to Chagres. Gonzales said that Henderson and others had consented to this scheme, as it did not violate U.S. law. He further stated that Cuban flags were provided for the vessels and for each regiment, but were unfurled only after leaving the United States. Gonzales reaffirmed that Henderson did not participate in the Cárdenas expedition. He tried to arouse juror sympathy by stating that he and all the Junta members had been sentenced to death by the Spanish government.[16]

Gonzales concluded his testimony for the defense after being on the stand for almost three hours. The judge called a recess until 5:00 P.M., when a rigorous cross-examination began. The shrewd Benjamin, later known as the "Brains of the Confederacy," proved a formidable opponent. Gonzales utilized his law training to withstand a "very minute and strict" interrogation and points "were frequently raised and debated with great warmth by the prosecution and the accused." The prosecutor, trying to discredit the witness, focused on his colossal admission of knowing more than anyone else all of the details of the expedition. When pressed by Benjamin for the particulars of the origin of the ammunition and the weapons, Gonzales testified that he could not recall or was not certain. He even went as far as saying that he was uncertain if his name appeared on the Cuban bonds, many of which had his signature along with that of López and others. The witness acknowledged that Cuban bonds worth one hundred thousand to one hundred fifty thousand dollars were sold in New Orleans at ten cents on the dollar, and that about forty thousand to fifty thousand dollars was raised by Henderson.[17]

On the redirect examination by Henderson, Gonzales claimed that the accused probably knew nothing of the weapons or how they were procured. He suddenly recollected, after Henderson refreshed his memory, that a limited number of the bonds were issued and signed in Jackson, Mississippi, and given to the defendant to sell. After answering a few more questions regarding the *Georgiana* and the *Susan Loud*, he stated that all who chose to return on those vessels had been free to do so, and that several did. The next witness was codefendant Laurent Sigur, who admitted participating in various expedition preparation meetings. He defined as ridiculous the precautions taken by Henderson not to violate the Neutrality Act, which he deemed unconstitutional.[18]

Henderson then offered into evidence the correspondence exchanged between Hunton and Secretary of State Clayton the previous year regarding the expedition. The Clayton letter to Hunton of January 22, 1850, showed that the prosecution was a result of orders from President Taylor, responding to complaints from Spanish minister Angel Calderón de la Barca. Henderson read to the jury Hunton's response four months later, assuring that "the leaders of the enterprise have had good *legal advisers* and have not rendered themselves amenable to our laws." The defense rested its case at 8:30 P.M. on Saturday and the court adjourned. Two days later, former Texas Republic president Mirabeau Buonaparte Lamar arrived at the St. Charles Hotel. He provided Gonzales with letters of recommendation for Georgia governor George Washington Towns, officers of the state militia, and other Masonic contacts. Gonzales quickly departed for Georgia to organize the next expedition before the trial concluded.[19]

On Monday morning, January 13, Benjamin spent all day giving the prosecution's final argument, pointing out the "the subterfuges to evade the law" to effect a "foray for the purpose of plunder." Henderson then addressed the jury for two days on what he perceived as the illegalities of his arrest and indictment. He insisted that the expedition was not formally organized until it reached Yucatán, and that Mexican neutrality, not American neutrality, was violated. The case ended on Monday, January 20 at 1:00 P.M. after Hunton presented his closing statement. The jury began deliberations and informed Judge McCaleb the next afternoon that they were deadlocked eight for conviction and four for acquittal.[20]

Another trial was set for Monday, January 27, 1851. It took three days to empanel a jury. Afterward, district attorney Hunton gave his opening remarks and Henderson presented the defense statement. The first witness, Archibald B. Moore, gave the same testimony as during the previous trial, implicating himself and others in the Cuban expedition. The prosecution accused Henderson of being responsible for the "murder, arson and rapine" committed at Cárdenas. The defendant angrily objected that he was not on trial for those offenses and that "the charge is most monstrous and atrocious." Henderson argued that "no evidence could be admitted to show that the acts of others were to be traced to his instigation." The prosecution wrapped up its testimony the following day and there was a four-day postponement due to a juror's illness. When the trial resumed, Gen. John Quitman, who had resigned as Mississippi governor four days earlier to answer the summons, pleaded not guilty and demanded a speedy trial.[21]

On Saturday, February 8, Henderson gave a four-hour closing statement to the jury. District attorney Hunton followed on Monday with a three-hour closing argument, and the jury was instructed. The next evening, the foreman told the court that they were equally divided on the verdict, six to six, resulting in their discharge. Eight days after the second hung jury, López felt very optimistic due to "the good results obtained by Gonzales in raising funds for the expedition in the state of Georgia." On February 20, Gonzales wrote from Georgia that Gov. George Washington Towns had given him 2 cannons, 400 muskets, pistols, and sabers. The adjutant general was expanding a network of collaborators in Savannah. He gave Cuban cigar maker Tomás M. Rosis a package with 150 photographs of General López to send to Macías in Charleston, so that it could be smuggled to Havana by Miguel F. Porto Zarazate, the purser of the steamer *Isabel*.[22]

The third filibuster trial began on March 3. Henderson was so confident that, in contrast to his lengthy presentation during the first trial, he did not reply to the prosecutor's opening statement. The testimony of the government witnesses concluded the next afternoon. Henderson then presented only two defense witnesses. Laurent Sigur stated that General Henderson had taken all precautions not to violate the Neutrality Act and that military commissions were issued at sea. General Quitman then described what constituted a military expedition, such as the one he led during the Mexican War. He said there had to be an organization, a contract between the parties assigning commands, and an obligation from others to obey, none of which had been effected by López in the United States. Special prosecutor Benjamin then gave his closing statement and the court adjourned at 4:00 P.M. When the trial resumed on March 6, Henderson summarized his defense in a few hours. Hunton presented his final argument in the afternoon. A third mistrial was declared the next day, when the jury deadlocked eleven to one for acquittal. Hunton then filed a *nolle prosequi,* dismissing all charges against Henderson, Gonzales, and all codefendants.[23]

That same day, General López complemented the organizational work that Gonzales was doing in Georgia by authorizing Henry H. Robinson and William A. McEwen in Cincinnati to sell one hundred thousand dollars in Cuban bonds and to procure twelve hundred muskets, one thousand swords, one hundred kegs of gunpowder, and five hundred uniformed men for a new expedition. William and Richardson Hardy, the former local recruiters, were excluded from the new endeavor. Gonzales was making good progress in Georgia, where on March 14 he had located fifty-two-year-old Gen. Mirabeau Buonaparte Lamar in Macon. Gonzales informed Lamar that thanks to his letters of introduction, Governor Towns had been "very friendly & *liberal*" toward him. This influence allowed Gonzales to sell many Cuban bonds and to acquire ten pieces of brass artillery. Several state militia cavalry and infantry companies promised to provide their armaments, and he "made arrangements to collect one thousand men and from 200 to 300 horses."[24]

Gonzales told Lamar that a steamer had already been procured that could carry up to seven hundred men and two hundred horses, and that he was "using every effort" to raise ten thousand dollars to send to López to complete the purchase of another steamer, capable of running sixteen miles an hour with the remainder of the expedition. The second vessel would allow passage for the eighteen hundred recruits in the Southwest ready to follow Mexican War veterans Col. Robert L. Downman, former worshipful master of Masonic Halo Lodge No. 5 in Cahaba, Alabama, and Col. W. S. Clendenin, a Jackson, Mississippi, attorney who participated in the battles of Monterrey and Buena Vista. Described as "a gentleman of great moral worth," Clendenin was appointed by Polk's Treasury secretary Robert J. Walker to the New Orleans Customhouse, but was later dismissed by President Taylor. To complete his expedition plans, Gonzales needed from Lamar "very especially your moral influence" with friends in Macon and Columbus, and with his cousin Charles Augustus Lafayette Lamar in Savannah.[25]

In Columbus, Col. John Forsyth, a thirty-seven-year-old newspaper editor and commander of the Columbus Guards Company of the Georgia Militia, was involved "to a very great extent" in the secret operations, and Capt. Ezekiel C. Davis, a twenty-seven-year-old merchant, unloaded

Cuban bonds at ten cents on the dollar. Forsyth served as adjutant of the Georgia Regiment in the Mexican War and was later the local postmaster, until he was turned out of office by the Taylor administration. In Macon, Gonzales had the collaboration of Robert A. Smith, Sidney Lanier (uncle of the renowned poet with the same name), and Capt. Isaac Scott, a wealthy forty-year-old merchant and president of the Macon and Western Railroad Company, later a Department of State informant. Lanier and Scott had returned in March from a trip to Havana, with messages of the "perfect unanimity of the Cubans" against Spain and of "the acknowledged fact that one half of the troops" on the island would join the revolutionaries.[26]

López wrote to his adjutant general on April 3 asking how many men and weapons were available and instructing to be ready to sail on April 16. Gonzales had mustered in Georgia a skeleton regiment of four hundred men, along with artillery and more than seventeen hundred rifles, to rendezvous with other invasion forces. He quickly learned to manipulate and dismantle the field guns from an instruction book and the practical lessons of the Georgia Militia artillerists. The Cuban already possessed the mathematical knowledge needed to target the cannons accurately. He telegraphed Laurent Sigur in New Orleans on the fifth, using the alias A. Herrmany (combining Ambrosio with his other code name Germán), indicating that his earlier suggestions to a Mr. Dixon should be implemented. Gonzales had already sent ten thousand dollars to O'Sullivan in New York and would send the rest in a week.[27]

The New York branch of the operations had been penetrated by informant Daniel Henry Burtnett, who had infiltrated the filibuster group in New Orleans the previous October. He was described as "a young man about 26 or 28 years of age, tall in stature, long features, wearing his whiskers full, and rather animated in appearance." Burtnett used the alias Dr. Duncan P. Smith and was also known as "Three Fingered Jack," because on one hand he "lost two fingers, and the first finger has the appearance of being broken in two places." He was introduced on April 13 to José Sánchez Yznaga and John O'Sullivan. The gullible conspirators informed the spy that they were going to purchase the paddle-wheel steamer *Cleopatra* the following day. They asked him to acquire another steamer within seventy-two hours, for which he would be reimbursed. Burtnett charted a schooner the next day, and met that evening at the residence of Sánchez Yznaga with O'Sullivan, *Creole* captain Armstrong Lewis, *Cleopatra* captain Phineas O. Wilson, Hungarian revolutionary Louis Schlesinger, William T. Rogers (O'Sullivan's business agent and cousin to his wife), and a Mr. Pittfield, a filibuster courier. The informant learned that four hundred expeditionaries awaited in New York and that the plotters had contacts at the telegraph stations to inform them of any government communications for foiling their plans.[28]

Burtnett then sold his information to Francis Stoughton, the Spanish consul in New York. Burtnett was shown a letter sent to Gonzales from Havana, warning that local conspirators were being arrested and begging not to delay further. Through O'Sullivan, the spy learned that three artillery pieces and seventy-one cases of rifles, containing two dozen each, were shipped by O'Sullivan to Gonzales in Savannah. The armament was secretly forwarded to the Woodbine Plantation in Camden County, Georgia, on the Satilla River. The property was part of the John Bailey estate, administered by his eldest son, twenty-eight-year-old David Bailey, a Georgia Militia major. He was married to Isabella Lang, of the neighboring Cambray plantation, and they had two children. Gonzales, from his Pulaski House headquarters in Savannah, recruited filibusters in coastal McIntosh and Camden counties and in northern Florida. A Northerner sojourning through the South heard of the filibuster plot and denounced it to New York senator Hamilton Fish, who relayed it to president Millard Fillmore. The informant correctly indicated that "their organization is very extensive, branching into Georgia, Alabama, Jacksonville and West Florida, Louisiana and Mississippi." He provided details of the invasion plans, in which a New Orleans municipal police chief was involved, and claimed that Gonzales "is of good dress, class, and accomplished, circulates in good society in and near Savannah, and is probably favored by some of the large planters. His instrument with the rank and file, is a man known as Harry Titus a celebrated fighting man." Gonzales was "said to have much money at his command, and to have enrolled as many men as he wants for his purpose." The Spanish vice consul in Savannah was monitoring Gonzales's activities and telegraphing the information to the Spanish minister in Washington, D.C.[29]

The Gonzales preparations in Georgia had begun to unravel on April 5. Warren R. Akin, residing in Cassville, forty-five miles northeast of Atlanta, informed the government that efforts were being made there and in adjoining counties "to get volunteers to make another attack on the island of Cuba." Five days later, Allen Ferdinand Owen, a former Whig representative and newly appointed American consul to Havana, was manifesting from Columbus, Georgia, that "men and money has been raised in Georgia" for a contemplated invasion of Cuba with "1600 men." That same day, the Georgia newspapers were reporting that 120 "enterprising" young men from the Cherokee region of Floyd and Cass counties, and others from Atlanta and Griffin, had departed on the Macon and Western Railroad on April 8 destined for Cuba. The *Rome Courier* declared on the tenth that "this new movement is sanctioned, if not promoted by men of influence and official standing in Georgia and other Southern States," and that the expedition would sail from Savannah in two days. These activities prompted J. Renean, editor of the *Atlanta Republican*, to telegraph president Millard Fillmore on the tenth: "Our rail-roads are crowded with an army of adventurers destined for Cuba—by way of Savannah beyond all doubt." The expeditionaries were going "to tender their services to Gen. Gonzales." Fillmore ordered the next day that the federal authorities in Savannah and the Navy intercept the expedition.[30]

A classified report sent from Charleston on April 12 to the British minister in Washington, Henry Lytton Bulwer, stated that Gonzales had organized the expedition in Georgia and Tennessee. The note stated that groups of filibusters were headed for various departure points on the coast and implicated Mirabeau Lamar. That same day, Lamar wrote to Gonzales from Macon, saying that the next time they personally met he would explain "the imperious circumstances which place it entirely out of my power to cooperate with you in your noble endeavors for the good and glory of your deeply injured and oppressed country." General Lamar's acquaintance, Macon merchant Isaac Scott, the Macon and Western Railroad Company president, informed Secretary of State Daniel Webster on April 13 that a company of forty expeditionaries had just passed through Macon, with their weapons on another train in merchandise crates consigned to Capt. John Lama in Savannah.[31]

The next day, Minister Calderón de la Barca informed Secretary Webster that he was notified by the Spanish vice consul in Savannah that "Antonio Gonzalez [sic], who is called the General, and who was wounded in the route of Cardenas, is now engaged . . . in planning these infamous schemes in that City, and, from a telegraphic despatch which reached this Legation yesterday, Sunday, it appears that the aforesaid Gonzalez, and other persons, are busy in enticing, and have actually collected together, a number of people at a point twenty miles distant from the aforesaid port of Savannah." On April 15, the minister met with Fillmore, who apologized that "our people are difficult to control" and vowed to suppress the expedition. When Gonzales concluded preparations in Savannah, he sent a cryptic telegram to López in New Orleans. Eight "Cuba Emigrants" from Alabama, led by Robert Downman, arrived there on April 18, and were joined days later by others. Another group was headed by Hungarian revolutionary Janos Pragay, former adjutant general to Louis Kossuth, who checked into the St. Louis Hotel on April 21.[32]

The next day, expedition preparations in New York were further undermined when the informant Burtnett, obeying the Spanish consul, denounced the plot to U.S. attorney J. Prescott Hall. The U.S. marshal was ordered to proceed with a force to Sandy Hook, New Jersey, on a revenue cutter to intercept the *Cleopatra*. The steamer had not left its New York City wharf for the rendevous point because of a pending debt of four thousand dollars. That evening, forty-five "Spaniards, French, Germans and Americans" landed at Perth Amboy on a vessel from New York and boarded the chartered sloop *William Roe*. The port collector, forewarned of their plans by informant Burtnett, ordered the vessel detained until morning. The *Cleopatra* was "getting ready to leave" on the morning of the twenty-fourth, with "supplies of water and provisions for at least five hundred men," when it was detained by federal authorities.[33]

Meanwhile, Gonzales met with López, two staff members, and the Georgia companies in Macon. As they prepared to take the Georgia Central Railroad train to Savannah at 7:00 A.M. on Friday, April 25, Gonzales received a telegram from Lama in Savannah warning that the customs collector, Hiram Roberts, and the federal marshal were waiting at the railroad depot to arrest them. López immediately wrote to Sigur desperately proposing to sail to Cuba with only the two hundred Georgians accompanying him. Gonzales wired Lama to have a carriage waiting at railroad Station No. 1, ten miles outside Savannah. The filibusters who boarded the train with them, disembarked at railroad Station No. 2, twenty miles from Savannah. They divided into many squads, one group going to Jacksonville, and the other to Burnt Fort, the departure point on the Satilla River, upstream from Woodbine plantation, in Camden County.

Gonzales and López continued on the train to the next station, where they boarded a carriage that took them by country road into Savannah around 6:00 P.M. One account claimed that Gonzales and López were "dressed as women and wore green face veils" to evade detection.[34]

The departure point in Jacksonville was the Henry Titus sawmill on the St. Johns River. Col. Theodore O'Hara, who commanded the Kentucky Regiment at Cárdenas, arrived with Titus on April 24 at the Jacksonville Hotel. The next day, the *Newark Advertiser's* "Jacksonville correspondent" was writing that within thirty-six hours the expedition was to sail from Jacksonville, St. Marys, and New Orleans. The reporter claimed seeing a letter from General Gonzales a week earlier, dated at Savannah, the headquarters of operations, "directing the movements of men and military stores." He also saw in a storehouse in Jacksonville "cannon, gun carriages, rifles, muskets, ammunition, and the furniture of an army equipment to a very large amount." The correspondent indicated that there were "some 600 men, 50 of whom were to be mounted." There were Creoles, Cubans a few Cárdenas veterans, and most of the privates were Mexican War veterans. The departure from Jacksonville had been scheduled ten days earlier, for the twenty-sixth, but "Gonzales has been threatened with arrest at Savannah."[35]

After arriving in Savannah on Friday evening, April 25, Gonzales and López were driven at midnight by twenty-eight-year-old hardware merchant George A. McCleskey to the town of Thunderbolt, five miles east of the city. His father, Thomas J. McCleskey, was the senior warden of Masonic Clinton Lodge No. 54 in Savannah. The younger McCleskey took the Cubans in a rowboat down the Augustine River to the plantation of thirty-two-year-old Elias Butts Barstow on Wilmington Island. Barstow, a Massachusetts native, apparently was a Freemason whose fraternal vows did not allow him to forsake his distressed brethren. He told them that he was a Whig, opposed to the expedition, "but as my guests you are welcome, and my house is yours." The next morning, the conspirators received a dispatch from John O'Sullivan informing that the *Cleopatra* had been seized and the New York expedition disbanded. Hours later, O'Sullivan, José Sánchez Yznaga, Louis Schlesinger, Armstrong Lewis, and William Rogers were arrested by the U.S. marshal. Gonzales was implicated by Burtnett's grand jury testimony, resulting in a forty-page, ninety-seven–count indictment against O'Sullivan, Lewis, and Schlesinger for violating the Neutrality Act. After the New York conspirators were released on three-thousand-dollars bail, O'Sullivan applied with the U.S. Attorney's office for the release of the *Cleopatra* and its cargo, to employ the vessel in "lawful trade." The issue was referred for action to the secretary of state, who returned the vessel to its "lawful owner and claimant" a month later.[36]

Savannah customs collector Hiram Roberts was informed at 9:30 A.M. on May 1 that Gonzales and López were on Wilmington Island and "would probably endeavor to take the Florida boat which left that morning at 10 o'clock." The collector and the marshal boarded the steamer *Magnolia,* which stopped on the Augustine River opposite the Barstow residence and rang the signal bell. Roberts noticed that a number of people approached the plantation landing, "where a boat appeared to be in readiness and waiting to convey passengers to the Steamer." In a few minutes, the collector spotted the steamer's mate rowing in a small boat toward the dock and speaking to the people there, who "immediately dispersed." Roberts, the marshal, and another companion pursued them in a second bateau. At the plantation residence they inquired for the gentlemen seen with Barstow at the landing. Barstow asked if they "had any authority to enter his private residence, to which the Marshal answered he had none." Roberts noticed place settings at the table where the guests had just finished lunch. He also considered peculiar that Barstow, whom he had seen in Savannah a few hours earlier, had returned home so quickly. The revolutionary leaders had fled into the surrounding woods guided by a slave. According to local legend a century later, Gonzales and López were concealed in the Barnard family crypt while the marshal searched Wilmington Island. The semisubterranean tabby brick burial vault measured some five feet in height, fifteen feet in length and eight feet in width, with a slate-shingle pitched roof. The ground entrance, with a short descending stairway, was covered by a stone slab with an iron ring. The spacious sepulcher accommodated six coffins, and its porous structure allowed light, air and water to seep in. After the Cubans entered the vault, their guide set the heavy stone back in place and later delivered their meals.[37]

The following day, Barstow departed in his boat with the fugitives. Leaving López in Savannah with John Lama, he took Gonzales up the Savannah River to Screven's Ferry, South Carolina, opposite the city, and left him with Freemason Samuel Prioleau Hamilton, whose father, Gen. James Hamilton, had been a War of 1812 veteran, a mayor of Charleston, a South Carolina Nullification Democratic legislator and governor, and a Texas Republic diplomat, instrumental in its annexation into the Union. Hamilton decided to take Gonzales to his father's cotton plantation on the Colleton River, across from Callawassie Island, Beaufort District, South Carolina. They headed back down the Savannah River, toward Wilmington Island, passing Collector Roberts on the customhouse boat as he went to look for Gonzales on the plantation of William Henry Mongin on Daufuskie Island, South Carolina. Roberts and Mongin were privates in the Georgia Hussars cavalry of the First Regiment, Georgia Militia. Gonzales went unnoticed after Hamilton had him lie on the bottom of the open boat covered with an overcoat. At dusk they reached the landing place at the mouth of the river on the South Carolina shore and proceeded some twenty miles on horseback until arriving at the Hamilton plantation at midnight. Gonzales was secreted for a month in a pinewoods shack one mile from the rice fields, under the care of the overseer.[38]

López left Savannah on May 2, assisted by John Lama, who telegraphed the conspirators in New Orleans: "Mr. Souverville [López] cannot recover his health and returns to that city." Collector Roberts cabled the acting secretary of the treasury that "General Lopez & Gonzales it is believed are on one of the Islands below. Efforts are making to arrest them." The message was immediately forwarded to President Fillmore. That same day, the Spanish vice consul in Savannah, John G. Doon, filed a complaint in federal court against the Georgia filibuster leadership for violation of the Neutrality Act. His statement was accompanied by a deposition from informant William E. Oliveira, who had infiltrated the Macon expeditionary contingent in mid-April. An arrest warrant was issued that day for Gonzales, López, Captains Robert Young, Samuel J. Koockogey and William B. McLean, Lieutenants Richard Ralston, Samuel St. George Rogers and Chapman, Sergeants Brown and Howard, and others. U.S. deputy marshal Michael Finney, pursuing the informant's lead, took the revenue cutter *Taney* up the Satilla River to the Waverly plantation in Camden County. Finney, assisted by naval officers, arrested Columbus contingent leaders Samuel Koockogey, Richard Ralston, and two other officers, rounded up twenty-seven other filibusters, and escorted them back to Savannah. All were released except the four leaders, who gave recognizance to appear before the November term of the U.S. District Court.³⁹

On May 5, Collector Roberts was informing the acting treasury secretary that the Cuba expedition had ended and that López had "gone to Mobile, and Gonzalles [sic] 'tis likely is wending his way towards Florida, of his whereabouts however there is nothing certain." Two days later, López was back in New Orleans with a beard to disguise his appearance. He told Cirilo Villaverde that when he arrived in Savannah there was an arrest order against him but because he had "Masonic protection," the Savannah City marshal refused to enforce his federal arrest warrant. The city marshal was Daniel H. Stewart, junior warden of Clinton Lodge No. 54. John Henderson returned to New Orleans on May 9 to meet with López, and checked into the St. Louis Hotel. A meeting was held to device damage control and renew efforts for another expedition.⁴⁰

In early June, Gonzales went to inspect the armament left on the Woodbine plantation, sailing from his South Carolina hideout to St. Marys, Georgia. The port was a bustling lumber and shipbuilding town on the St. Marys River, facing the Florida border. St. Marys had a courthouse, five churches, three schools, a market, nine dry-goods and grocery stores, three physicians, three lawyers and a federal collector. Gonzales hired a buggy and a black guide for the twenty-five mile journey on the Old Post Road across the Camden County pine lands to Woodbine plantation. They had proceeded less than ten miles when a midday thunderstorm left their buggy bogged in the mud. The driver suggested seeking help at the nearby Laurel Isle cotton plantation. As Gonzales approached the residence, he flashed the Masonic Grand Hailing Sign of Distress to the tall gentleman standing on the piazza.⁴¹

The Cuban introduced himself and described his situation. "It so happens," said the stranger, "that I have an order for your arrest from the president of the United States." The plantation owner identified himself as John Hardee Dilworth, the St. Marys collector of customs, and assured his stunned visitor, "I am most happy to make your acquaintance and to welcome you to my house. I sympathize deeply with your cause; had I seen you at St. Marys I would certainly have arrested you." The thirty-nine-year-old Dilworth, a founding member of Masonic St. Marys Lodge No. 126, had studied at the University of Georgia and was a colonel in the Third Regiment, Georgia Militia. A Whig, he had been appointed to his post by President Taylor in August 1849. Dilworth invited Gonzales to dine with his family and to spend the night in his home. After continuing to the Woodbine plantation the next day and completing his task, Gonzales returned to the South Carolina lowcountry.[42]

While staying at the home of Col. Gaston Allen in Bluffton, South Carolina, four miles from the Hamilton's Callawassie Island plantation, Gonzales contracted a "severe bilious fever," apparently malaria, which left him prostrated and under a physician's care. He took quinine pills to alleviate the fever and returned to the Barstow plantation on Wilmington Island to further expedition preparations. While there, he received a letter from Villaverde in New Orleans, indicating that since June 12 a group of movement dissidents had been invoking Gonzales in their campaign to discredit López. Gonzales replied to Villaverde on June 23 and also wrote López, denouncing the "infamous slyness and machinations" against him and the prestige of the general.[43]

Meanwhile, López had written to Gonzales on June 23, concerned that he had not received a reply to his previous letter. The general indicated that he had modified his plans and wanted Gonzales to swiftly return to New Orleans with a Florida coast pilot and sell a few bonds to cover expenses. Nine days later, when López received the latest Gonzales communication, he replied that he was going to ignore "the madness and chicanery" of José Hernández Canalejos and the six or seven "tricksters and fools" who sustain him. Upon learning that Gonzales was still recuperating from malaria, López countermanded his previous order to meet in New Orleans, and instead told Gonzales to reestablish his health at the mineral spring baths in Virginia. López wanted Gonzales to "take part in the fight that will occur, at least departing with the second push," of expeditionary reinforcements. López reiterated: "To the baths then, Mr. G., and without thinking of anything else until you receive my next letter in which I will manifest what I will want you to do from there." Gonzales went on June 26 to Jacksonville, informed Henry Titus of his plans, and continued to Fauquier White Sulphur Springs, in Warrenton, Virginia. Rumors originating in Savannah in early July purported that "Gen. Gonzales had left the sea-coast of Georgia with troops in three steamboats."[44]

López received a telegram on July 1 from Gaspar Betancourt in New York, announcing that an uprising in his native Camagüey was imminent. The general was outraged at being excluded and at Betancourt's lack of assistance toward his own effort. The annexationist insurrection led by Joaquín de Agüero, a thirty-four-year-old Freemason, was scheduled for the July 4 anniversary of American independence. Two days earlier, his wife confessed the plot to a local priest and gave him a Cuban flag that she had sewn, to be placed on the church altar during a special mass. When the service ended, the priest returned the emblem and quickly informed the authorities. Agüero, accompanied by forty-four men, many of them members of his Masonic lodge, issued a proclamation of independence on July 4 in a farm on the outskirts of Puerto Príncipe. After various skirmishes with Spanish troops, Agüero and his remaining followers were captured on July 23. The news of this defeat was not published in the United States until the following month, because the government suspended postal service in Puerto Príncipe. The day after Agüero was captured, militia cavalry lieutenant colonel Isidoro Armenteros Muñoz, a leader of the 1848 López conspiracy, revolted with twenty-four men. They were dispersed after engaging the Spanish army on the twenty-fifth at Guayabo. Armenteros was captured five days later and executed in August.[45]

The lack of information from Cuba on the impending insurrection made López restless. He sent his agent Paul Francis de Gournay to Havana on July 13, on the steamer *Cherokee,* bearing a false passport under the name O'Callaghan. De Gournay carried a letter from López to his followers on the island, telling them to revolt as soon as they heard of his landing, to detain all government officials, to form a council, and name a leader among them who would rule until he contacted them. The next day, the *New Orleans Picayune* announced that four Puerto Príncipe rebels, pursued by the government, had "escaped to the mountains, where they have since been joined by upwards of twenty of their comrades, who are implicated." The impulsive López decided to hasten his plans, and the next day sent Cirilo Villaverde to Savannah with a note of instructions addressed to Gonzales, John Lama, Henry Titus, and Juan Macías.[46]

On July 28, five days after Villaverde arrived in Savannah, he proceeded to Florida with Leopoldo Turla, and they checked into the Jacksonville Hotel. The Cubans later met with Titus at Empire Mills across the St. Johns River, where they found Col. Theodore O'Hara, who had been in Jacksonville since May. Six days later, Titus sold Empire Mills and invested his proceeds in equipment for the expedition. He then invited Villaverde to join Solomon's Lodge No. 20 free of charge. Other volunteers, including Cárdenas expeditionary Juan Manuel Macías and twenty-five-year-old Alejandro Angulo Guridi, reached Jacksonville on July 30. They were joined four days later by José Sánchez Yznaga, proceeding from New York via Savannah. The same day, Agustín Manresa arrived from New Orleans with the last letter from López, stating that he would depart on the steamer *Pampero* on the thirty-first and would be in Jacksonville four days later. López had also notified the Kentuckians, who were rushing south. Maj. Thomas T. Hawkins arrived in Louisville from Newport, Kentucky, on July 24, where he was joined by Col. John T. Pickett three days later.[47]

Hundreds of filibusters descended on New Orleans as the news of the Cuban uprising spread. Louis Schlesinger and *Creole* captain Armstrong Irvine Lewis arrived on July 28, two weeks after appearing with O'Sullivan before a federal district court in New York on charges stemming from the *Cleopatra* expedition. O'Sullivan, as usual, stayed behind during the call to arms. That same day, two contingents of Cincinnati youths commanded by Cárdenas veteran Capt. Henry H. Robinson departed for New Orleans, which was "all in a blaze of sympathizing excitement about Cuba and for Cuba." Lewis inspected the expeditionary steamer *Pampero* after its arrival on July 29, discovering that the "boilers were burnt out" from a collapsed fifteen-inch diameter exhaust pipe. The *Picayune* reported on July 31 that Trinidad and Villa Clara, where López had organized the 1848 insurrection, had also revolted. The next afternoon, filibuster leader William Logan Crittenden, a Customhouse officer, informed López that due to a complaint from the Spanish consul, the *Pampero* would be impounded Monday morning. The general then ordered the expedition to sail Saturday at midnight, doing the required repairs en route.[48]

López had previously made arrangements with Major Hawkins to raise a regiment of six hundred carefully chosen Kentuckians and Hoosiers to serve as his main force. The Trinidad news made López hasten his departure; hence, instead of waiting for this elite force to arrive, he took an inexperienced group raised within forty-eight hours in the streets of New Orleans. López knew that the riff-raff that Col. Chatham Roberdeau Wheat had mustered under the same circumstances the previous year had greatly contributed to the Cárdenas failure. Gonzales attributed the hasty departure to López's believing the newspaper accounts fabricated by the Spanish government of widespread insurrection on the island, and to his not wanting to "miss the opportunity of marching in triumph into Havana." Crittenden, assisted by Customhouse official Victor Kerr, hastily raised a regiment of artillery composed of 114 novices to operate the Georgia Militia cannons procured by Gonzales. López left instructions for Col. Chatham Roberdeau Wheat, who was in North Carolina, and the one-armed Lt. Col. William H. Bell to form other regiments. Cárdenas veterans Hawkins and Pickett, unaware that López had again changed plans, reached New Orleans the day after the *Pampero* sailed, and the Kentucky Regiment arrived a week later.[49]

The *Pampero* was moored at the St. Marys Street wharf, where López arrived at midnight on Saturday, August 2, accompanied by his chief of staff Janos Pragay, Louis Schlesinger, and other staff members. The loading of passengers and provisions lasted until daybreak, when a towboat pulled the *Pampero* downstream. A shoddy replacement exhaust pipe was installed while the steamer was heading out to sea. One expeditionary calculated there were 450 men on board, including the First Regiment of Cuban Patriots, commanded by Capt. Ildefonso Oberto Urdaneta, a Venezuelan-born former Spanish soldier. It contained forty-nine Cubans and Spaniards, including a dozen of the Cárdenas company deserters. Schlesinger estimated the volunteers at "a little over 400." Another participant gave an exact number of 434 men. The expedition muster roll was later seized by the Spanish authorities.[50]

After five days at sea, the *Pampero* stopped at Key West. López was told that most of the Spanish troops in Havana had gone to suppress the revolt in the central region, that the insurrection had extended to thirteen towns, pushing westward to Vuelta Abajo, and that most of the Cuban people anxiously awaited his arrival. The *Pampero* quickly departed for western Cuba, ninety miles distant, instead of making the twelve-hundred-mile round trip to pick up Gonzales and the Jacksonville and Savannah regiments. Near midnight on August 11 the *Pampero* reached Morrillo, a small hamlet sixty miles west of Havana, and unloaded the expedition. Captain Lewis and Callender Fayssoux then took the steamer to Jacksonville with two sick men and a letter from López to "Sánchez Yznaga, Gonzales, Macías, and others," indicating his change of plans and ordering them to proceed at once with the secondary expedition to Puerto Príncipe.[51]

Morrillo, containing only four houses, lacked transportation. Pragay persuaded López to split his force, proceed ten miles to the next village of Las Pozas to obtain carts and horses, and send them back to Colonel Crittenden, who would remain at the beach with 120 men guarding the supplies. These consisted of "four barrels of powder, two of cartridges, about 150 muskets, the flag of the Expedition, and the officers' luggage." The main force departed around 9:00 A.M. and reached Las Pozas at noon. The community, located on a ridge, contained some fifty houses astride the main road from Morrillo. It was deserted except for the Spanish owners of the only two stores and some Afro-Cubans. López issued a liberating proclamation and forwarded to Crittenden a few ox carts with black teamsters, each accompanied by five to eight expeditionaries. Shortly after they departed, a peasant brought news from Bahía Honda that a division of Spanish troops had just arrived there from Havana by steamer. López then sent a courier to Crittenden with orders to abandon the heavy supplies and leave for Las Pozas immediately. Schlesinger stated that Crittenden got the message one hour before the carts arrived, and that these "could not have reached him later than eight o'clock" that evening. Crittenden hesitated and did not move out until 11:00 P.M.[52]

At 8:00 A.M. on August 13, before Crittenden's expected arrival, a Spanish force of eight hundred men from Havana attacked Las Pozas. Ildefonso Oberto was mortally wounded while leading the Cuban company in a countercharge that dislodged the enemy from the left ridge leading into town. The remaining expeditionaries occupied the right ridge and Col. Robert L. Downman was killed while repulsing two enemy column advances. The Spaniards retreated after half an hour, leaving behind 180 dead. The 280 expeditionaries had thirty-five casualties, with Adjutant Felix Huston Jr. among the dead, Pragay and Gotay mortally wounded, and another ten slightly hurt.[53]

López dispatched five men to find Crittenden, but this search party returned about 9:00 P.M., saying that a large enemy force was blocking the road to Morrillo. Ninety minutes later, forty of Crittenden's force arrived at Las Pozas led by Capt. James A. Kelly. They informed López that at 8:00 A.M., halfway to joining López at Las Pozas, they were attacked by a Spanish unit while having breakfast. After repulsing the enemy, Crittenden chased them into the woods with some eighty men and never returned. Kelly and his men were unaware that Crittenden, unhappy with the turn of events, went back to Morrillo with fifty followers, seized four launches, and departed for the United States. Crittenden and his men were later captured by a Spanish warship while resting in a coast mangrove eighteen miles away and given a summary court-martial on the voyage to Havana. The fifty prisoners were executed ten at a time by firing squad at Atarés fortress before sixteen thousand spectators. The bodies were mutilated by the crowd and thrown in a garbage dump. López, unaware of Crittenden's fate, held a war council as the Spanish army once again approached Las Pozas. The officers agreed not to wait for Crittenden but to head for the countryside, where the general "believed it would not be long before they would unite with insurgent parties of the Cubans."[54]

López arrived at noon on August 18 at his brother-in-law's Cafetal de Frías coffee estate. While the 250 filibusters prepared lunch, Spanish general Manuel Enna attacked with twelve hundred infantry, 120 cavalry, and four howitzers. Enna was mortally wounded while leading a cavalry charge, and the entire force broke in a rout. López tried to pursue them, but few obeyed, disheartened by the lack of popular support since landing. The expeditionary force left the next day, camping at the Candelaria coffee plantation, where the enemy attacked them at daybreak on August 19. The filibusters, "with a trifling number of effective rifles," disbanded and fled into the woods. López regrouped a large number and led them up a mountain under a tropical storm. The invaders had been in Cuba for over a week, and only López and a few others still retained hope that Gonzales would arrive with an auxiliary expedition.[55]

Accounts of the López landing began appearing in American newspapers on August 20, when Laurent Sigur arrived in Savannah. The *Pampero* had not been heard from, prompting Sigur to go to Charleston, South Carolina, on August 23 to hire another steamer. That day, various newspapers, including the *Washington Union* and the *Louisville Courier*, published a dispatch indicating that the *Pampero* had "returned to Key West, and then proceeded to Jacksonville for Gen. Gonzales and his regiment." Gonzales registered at the Charleston Hotel on August 23 and apparently met with Sigur. López failed to send word to the convalescent

Gonzales of the expedition's early departure, so the Cuban was surprised to hear of the Morrillo landing. Upon receiving this news, he immediately traveled by stagecoach and train to Wilmington, North Carolina, where he took a steamer to Charleston.[56]

Gonzales did not return with Sigur to Savannah, where he still had an arrest warrant pending since April. He wrote on August 24 to the editor of the secessionist *Charleston Mercury* to rebut an editorial citing the recently deceased John Calhoun's opposition to the acquisition of all of Mexico, inferring from this that he would have opposed the annexation of Cuba. Gonzales objected to seeing "the great name of the South Carolina Statesman thrown into our adverse scale," and recalled that when he and General López arrived in Washington, D.C., in 1849, Calhoun had paid them a visit and "expressed himself as warmly in behalf of Cuba and her annexation." According to Gonzales, the sectional issue was then revived in Congress with the debate over the admission of California, forcing Calhoun to procrastinate on Cuban statehood. Had Calhoun lived to see California admitted to the union as a free state, Gonzales argued, he would have striven for Cuban annexation to obtain "that 'equilibrium' with which alone can this Union be preserved, through the Union of the South."[57]

The Gonzales letter was an attempt to elicit sympathy from South Carolinians for the cause of Cuban annexation, as he believed Calhoun would have done, at a desperately needed moment. Gonzales departed on the twenty-fifth by train for Columbia, the state capital, appealing for assistance to Mexican War veterans general James Hopkins Adams and colonels Maxcy Gregg, Adley Hogan Gladden, and D. J. McCord. He returned to Charleston with them the next day and then traveled alone to Wilmington Island, Georgia, avoiding the U.S. marshal. Two days later, the Jacksonville *Florida Republican* reported that the *Pampero* was "now in the waters of Georgia, to receive reinforcements, and may momentarily pay this port a visit." The *New Orleans Picayune* announced on the twenty-ninth that General Gonzales commanded the Jacksonville expedition.[58]

The *Pampero* arrived in Jacksonville at 10:00 A.M. on Sunday, August 31. Macías immediately left there for Savannah with the news. During Gonzales's absence, Henry Titus had usurped the leadership of the Florida contingent, while Col. Theodore O'Hara, his former Kentucky Regiment commander, "did not hide his resentment" at being demoted to lieutenant colonel and second in command. The *Pampero* engine was repaired by Tuesday afternoon, September 2. Titus loaded the steamer with wood fuel, provisions, and "two [twelve-pounder brass] cannons, two howitzers, 5 or 600 muskets, about 150 Yauger Rifles, about 150 cutlasses, 10 or 15 kegs of Powder, some Bombs and 50 or 60 kegs of cartridges and some [thirty] saddles and also about 75 men."[59]

The *Pampero* went to Wilmington Island to retrieve some seventy men of the Georgia contingent, led by Captain Williamson, who arrived from Savannah on the steamer *Jasper* with Sigur and a stock of provisions. Gonzales, who had been once again hiding at the Barstow plantation, was dismayed to see that the secondary expedition had been delayed for more than a week, that the engine continued to malfunction, and that there was no coal or water on board for a sea voyage to Cuba. A dispute arose when Gonzales claimed the expedition leadership and tried demoting Titus to a lieutenant colonelcy. Titus, who invested heavily in the affair, argued that he had received a letter from López in July, when Gonzales was ill in Virginia, giving him command of the Jacksonville Regiment. Sigur, the *Pampero* owner, settled the controversy by siding with Titus. Gonzales later wrote that "without interfering with that movement already in the hands of others, I at once proceeded to raise the promised reinforcement," and returned on September 3 to Charleston. Titus then refused to follow Sigur's instructions, creating a schism that prompted the latter to take one-third of the *Pampero* cargo on the *Jasper* back to Savannah. Adjutant John L. Hopkins, Captain Andrew V. Colvin, and others departed on the *Jasper,* after Sigur promised to charter the steamer *Monmouth* for them. The *Pampero* left before dawn on September 4, stopping at a Nassau River plantation, where the crated arms were distributed and the men drilled for about three days, while waiting for more volunteers to go join López in Cuba.[60]

By that time, however, López was already dead. He had been taken prisoner on August 29 and sent to Havana, where he was publicly executed at 7:00 A.M. on September 1 as a common criminal on the *garrote vil,* a medieval strangulation contraption. His last words were: "My fate will not change thy destinies; adieu, dear Cuba." López was interred in an unmarked pauper's grave in Espada Cemetery. The filibuster tally published in the *Gaceta de Gobierno* was 106 killed in action; 136 executed in the countryside and another 50 in Havana; with 176 captured, totaling 468. The captain general freed 4, who returned to the United States with inconsistent accounts of events, and sent the others to prison in Ceuta to labor in the quicksilver mines. He provided a list with the names of only the prisoners and those executed with Crittenden. Relatives of missing expeditionaries were still inquiring about their fate two years later.[61]

Accounts of the López defeat reached New Orleans on the steamer *Cherokee* on September 4. The following day, Gonzales read in the *Charleston Courier* that López "had been captured and shot," and that "all who left on the Pampero had been killed or made prisoners." Gonzales then ceased his recruitment efforts in Charleston. Two days later, on the seventh, Villaverde, Sánchez Yznaga, Alonzo B. Luce, and four other Cubans departed from Savannah for New York on the steamship *Florida*. Meanwhile, John L. O'Sullivan had arrived in Jacksonville from New York on September 10 to arrange with Sigur the surrender of the *Pampero*, hidden at Dunn's Lake, near Palatka on the St. Johns River. Captain Lewis, accompanied by first mate Callender Fayssoux, Macías, and O'Sullivan, afterward left Jacksonville. The U.S. marshal's office in New Orleans informed President Fillmore on the eleventh that the multitude of filibuster recruits in their city had disbanded. The *Pampero* was adjudicated for libel in federal court and auctioned in Jacksonville four months later. Titus eventually sold the expedition's equipment to the Florida Militia to recoup his losses.[62]

After six months as a federal fugitive, Gonzales surrendered on October 17 to Savannah U.S. marshal William H. C. Mills and "entered into recognizance returnable likewise to the Circuit Court." The following week, Savannah U.S. district attorney Henry Williams requested instructions from Secretary of State Daniel Webster on the course to pursue against Gonzales and the four *Cleopatra* expedition defendants from Georgia. He indicated that this was not a case of "offenders against local criminal laws," insinuating that conviction would be difficult. Williams suggested that any action should be dictated by "the policy of the General Government," asking "whether prosecutions shall be instituted and pressed against them." The news of Gonzales's arrest was published in the Savannah, New Orleans, New York, Jacksonville, and Cincinnati newspapers a month later. The Fillmore administration also went after the Cincinnati filibuster leaders. William Hardy, Henry H. Robinson, and William A. McEwen were indicted for violation of the Neutrality Act on November 1 by a federal grand jury in Columbus, Ohio. Gonzales remained in Savannah, where on December 15 he answered a sworn written deposition in the pending case against O'Sullivan, Lewis, and Schlesinger.[63]

Gonzales was devastated by the execution of López and the swiftness with which the internal revolts had been crushed. He had demonstrated a great ability at organizing and equipping an entire expedition in three states, by swaying influential people like Governor Towns. The Cuban was often seen in good society, who were impressed by his manner, intelligence and attire. The Masonic brotherhood assisted in his endeavors, which were destroyed by the indiscretions of López and O'Sullivan, prompting federal government intervention. Insurrections had occurred in Cuba for the first time, but they were small, isolated, and quickly suppressed. Once again, fate had been favorable to Gonzales, when his illness prevented sailing on the *Pampero*. The expedition may have succeeded if López had not been so restless.

The first López internal conspiracy was better planned than his subsequent ones: it had greater appeal to the Creoles, since it lacked the massive foreign and mostly rowdy element employed in later enterprises. We can only speculate, had independence been achieved in 1848, whether the Cuban people would have chosen the Texas model for annexation. Perhaps López would have turned the island into his own *caudillo* fiefdom, similar to those ruling other Latin American countries. López was impulsive and never learned from his mistakes. He had failed at Cárdenas and again at Las Pozas largely because of the inexperienced and undisciplined recruits hastily scooped up in New Orleans. López desired to land in Cuba before the Agüero uprising triumphed so that he could share in the military glory and power. His uncontrollable impetus wrecked the enterprise and cost him his life. The future had never looked so bleak for Gonzales now that López was gone; he faced another federal trial, and a death penalty awaited him in his homeland.

## Four

# PATRONAGE PURSUER

Gonzales remained in Savannah after the *Pampero* fiasco, surviving on the economic assistance forwarded by his family and the generosity of his Southern friends. The handsome Cuban embodied a romantic cavalier figure during the era of Southern chivalry. He had received an early education in the United States, had traveled the country widely, and spoke English well. Flaunting the title of "General" in hotel registers and personal cards, the enthusiastic Gonzales wore an impeccably tailored uniform and a sword with a gold-plated scabbard. His refined upbringing was evident in his elegance and manners. His combat wounds, the notoriety of the New Orleans trial, recognition in national newspapers, published biographical sketches in *The History of the Late Expedition to Cuba* and *The History and Adventures of the Cuban Expedition,* and his political writings added to his revolutionary mystique. He attended Episcopal Christ Church at Monument Square with his filibuster and Masonic associates. These included Mayor Richard Wayne; aldermen Richard R. Cuyler and James P. Screven; attorney Robert Milledge Charlton; planter William Henry Mongin; and Solomon's Lodge members John G. Falligant, John Hunter, George S. Nichols, and William P. White. Gonzales performed in the choir a repertoire of sacred music, notably a solo of Gioacchino Rossini's *Stabat Mater.*[1]

Another member of Christ Church was fifty-four-year-old Ralph Emms Elliott, who most likely introduced Gonzales to his brother, William Elliott, a sixty-three-year-old South Carolina planter, politician, and writer. Precisely how or when the two met is not clear. One writer claimed that Gonzales went to Charleston with letters of introduction from John Quitman and P. G. T. Beauregard and was Elliott's guest at Oak Lawn plantation after his arrest warrant was issued. Another author believed that Gonzales appeared at Oak Lawn during "his recruiting in the Savannah area." The Cuban and the Carolinian had mutual acquaintances, including William Henry Mongin and the Hamiltons. Steamer passenger and hotel records show that Gonzales left Savannah for Charleston on February 3, 1852, staying overnight at the Charleston Hotel. He returned to Savannah on the steamer *Calhoun* on March 2. It was probably during this month that he initially stayed at Oak Lawn with the Elliotts. The Elliott family correspondence mentions the Cuban for the first time in the summer of 1852.[2]

The Elliotts could trace their Hutchinson lineage to the twelfth century. It was originally "Vitonesis," modernized into Hutchinson. Col. Thomas Hutchinson was governor of Nothingham Castle during the reign of Charles I. His grandson emigrated to the Massachusetts Bay Colony and his great-grandson settled permanently on the Chehaw River bank in South Carolina. William Elliott was born in Beaufort, South Carolina, on April 27, 1788. His father, William Elliott (1761–1808), was a state senator and the first Carolinian to experiment successfully with the high-quality long staple (sea island) cotton. After his father's death, Elliott received a B.A. degree from Harvard College in 1810 and an M.A. from the same institution five years later. He married Ann Hutchinson Smith, daughter of Thomas Rhett Smith and Ann Rebecca Skirving, at Oak Lawn on May 23, 1817, when he was twenty-nine years old and she had just turned fifteen. His bride's parents had also married at Oak Lawn on the same date twenty-three years earlier. Ann Skirving Smith's grandfather was the physician and planter James Skirving, a militia colonel of the American Revolution who commanded a regiment at the Battle of Port Royal on February 3, 1779. Skirving owned over 4,400 acres of land between the Chehaw and Ashepoo Rivers, along with other properties, including Oak Lawn. William Elliott acquired most of these lands through marriage settlement and purchase, which along with his own inheritance, amounted to large tracts in Beaufort and Colleton districts in South Carolina and on the Ogeechee River in Georgia. These plantations were Myrtle Bank, on Hilton Head Island; Balls (1,083 acres) in Colleton District, on the Chehaw River; Social Hall, The Bluff, and Middle Place (3,400 acres), between the Ashepoo and Chehaw Rivers; Oak Lawn (1,750 acres), Bee Hive and Hope on the Edisto River; Ellis, Shell Point, and the Grove on Port Royal Island, and Bay Point on Phillips Island. Elliott also owned summer residences in Adams Run, Anchorage House in Beaufort, and *Farniente* estate in Flat Rock, North Carolina. The 1860 slave census indicates that he possessed 217 slaves in St. Helena and St. Paul Parishes. The two-story mansion at Oak Lawn served as the family residence from November to May. A wide avenue of interlocking live oaks, festooned with Spanish moss, stretched from the highway to the main entrance. The interior decoration depicted the wealth and refinement of the planter, writer, and public servant. The furniture was expensive American traditional and the paintings portrayed extensive travels and ancestral pride. The portraits of Colonel Skirving and his family hung in a prominent place. The chandeliers, clocks, vases, silver tea service, and glassware were of exquisite quality. The garden contained a profusion of roses, jasmine, japonicas, and other plants.[3]

Gonzales and William Elliott eventually formed a strong bond through their mutual interests: both proceeded from aristocratic planter families; both were college graduates; both were Unionists who considered slavery as an economic necessity; both favored American expansionism, especially the annexation of Cuba to the United States; both were impressive orators; both spoke French and had visited Europe; both were outstanding sharpshooters; both played the piano and had a taste for classical music; both frequented the opera; both were voracious readers and Elliott's voluminous Oak Lawn library allowed them much literary conversation; both were good billiard players and the Oak Lawn pool table was a source of shared recreation; both suffered from asthma; both had a similar obstinate temperament and a streak of vanity that surfaced under stress; and both were Freemasons. Elliott had been initiated in 1808 in the Charleston Orange Lodge No. 14, and his sons Ralph Emms Elliott and Thomas Rhett Smith Elliott, along with their cousin Stephen Elliott Jr., were members of Harmony Lodge No. 22, in Beaufort.[4]

The thirty-four-year-old Gonzales regarded the Carolinian as the fatherly figure that he missed dearly. Elliott, the same age as the Cuban's deceased progenitor, extended his paternal affection to the dashing swashbuckler, who had exceeded the expectations he had of his son, William Jr., a recent Harvard dropout. Gonzales also became close to one of the Elliott sons, and after his departure he spoke kindly of the Elliott family. The *New Orleans Picayune* had described him as "thoroughly Americanized," having "on all occasions exhibited courage and talents of a high order, with fine manners and courteous address." These characteristics allowed him to glide easily into the Elliott world, instead of having "abruptly crashed" it, as has been suggested by a Carolinian historian.[5]

Gonzales returned to Charleston from Savannah on March 26, 1852, accompanied on the steamer by Georgia Democratic representative David Jackson Bailey, whose name appears contiguously in the Charleston Hotel register. Gonzales stayed in the area another month, probably a guest again at Oak Lawn, before going back to Savannah. He then heard the good news that the New York filibuster trial against John Louis O'Sullivan and filibuster steamer pilot Armstrong Irvine Lewis, for which his deposition had been taken in December, had ended in a hung jury. The case against O'Sullivan was eventually dismissed. Lewis died of "malignant fever" in Mobile a few months later. In June, Gonzales wrote an article for the *Savannah Georgian*, denouncing the "ignorance" of Havana's *Diario de la Marina* newspaper regarding the U.S. presidential campaign. The Cuban praised Democratic candidate Franklin Pierce as a warrior and statesman "quite competent to fulfil that call of destiny" of American expansionism. The following month, Gonzales went to Washington, D.C., for two weeks and proceeded to the summer resort at Warrenton Springs. He was incensed to read in the *Washington Union*, the Democratic Party organ, an editorial reprinted from the *New York Courier and Enquirer* stating that "the Cuba revolution was a mere piratical foray, having no real origin among the residents of the island." Gonzales replied that during the previous four years, the Cuban independence movement had yielded a clandestine press, political prisoners, exiles, four expeditions, martyrs, and "the blood of Cubans has mingled with the American," to "bring forth the aspiring tree of liberty."[6]

During his summer stay at Warrenton Springs, Gonzales wrote the 10,385-word *Manifesto on Cuban Affairs Addressed to the People of the United States,* on the first anniversary of the López execution. The document was meant to correct misled American public opinion as to the causes of the Cuban revolution, as to the noble character of the Cuban movement in the United States, and to "unfold, in self-vindication, the double-dealing arts of our detractors." The author indicated that the question of the independence of Cuba deserved the special consideration of the American people, as the island's "close proximity, controlling position, domestic institutions [euphemism for slavery] and commercial wants, bring home to the safety, the peace, the welfare and development of America, all incidents likely to bear upon her future."[7]

Gonzales outlined the grievances by which the Cubans, "groaning under oppression a thousand times more galling" than those of the thirteen North American colonies, were seeking the aid of "a neighboring republican people." Cuba's white population paid a tax of nearly $40 each, while the Spaniards paid an average of $9, and the Americans $2.39. Gonzales listed a number of highly taxable commodities, such as foodstuff, salt, agricultural exports, all property, including slaves, and even stamp paper, required for all legal documents. These taxes were not being used by the Spanish Crown for the benefit of the Cuban people. Only one child in every eighteen was in school, due mostly to private exertions, and since 1849 the Spanish government had prohibited its citizens from studying in the United States. The Cuban captain general, "a mere soldier," was an autocrat who even presided over the Supreme Court of Justice. He highly benefitted from the "horrid traffic" of slavery, along with the Queen Mother, by a special importation fee of $51 per head. Cubans had no political representation in the Spanish Cortes parliament and they were "deprived of all liberty of conscience, of speech, or of the press." Cubans could not own firearms, nor assemble in public, nor have a trial by jury. They needed a permit to have company at home, to change domicile, to leave town, or to travel abroad. The judicial *habeas corpus* did not exist, and Cubans could be incarcerated for months due to secret denunciation. All public offices were controlled by Spaniards, except "a few of the lowest order." There was a glut in the medical and law profession, the latter "under the pervading influence of a corrupt system." The manifesto denounced that "Cubans of high intelligence and education, every avenue of distinction and emolument being closed to them, are constrained to discharge the duty of overseers to planters, machinists, etc., in order to earn a livelihood." Gonzales criticized the colonial government, which sentenced him to death for a political document published in U.S. newspapers, as "a hideous compound of base rapacity, wanton insult, and dire oppression."[8]

The manifesto contrasted the favorable conditions of the American Revolution and the unfavorable insurrectional situation on the island, to answer those who questioned the Cuban failures. While the American colonists were surrounded by the wilderness of a vast continent, the island of Cuba offered no possibility of retreat. The Minutemen evolved from a militia, possessed firearms and knew how to use them, while Cubans lacked military experience and were prohibited weapons. The American colonists "met, discussed, and resolved, printed and spoke and went about freely and unshackled," but the same liberties were denied to the Cubans, whose movements were constantly watched. Gonzales pointed out that the American revolutionaries "had on their side the fleets and armies of France, the chivalry of Europe, the financial aid of Spain, and the moral countenance of all nations. We have against us not only Spain, but that very France and England, and the menace of the blacks, the squadrons of the United States, and the denunciation of the republican government as pirates and freebooters, to draw from our feet our only plank of support, with the world against us; the moral aid of this free country."[9]

Gonzales sent copies of the manifesto to Democratic newspaper editors and political allies, including one on September 10 to *New York Herald* editor James Gordon Bennett, enclosing a "list of persons to whom the Herald containing the manifesto is requested to be sent." These were forty-seven people Gonzales considered "friendly, indifferent or hostile" to the Cuban cause. Bennett published the roster with an editorial on the twenty-second, indicating that these were "sympathizers of rank and position in the Southern States, who are supposed to be thoroughly enlisted in the cause of Cuban emancipation." This elicited public corrections from various people mentioned, including Henry D. Holland, the Jacksonville customs inspector who had seized the *Pampero*; Charleston commission merchant Henry Gourdin, who was part owner of the steamship *Isabel*, and Savannah banker Richard R. Cuyler, a Whig activist and the father of George A. Cuyler. Gonzales was obliged to send letters to the editors of various newspapers correcting the error.[10]

The manifesto had a strong impact on friend and foe alike. The *Washington Union* responded to the Gonzales document on September 24 stating that "Spanish rule in Cuba is an outrage on the sentiments of the civilized world." Although the newspaper deemed the López expeditions as "ill-timed and mismanaged," it blamed President Fillmore's policy for "the murder of our citizens at Havana," the "usurpation of power to use our navy to defend and protect the coast of Cuba," and for abandoning "the rights and honor of the nation" by giving "permission to England and France to establish a police in our waters." A week later, Spanish minister Angel Calderón de la Barca sent a letter of protest to the acting secretary of state, Charles M. Conrad, denouncing the activities of the Cuban exiles, especially, "a Creole, without any known profession or respectability, a native of Matanzas, named *José Ambrosio Gonzalez* [sic], has published what he calls a manifesto to the American people, in which he dares to make insidious insinuations, and to cast censure, upon the conduct of the enlightened and upright Chief Magistrate of the Republic. It is not difficult to perceive, that another pen, more able than his own, has indited that slanderous document."[11]

Gonzales had returned to Washington, D.C., from Warrenton Springs after October 5. He consulted with the editors of the *Washington Union* and Louisiana Democratic representative Alexander Gordon Penn, a planter and former New Orleans postmaster (1843–49), on promoting Cuban annexation during the Democratic presidential campaign. The Cuban had worked "earnestly and diligently" wherever he could exert his "little influence" toward the Democratic campaign, prompted by his "democratic faith" and the favorable results it would have toward the welfare of Cuba. He traveled with Florida senator Stephen Mallory to New York City, where they lodged at the Irving House on October 26. Gonzales had written a speech in favor of Franklin Pierce to present at a Tammany Hall political rally that evening where Mallory was scheduled to speak. Filibuster colonel John T. Pickett was also present. The event was "one of the largest and most enthusiastic political demonstrations ever witnessed" in New York City. Tammany Hall was brilliantly illuminated, there was a music band, a thirty-one-gun salute, a series of rockets and other projectiles, and two fireworks displays. One was "a star—a lone star—with the word 'Cuba' inscribed in red, the star itself being white, and having several splendid jets." The Cuba issue was not addressed by any of the orators, who included Daniel S. Dickinson and Gov. Howell Cobb of Georgia. Gonzales was dissuaded from giving his prepared speech by filibuster attorney Theodore Sedgwick, who insisted that "it were better not to bring the Cuban question into the contest there." After the rally, Gonzales left for Richmond, Virginia, where two days later he served as best man in the St. Paul's Church wedding of thirty-one-year-old Savannah banker George A. Cuyler and Bessie B. Steenbergen. Cuyler was the Worshipful Master of Zerubbabel Lodge No. 15, whose members included some filibusters. Gonzales then returned to New York City with the newlyweds, being unable to vote in the national election because he was absent from his registered district in Georgia.[12]

The trio departed for Savannah on November 20 on the steamship *Florida,* checking into the Pulaski House three days later. After a short stay, Gonzales left for New Orleans, rooming in the St. Louis Hotel on December 2, where Jefferson Davis had also lodged that day. Gonzales apparently spoke with Davis about the possibility of obtaining patronage in the new Pierce administration. The Cuban met with John Henderson before briefly leaving New Orleans, probably searching for Gen. John Quitman, returning to the same hotel on the seventh. Domingo de Goicouría checked into the St. Louis Hotel on the seventeenth, and the next day Mirabeau B. Lamar and a Colonel White arrived at the Veranda Hotel from Texas. Other filibuster leaders in New Orleans at that time included Cárdenas veterans William Bell of Mississippi, J. C. Davis of Kentucky, and Chatham Roberdeau Wheat of North Carolina. The last was an organizer of the massive public funeral ceremonies in honor of Calhoun, Clay, and Webster. During this time, Spanish minister Calderón de la Barca notified his consulate spy network to be on the lookout for the filibuster leadership, which seemed to have gone underground. The Spanish consul in Charleston replied on December 22, "With regard to Macías, Yznaga, Gonzalez and company I have not been able to find out anything yet, but I will continue my inquiries with determination."[13]

Gonzales lodged in New Orleans with haberdasher James O. Nixon, the secretary of the Southern Yacht Club, and a member of the Joint Committee on the recent funeral memorial. After Quitman's son arrived at the City Hotel on December 22, Gonzales was told to await the general after Christmas. He wrote to Southern solons influential with Pierce, asking to be recommended for "an independent and influential post" advantageous to the party and the Cuban annexation cause, especially in a diplomatic position. Gonzales felt suited for the job, saying that he was conversant in English, Spanish, French, Italian, could translate Portuguese, knew some German, had studied law, and possessed a classical education. He believed that his appointment would be "pleasing to both branches of the democratic party, the progressive and the conservative, besides a large number of pro-Cuba men not affiliated with any party but would be more especially so to my numerous friends in Virginia, South Carolina, Georgia, Florida & Louisiana." Gonzales indicated that his post "would have a significance to Spain which to some extent would tell to advantage in her future course concerning Cuba." He stressed, "My character, I trust, is sufficiently well established in the country to require any remark of my own. Genl. Pierce will, I believe, know me well upon that score."[14]

Gonzales wrote from New Orleans to Quitman on January 25, 1853, that he was anxiously expecting his arrival, and not knowing when it would occur, he was instead writing "to give you some idea of my situation, my fears and my hopes." He enclosed a copy of his letter requesting a diplomatic post and asked Quitman to help him obtain the position directly and through his friends Jefferson Davis, John F. H. Claiborne, Emil La Sere and "any other person likely to have an influence with Genl. Pierce." Gonzales feared that recent dissension among the Democrats had considerably weakened their belligerent Cuba position. Since a "great majority" of people in Cuba and the United States now favored "the possibility or practicability of acquiring the island by purchase," sufficient means for an expedition would not be available. He thought it might take two years to undeceive these people, but hoped to be proven wrong, especially by "some unexpected favorable incident" that would allow "asserting our own independence by force of arms." The Cuban described himself as "a man of action and not of words," who would not limit himself like other exiles merely to conspire by correspondence with dissidents on the island. He told Quitman that whenever the opportunity arose to strike for the redemption of Cuba, no matter what his occupation, he would serve under his orders "upon the same principle that Genl. Lopez acted in our joint proposition to you of 1850." Gonzales wanted the Pierce diplomatic post to further the cause of Cuban independence, as he stressed to Quitman, "Any consideration which I may acquire in the meantime, will but render it more in my power to aid you in such a work." Gonzales claimed that he could probably "raise under favorable circumstances in South Carolina, Georgia, Florida & East Tennessee a force of from 2 to 3,000 men, to operate in concert with another from the Southwest." He reiterated his "unlimited confidence" in Quitman "as a man, a soldier and a republican."[15]

Quitman traveled to New Orleans on February 7, 1853, and checked into the City Hotel along with brother-in-law Henry Turner. Col. John T. Pickett arrived the next day from Kentucky and settled in the St. Charles Hotel. Quitman met with Gonzales and some conspirators, including William Scott Haynes, but requested to confer with other Cuban leaders before making a final decision on leading the next expedition. Haynes immediately wrote to plotters in New York that they would soon hear from Quitman. He requested that Gaspar Betancourt or Domingo de Goicouría travel south to meet Quitman, or the general could meet them during his scheduled trip to New York City in mid-March. When this was accomplished, recruitment efforts for another expedition commenced in the South and secret societies flourished in various Cuban towns for collecting funds to finance the next invasion. After seeing Quitman, Gonzales departed for Washington, D.C. He stopped in Montgomery, Alabama, where he wrote to Mississippi senator Henry S. Foote and Massachusetts Supreme Court judge Caleb Cushing, requesting their influence with Pierce, and enumerating a dozen reasons for his political appointment. The Cuban arrived at the Irving Hotel in the capital on February 18 to await the patronage distribution of the new Democratic administration.[16]

The day before Pierce was sworn into office, Gonzales sent a memorandum to Cushing who, like Pierce, was staying at Willard's Hotel. It gave the reasons why he should be appointed chargé d'affaires to Venezuela. Gonzales at times was contradictory in his desperate arguments. He twice stressed that by sending him abroad, the United States would demonstrate to Spain that it was not encouraging private expeditions, but then stated that if the island were "to be wrested from Spain," his military services "will be far more needed." If "the island had to be annexed either by peaceable or forcible means," Gonzales claimed to be the perfect candidate to influence Cubans "for the interests of the democratic party in the subsequent formation of political lines." He admitted his lack of experience for the post, but would thrive on the "opportunity of acquiring knowledge & experience" after his appointment. If "some inferior post" was offered him, his prestige among Cubans, acquired "through long years of privation & labor would be totally lost."[17]

The next day, March 4, Pierce emphasized in his inaugural address that his administration's policy would not be "controlled by any timid forebodings of evil from expansion," marking a stark contrast with former Whig practice. In apparent reference to Cuba, the new president stated that the undisguised national attitude was "the acquisition of certain possessions not within our jurisdiction, eminently important for our protection." Pierce seemed to be thinking of reviving the Polk administration policy of purchasing Cuba and discouraging filibuster attempts when he added that "should they be obtained, it will be through no grasping spirit. . . . We have nothing in our history or position to invite aggression—we have everything to beckon us to the cultivation of peace and amity with all nations. Purposes, therefore, at once just and pacific, will be significantly marked in the conduct of our foreign affairs." Pierce was adopting a compromise position on acquiring Cuba which would please the various factions of his new seven-member cabinet.[18]

Top cabinet posts went to William Learned Marcy, secretary of state; Jefferson Davis, secretary of war; and Caleb Cushing, attorney general. The trio favored territorial expansion, but strongly disagreed on other matters. Marcy, a former New York governor, senator, and Polk's secretary of war, detested the Southern "fire-eaters" and the Young America radicals, who considered him as the worst of the "old fogies." Davis was the leading exponent of Southern rights after Calhoun's death. Cushing, a Whig renegade, was regarded as unstable, unreliable, inconsistent, void of political judgment, and lacking a sense of humor. The cabinet also had divided views on slavery. Davis became the sentinel of the institution while Marcy and Cushing favored popular sovereignty. Davis and Cushing had previously discussed Cuban annexation with Gonzales.[19]

The Pierce administration spent three months selecting candidates for more than seven hundred appointive government posts from thousands of applicants. On the evening of March 21, Democratic senators Stephen R. Mallory of Florida and Pierre Soulé of Louisiana presented Gonzales's letters of introduction to Pierce and endorsed his appointment to the

Venezuelan post. Mallory had assisted him in Key West after the Cárdenas fiasco. Soulé, a French republican refugee, had praised in the Senate the morality of the actions of López and the filibusters. The president "expressed himself in very kind terms as to his disposition" toward Gonzales, but disassociated himself by claiming that the matter rested "chiefly with the Secretary of State." The next day, Gonzales again wrote to Cushing, asking him to exert his "powerful influence" on his behalf. The great stress that Gonzales was under forced his vanity to surface, by exaggerating, "I happen to be better qualified for a mission to that place than those who would apply for it. The administration could hardly give me a position inferior to that of chargé without impairing my prestige and condemning me to political obscurity to the obvious detriment of the interests of the party both with reference to Cuba & the United States."[20]

Secretary of State Marcy was being badgered by job seekers at his Willard's Hotel quarters when Gonzales went to see him with a March 24 letter of introduction from Alexander Gordon Penn. Gonzales also carried written recommendations from Sen. John Slidell and other distinguished Louisianians, who described him as "not only a Gentleman of great intelligence and individualism, but one who is entitled to the Strong Sympathies of our friends." Gonzales expressed to Marcy his qualifications for the job, but like hundreds of other patronage pursuers was left waiting indefinitely for a reply. Two months later, Gonzales relayed to Cushing what he apprehended could be grounds against his nomination: that he was Catholic; that he would promote Cuban revolutionary movements while in Venezuela; and "that a general outcry will be raised against it by the opposition press." In replying to these points, Gonzales indicated that he was Protestant, enclosing wardens and vestry resolutions he received from Christ Church in Savannah; emphasized that he would not betray his oath of office; and believed that the majority of the Whig press "will not be blind to the policy and reasonableness of my appointment."[21]

Twelve days later, on May 27, Gonzales again addressed Cushing "for the last time upon this subject." He gave an account of his activities during the previous year which he deemed beneficial to the Democratic Party. The Cuban also enclosed various newspaper clippings of his political articles and a letter of recommendation, circulated in Savannah by

Alonzo B. Luce and John Lama, signed by Mayor Richard Wayne and a few prominent citizens. He concluded by adducing "these facts merely as an answer to those invidious & malignant persons, who, on the plea that I have done nothing for the party pretend that I shall not be the recipient of its favors." The next month, Gonzales informed Cushing that he would accept the job of live-oak agent, if the government consolidated the two live-oak regions of middle and west Florida, similar to the arrangement made for John F. H. Claiborne of Louisiana, with an annual salary of two thousand dollars. The agencies of eight Southern live-oak regions were still occupied by Whig appointees, soon to be replaced by Democratic patronage.[22]

The relentless Gonzales again wrote to Cushing six days later, on July 6, enclosing a newspaper article that he believed would strengthen his chance of being appointed chargé d'affaires to Venezuela. The *Baltimore Sun* clipping of the previous day stated that the Mexican minister to New Granada would visit Venezuela and Ecuador to propose, in alliance with Spain, a Latin American Congress to oppose United States policy in the region. Gonzales recommended a counterplan "having Cuba, morally & physically as a 'point d'epee' & a connecting link . . . in any contingency affecting the interests or the safety of the United States." In such a case, Gonzales had also informed Pierce in his letter of application, he would raise ten thousand allies in Venezuela with the aid of exiled Gen. José Antonio Páez, for "any struggle upon the Gulf or the Caribbean sea." Gonzales indicated that he had political enemies, including Cuban exiles, "who will resort to underhanded & insidious means to thwart my purposes & clip the wings of my ascent." His repeated references to expansionist belligerency for Cuba, which he mistakenly assumed was the ideology of the Pierce administration, was probably one of the reasons why he did not receive a diplomatic post. Pierce never supported the filibusters, did not perceive a Mexican threat, and achieved during his presidency only one expansion: the $10 million Gadsden Purchase from Mexico six months later.[23]

Gonzales had become estranged from the filibuster movement after General Quitman had accepted nine weeks earlier, from the recently organized New York Cuban Junta, the exclusive military and civil command of the next filibuster expedition. Junta leaders José Elías Hernández, Porfirio Valiente, and Gaspar Betancourt Cisneros, had gone to Natchez, via New Orleans, to personally proposition Quitman. The former governor accepted under the stipulation that the Junta would provide the adequate means for the venture and secure the support and unity of all Cuban exiles. The slavery issue by then was increasingly dividing the Cuban émigré community, a reflection of American society after the Compromise of 1850. Gonzales had "kept aloof" of Junta activities because of "the antislavery notions of some of its members" and his "utter dislike of the personal character of others" like Betancourt, who had undermined General López. As a result, he lost communication with his former Southern filibuster associates and dedicated his energies to serving the Cuban cause from a diplomatic position.[24]

By mid-July, the Pierce administration had appointed ten foreign ministers and ten chargés d'affaires. Some important foreign posts went to Young America adherents. Louisiana senator Pierre Soulé became minister to Spain; German-born Wall Street banker August Belmont was sent to The Hague; John L. O'Sullivan, who now favored the acquisition of Cuba through purchase after being twice indicted for filibuster activities, was appointed minister to Portugal; and Cárdenas veteran John T. Pickett, became consul at Veracruz, Mexico. Soulé "personally and strenuously recommended" Gonzales as secretary of legation to France when Gen. John Adams Dix, the U.S. assistant treasurer in New York, was being considered as minister. Dix had left the Democrats in 1848 to run unsuccessfully as the Free Soil Party candidate for governor of New York. His appointment was postponed until after the summer elections in the South because of strong opposition from that area. O'Sullivan, a former "barnburner" with Dix, discouraged him from taking the ministry.[25]

During the first days of August, Gonzales accompanied Soulé from Washington to New York City. The Cuban exiles celebrated an act on September 2 commemorating López and the *Pampero* martyrs. Three days later, they feted Soulé with a torch light procession, a serenade, and a patriotic speech by Miguel Teurbe Tolón. Soulé responded with assurances that in spite of his official position, he would carry to the tyrants of Europe the throbbings of those people who express such tremendous truths. Before departing for Spain, Soulé introduced Gonzales to General Dix, the unpopular Northern political heretic to whom the Cuban's destiny had been tied. Unfortunately, Dix did not receive the pending mission to France. Gonzales then left for the summer resorts at Newport and

Bristol, Rhode Island. While there, he wrote to Cushing on October 20 that he was not involved with "the late action of the Cuban Junta in New York in connection with the Ingraham demonstration & in furtherance of party dissentions." The following day, Cushing's Justice Department quietly dropped all charges pending for two years against Cincinnati filibuster leaders William Hardy, Henry H. Robinson, and William A. McEwen. This was not an indication of the Pierce administration sanctioning filibuster activities. Secretary of State Marcy was just as vigilant of Cuba movements as his Whig predecessor. In November, he wrote to the New Orleans U.S. attorney and port collector and to the U.S. consulate in Havana, inquiring about rumors of a projected Cuban expedition. Acting consul William H. Robertson replied that "there has not been a day since I have been in the Island, now upwards of two years, that there have not been rumors of expeditions prepared or preparing in the United States, and frequently that they were already landing."[26]

In mid-December, Gonzales wrote from New York City to Attorney General Cushing and Secretary of War Jefferson Davis indicating that references in the president's first State of the Union message on December 5 were propitious regarding his pending appointment. Pierce had alluded to U.S. interests in acquiring territory that would be important to the country's security and commerce. Gonzales and most Cuban annexationists erroneously assumed that he was referring to their homeland, but it turned out to be the Gadsden Purchase signed on December 30. Gonzales asked Davis for help in getting appointed secretary to the minister to France or to the more desirable post of chargé d'affaires to Venezuela. The Cuban returned to Washington, D.C., on January 4, 1854, lodged at Willard's Hotel, and continued lobbying for patronage.[27]

Two weeks later, on the day Pierce issued a proclamation against William Walker's planned filibuster expedition against Mexico, Senator Mallory met with the president and stated that Cuba was "on the verge of a change" regarding slavery that would "take place *with* or *without* the consonance of the U.S." Mallory believed that "in such a state of things the abilities, and qualifications of such men as Gonzales will become of the very first consequence, in fact indispensable." He argued that the Venezuelan position was "not much sought as a political appointment," and that although "Gen. Gonzales has not the united political influence of any state," he did have recommendation letters "of the most satisfactory character" from the leading men of several states. Mallory further indicated to Pierce that "no objection to him can be raised on the score of his connection with Cuban expeditions" because since his involvement with the first one, "he has done all in his power to suppress such expeditions," and John T. Pickett, a member of the first expedition, was made consul at Veracruz. Furthermore, "his views on Cuban matters . . . are just such as are entertained by Soulé, [Romulus M.] Saunders, [John] O'Sullivan, [Solon] Borland & a host of others." The next day, Mallory wrote to Attorney General Cushing, stressing the same points on behalf of Gonzales.[28]

One of Pierce's recent concerns was the growing agitation over rumors that Spain, pressured by the abolitionist British and French governments, would "Africanize" Cuba through the introduction of the African apprentice system. This prompted him on April 3, 1854, to authorize the American minister in Spain, Pierre Soulé, to offer the Spanish Crown up to $130 million for the purchase of the island. If the sale of Cuba was rejected, Soulé was to direct his efforts toward inducing Spain "to consent to its independence." The slave apprenticeship system was supported by Cuban émigré abolitionists in New York with a new publication, El Mulato, edited by attorney Santiago Bombalier. Gonzales perceived how some Whig journals, "mistaking the origin, nature, and bearing" of the paper, were warning the South "against incautious sympathy in behalf of Cuba." He also objected to how the opposition papers were incessantly coupling the patriotic acts of López with those of filibuster William Walker and "the band which is now foraging Lower California." Although Gonzales himself did not sympathize with Walker, at least ten former Cuban filibuster officers joined him in Nicaragua. Gonzales sent a letter to the editor of the Washington Union on April 13, pointing out that the Cuban exile publications El Cubano, the Correo de Ambos Mundos, and El Filibustero in New York, along with The Beacon and the Independiente in New Orleans, "have abstained from meddling with the question of domestic slavery, which, Cuba being free, will, as naturally as in South Carolina, take good care of itself." Gonzales derided El Mulato as "an abolition sheet of very small dimensions and much smaller influence," hooted "as a political hat by the great majority of the resident Cubans at a meeting held for the purpose, in which appropriate resolutions were passed." He correctly predicted that the newspaper, supported by British or American abolitionism, would soon expire.[29]

The Africanization of Cuba scare was heightened after Cuban captain general Juan Manuel de la Pezuela decreed a yearly slave census that would grant freedom to those whose masters lacked legal ownership title. Pezuela also proclaimed freedom for all *emancipados* arrived prior to 1835; the banishment of slave importers; the encouragement of racial intermarriage; and the dismissal of governors who did not denounce the slavers. The measure had the support of the Cuban Catholic Church, but the aristocracy panicked, and again looked toward American annexation to preserve the status quo. Since most Africans had been clandestinely imported to avoid paying the government tax, many Cubans reeled at the thought of thousands of emancipated slaves roaming the island. When the order inaugurating the African apprenticeship system was printed in the *Washington Union,* Gonzales responded on April 25 with the article "Cuba—The Turning Point." He recalled how Spain had vowed that the island would "be African when it lapses from her grasp" and warned that five hundred Africans confiscated by the British navy from slave ships the previous month had been deposited in Trinidad, Cuba, to begin one-year terms as apprentices before being freed. Gonzales feared that these manumitted "savages" and other "wild, untutored, and ferocious Africans" to be dumped on Cuba by the British, with the consenting "folly, the temerity, or the wickedness of Spain," could "produce a thirst and longing for self-control in half a million slaves" that would result in racial massacres as previously occurred in Haiti. Gonzales warned that if Cuba "be not drawn at once, by the eddy of the popular will of America, into the arms of the Union," it would result in dire consequences for the island and the best interests of the United States. He concluded that statehood would make all proclaim the political axiom that "Cuba has become the bond of the Union, as she was made by Nature the CLASP of America."[30]

A week later, Gonzales published another article in the *Union,* titled "Cuba—Our Duty to Ourselves," proposing that just as the Monroe Doctrine had impeded the European schemes of white colonization in Latin America, a new "Pierce Doctrine" should prohibit Spain's "evil" plan to increase African migration to Cuba as either slaves or under the apprenticeship system, "with all its barbarism and concomitant results." A black colony would be formed within the Spanish colony and eventually "turned against us." The result would be "ruin to our commerce, destruction to our interests, and the obliteration of our race in Cuba and Porto Rico." Gonzales concluded by asking the "enlightened, the free, the powerful, and generous American people" to say "No!" to those plans through a Declaration that would safeguard the continent.[31]

Four days later, on May 7, Gonzales sent another letter to the editor of the *Union*, replying to a *Washington National Intelligencer* article that claimed that "Cubans are lily-livered because they did not join General López and his companions, and for this reason [are] not deserving of freedom." He outlined a history of Cuban independence movements since 1823 that resulted in "a series of imprisonments, deportations, and executions" and "*hundreds* of exiles now in the United States." Gonzales cited from his manifesto the overwhelming advantages the American Revolution had over the Cuban Revolution, and stressed that the failure of General López "was merely the result of circumstances. Bolivar made several trials before he established liberty in Colombia, Bruce made nine attempts for the independence of Scotland." The Cuban lethargy was because "three centuries of unheard tyranny, corruption, and exactions, do not weigh in vain upon a miserable people."[32]

His fiery rhetoric in the Pierce administration's organ was read by American government officials and the Spanish minister in Washington, who were also receiving reports of the filibuster organizing activities of General Quitman and the Cuban Junta. President Pierce issued on May 31, 1854, a proclamation against persons "engaged in organizing and fitting out a military expedition for the invasion of the island of Cuba." He threatened to "not fail to prosecute with due energy all those who, unmindful of their own and their country's fame, presume thus to disregard the laws of the land and our treaty obligations." When Gonzales called on Cushing on June 20, the attorney general, who was planning to arrest Quitman and erroneously suspected that Gonzales was involved, abruptly terminated their conversation when it turned to "the subject of Cuban movements." After departing, Gonzales penned a scathing note, the last communication he ever had with Cushing, accusing him of being discourteous and reversing his views on Cuban annexation after taking office. Two days later, the *Washington National Intelligencer* denounced the Quitman plans, to be seconded by a Mexican War brigadier general, and purported that "General Gonzales, whose name has been associated with this movement from the first, will without doubt be third in command."

Gonzales quickly replied with a letter to the *Union*, denying any involvement "with a Cuban expedition at the Southwest. I have full confidence in the president and in his ability to attain success. I shall abide by his action as long as that confidence is unimpaired." He also expressed to President Pierce, "As this action of mine might subject me to the imputation of interested motives, I desire to state that I withdraw the application that I have made & which my friends have urged." The position of chargé d'affaires to Venezuela later went to Charles Eames, who had been managing the *Washington Union*.[33]

That same day, the *New York Times* also published an account of the preparations of the Cuban Junta, whose "departure of the first expeditionary corps for Cuban emancipation is fixed for an early day in July." The article claimed that fifty thousand "robust and thoroughly disciplined" filibusters had secured fifty thousand uniforms, eight staunch steamers, over fifty thousand arms and ninety pieces of artillery. The newspaper alluded to Quitman as the leader of the movement, and that "General Gonzales, whose steadfast devotion to the Cuban cause has won for him the confidence of all parties of his own countrymen, and of their sympathizers in the United States, will also be placed in prominent command." The report indicated that "Colonels Wheat, Pickett and Bell, Majors J. A. Kelly, and Moore, Captain W. S. E. King and J. W. Denient, with Lieut. Frank Oilnette, are of the officers pledged to the cause." Gonzales replied with a letter to the *New York Times* editor disavowing any involvement and stating, "I prefer, for the present, the action of the Government to that of individuals. Should my confidence in the former ever be shaken, I shall go with the foremost for that 'ultima ratio *civis*,' in favor of oppressed humanity: *the right of expatriation,* and of meeting the tyrant hand to hand upon the soil he desecrates."[34]

This letter, reprinted in the *New Orleans Picayune* on July 4, had unfortunate and unforeseen consequences for Gonzales. The same page of that *Picayune* edition contained an article indicating that General Quitman, John Sidney Thrasher, and Dr. A. L. Saunders had each been fined a three thousand dollar surety bond for the term of nine months by U.S. Circuit Court judge John A. Campbell "for their observance of the laws of the United States in general, and the Neutrality Act of 1818 in particular." The trio had refused to testify before a New Orleans federal grand jury investigating rumors of preparations for a filibuster expedition. The grand jury report concluded that while there was evidence that Cuban meetings were frequently held, Cuban bonds were issued and funds had been raised, there was no proof of "military organization or preparation." Quitman, the Cuban Junta, and their supporters interpreted the Gonzales letter to the *New York Times* as his definite break with their cause.[35]

In early August 1854, Gonzales was hired as "occasional" translator in the Department of State with a "moderate income." He received a batch of foreign-language documents that he took to the summer resort in Warrenton Springs, Virginia. Gonzales translated for the use of Congress, at the request of Speaker of the House Linn Boyd, a French-language work by officers of Napoleon III on the Panama Canal and a Spanish-language pamphlet by Mexican general Juan de Orbegoso on his 1825 survey for a Tehuantepec Isthmus Canal. The Cuban also kept in frequent correspondence with the Elliott and the Wayne families of South Carolina. Mrs. Mary W. Wayne, a "venerable cousin" of the Elliotts, informed Mrs. Elliott that month that she and her husband, U.S. Supreme Court associate justice James Moore Wayne, were looking forward to seeing Gonzales again and hoped "that he may soon recover all his property in Cuba." When Gonzales returned to the capital in late October, he called on Secretary of War Jefferson Davis, who was not at home. The congressional and state mid-term elections were a week away, and Davis and the Pierce administration had come under fire in Mississippi because of the government's opposition to the Quitman expedition. Davis asked Gonzales on August 28 to provide a translated copy of a *La Verdad* article that was "essential to the proper defense of the administration in its Cuban policy." In spite of the presidential proclamation against Cuban filibustering, Davis was still "always hoping for success and trusting with sanctity of a cause which has for its end the prosperity of our own country and the regeneration of an oppressed neighbor. I have not despaired through the vast and the bad each in his own way contribute to defeat our efforts." Gonzales promptly did the translation request, which Davis forwarded to Mississippi, hoping it would "correct some errors." Davis highly praised Gonzales for "the patriotism and discretion which has characterized your course."[36]

After the November election, Gonzales started an article for the *Union* on the adaptability of light-draft war steamers for American national defense. He left the capital during a few days, possibly visiting the U.S.

Naval Academy at Annapolis. The return of "the well-known Cuban patriot" to Washington, D.C., on November 18 was heralded in the *Union*. Gonzales sent his manuscript to Secretary of War Davis, who approved most of it, and provided some suggestions on harbor defenses. The article argued that America, with a feeble navy, was threatened by the recent Anglo-French alliance. Since the national debt could not afford "scores of millions" for new warships, Gonzales suggested constructing a cheaper fleet of "vessels propelled by steam, of from four to six feet draught, mounting each two of the heaviest columbiads which they can be made to carry." These armed steamers, plying shallow coastal waters "inaccessible to larger ships," would be "invaluable, as they could avoid or give battle at will, thus baffling intervening squadrons, *land troops with ease,* and afford them, whilst they landed, the very best protection." In battle, "the fire of 50 or 100 such gun-boats, concentrated upon the advanced ship of the enemy's squadron, would sink her before she could bring more than her chase guns to bear upon them." Seven years later, during the Civil War, Gonzales proposed a similar plan to the Confederacy for the coastal defense of South Carolina.[37]

Eleven days after his article appeared, Gonzales wrote from Washington, D.C., to Laurent Sigur regarding a "derogatory" article that had recently appeared in the *New Orleans Delta,* "in a manner totally unwarranted by facts," assailing his motives in keeping aloof from Junta activities. The Cuban stated that he wanted Sigur and John Henderson to know that the obnoxious character of some Junta members and the abolitionist ideals of others is what made him disassociate himself from that group. He regretted that this had caused a separation from his Southern friends after their association with the Cuban Junta. Gonzales reiterated his ideals by emphasizing that he had "advised *no* purchase of Cuba" to the administration and that, even so, it was preposterous to believe that they would have heeded his words. Although he had asked for government employment, he did not consider it "impending to *proper* action when its proper moment came." Gonzales indicated that Sigur and other filibusters had held public office, which did not interfere with their activities. He stressed that he had stated distinctly that his confidence in the administration was only as long as "*its ability to attain success* remained unimpaired," which did not imply "a shackled judgement." Gonzales informed Sigur that he had withdrawn his patronage application in June and was "entirely destitute & depending upon distant friends for a modest & precarious support." He concluded that "if after this, justice be denied me, let it be so; censure has no darts for me."[38]

The whereabouts of Gonzales during the next two months are obscure. José de la Concha, who sentenced Gonzales to death, had returned as captain general of Cuba in late December for a five-year period, intent on again crushing the filibuster movement. A spy report from New Orleans filed with the Spanish Navy General Headquarters in Havana on March 4, 1855, and dated the "23rd," purported that Gonzales was acting as adjutant general and private secretary to Quitman. The document stated: "Said Ambrosio Gonzales joined Quitman in Alabama and gave him communications from the Junta (Cuban) in New York. . . . General Gonzales claims $3,000 in salaries and expenses in the North." The report warned that "tomorrow 5 members of the Junta leave for Savannah, and they are Porfirio Valiente, Goicouría, Betancourt, Duanes [Duany] and [John] Trasher from where they will go to New York." But the Spanish spy's report was unreliable. Goicouría had left the Junta in 1854. New Orleans hotel arrival notices indicate that since New Year, the only Junta member who visited the city was Valiente. He stayed at the City Hotel on January 18, where General Quitman arrived four days later. Quitman and Gustavus Woodson Smith later checked into the City Hotel on February 16, where Felix Huston, who lost a nephew in the *Pampero* expedition, and Chatham R. Wheat had lodged a few days earlier, but there is no mention of any other Junta leaders there or arriving in Savannah. During the two months in question, Gonzales did not get involved with Quitman or the Junta, and he probably remained in Washington, D.C.[39]

When Carolinian planter William Elliott visited the capital in late February 1855 with his daughters Anne and Emily, they frequently saw Gonzales "in the best circles" and noticed that he "appears to have the respect of the principal people." The Elliotts were staying at Willard's Hotel while applying for passports. The ladies were "much pleased" to attend various splendid parties where they met "gentlemen who could appreciate them." Elliott was trying to get his two eldest daughters betrothed into high society, but their age made them spinsters by Southern standards. Thirty-three-year-old "Annie," who weighed, "to the astonishment of all who have heard it 173 pounds," and twenty-five-year-old "Emmie," who weighed ninety-nine pounds and had an irascible character, never married.⁴⁰

Emmie was smitten by the Cuban's attentions, charming manners, and years of correspondence with her, and she misunderstood their platonic relationship. In mid-May 1855, she wrote from Oak Lawn to her youngest sister, fifteen-year-old Harriett Rutledge Elliott, affectionately called "Hattie" and "Little Girl," who since January was boarding at Madame Rosalie Acelie Togno's School for Girls in Charleston. Emmie boasted receiving a twelve-page letter from Gonzales, who had been delighted to hear that Hattie had grown taller than her, and inquired what pieces the school girl was playing on the piano. A streak of jealousy surfaced in Emmie's own response: "Shall I tell him—Variations on the air 'An Old Man will never do for me'?" Emmie added that Gonzales "wants to know our plans as his own will be contingent upon them—if we are to be at Flat [Rock] so will he." The aging Southern belle, sensing a growing interest in the Cuban toward her teenage sister, questioned Hattie's sentiments toward him: "Shall I tell him that little Annie [Johnstone], Mama & yourself, will greet him with gladness?" Emmie also told Hattie that a "Col. Morris has gone without proposing" to her, and that Gonzales's friend, forty-one-year-old Pedro José Guiteras Font "intends to send me his likeness." Emmie and her sister planned spending the summer at the Elliott's new seasonal house in Adams Run, South Carolina, which was being renovated with an additional wing to accommodate the entire family.⁴¹

The Elliott family split up for the summer. William Elliott sailed on May 30 on the steamship *Baltic* from New York to Liverpool with his daughters Caroline and Mary. He then proceeded to France as South Carolina's commissioner to the Paris Exhibition. Days later, Mrs. Ann Elliott and Hattie moved to Mary's *Beaumont* summer residence in Flat Rock, North Carolina, to care for her five children. Hattie considered the place dreadfully "awful" and preferred to have stayed at their nearby *Farniente* estate, which had been rented for the summer. She indulged in sleeping "abundantly," walking, riding, studying, playing polkas on the piano, and teaching her nieces and nephews. Hattie and her mother went on carriage rides to the French Broad River valley and attended services at St. John in the Wilderness Episcopal Church.⁴²

Emmie and Anne got bored in the village of Adams Run and in late July departed with their ill brother William, by railroad and stagecoach, for the health resorts in Virginia. The Elliotts went to White Sulphur Springs, where they encountered Gonzales, who had been working as a temporary clerk in the Patent Office in Washington, D.C. After a few weeks of "dirty people and bad fare," the Elliotts left with Gonzales for Salt Sulphur Springs, in present-day Monroe County, West Virginia, a more "quiet" and "quite comfortable" place. The resort was a favorite of South Carolinians, who occupied a row of brick cottages called "Nullification Row." William wrote to his brother Ralph, who had remained at Chehaw plantation tending to his crops, that good food, tranquility, and "the man they call Gonzales" had "done much towards restoring us to equanimity." The Elliotts and Gonzales, departed by stagecoach on September 3 for Sweet Springs, Monroe County, Virginia, to enjoy the sulphur baths. In mid-September, Gonzales apparently accepted their invitation to visit Flat Rock. On October 2, Hattie celebrated her sixteenth birthday, while Gonzales had his thirty-seventh birthday the following day. During this prolonged stay, the Cuban patriot and Hattie fell in love.[43]

After a whirlwind two-month romance, during which Gonzales accompanied the Elliotts on their return to Oak Lawn, the couple celebrated their engagement on December 11, 1855. Hattie's holiday gift to her fiancé was a Bible with a dedication and a prayer book. Her sister Mary wondered if Hattie would finish her education, or if she was "going to be romantic and settle down at Myrtle Bank? or is she going to be a public character and live in Washington?" Gonzales spent the Christmas and New Year's holidays in South Carolina, before returning by train and steamer to the capital, via Baltimore. His vessel grounded on the Chesapeake Bay ice two miles from shore, and he had to walk the remaining distance with his carpetbag and trunk, plus another three miles on land, to reach the Baltimore railroad station. The ten-degree-below-zero weather left icicles on his moustache. When Gonzales reached a pub, there being no wine, he took whiskey and water to keep warm. He reached Willard's Hotel, in Washington, D.C., on January 11, 1856. The Cuban shaved off a long beard, arranged returning to his former rooming house, and wrote to Pedro José Guiteras Font and a Miss White of his engagement. Gonzales later forwarded their replies to Hattie, referring to Miss

White as "your 'illustrious *teasing* predecessor.'" The next morning, he returned to work as a temporary clerk in the Patent Office, ruling on three hundred patents that day. The pay scale was ten cents per hundred words. Gonzales worked overtime by evening candlelight to save for his next trip south.[44]

That week, the former filibuster spoke with Secretary of War Jefferson Davis, who had previously agreed to talk to President Pierce about a diplomatic post for him. He spent the evening of January 22 at the Davis home on the corner of 18th and G Streets, but family distractions prevented Gonzales from fully discussing his future. Davis told Gonzales that he and President Pierce shared the same "great interest" toward him. Three days later, Gonzales went to see Pierce, who said that he had not addressed Secretary Marcy yet on the Cuban's petition, but would do so as soon as possible. The next evening, Saturday, January 26, Gonzales had an appointment with Mrs. Pierce at the White House. She asked if he had heard from Hattie and commented on how happy he must feel upon receiving her missives. The weekly love letters included an exchange of hair locks, to which the Cuban pointed out, "mine is silver and yours is gold," closing with "Gonzie," the sobriquet Hattie gave him. Emmie showed the first signs of jealousy toward her engaged sister by being sarcastic about the extent of Gonzales's friendship with Miss White. The Cuban tried to neutralize the situation by sending Hattie the next note he received "from my good friend Miss White, *our* good friend, I should have said." He did so "even at the risk of demolishing Emmie's satire which I knew never could reach the sublimity of ever a 'ferocious instinct.'" Thus began Emily's perpetual love-hate attitude toward Gonzales that lasted a lifetime.[45]

Gonzales and Hattie were married on April 17, 1856, at Oak Lawn, where her mother and grandmother had also contracted nuptials. The ring bearer was Hattie's seven-year-old nephew Elliott Johnstone. Her sisters Anne and Emily were bridesmaids. Pedro José Guiteras Font traveled from Philadelphia to serve as best man. He gave Emmie two Spanish madrigals, dedicated "as a testimony of tender friendship." Another guest was Florida senator Stephen R. Mallory, who arrived by steamer from Key West. The South Carolina lowcountry aristocracy turned out, with few exceptions. Statesman James Louis Petigru, a Whig Unionist, wrote to William Elliott, regretting not being able to attend because his law practice kept him that week "tied down" before the docket. Judge John B. O'Neall was also kept away by his court calendar. The lavish ceremony was performed by Episcopalian Reverend Charles Cotesworth Pinckney, a nephew of William Elliott. Hattie brought a twenty-thousand-dollar dowry with seven hundred dollars' annual interest income into the marriage, her brother Ralph being appointed trustee. Although the husband customarily administered the dowry and maintained its value, it remained the wife's property and future inheritance of their offspring. This was a common arrangement among the wealthy, as it would protect the woman from her husband's creditors. The wedding notice appeared in the *Charleston Courier* and the *Savannah Morning News*.[46]

The newlyweds stayed at Oak Lawn, Charleston, and Adams Run after the ceremony. The Reverend Pinckney gave them an elegant reception in his Charleston residence at 46 Beaufain Street. The couple left Charleston on June 29 on the steamship *State of Georgia,* arriving in Philadelphia fifty-four hours later. They continued to Washington, D.C., and checked into Willard's Hotel on July 1. Gonzales visited the White House to introduce Hattie to the president and his wife, but was told by the porter that they were both ill. The Cuban later went by himself to visit Jefferson Davis, whose wife admonished him for not bringing Hattie. Mrs. Varina Davis already had a detailed account of the bride and the wedding from a letter written to U.S. Supreme Court associate justice James Moore Wayne from his wife Mary W. Wayne in South Carolina.[47]

A few days later, Gonzales and his wife went to spend the summer at the Rockbridge Alum health resort in the mountains near Lexington, Virginia. The place was considered "second only in fashion and elegance to Fauquier White Sulphur Springs," containing five hotels and two rows of cottages, that accommodated up to eight hundred guests. Hattie wrote home on July 11: "The life we lead is funny, immediately after breakfast, dinner, & supper, Gonzie brings me to our room & there we sit playing or reading Spanish or sleeping & not making a single acquaintance." The Cuban would rise early, get plenty of exercise, avoid wine, and take long baths with his wife in the sulphur waters. He shaved every other day, spent hours reading a book of fables by French poet Jean de La Fontaine dealing with human behavior, while Hattie dyed his hair black. The resort included two bowling alleys, a billiard saloon, a bar room, a music band, and the services of a physician and a dentist. Hattie told her family how

"Gonzie went today to see some men play billiards & ended of course in playing himself & winning too." In closing her letter, she sent her remembrances "to all the servants." Mrs. Ann Elliott then decided that she would depart on July 26 with her family to join the newlyweds. Her husband, settled at Bay Point plantation on Phillips Island, not being "strong enough in mind or body," declined to go. Mrs. Elliott refused to take the steamer for the longer part of the trip due to her "foolish dread of the Sea." Gonzales looked forward to the visit, planning to leave his wife with her family for a day or two while he returned to Washington and tried to see President Pierce regarding his pending position. He asked Annie to bring from the Oak Lawn library a volume of the Boston journal *Littell's Living Age,* which contained the article "Political Tendencies of America."[48]

Mrs. Ann Elliott, Ralph, Caroline, Annie and Emmie joined the honeymooners in Virginia. Gonzales returned to Washington, D.C., on October 14, checking into Willard's Hotel with his wife and in-laws. Before the Elliotts departed for South Carolina via New York, the Cuban took them to the White House, and Mrs. Elliott was "much pleased" with the green houses and gardens. Gonzales and Hattie moved into a rooming house run by a Mrs. Paine, who prepared them "delightful" meals. The former teacher gave his wife Spanish lessons, which included rolling her r's "in a most Cuban manner." In the evenings they took long walks across town to the Capitol. At the time, there was an ebullient presidential election campaign between Democrat James Buchanan, Republican John C. Frémont, and Millard Fillmore of the Whig-American Party alliance. When it appeared that Buchanan was going to win, Hattie got excited when a Mrs. Ashton White expressed her strong belief that Gonzales would be governor of Cuba. The Spanish Crown had granted that year a political amnesty that rescinded Gonzales's death sentence. Many former filibuster activists returned to Cuba, but Gonzales made no plans to join them. In December, when Hattie was three months pregnant, they started going to church, which she described as "quite an event in our lives & the first time I have been for more than a year. Gonzie was so devout, it was too funny." The couple also visited the White House and attended social parties, which Hattie did not enjoy. The teenager recalled how during one concert, "while crowds of persons, foreign ministers, Senators & members of Congress, were waiting to see the president and

Mrs. Pierce pass, as they approached us, bowed several times, stopped, shook hands & began to talk." Just then, Hattie saw the surprised look on South Carolina Democratic representative William Aiken of Charleston, and "could scarcely keep from laughing" at his astonished expressions. She wrote Annie and Emmie of being anxious for them to visit the capital again, so that they could accompany Gonzales to the parties while she stayed home.[49]

Emmie's resentment at being single surfaced in a letter to Hattie a few days before Christmas. She enclosed a previous missive from Gonzales that she had just found in her desk, jealously claiming, "it reminds me of when 'I too once had a Gonzie.' Are you not sorry for me? Don't cry. I am perfectly content to let you have him. . . . I do wonder he did not think me young enough for himself I never thought of it before." Emmie also sent a bouquet of flowers from the Oak Lawn garden for Pedro José Guiteras Font, and asked that Gonzales persuade his friend to visit Washington for the inauguration, which she planned to attend. She complained of "growing fat" and indicated that on her upcoming visit to Santee the next month, "perhaps I shall flirt with Parson [Thomas] Gerardeau." Nine days later, Emmie wrote to her sister again deriding the forty-two-year-old "Guiteras's age & infirmities." She rudely asked if her husband still corresponded with Miss White, sarcastically wondering "how does she stand the separation?" Hattie, for her part, considered herself fortunate to be the Cuban's wife and never doubted his love for her. She had written Annie the previous week that two Washington socialites, "Jehosey & Etta might have been Gonzie's, it is my opinion, if he had been calling," and that Warrenton Springs proprietor Thomas Green "seems to truly love Gonzie & not to feel vexed with him because he did *not marry* one of his nieces." In contrast to Emmie, her sister Annie was grateful to receive "sweet" letters from Gonzales, whose sentiment "brought tears to the eyes of our old nurse who can judge of heart— if not of styles."[50]

The Elliott sisters, escorted by their brother William, a recent planter after acquiring the family's Myrtle Bank plantation on Hilton Head, arrived at the capital in mid-February. They gave Hattie a pink French bonnet and a Canton crepe shawl and lodged in a rooming house procured by Gonzales. Their sister Caroline, suffering from bronchitis, sought the warmer Cuban climate with her father, who was afflicted with asthma.

William Elliott traveled on the steamer *Isabel* on March 4, with letters of introduction from Secretary of the Navy James C. Dobbin for U.S. Navy commander Hiram Paulding, describing him as "a gentleman of fortune and character," and one from Gonzales for his Cuban friends. The Elliotts stayed a week in Havana, attending a reception at the "princely mansion" of Miguel Aldama Alfonso. His wife, Hilaria Fonts Palma, was anxious to know Gonzales's response to her husband's request that they be the godparents of his first-born. The Elliotts visited the fortifications of El Morro, La Cabaña, and El Príncipe prison, and at the Tacón Theater saw Capt. Gen. José de la Concha, who had signed Gonzales's 1850 death warrant. They also visited La Punta and Atarés fortresses, where López and fifty American filibusters had been executed. The Elliotts later went to the city cemetery, erroneously assuming Crittenden and his men were interred there, but found no trace of their graves. Benigno Gener Junco took them sightseeing in Matanzas to various places, including the Cumbre, the Yumurí Valley and Cárdenas, the scene of the filibuster landing. Pedro José Guiteras's brother lodged the Elliotts at the sugar plantation owned by Aldama and his brother-in-law, José Luis Alfonso García de Medina, a. former Havana Club conspirator. The Carolinian was fascinated with the efficiency of the mill, the task organization of the 350 slaves, and the immense net annual profit of $150,000. Elliott later published in *Russell's Magazine* seven consecutive articles detailing his visit. Upon returning on the *Isabel* to Charleston on March 28, Callie wrote to Emmie in Washington, D.C., about having "never enjoyed myself as much in my life" and the island being "magnificent, the climate delightful & the people (the Cubans) the nicest in the whole world. I would prefer returning there to going to Paris." Callie reiterated her desire and that of her parents that the entire family reunite soon in South Carolina.[51]

The following month, Gonzales, Hattie, and her siblings returned home. William went to his Hilton Head plantation and the others settled at the Adams Run summer residence. While there, Hattie gave birth on May 29, 1857, to Ambrosio José Gonzales Jr. Five days later, Emmie described the baby to William, as "a Little Spanish American" of "uncommon size, shape, intelligence & appetite." She said that the mother and child were well, the grandfather was "delighted," and that Gonzales was "radiant" and had written the news to his Cuban friends. In late July, the Elliotts again split up for the summer. The patriarch and Callie headed for Bay Point, Mrs. Elliott and Emmie departed for Flat Rock, and Ralph stayed with Annie and the Gonzales family at Adams Run.[52]

That summer, Gonzales returned to Washington to lobby the new Buchanan administration for a diplomatic post. He went to Warrenton Springs for recommendations from politicians vacationing there. When his absence extended from weeks to months, his pessimistic wife, probably agitated by Emmie, wrote him a hasty letter in mid-August desperately asking what would become of them if he did not get a position, since she and the baby would not be satisfied with eating "wild rabbits & blackberries." The distraught Gonzales advised her to be more considerate of his feelings, as he was toward others, because since the spring she had been too often acting "hastily and *under excitement*." He was "most anxious" about resolving the matter and what he needed from her was "comfort which strengthens and bears up & not doubt which weakens & pulls down." Gonzales thought it was "improbable" that Buchanan "will not give me what I most desire. I consider it *next to impossible* that he will not do something for me. But the time has not come for any body to decide about the result, in my case no more than in those of a thousand more who are as expectant as I am."[53]

The Cuban had been strongly endorsed for the Chilean mission by sixteen actual and former Southern senators and representatives who sympathized with Cuban annexation: John Quitman, John Henderson and Jefferson Davis of Mississippi; James H. Hammond, Josiah J. Evans, William Waters Boyce, Lawrence Keitt, and John McQueen of South Carolina; Robert Toombs, Alfred Iverson, Martin J. Crawford, and John H. Lumpkin of Georgia; Alfred Nicholson of Tennessee; Stephen R. Mallory of Florida; Clement Claiborne Clay of Alabama; John Slidell of Louisiana; Thomas J. Rusk of Texas; William K. Sebastian of Arkansas; and in addition, John Forsyth, of the *Mobile Register*; Mayor James P. Screven of Savannah, and a dozen aldermen, including Freemason and filibuster supporter Richard D. Arnold; former schoolmate P. G. T. Beauregard and others.[54]

William Elliott then traveled to Washington, D.C., to exert influence for his son-in-law. He arrived at Willard's Hotel on September 24 with his seventeen-year-old grandson William, the second of thirteen children of Thomas Rhett Smith Elliott. Gonzales joined them at Willard's three days later. The patriarch obtained a letter of introduction from Secretary of the

Navy Isaac Toucey for the superintendent of the U.S. Naval Academy at Annapolis, where he tried to enroll his grandson. The teenager failed the entrance examination due to his "absolute ignorance of grammar and spelling," and returned to South Carolina with Gonzales. Elliott went to New York, to negotiate publishing an illustrated edition of his book *Carolina Sports,* before returning to Willard's on October 26.⁵⁵

Upon arriving at the hotel, Elliott met William Cazneau, husband of filibuster activist Jane McManus Storm, who wanted to introduce him to Mirabeau Lamar. The former Texas president had just been appointed minister to Argentina and wanted to offer Gonzales the post of secretary of the legation to that mission. Elliott replied to Cazneau that his son-in-law was expecting a mission and that the position offered would not suit him. The next morning, the Carolinian left his calling card at the office of Secretary of State Lewis Cass, who was unable to receive him. Cass was a political expansionist who during his failed 1848 presidential campaign had favored the purchase of Cuba to safeguard national security interests. Elliott visited the White House that evening, and was surprised to see that Buchanan had "grown very old" since he last saw him in London. The president was attending other guests "from distant portions of the country," including Lamar, and Elliott "could not get a single opportunity for conversation" on the Gonzales appointment. The president's niece, Harriet Lane, indicated that she expected Gonzales and Hattie to spend the winter in Washington. A day or two later, Elliott had a five-minute interview with Secretary Cass, in which he pointed out the strong recommendations that Gonzales had received from prominent Democratic Party leaders for an appointment to a South America mission. Cass alleged having a favorable recollection of Gonzales but added that "he could not say that at present there was any opening for an appointment but that he would bring his name again to the notice of the President." Elliott believed that this meant very little since "the same things are said to hundreds in Washington who never get appointments."⁵⁶

Gonzales had recently returned empty handed to join his wife at Flat Rock, North Carolina, for the christening of their son at St. John in the Wilderness Episcopal Church on November 1. He named Benigno Gener Junco, Miguel Aldama Alfonso, and his wife Hilaria Fonts Palma, as sponsors or godparents of the child. Hattie conceived a second son before the restless patronage pursuer returned to Washington, D.C. Gonzales arrived at Willard's Hotel with Florida senator Stephen R. Mallory on December 18, eleven days after the opening of the thirty-fifth Congress, in which Mallory was chairman of the Senate Committee on Naval Affairs. The Cuban spent the next four months in the capital, relentlessly lobbying politicians for a diplomatic post and Cuban annexation. On March 31, 1858, Gonzales sent his father-in-law a *New York Times* clipping from the previous day, "relating to the President's anticipated action in regard both to Cuba and Mexico," which he attributed to its editor, who was visiting the capital. The optimistic Gonzales believed that the report, and a similar rumor that he had heard, would initiate the "dawning of a result so long prayed for and so long expected." He considered Elliott's recent article on Cuba in *Russell's Magazine* as "both timely and forcible," and had delivered copies of it "to the fountain heads of power." Gonzales informed that the capital was convulsed with "the vexed Kansas question" of statehood and its Lecompton constitution, which he regarded as delaying his personal pursuits. He wanted to take home "some welcoming news to Hattie" before their second wedding anniversary. Gonzales wrote that if Buchanan's policy was not quickly developed, he would go to Oak Lawn and then take his "chances for another trip to Washington which I trust will not be again either so long or so very disagreeable."[57]

After returning to Oak Lawn, Gonzales continued his political lobbying efforts by mail, sending copies of his father-in-law's last Cuba article in *Russell's Magazine*. He forwarded a sample on May 9 to secessionist South Carolina Democratic senator James Henry Hammond, indicating that the article "makes it extremely plain that Cuba cannot, under any circumstances, be anything but a *Slave State;* a fact not unimportant to Southern Statesmen." Elliott had written that Cuba was "under the worst form of government—an unchecked despotism—exercised by deputy—enjoying an extraordinary degree of prosperity" from "the agricultural and commercial classes." Yet "the property holders are unrepresented, and are subject to all kinds of misgovernment and exaction." The article denoted that the agricultural production of Cuba for 1855 amounted to $77.9 million and its exports to the United States were valued at $45 million. Elliott agreed with his son-in-law and most Cuban planters that abolition would ruin the island's economy, creating an agricultural disaster similar to what had occurred in Haiti and Jamaica after emancipation.[58]

Gonzales moved with his family during early June into the summer home of Mrs. Mary Wayne, which his father-in-law had rented for $150, at Eddingsville, Edisto Island, forty miles south of Charleston. For over thirty years, the seaside village on a sandy ridge had been the summer resort of the wealthy lowcountry rice planters, who migrated there from May to November. Eddingsville had a church, a ballroom, and forty-two two-story dwellings with tabby chimneys, lined in dual rows, one facing the surf and the other on the back beach, separated by a main street. The homes had vegetable and flower gardens, carriage houses and slave quarters. Groves with palmettos and live oaks laden with Spanish moss lined the streets. The residents spent their time entertaining friends from Charleston and nearby areas, giving teas, having balls, sailing, fishing and swimming. The Cuban went with friends on fishing excursions, while his father-in-law advised that he use "less patience" during bass fishing. Gonzales could not swim with Hattie, since mixed bathing was not permitted, and the ladies were segregated on the eastern beach. Swimming was considered masculine, and the women would bob up and down on the water, shrieking and holding hands for support. William Elliott found that sometimes the ladies would "scream so loud as to be in no danger from sharks."[59]

During this retreat, Gonzales wrote a letter on July 29 to the *New York Herald* responding to their article linking the López movements with Southern secession. He claimed that López's "only purpose was a change of rule in Cuba, whereby she might become incorporated in this sisterhood of States, following the annexation lead of such Spanish colonies as Louisiana, Florida and Texas, retaining, as they did, her domestic institutions and coming, as in their case, to the enjoyment of civil and religious liberty, and to the protection extended to those institutions and those liberties of our federal compact." Gonzales quoted his April 25, 1854, article in the *Washington Union* on the benefits of "commerce, security, civilization and future greatness of America" that annexation would bring, with Cuba serving as "the bond of the Union as she was made by nature the clasp of America." The Cuban remained a strong Unionist, like his father-in-law, during the whirlwind political controversy over Southern secession.[60]

A week later, at 5:55 A.M. on August 5, 1858, Hattie gave birth to Narciso Gener Gonzales in Mrs. Wayne's home. He was named after Cuban revolutionary leaders Narciso López and Benigno Gener Junco. Gonzales immediately wrote about it to family and friends, including brother-in-law Ralph Elliott in Adams Run, babysitting Brosie, the "Little General," whose delicate health was under a physician's attention. Hattie was disappointed that the second child was not a girl.[61]

During the last week of August, William Elliott stopped in Washington, D.C., on his way to Saratoga Springs, New York, to again lobby on behalf of his son-in-law for a diplomatic post. He called on Secretary of State Lewis Cass, who was absent, and then went to see President Buchanan, who responded "rather formal, and invited no communication." Elliott met with former minister to Spain, Pierre Soulé, and drew the conversation to the annexation of Cuba. Soulé bitterly decried that Spain would never give up the island and that the Cubans no longer desired annexation because the abolitionists would tamper with their slaves, and the Know-Nothing Party would despise them as Catholics and foreigners. Elliott believed that Gonzales had failed to receive a mission to South America, "though qualified by nationality and suitableness in other respects," with high recommendations from prominent congressmen, because "his appointment would interfere with the acquisition [of Cuba]—from the offense which Spain would take at it." Elliott also asked Louisiana senator John Slidell to intercede with the president and the cabinet on behalf of Gonzales.[62]

During fall 1858, South Carolina Democrat James L. Orr, Speaker of the House of Representatives, requested that Gonzales write a series of articles on Cuban annexation for the *Detroit Daily Free Press* "to help the action of the Government in that regard." Ten consecutive articles totaling 25,700 words, particularly directed at Northerners, were published between November 25, 1858, and January 4, 1859, "and circulated all over the United States." The editor identified Gonzales as "one of the best Cuban minds. . . . He is an undoubted patriot, whose chief desire is to confer benefits upon his native country. In its annexation to this country he beholds the highest welfare of the island. Aside from the question of annexation, the facts and statistics of the articles are very interesting and instructive." Gonzales relied on Maturin Ballou's *History of Cuba or Notes of a Traveller in the Tropics,* published in Boston in 1854, for data and a synopsis of Cuban history. His newspaper articles denoted the island's physical description, its natural wealth, derided the colonial government and its taxation, described the system of labor and its relation to the United States, and highlighted the geographical and strategical position of the island, all of which, according to John Quincy Adams, "were indispensable to the continuance and integrity of the Union itself."[63]

These writings were the last effort Gonzales made for Cuban annexation prior to the Civil War, when the issue became inextricably entangled with the secession crisis. Since the previous year, Buchanan had been secretly negotiating the purchase of Cuba with Christopher Fallon, an American representing the financial interests of the Spanish queen. Senator Slidell introduced a bill on January 1, 1859, to appropriate $30 million toward purchasing Cuba. The proposition divided legislators along sectional lines. The Republicans refused to debate the merits of annexation, which was opposed by the entire free-soil press, and on February 25, the Cuba bill was defeated thirty to eighteen. Newspaper hotel arrival notices indicate that Gonzales did not return to Washington, D.C., in 1859. After six years of lobbying the Pierce and Buchanan administrations for a diplomatic post, the perennial optimist decided that it was time to look for another occupation.[64]

That summer, Gonzales and his family accompanied Mrs. Elliott, Annie, and Emmie, on their seasonal jaunt to Flat Rock. The Cuban mulled over becoming a planter, and inspected some farms in the nearby French Broad River valley. He also planned "going to Texas this winter to look for lands." East Texas was one of the most prosperous and fastest growing plantation regions, prompting a large southern migration there, including more than a quarter million South Carolinians. Gonzales spent part of his time at Flat Rock bear hunting with a new model Maynard 90-gauge rifle that had a custom-made barrel. He later acquired a special shade for its sight, using it during a hunting party to kill a deer at 160 yards, as it ran through a cluster of large pines. This was such an astonishing feat that the Maynard Arms Company of Washington, D.C., published his letter in their promotional pamphlet. Gonzales returned to Oak Lawn with his family and in-laws on November 8, on the Elliott carriage and the mail stagecoach. The next month, he accompanied his father-in-law to supervise the rice harvesting at Social Hall. They also went deer hunting on horseback, with a servant handling the bloodhounds, while Gonzales became thoroughly accustomed with the Maynard rifle. In February 1860, twenty-four-year-old Ralph Elliott became a planter after his father bought him a seven-thousand-dollar plantation on the Pon Pon River. Ralph, a captain in the Willtown Beat Company of the South Carolina Militia 13th Regiment, was elected to the state house of representatives that year.[65]

Gonzales then wrote on March 5, 1860, to W. C. Bestor, secretary of the Maynard Arms Company, offering to serve as an agent for their rifle and shotgun in South Carolina, Georgia—where he had "many friends and connections"—and in Florida, where he planned to reside. He claimed that "for these reasons as well as for my being known, to some extent, in connection with military matters, and for my greater experience of Maynard rifle & shot gun than any other purchaser of that weapon, I believe that I could serve the interests of your company in the capacity aforesaid, probably to a greater extent than most persons at the South." His vanity again flourished under the stress of acquiring a permanent occupation after three years of unemployment.[66]

That month, Gonzales traveled to Florida to assess the advantages of settling there, after deciding not to move to Texas because it was too far from his wife's family. Leaving Charleston on the steamer *Darlington*, he traveled to the head of navigation on the St. Johns River at Enterprise, to New Smyrna on the southeast coast of Florida, and to St. Augustine. During the voyage, Gonzales joined eight or ten sportsmen on the ship's deck who were shooting at wildlife with older model rifles and shotguns, using his Maynard rifle with amazing results. He later informed William P. McFarland, the Maynard agent at Chicopee Falls, Massachusetts: "I not only killed alligators that were missed by the rest, but some at a distance, say of one hundred and thirty yards, long before the rest could get within range, and having their heads only partially out of the water; white cranes at 150 yards, water turkeys & cormorants at 100, even 'divers,' single birds, three almost in succession, at from 60 to 130 yards, severally; a blue heron at 200 yards and finally an eagle in his nest on the summit of a swamp cypress tree on the shore. I, in all these shots, standing, & the boat in motion." When Gonzales arrived in Savannah on May 6, on the steamer *St. Mary's* from Palatka, he sent to McFarland eight Maynard orders for the excursionists and one for the steamer's captain. He boasted of being able to sell "at least *one hundred* more" rifles if the recent purchasers were satisfied.[67]

Gonzales returned to Oak Lawn to await the May 30 birth of his third son, Alfonso Beauregard Gonzales, named after his friends José Luis Alfonso García de Medina, the Cuban aristocrat and Havana Club conspirator, and Pierre Gustave Toutant Beauregard, his childhood classmate and Mexican War veteran. Gonzales named Stephen R. Mallory and Emily Elliott as godparents. At home he found a letter from Bestor, enclosing a Maynard Rifle pamphlet. The Maynard secretary indicated that he would present the Gonzales petition to be a sales agent before the company board of directors. Gonzales replied on May 10 that after giving up previous business plans, a family member suggested he become a Maynard agent due to the great interest he had in the weapon. He also indicated that his brother-in-law, Andrew Johnstone, was aide-de-camp to Gov. William Gist and that both had been the guests of William Elliott at Oak Lawn while Gonzales was in Florida. The Cuban indicated that, had he been empowered by the Maynard Arms Company, he would have delayed his trip to Florida and proposed to Gist the adoption of the Maynard rifle for South Carolina. Gonzales informed Bestor that he had already placed an order for nine rifles with McFarland, and that during his stay at Flat Rock the previous year, he had ordered four Maynards for his friends. He concluded the letter saying that he had "seen a desirable location for myself and family in East Florida and may have soon to go there, passing through Georgia."[68]

Bestor replied on June 15 with an offer of a twenty per cent commission on articles sold to sportsmen and three dollars on each rifle sold to companies or state governments. Gonzales, not fully pleased, responded a month later that he intended on going to Washington, D.C., shortly and would then make an arrangement satisfactory to both parties. In late July, Gonzales accompanied eleven relatives to Saratoga Springs, New York, via the ocean route from Charleston to New York City, bypassing the capital. He wrote to Bestor on August 20, indicating that he still planned on going to Washington as soon as possible to finalize their business deal. Gonzales concluded that a substantial sale of Maynards would not be made in South Carolina until the sporting season began in November. A month later, Gonzales went to Chicopee Falls, Massachusetts, where the Maynard rifles were manufactured by Springfield Arms, and met with company agent William P. McFarland.[69]

The Gonzales-Elliott family left Saratoga Springs for home in October. They stopped at the St. Nicholas Hotel in New York City, on Broadway, between Broome and Spring streets. The six-story structure had elegant marble interiors, carved white-oak staircase and fresco ceilings. The streets were crowded with marching clubs participating in the presidential campaign of 1860. From their hotel window, the Carolinians watched the Republican "Wide-Awakes" parading in uniform with tin torches on poles, banners and martial bands, drumming up support for Abraham Lincoln.[70]

When Lincoln won the presidency, the South Carolina legislature called for a state convention to discuss seceding from the Union. After returning to Oak Lawn, Gonzales read in the *Charleston Mercury* on November 19 that President Buchanan would enforce the federal laws against all attempts at secession by those states which objected to the Lincoln victory. The War Department was sending Maj. John L. Gardner to his newly appointed command at Fort Moultrie on Sullivan's Island, near the mouth of Charleston harbor. The dispatch from Washington concluded, "The news of the demonstrations at Charleston have at last aroused people here. Everybody now believes that South Carolina will go out, and there is great consternation in consequence."[71]

That morning, Gonzales penned a hurried letter to Gov. William Gist, stating, "It is reported, by telegraph, from Washington, as I see in to-day's 'Mercury,' that it is the intention of the Federal Government to coerce South Carolina into submission. In such a contingency I beg respectfully to proffer my humble services to the State of my adoption. I have served as adjutant Genl. & Chief of the Staff to the lamented patriot, Genl. Narciso Lopez." Thus, Gonzales has been referred to as "perhaps the earliest volunteer" to the Confederate cause and "among the last in service—three weeks after Appomattox." Gist replied, "Should the Genl. Government wage war upon this state, I will, certainly, accept your services in some position suitable to your former rank." The Cuban then began procuring weapons for South Carolina. On December 11, Maynard Arms Company agent William P. McFarland received a letter from Gonzales, saying he had arranged with W. T. J. O. Woodward, the agent for Adams Express Company in Charleston, to take all the guns sent to them. McFarland reported to Bestor on the twenty-ninth that the net amount of the Gonzales orders were $1,978.45. This amounted to about sixty-five rifles, at $30 each, and their appendages.[72]

South Carolinians had been discussing nullification from the United States since the 1830s. It began as a protest against the 1828 "Tariff of Abominations" by the state legislature. It declared the new duties unconstitutional and endorsed the right of a state to nullify federal law. The crisis subsided after the Compromise Tariff of 1833. Lincoln's electoral victory prompted the secession convention. The South Carolina State Convention met in Columbia on December 17, 1860, and then adjourned to continue in Charleston's Secession Hall. They discussed the passage of an Ordinance of Secession and approved an "Act to Provide an Armed Military Force." That same day, Francis Wilkinson Pickens was inaugurated governor of South Carolina and immediately got involved in the feud over the three federal forts in Charleston harbor. Two days later, he arrived at the Mills House in Charleston to attend the secession convention. Gonzales checked into the same hotel that day, and the *Charleston Courier* announced his presence as "agent for the Maynard arms, and for the Le Mot [sic] grape shot revolver." Also hoping to cash in on the military fervor was Capt. A. H. Colt, agent for Colt arms, who was passing out information at the Charleston Hotel and had his weapons exhibited in local hardware stores. Gonzales had an interview with Governor Pickens, and proffered his services. Pickens asked how far he lived from Beaufort, and how long it would take him to reach there. "At the tap of the drum," replied Gonzales. The next day, the Ordinance of Secession was approved by the convention, declaring that the "union now subsisting between South Carolina and other States, under the name of 'The United States of America' is hereby dissolved."[73]

The Convention sent a commission to Washington, D.C., "to treat with the Government of the United states for the delivery of the forts, magazines, light houses, and other real estate, with their appurtenances, within the limits of South Carolina." Meanwhile, U.S. Army major Robert Anderson evacuated Fort Moultrie on the evening of December 26, spiking the cannons and burning the carriages of those aimed at Fort Sumter, where he retrenched his forces. The Carolinians claimed that "Major Anderson waged war," and proceeded to take by force Castle Pinckney, an old brick structure on Shute's Folly Island, one mile east of Charleston, while its federal garrison retreated to Fort Sumter. The military forces of South Carolina then started to build up a ring of artillery positions around Fort Sumter to force its surrender. The five-sided brick fortification was located on a shoal, near the harbor entrance, three-and-a-half miles from Charleston. It was designed to mount 135 guns on levels of embrasure and in barbette.[74]

Gonzales now supplanted the fight for Cuban freedom with the independence struggle of his adopted state. The growing sectional squabble over slavery had immobilized the two previous Democratic administrations from actively pursuing Cuban annexation. Filibuster activities were denounced and suppressed by the Pierce administration with the same vigor as its Whig predecessor. Gonzales, a Union Democrat, carried out a propaganda campaign with a manifesto and newspaper articles, denouncing Spanish colonial abuses, defending López, and agitating on behalf of Cuban statehood. As a Protestant, he wanted religious liberty for his homeland, where the Spanish Crown imposed Catholicism as the official religion. Gonzales abandoned the filibuster movement when it was co-opted by the enemies of López and abolitionists, who foundered in 1855.

During six years, Gonzales had obtained dozens of recommendations for a diplomatic post from prestigious statesmen, including cabinet members, who praised his capabilities. The Cuban optimist, a fixture in aristocratic circles, had frequently visited the White House and enjoyed cordial relations with President Pierce and Secretary of War Davis. He had been endorsed as chargé d'affaires to Venezuela, secretary to the minister to France, and head of the Chilean mission, but Pierce and Buchanan rejected him. Both presidents had been secretly negotiating the purchase of Cuba and probably felt that a Gonzales appointment would antagonize the Spanish Crown, which regarded him as an insurrectional fugitive under a death sentence. Gonzales acknowledged that he lacked expertise and that his appointment would require on-the-job training. He refused to accept an inferior diplomatic post, but ended up as an occasional translator in the Department of State and a temporary clerk in the Patent Office. Gonzales became an agent for the Maynard Arms Company on the eve of the Civil War. He believed that he could do more for South Carolina than for Cuba. If he had any notions about Cuban annexation to the Confederacy, he never wrote about them, as the "irrepressible conflict" finally loomed on the horizon.

## Five

## DAVIS DISPUTE

Gonzales was ebullient when his former schoolmate Pierre Gustave Toutant Beauregard arrived at the Charleston Hotel on March 3, 1861, two days after being appointed the first brigadier general of the Provisional Army of the Confederate States, and reported to Gov. Francis Wilkinson Pickens for military duty in South Carolina. A subaltern recalled: "His uniform fitted to perfection, he was always punctiliously neat, his manners were faultless and deferential. His voice was pleasant and insinuating, with a perceptible foreign accent. His apprehension was quick, his observation and judgement alert, his expressions tense and vigorous." Beauregard was five feet seven inches tall, weighed 150 pounds, had an ample brow, high cheekbones, dark complexion, gray hair, a moustache slightly drooping on the ends, and a protruding chin. Gonzales closely resembled Beauregard in physique and features, including the dimpled chin and the slight accent. His 1865 Amnesty Oath and his Cuban passport described him as five feet eight inches tall, with a dark complexion, gray hair, and hazel eyes. Mary Boykin Chesnut noted in her subsequently famous diary that she found Gonzales handsome, a "fine person," and "so like Beauregard as to be mistaken for him." Gonzales idolized his Creole friend and brother Freemason and tendered his services as a volunteer aide. Beauregard, who already had a military staff, was being besieged with similar assistance offers from former governors and senators and did not see the need to keep Gonzales in Charleston permanently.[1]

General Beauregard, a West Point engineer, spent the first three days with Governor Pickens inspecting the artillery batteries on Charleston harbor encircling the federal garrison in Fort Sumter. He set up headquarters at 107 Meeting Street, then issued a general order assuming command of all militia, volunteer, and regular troops in the area. Confederate secretary of war Leroy Pope Walker informed Beauregard on March 9 that a U.S. naval expedition was underway to reenforce Fort Sumter. This included a Union plan by Capt. John Rodgers "to force a landing on the southern extreme of Morris Island; to carry the [Cummings Point] batteries by the rear and destroy the channel." On April 8, U.S. State Department official Robert S. Chew met with Pickens and Beauregard in Charleston. He notified them on behalf of president Abraham Lincoln that soon "provisions would be sent to Sumter peaceably, otherwise by force." The expeditionary vessels began arriving three days later off Morris Island at the mouth of the harbor. That same day, Beauregard was informed that the *New York Tribune* declared "the main object of the expedition to be the relief of Sumter, and that a force will be landed which will overcome all opposition." Beauregard claimed that due to these threats, "we were compelled by self defence" to fire on the garrison, after Maj. Robert Anderson had "refused most honorable terms not of surrender but of evacuation."[2]

The boom of Confederate cannons firing on Fort Sumter awoke Gonzales at Oak Lawn at 4:30 A.M. on Friday, April 12. He donned a military uniform, packed a white canvas trunk, strapped on his sword, and grabbed his Maynard rifle. The Cuban boarded the Charleston and Savannah Railroad train that stopped at Adams Run for the two-hour, twenty-six-mile trip to St. Andrews depot, across the Ashley River from Charleston. Upon arriving, the citizenry were exuberant and the tempo of the cannonade had doubled. The *Mercury* reported that "the Battery, the wharves and shipping in the harbor, and every steeple and cupola in the city, were crowded with anxious spectators of the great drama." The former filibuster offered his services "as an officer or as a soldier" to Beauregard, who immediately appointed him as one of his seven volunteer aides-de-camp. The others were prominent Southern politicians and fire-eaters, including former South Carolina governor John Laurence Manning, Senators Louis Trezevant Wigfall and James Chesnut Jr., Representatives William Porcher Miles and Roger Atkinson Pryor and Beaufort planter

Alexander Robert Chisolm. Gonzales's appointment was heralded in the *Charleston Courier* the next day, identifying him as "a class-mate and friend of Gen. Beauregard" who had tendered his services to Gov. William H. Gist "on the first intimation of the secession of South Carolina," and was now "assigned to important duty." When sister-in-law Mary Barnwell Johnstone read the article in Georgetown, South Carolina, she wrote to her mother that she was "so very glad to see the General's name on Beauregard's Staff." She and her husband, fifty miles away, had "heard the firing of the batteries, and we were all so anxious to know what it meant. On Saturday the rapidity of the guns made us imagine something besides Sumter was being attacked." Mary's cousin, Dr. Ralph E. Elliott, and her twenty-year-old nephew, William Elliott, son of Thomas Rhett Smith Elliott, were members of the Beaufort Volunteer Artillery, firing on Sumter from the Iron Battery on Morris Island.[3]

Beauregard expected a Union disembarkment in the center of Morris Island, on the low ground between Vinegar Hill and Lighthouse Hill, five miles directly southeast of Charleston. The four-hundred-acre island was nearly four miles long and up to one thousand yards wide, depending on the tide. Beauregard had sent two twelve-pounder guns and a battery of six-pounders to defend the position and ordered Col. Maxcy Gregg, commanding two thousand men on the island, to "make a desperate stand there" if challenged. On the thirteenth, Beauregard gave Brig. Gen. James Simons of the South Carolina Volunteers command of the northern half of Morris Island, from Vinegar Hill to Cummings Point, and ordered Maj. Gen. Milledge Luke Bonham of the Army of South Carolina to dominate the Southern half. Beauregard temporarily detached Gonzales as assistant adjutant and inspector general on Bonham's staff and told him that "the first fight was to be expected" on Morris Island. The forty-seven-year-old Bonham, an attorney and former state and U.S. representative as a states' rights Democrat, was a veteran of the 1836 Seminole War and had been a colonel of the Twelfth Infantry Regiment during the Mexican War. Gonzales and Bonham were taken across the harbor in a rowboat during the bombardment, landed by the active Cummings Point gun batteries, and walked in inclement weather a few miles to headquarters behind Vinegar Hill.[4]

After Gonzales and Bonham reached headquarters, Maj. William H. C. Whiting, Confederate chief engineer and adjutant and inspector general on Beauregard's staff, informed his superior: "The gentlemen you sent are very efficient. Quartermaster Hatch should send down tents for the general and his staff." Whiting asked for at least a thousand men to reinforce his position, from where he observed "six of the hostile ships. . . .With these vessels so close to us, I cannot relax my vigilance without such a force as would render a *coup de main* impossible." Robert Barnwell Rhett Jr., editor of the *Charleston Mercury*, telegraphed Confederate secretary of war Walker on April 13: "Fight expected on Morris Island to-night." Gonzales and many volunteers at Vinegar Hill spent a sleepless night, although the terms of evacuating Fort Sumter had been accepted that Saturday at 8:00 P.M.[5]

Beauregard's official report on the Fort Sumter attack included Gonzales among those in his volunteer staff to whom he was "much indebted for their indefatigable and valuable assistance." The forces on Morris Island were reorganized on April 15 into one division, commanded by Maj. Gen. Bonham, with Gonzales as his acting inspector general. The division embodied two brigades. The brigade on the north side of the island was led by Brigadier General Simons and included the militia regiments of Colonels Johnson Hagood and John Cunningham and an artillery battalion under Lt. Col. Wilmot Gibbes DeSaussure. The brigade on the south side, headed by Brig. Gen. Patrick H. Nelson of the South Carolina Volunteers, contained the regiments of Colonels Maxcy Gregg and Joseph B. Kershaw and an artillery battalion under Col. James Henry Rion. That day, according to one soldier, the regiments changed positions a dozen times, during "the most violent storm of wind and rain that I have witnessed in many a day, and at present the tents afford but little protection in the driving rain which still continues." Gonzales shared a tent with a volunteer from Winnsboro, who the next year recalled having "rarely been more impressed than then, by the evidence he gave of the qualities which make the soldier, and the modest demeanor which marks the gentleman." The Cuban was also remembered for "his valuable services, varied knowledge and experience."[6]

Gonzales applied on April 17 for the command of militia general Alfred Moore Rhett's Brigade of Volunteers after Rhett was appointed quartermaster general. He asked Beauregard, the most idolized contemporary Confederate, to present his claim to Governor Pickens or to president Jefferson Davis, as the case may be. Beauregard's endorsement affirmed:

"I take much pleasure in recommending the above application of General Gonzales to the favorable consideration of his Excellency Gov. Pickens. Genl. G. has been very active & intelligent in the discharge of the duties referred to by him." Two weeks later, Gonzales personally petitioned his old friend Jefferson Davis for command of Rhett's brigade. The Confederate president replied: "I have no power to act in the case. The commission is under the State authority alone." This was the first of many efforts Gonzales made during the war to obtain the rank of brigadier general.[7]

More than seventy Cubans participated in the Civil War. Sixty-two New Orleans residents organized the Cuban Rifles Company, under Capt. Antonio González Vigil. They served as Company C of the Jackson Rifle Battalion, in the Provisional Regiment Louisiana Legion.[8] Prominent Cubans embracing the Southern cause included Confederate diplomat José Agustín Quintero,[9] physician Col. Ramón T. de Aragón, Lt. Col. Paul Francis de Gournay,[10] First Lt. Enrique B. D'Hamel[11] and Loreta Janeta Velázquez, who used her husband's uniform as a disguise to fight with the Confederates under the name Lt. Harry T. Buford.[12] Cubans in the Union ranks included physician Col. Antonio Luaces, Lt. Col. Julius Peter Garesché du Rocher,[13] and the brothers Federico and Adolfo Fernández Cavada, also lieutenant colonels. Gonzales eventually outranked his compatriots after attaining the rank of colonel.[14]

On April 18, Gonzales wrote to Beauregard from Morris Island headquarters that Colonel Hagood proposed strengthening his position at Vinegar Hill. The general, following Major Whiting's advice that "Fort Sumter cannot be retaken from Morris Island alone," decided to dismantle the Cummings Point batteries and gradually reduce troop strength. Beauregard told Gonzales that he would instead fortify Fort Palmetto, eight miles west of the Charleston bar on Cole's Island at the Stono Inlet, and Fort Pickens, a slightly built *barbette* work on Battery Island three and a half miles up the Stono River. This would block the Stono River entrance and protect James Island, considered by Beauregard the key to Charleston's defense. Colonel Rion's forces were ordered to Fort Pickens, and General Bonham, appointed Confederate brigadier general by President Davis, departed on April 21 with parts of Gregg's and Kershaw's regiments for Virginia. The next day, Gonzales returned to the Mills House in Charleston, searching for his white canvas trunk, which had not been forwarded to Morris Island. Hearing that it was "supposed to be at some Rail Road Station," he advertised in the *Courier* a reward for its return. William Elliott arrived on May 7 at the Mills House, where he found his son Ralph with Gonzales. He brought from Oak Lawn a parcel of clothes for his son-in-law, whose trunk was never recovered. Elliott also met there an old Boston friend, William Appleton, accompanied by William Amory and his wife, and invited them to dine at Oak Lawn.[15]

Gonzales returned to Morris Island the next day and filed an inspector general's report on the needs of the state regiments stationed there. These were the 832 men in Colonel Hagood's First Regiment South Carolina Volunteers and the 541 soldiers in Lt. Col. James Douglass Blanding's Second Regiment South Carolina Volunteers, who arrived on the island with only their arms, ammunition and cooking utensils. Gonzales, doing a characteristically thorough job, requested for each soldier waterproof knapsacks and haversacks, blankets, broad-brimmed felt hats, an extra flannel shirt and two pair of socks. He suggested that the men on the seaboard should be armed with rifles instead of muskets and given extra ammunition for target practice, along with a succinct marksmanship manual. The field hospital lacked a surgical instruments case, ice, and ten dozen sheets and pillow cases. Headquarters needed company ledger books and blank returns, arms racks, and punches and dies for marking the muskets and accoutrements. Both regiments were later evacuated from Morris Island by the end of the month.[16]

Beauregard returned to Charleston on May 10 from a one-week trip to the Confederate capital at Montgomery, Alabama. The next day, the first federal blockading vessel, the frigate *Niagara*, appeared before Charleston harbor. Beauregard quickly organized a tour of inspection of the coastal defense lines from North Edisto River to Broad River. The general departed from Market wharf on May 12 on the steamer *General Clinch*, accompanied by Gonzales and his father-in-law William Elliott; Lt. Col. George P. Elliott; aide-de-camp Col. Alexander Robert Chisolm; Majors John G. Barnwell and W. M. Murray; and Lt. John W. Gregorie of the Corps of Engineers. They visited eight artillery emplacements, including Fort Elliott and Fort Schnierle in Beaufort. Returning to Charleston three days later, Beauregard recommended to Governor Pickens the construction of fortifications on opposite sides of the Port Royal bay entrance on Phillips and Hilton Head Islands, on Mackays Point, where the Tullifinny and Pocotaligo Rivers converge on the headwaters of the Broad River, and works on North and South Edisto Rivers, eighteen miles southwest of the Charleston bar.[17]

Two days later, on May 17, Beauregard ordered Gonzales on a three-day coastal voyage with engineer Capt. Francis D. Lee and Colonels Arthur Middleton Manigault and Benjamin Allston to select gun emplacement sites along the coast north of Charleston at Bull's Bay, on the South and North Santee Rivers, and at Georgetown. That same day, Beauregard informed President Davis that defense preparations in Charleston were nearly completed and that he was ready for another assignment. The following morning, Gonzales sent Davis a telegram requesting to command, with the appropriate general's rank, the three regiments the president ordered raised for the defense of the South Carolina coast. The Cuban indicated, "I have for this post the strong preference of Gen. Beauregard, of the Colonel of the Regiment likely to be assigned on that duty and of very many of the inhabitants of the sea board. Will you intrust me with it in the absence of Gen. Beauregard?" Once again, Davis turned him down, but the reply has not been found.[18]

When Beauregard was transferred a week later to Virginia, leaving Col. Richard H. Anderson in charge of Confederate forces in Charleston, he wrote to Gonzales with gratitude for "coming, as you did, so efficiently to my assistance at such an important moment and lending to the time all your untiring energy and perseverance. Such services, past and present, I assure you, will be recalled in my mind with constant pleasure hereafter." A few days later, Governor Pickens appointed Gonzales special aide-de-camp, with the rank of lieutenant colonel, in charge of inspecting the coastal defenses of South Carolina. The Cuban had acquired this expertise from his father, who had been the Matanzas coastal inspector. Coastal defense was initially assumed by state governors due to the weakness of the temporary Provisional Government of the Confederate States. A coastal defense strategy for the Southern states was not formulated until the following year.[19]

On his next tour of inspection, Gonzales visited the points between Edisto and Hilton Head that needed strengthening, before returning to Charleston on June 9 on the steamer *Cecile* with Capt. N. L. Coste of the South Carolina Navy. Coste had previously commanded the U.S. revenue cutter *Jackson* at Savannah. They had no difficulty getting past the four

Union blockaders outside the harbor. The next day, Gonzales requisitioned artillery and ammunition for three seaboard defense positions: two twenty-four-pounder cannons with two hundred rounds of ammunition, and two six-pounder iron guns with shot for South Edisto Inlet; one twenty-four-pounder gun and two twenty-four-pounder howitzers with corresponding ammunition for Fort Beauregard at Bay Point; and three eighteen-pounders or two twenty-four-pounder guns with shot for Braddock's Point on Hilton Head, facing Daufuskie Island. On the twelfth, Gonzales started a tour of inspection on the steamer *Firefly*, piloted by the newly appointed secretary of the Coast Police, Col. Robert S. Duryea, "for the purpose of determining suitable locations for erecting fortifications, and preventing the entrance of the enemy."[20]

Their first task was to relieve Maj. W. M. Murray and the Calhoun Artillery at the redoubt on South Edisto Inlet. Redoubts were enclosed earthworks of various shapes, most commonly a square or triangle. The unit was transferred with their two field pieces to Eddings Bay, midway between North and South Edisto. Gonzales recognized that in this region "the shallowness of the water opposite the coast, extending for several miles, would prevent any but the lightest vessels from approaching it." He therefore worked on strengthening the defenses on the upper and lower inlets of Edisto Island. When the *Firefly* presented mechanical problems, Gonzales devised a plan where the steamer *General Clinch*, piloted by Captain Coste, transferred the newly organized St. Paul's Rifles and the Colleton Rifles, with two twenty-four-pounder guns, from North to South Edisto. The sixty-four recruits of the St. Paul's Rifles were led by Capt. B. Burgh Smith, a civil engineer and graduate of the Charleston medical college. The seventy-eight men of the Colleton Rifles, under Capt. J. D. Edwards, were attached to the Ninth Regiment South Carolina Volunteers, commanded by Col. William C. Heyward. The open sand work at North Edisto was left operational with six guns and a garrison. It consisted of "two redoubts for five guns each, connected by a long curtain, and protected in the rear by a double fence of thick plank, with earth between and loopholed." The schooner *Helen*, towed by the *General Clinch*, was used to land the men and supplies on South Edisto, but they had to beach the steamer to unload the cannons. The artillery was placed in one of the redoubts, tents were pitched, and in the evening, before departing, Gonzales addressed the troops on behalf of Governor

Pickens. "He thanked them for their voluntary services, the sacrifices they had made of their planting interest to protect the seaboard, their generous contributions of labor for the works laid out by Gen. Beauregard; their liberality in furnishing themselves with arms, tents, accouterments and ammunition free of charge to the State, their constant readiness to meet the enemy, should they have attempted to tread the soil of Carolina." Gonzales also encouraged the recruits "to pursue their instruction in heavy as well as in field artillery." The troops then "gave three hearty cheers for Gen. Gonzales," after which Colonel Heyward "address[ed] them in a few appropriate remarks." As Gonzales left the fort, the Colleton Rifles fired a gunnery salute.[21]

The Cuban returned to Charleston in the malfunctioning *Firefly*. He wrote an anonymous article on his latest coastal defense activities that was published in the *Courier* and the *Mercury* on June 24. He enclosed a copy of it with a detailed report to Governor Pickens that same day. This indicates that Gonzales, like Beauregard, sometimes wrote his own favorable articles in the press with the approval of the editors. The coastal inspector recommended to Pickens the occupation of Daufuskie Island as protection for Hilton Head Island and the rear guard of Fort Pulaski. He suggested the deployment of cavalry patrols on Hilton Head and Eddings Island. Gonzales then placed advertisements in the *Courier* and the *Mercury* indicating that although he continued to receive orders for the purchase of Maynard rifles, he had long since stopped being their agent. The next day, Gonzales departed for Beaufort with Capt. Henry Middleton Stuart and the Hamilton Guards, and a dozen thirty-two-pounder guns received from the Norfolk Navy Yard for Fort Beauregard at Bay Point. This artillery was mounted on permanently fixed naval carriages, which Gonzales considered vulnerable to being overrun by the enemy. He therefore requested smaller guns on siege carriages to protect the laborers and garrisons at Port Royal Sound.[22]

At Bay Point, Gonzales encountered his brother-in-law Ralph Elliott, serving as a private in the Beaufort Volunteer Artillery, captained by his cousin Stephen Elliott Jr. Ralph wrote his family that he was "mounting heavy guns, building hot shot furnaces, digging trenches, moving heavy timbers, hauling bricks, mortar & sand in wheelbarrows, & laboring harder than any negro on my plantation." The Fort Beauregard embankment ran back up to the house of his older brother Tom. The summer home was later sold to be used as officers' quarters. Younger brother William, a private in the same outfit, also labored on the fortification, "painting guns and blacking balls and other duty work interrupting between morning & evening drills." There was a real urgency in completing the defenses. In July, the *New York Herald* called for an invasion of South Carolina through Port Royal harbor "[l]ate in the fall, when the warm weather moderates and the region becomes healthy for Northern troops." The *Charleston Mercury* editorialized that "unless the Northern forces are driven out of Washington before frost, it is a moral certainty that just such an attack will inevitably be made upon the coast of South Carolina." That same month, the U.S. Navy Blockade Strategy Board, presided over by Capt. Samuel Francis Du Pont, presented a memoir to Navy secretary Gideon Welles calling for "the occupation of Bull's Bay, St. Helena Sound and Port Royal Bay," to secure the South Carolina coast.[23]

In late July, the *Courier* announced that Gonzales, on "our sea coast defences, has been industriously engaged, and has lately visited several important points." These included Georgetown and Bull's Bay, which had been fortified with the aid of the steamer *The Planter*. The *Mercury* simultaneously declared that "at all the available approaches to our coast batteries have been erected; but these have been furnished for the most part with guns of comparative small calibre." It called for the deployment of long-range rifled cannons on all seacoast defenses without delay.[24]

Gonzales was in Charleston, making arrangements to obtain heavy artillery, when he heard of the Confederate victory on July 21 at the Battle of Manassas, Virginia. Five days later, the remains of Gen. Barnard Elliott Bee and Lt. Col. Benjamin Jenkins Johnson, killed in action, arrived in Charleston at 1:00 P.M. by special railroad car. The bodies lay in state at City Hall. A funeral procession was organized during the late afternoon. Gonzales served as a pallbearer to General Bee, along with four other officers and Colonel Roswell Sabine Ripley, who the following month was appointed Confederate brigadier general and given command of South Carolina. The Charleston Brass Band, playing Beethoven's dead march, led the way up Meeting Street. The hearses were accompanied by twenty-one military companies, the mayor and city council, clergymen, and other dignitaries. The stores were closed, flags at half-mast were shrouded in mourning, church bells tolled, and the heavy beat of the muffled drum and the melancholy dirge resounded throughout the streets.

Gonzales accompanied the bier to St. Paul's Church on Coming Street. The Battle of Manassas also brought the war home to the Elliott family. Their nephew, William Elliott, a private with the Palmetto Guard, had been slightly wounded when "struck by a piece of shell on the breast."[25]

In early August, Gonzales went by railroad to Richmond, Virginia, to procure artillery for South Carolina from the Tredegar Foundry. He remained for two months while supervising their manufacture and periodic shipment. In the Confederate capital, Gonzales often called on James Chesnut Jr., who had also been on Beauregard's volunteer staff and had gone to Richmond with his wife Mary Boykin Chesnut, seeking the Confederate ministry to France. Gonzales participated in a soiree on August 13, and Mrs. Chesnut noted in her diary that he was "handsome" and "sings divinely." One evening, after Gonzales lit a candle for Mrs. Chesnut and gave her a bag of quinine pills for her ailing husband, she wrote, "These foreigners have a way of doing things—gracious and graceful—no native [of the] USA can approach." Her cousin, Mary Whitaker Boykin, found the Cuban "splendid in gray uniform."[26]

Gonzales was distraught by the surrender on August 28 of two Confederate forts at Cape Hatteras, North Carolina, to a Union expeditionary force. The following day, he telegraphed Port Royal, warning that it would probably be the next point of attack. His first shipment of heavy ordnance from Richmond arrived in Charleston on September 5 by railroad. It included "one 24-pounder rifled cannon, four eight-inch Columbiads, four ten-inch Columbiads, and several 32 and 42-pounders." A week later, a rifled nine-inch and an eleven-inch Dahlgren gun reached South Carolina from the Tredegar Foundry. The Palmetto coat of arms was emblazoned on all the cannons, which also had the foundry marks. Gonzales sent with these weapons an article titled "Big Guns" that appeared on September 12 in both the *Courier* and the *Mercury*. The Cuban then devised a plan for a system of coastal defense for the Atlantic and Gulf states. Two days later, after consulting with military engineers, he outlined his strategy to President Davis in an elaborate 2,800-word letter. Gonzales indicated that the guns on North and South Edisto Inlets could be overrun from behind if the enemy used barges to land a large force on the beach midway between both positions. The best protection against such a disembarkment was to create a "sea-coast flying artillery," composed of rifled twenty-four-pounder guns mounted on siege carriages that could be quickly moved to engage the enemy and sink approaching low-draft vessels. The object of this flying artillery was "to link the batteries at the inlets, to close and watch the space which intervenes between them, and to prevent their being taken in reverse."[27]

At the time, land based artillery had two basic classifications. Field artillery was ordnance small enough to freely move with the armies in the field. These pieces could be either smoothbore or rifled and generally were designated by the weight or caliber of their projectiles, for example, six-pounder, twelve-pounder, and three-inch. The field artillery included one of the most popular cannons of the war, the twelve-pounder 1857 model commonly called a Napoleon gun. This gun had an effective range of about 1,680 yards and was particularly effective at breaking up infantry assaults. The other classification was heavy artillery, divided into two classes—seacoast guns or siege and garrison guns. Seacoast guns were large, cast iron pieces that were used in permanent fortifications. They included twenty-four-pounders, thirty-two-pounders, forty-two-pounders, and various calibers of Columbiads and Parrott guns. Mounted on specialized, fixed carriages, seacoast artillery could fire projectiles up to three miles. Siege guns were massive but could be mounted on large, wheeled carriages so that they could accompany an army. The most popular of these cast iron cannon were the twenty-four-pounder (both smoothbore and rifled) and thirty-pounder Parrott rifles.[28]

Gonzales conceptualized combining these siege pieces into a "train" of flying artillery with ten horses pulling each cannon, while ten other horses were assigned to its ammunition wagon. A ten-man cavalry detachment would accompany each unit. During difficult traveling, the cavalry horses would temporarily be attached by the saddle breast-band to ten straps permanently affixed to the gun carriage, bringing a total of twenty horses to the draft of each cannon. Gonzales proposed arming the cavalry with sabers and double-barreled shotguns, "especially as in landing from boats the enemy must be crowded," and boasted that "nothing that the enemy can bring can neutralize this system." He indicated: "We have in South Carolina very nearly the number of 24-pounders on siege carriages required for the establishment of this system." Gonzales also suggested that "a central battery of sea-coast flying artillery of twelve rifled twenty-four-pounders" be stationed in Charleston, which could be quickly moved by railroad or steamer to any point on the coast.[29]

Almost fifty years later, former Confederate cavalryman Ulysses R. Brooks described the new invention: "The siege train originated and organized by Colonel Gonzales was here used for the first time in warfare. It was, in brief, the mounting of heavy siege guns on special artillery wheels, attaching ropes and moving them from point to point with hundreds of men and horses. This method of handling heavy artillery was used a generation later in the Boer war, and thereafter by the Japanese at the siege of Port Arthur. As in iron clads and submarines, the South was also a pioneer in thus mobilizing heavy guns."[30]

Gonzales hoped that his promising plan, along with his service on the South Carolina coast, his previous experience in the Cuban cause, and his friendship with Davis during the last twelve years, would earn him the rank of brigadier general, which had been bestowed on others with lesser military capabilities. The Confederate president had been badgered since he took office by prominent men seeking a generalship, but the positions available were limited. Most commands went to West Point graduates, even though some of the new generals lacked military experience. Episcopalian bishop Leonidas Polk, who befriended Davis at West Point, had resigned his army officer's commission to become a priest after graduating in 1827. When the war started, "he was hardly qualified to serve as a second lieutenant," but Davis made him major general in charge of Department No. 2, which included the defense of the Mississippi River. Gonzales believed, as did most European military tacticians, that political leaders with appropriate attributes, study, and experience, could be capable generals. He therefore felt that he deserved the rank of brigadier general in the Confederacy.[31]

Gonzales impatiently wrote again to Davis on the sixteenth, claiming that his plan, his "thorough knowledge" of the South Carolina and Georgia coast, along with his "energy activity combination & fertility of resources" in his prior military movements, made him "better qualified to command a Brigade of mobile troops, of the description I have planned, than perhaps, any other man." This self-aggrandizement usually appeared in his correspondence when he suffered great stress. Gonzales indicated that his command of a flying artillery column on the Atlantic Coast would meet with the approval of politicians and influential citizens. He reminded Davis that Governor Pickens had already stated that he had intended to give Gonzales command of the First Brigade of South Carolina Volunteers.

Pickens later offered to recommend him as a Confederate diplomat in Europe, but the Cuban refused, "for the reason that having borne for a long time a high military title I consider it my duty to serve my adopted State and Section where danger was to be met." Gonzales felt that he was "entitled from one who for so long has called himself my friend, to a place in this contest more in accordance with what I have understood to have been his estimate of my abilities than the one I have received. A bureau would cramp my energies and kill off my specialty, which is action."[32]

Gonzales wrote to Ralph Elliott on September 23 from Richmond, saying that he had been "very anxious about Port Royal" since the departure of the Union expedition from Cape Hatteras. The balance of the ordnance he recently acquired had been forwarded that day to South Carolina. The entire order amounted to thirty guns, including sixteen large Columbiads, six of those being rifled, all of which he regarded "the heaviest & most formidable Armament now existing in any one spot in the Southern Confederacy." Gonzales told Ralph that he would await the decision of the government on his plan of coastal defense but would rush home in case of imminent enemy attack. Gen. Joseph Reid Anderson, the former superintendent of the Tredegar Foundry, noticed the new artillery in railroad transit through his military district in Wilmington, North Carolina. He telegraphed President Davis for permission to seize two of the large cannons and their carriages, because he lacked proper artillery. Davis referred the matter to the secretary of war, suggesting that the "necessity justifies the change if only a brief delay will be involved" in replacing the guns.[33]

Gonzales returned to Charleston by October 8, when both local newspapers praised his dedicated work. The *Mercury* heralded how "the untiring patience and urgent and watchful zeal of Gen. Gonzales, have furnished us with artillery sufficient to our security." The secessionist newspaper, a bitter critic of President Davis and previously opposed to Cuban annexation, now declared that "the people of South Carolina owe a debt of gratitude to this gentleman, special aid to Governor Pickens, for his very efficient services in procuring arms, ammunition and equipments for the seacoast defence of the State. . . . Gen. Gonzales is well entitled to our warmest thanks for his very successful labors in the part he has undertaken to perform." The *Courier* announced that Gonzales had "employed his time and talents with his characteristic energy and devotion for the cause of the State and the South. He has done excellent service, the duties of which cannot be properly presented now." For all of his efforts, the Confederate adjutant general proffered Gonzales the position of adjutant and inspector general with the rank of major. Gonzales replied from Charleston on October 9, enclosing four prominent endorsements and declining the appointment.[34]

The Cuban then visited his family at their *Farniente* summer residence in Flat Rock, North Carolina. Tom Elliott, working on the defenses of Fort Beauregard, wrote to his father that praises for Gonzales "are sounded on all sides for his great energy in procuring the heavy Guns & Carriages, & the prevailing opinion is, that if it had not been for his exertions in the matter we would have been in a sad fix now." William Elliott Jr., informed his mother from Bay Point on October 7 that "in Beaufort too there is great alarm felt by all but old Mrs. Albergottie, who declared, when she heard that Ed[mund] Rhett was at Bay Point, she felt no apprehension whatever for he, Mr. Rhett, would *prohibit* the fleet from coming to Beaufort." The Elliott patriarch prophetically responded to his namesake, "I feel very unhappy when I contemplate the dangers to which you and Ralph and Tom too for aught I know may be exposed should this invasion occur as I have too much reason to fear it will. It will not be for the purpose of taking the Town of Beaufort. It is to possess and occupy Hilton Head as a strategic point—*for the war,* that they will assail Port Royal." Elliott advised his son to remove most of his slaves, provisions and cotton from his Myrtle Bank plantation on Hilton Head. Gonzales spent a few days with his family at their *Farniente* estate, before departing for an inspection tour of Port Royal Sound, where most of the artillery he procured had been forwarded.[35]

Prior to leaving, Gonzales wrote to President Davis on October 10, saying that he had turned down the offer made by the Confederate adjutant general and hoped that the reasons given would prove satisfactory. Since this letter has not been located, we can only assume its content from the scathing and sarcastic reply Davis wrote six days later. The Confederate president said that he was not informed of the response given to the adjutant general, that Gonzales had the absolute right to decline, and that the position conferred was meant to continue the duties he was already performing under State authority. He then admonished, "I object to your recital of services rendered to me; 1st, because personal obligations do not form my standard in the appointment of officers; 2d, because you have widely mistaken as well as what you gave, as what I received."[36]

Davis admitted that he was "slandered in relation to the Cuban expeditions" suppressed by the Pierce administration when he was secretary of war but stated that he did not recall the efforts on his behalf Gonzales claimed to have made in a letter to him during his 1857 campaign for the U.S. Senate. Gonzales had failed to specify that he was referring to the November 1854 Congressional election in Mississippi, for which Davis had requested a translated copy of a *La Verdad* article to defend the Pierce administration's Cuba policy. Gonzales manifested that his letter of support to Davis made him lose his friendship with the editors of the *New Orleans Delta* and Gen. John A. Quitman. The chief executive wondered how they could have been privy to the correspondence or why Gonzales's change of position on Cuban affairs was not revealed to the public and only proclaimed in the letter to him. Gonzales believed that his actions had benefitted Davis, but the Confederate president replied that he had been twice elected to the Senate before and the slander promulgated caused little disturbance in the public confidence. Davis emphasized, "The people of Mississippi would no doubt be as much surprised as I was to learn that after many years spent in their service, I had been compelled to seek your aid to save me from their condemnation. There never was a time when I would have sought the letters of any man to sustain my political fortune; it would have been absurd, for such a purpose, to have sought your endorsement."[37]

Davis also remonstrated that he had complied with the "frequent" applications Gonzales made to obtain employment abroad and at home. He sarcastically added that the little success of those efforts "may have been your measure of their value, and have led you to pay me in the form of your present tender. It is accepted." Gonzales had ended his October 10 missive to the Confederate president stating, "Had I been your 'classmate at West Point' my humble self could have done no more for you." Davis understood this to be a comparison of the services rendered to him by Thomas Fenwick Drayton, a West Point classmate and lifelong friend, appointed on September 25 Confederate brigadier general in charge of the Port Royal defenses, the position Gonzales coveted. Drayton, a Hilton

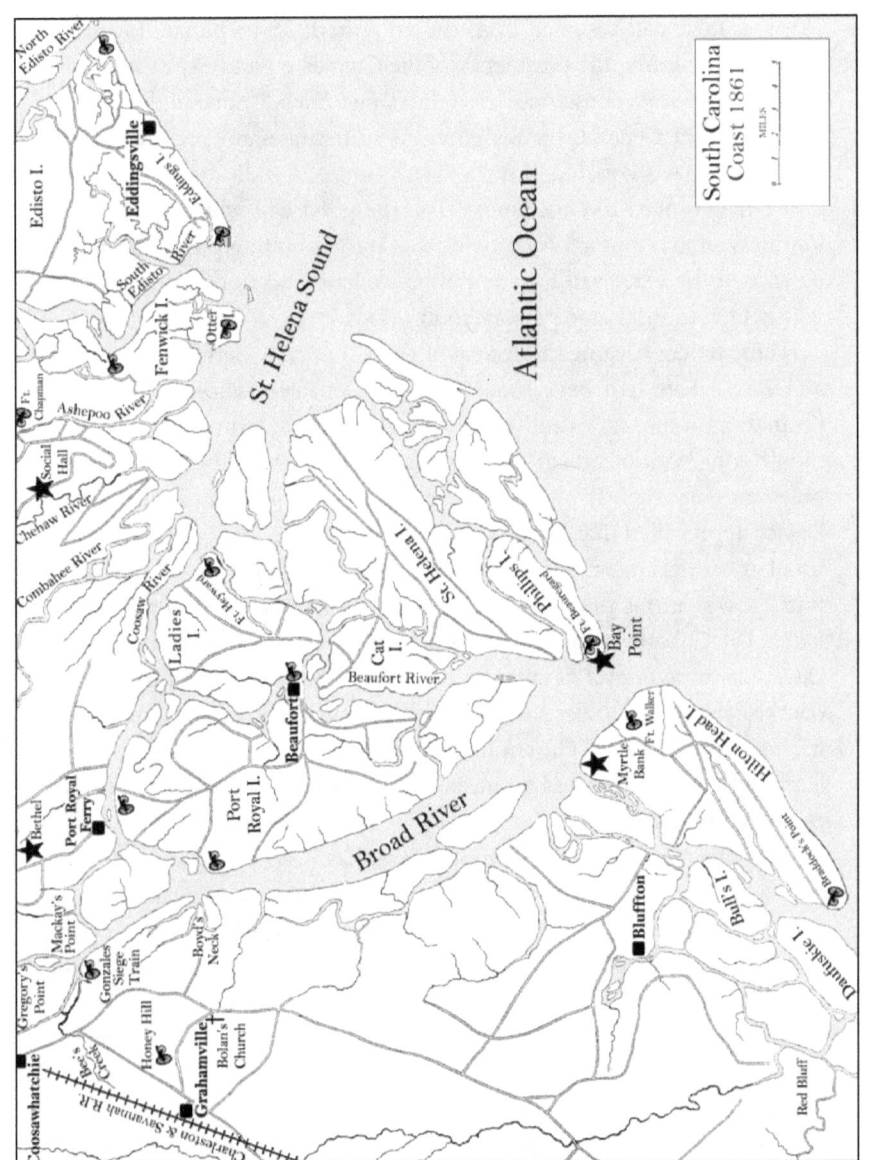

*The South Carolina Coast, 1861*

Head planter, had resigned from the army in 1836 to pursue business ventures, including the presidency of the Charleston and Savannah Railroad. Governor Pickens considered that Drayton had "none of the higher attributes of a General, for his nature is to hesitate and become rather confused." Davis concluded his letter to Gonzales indicating, "I am not aware that I owed to either more than the good will which answers to kindness shown; but am fully aware that the appointing power is a public trust to be exercised for the public welfare, and not a private fund with which to discharge personal obligations."[38]

These letters became the source of deep animosity between Gonzales and Davis. Fate had been unfavorable to Gonzales, whose request for promotion came at a time when Davis had just recuperated from a month-long bout of ague. Davis was also worried about the health of his pregnant wife, slightly injured in a recent carriage accident. Mrs. Varina Davis later recalled that her husband at the time was more sensitive than usual to criticism, and that "even a child's disapproval discomposed him." It was at this juncture that Davis lashed back at Gonzales over his letter. The Cuban was partly to blame because, having intimately known Davis for more than a decade, he needed to have considered the president's character. Gonzales should have been less impatient when requesting the promotion and chosen his written words more carefully. He later admitted to Davis that he had mistakenly referred to the wrong electoral campaign in his letter.[39]

The character and personality of Davis has been an ongoing debate among historians, ranging from high praise to extreme criticism. David M. Potter believed that the poor performance of Davis as president and commander in chief and his deficient relations with others led to Confederate defeat. Allan Nevins pointed to arrogance and pride among the most negative aspects of Davis. In the foreword to the latest edition of Davis's *The Rise and Fall of the Confederate Government,* James M. McPherson stated that Davis had "the type of personality that readily made enemies." Clement Eaton agreed that the Confederate president's "quick resentment of criticism interfered with the efficient carrying out of his duties." Eaton demonstrated that Davis was aware of his own shortcomings in dealing with people who snapped at him but was unable to overcome them. A recent biography of the Confederate leader describes him as "intensely critical of opposition, and [having] the tendency to react rather than act." Davis believed that "he and only he had the power to nominate officers, and he would not debate the propriety of his decisions." Steven E. Woodworth, in his analysis of Davis as commander in chief, points to "his almost compulsive drive to prove himself right and his subordinate wrong—and to demonstrate this to his subordinate with extensive documentation." He depicted Davis as insecure, "rarely willing to admit a mistake," and asserted that "his determination in all disagreements, past or present, trivial or significant, to prove himself right and others wrong made him many enemies." Those whom Davis cherished as intimate friends hardly noticed his personal defects. One of them, John H. Reagan, estimated that the Confederate president "had two characters—one for social and domestic life and the other for official life." Gonzales was unable to differentiate between the two, and it proved costly.[40]

The Cuban spent the last two weeks of October inspecting the South Carolina coast and watching the maneuvers of the blockading squadron. He apparently wrote an article that was simultaneously published in the *Courier* and the *Mercury*, regarding the movements of the enemy fleet. When Gonzales returned to Charleston on October 31 for a brief visit, the *Courier* reported that "he is engaged with characteristic zeal and energy in public defences, in which field he has done and is doing noble service." Five days later, the largest Union fleet ever assembled entered Port Royal Sound. It consisted of fifteen warships, with 148 large guns of various calibers, which left Hampton Roads under flag officer Samuel Francis Du Pont, commander of the South Atlantic Blockading Squadron. Brig. Gen. Thomas W. Sherman was in charge of 12,653 troops on thirty-six transports. In contrast, Confederate forces in South Carolina amounted to only 8,271 men. That same day, another note appeared in the *Courier* stating, "Gen. A. J. Gonzales is ready and prepared to do duty as a full private in one of the regiments now engaged in the defence of our coast. He has labored ably, perseveringly and zealously, as far as he has been permitted to do, for the public defence, and without adequate position or recognition." The notice was read by Gen. Roswell Sabine Ripley, who had accompanied Gonzales as a pall bearer during the Bee funeral. Ripley appointed Gonzales on November 5, under Special Orders No. 175, as volunteer aide-de-camp and assigned him to the command of an artillery siege train of "four eight-inch howitzers with a full complement of horses, ammunition wagons, traveling forge and equipments." When

Mary Boykin Chesnut heard the news, she believed that the Cuban "means to do his duty, and he is so clever, may he be a real aid to Ripley." Capt. Thomas R. S. Elliott, a volunteer aide-de-camp to Gen. Thomas Drayton, erroneously informed his mother and Hattie, who had just returned to Oak Lawn, that "Gonzales has been made a Brigadier General" and hoped that it would be a salaried position.[41]

Joined by a group of aristocratic Charlestonians, Gonzales appointed Theodore Dehon Jervey and William M. Sayer among his volunteer aides-de-camp. Jervey, a former banker and Charleston collector of customs, was a partner in the firm William C. Bee and Company, whose principal owner was the cousin and factor of William Elliott. Gonzales requisitioned on November 7, from the state ordnance officer at the Citadel, thirty-five howitzer cartridges, one hundred pounds of cannon powder, and a pair of cavalry holsters. The siege train was manned by the Palmetto Guard artillerists, under Capt. George Lamb Buist, along with the cavalry companies of the Charleston Light Dragoons, commanded by thirty-two-year-old attorney and Yale graduate Capt. Benjamin Huger Rutledge, and the Rutledge Mounted Riflemen, led by Capt. William L. Trenholm. The latter group, until then stationed at Sullivan's Island, were armed with Sharp's carbines, Colt's navy revolvers and sabers. Gonzales regrouped his force of eighty men in Charleston for the railroad trip to Port Royal.[42]

The Confederate defenses on Port Royal Sound started bombarding the approaching Union fleet at 9:25 A.M. on November 7. Fort Beauregard, on Bay Point, under Capt. Stephen Elliott Jr., had nineteen guns and 640 men. Fort Walker, on Hilton Head, was commanded by Col. John A. Wagner, and contained twenty *barbette*-mounted guns of various caliber and 622 men. Among them was filibuster activist Randolph Spalding, of Sapelo Island, Georgia, who had attached himself to the Fifteenth Regiment of South Carolina Volunteers. General Drayton witnessed from Hilton Head how "the fleet soon passed both batteries apparently unharmed, and then returning delivered in their changing rounds a terrific shower of shot and shell in flank and front." His brother, U.S. Navy captain Percival Drayton, commanded the armed steamer *Pocahontas* during the attack. Three Confederate steamers under the command of Commodore Josiah Tatnall briefly exchanged fire with the fleet before retiring. The Union ships fired a total of 2,461 shells. Five hours later, when most of the guns in Fort Walker were disabled, Drayton ordered evacuating Hilton Head and spiking of the guns in Braddock's Point. When a federal landing force took Fort Walker, to the tune of "Yankee Doodle," Captain Elliott was instructed to abandon Fort Beauregard. The fleet, more than 2,500 yards away, had remained out of the reach of most of Elliott's guns. Union forces captured 52 cannons, including 2 ten-inch Columbiads and 3 rifled guns procured by Gonzales in Richmond, and freed 5,000 slaves.[43]

Capt. Thomas R. S. Elliott assisted the Fort Beauregard garrison in passing White Hall Ferry during the retreat to Pocotaligo, where he returned to his Bethel plantation. It had one thousand acres, more than one hundred slaves, and was bounded on the east by the road from Pocotaligo to Salkehatchie Bridge, and on the south by the Pocotaligo River. His Beaufort home, worth $23,500 with furnishings, was "completely turned upside down, & inside out," by the freedmen after the town was abandoned. When his cousin went inside, he found the former slave Chloe "seated at Phoebe's Piano playing away like the very Devil & two damsels up stairs dancing away famously, all of them were from the Grove." The house on the Grove plantation was "also depredated upon & nearly every thing stolen." Tom also lost his carriage, a buggy, two boats, a horse and two mules, worth $1,755, and eleven prime slaves valued at $9,900. William Elliott Jr., lost the Myrtle Bank plantation on Hilton Head. His father lost the Grove, Shell Point and Ellis plantations on Port Royal Island, with 95 slaves, 9,700 pounds of sea island cotton, 115 head of cattle, 1,200 bushels of corn, 1,200 bushels of potatoes, 300 bushels of peas, 2 boats, two flats, wagons, mules, gins, and other farming implements. His $10,000, three-story Beaufort mansion was later "inhabited by the female abolition teachers." The seventy-three-year-old patriarch petitioned to command or to accompany a military force "to save property, and restrain the negroes," but Gen. Robert E. Lee told him that there was "no adequate force to hold the island."[44]

The evening of the defeat, Lee arrived at Coosawhatchie, the Charleston and Savannah Railroad station nearest Port Royal, and established his headquarters in an abandoned house. He had just been appointed to command the newly created Department of South Carolina, Georgia, and East Florida, where he found a "general scarcity of ammunition" and a "deficiency in powder." Lee was also given command of State troops and soon divided the two-hundred-mile South Carolina coast into five military districts: the first, from the North Carolina border to South Santee River, headquartered at Georgetown, under Col. Arthur Middleton Manigault; the second, from South Santee to the Stono River, headquartered at Charleston, under General Ripley; the third, from the Stono to the Ashepoo River, headquartered at Adams Run, under Gen. Nathan George "Shanks" Evans; the fourth, from the Ashepoo to Port Royal Sound, headquartered at Coosawhatchie, under Gen. John Clifford Pemberton; and the fifth, from Port Royal Sound to the Savannah River, headquartered at Hardeeville, under General Drayton. Lee began to concentrate his meager scattered forces into a line of defense, which would regroup and resist the enemy "at the strongest point on his line of advance." He decided to protect the mainland, evacuating the men and guns from the smaller forward island strongholds that were within range of enemy naval gunfire. These included the fortifications Gonzales had strengthened on Edisto Island, Fort Heyward at Sams Point on Ladies Island protecting the Coosaw River, the defenses on Otter Island at St. Helena Sound, and the redoubt on Fenwick Island on the Ashepoo River. Lee, a skilled engineer, also reinforced the defenses of Charleston, Fort Pulaski at Savannah, and built a deep defensive line in front of Savannah and the area between Port Royal Sound and the Charleston and Savannah Railroad. One of his tasks was to prevent the enemy from sailing their gunboats up the Coosawhatchie and Tullifinny Rivers to destroy the railroad bridges linking Savannah and Charleston. Throughout the war, the Union made repeated attempts to destroy this important interior communications line, which transported soldiers, armaments, victuals, along with cotton, rice, tobacco, and lumber exported by the blockade runners. The *Mercury* described the Charleston and Savannah Railroad as "the main artery, along which the never-ending supplies of Quartermaster's, Commissary and Ordnance stores, are delivered within easy access of the hundred camps which dot the seaboard strip." While the Confederates made an extraordinary effort to keep this railroad open, Union general Thomas W. Sherman repeatedly insisted that "some point on this line should be struck soon."[43]

The Gonzales artillery siege train was ordered from Charleston to a forward position at Port Royal Sound. They left Charleston on Friday evening, November 8, crossed the Ashley River, and went to the St.

Andrews railroad depot. A locomotive delay forced them to sleep that night on the ground beside campfires. The siege train departed by railroad at 7:00 A.M. the next day, reaching Pocotaligo in the afternoon. After unloading their "piles of baggage of every description, domestic and military," they pitched camp. At dawn, on the tenth, General Ripley ordered the Gonzales siege train to Port Royal Ferry, where they "took quarters in the comfortable dwelling of Mr. J. S. Chaplin." Five days later, the *Charleston Courier* and the *Mercury* announced that "a Special Mail will be made up daily at 9 A.M. at the Post Office, for Gen. Gonzales' Command, including Charleston Light Dragoons and Palmetto Guard. Address letters: 'Gen. Gonzales' Command.'"[46]

Port Royal Ferry was fortified with a new redoubt overlooking the river. The guns from the steamers *Lady Davis* and *Huntress*, which belonged to Tatnall's "mosquito fleet," were placed there. The Gonzales siege train was by the eighteenth stationed at Bee's Creek landing, on the Huguenin's Roseland plantation, seven miles below Coosawhatchie. The plantation had been built by David Huguenin at the close of the Revolutionary War. For the next three weeks, Gonzales supervised a slave force from the nearby plantations as they built two earthworks and barrier obstructions on the Coosawhatchie River southeast of the landing. The artillery redoubts, some fifty yards apart and connected by a six-foot-wide trench, still stand today. The northern one, for two gun emplacements separated by a traverse, is about six feet high and sixty yards wide, with a seven-yard-wide parapet and a ditch in front. The southern rectangular position for a single gun has a fifteen-yard front and flanks, surrounded by a ditch. It is six feet tall with an eight-yard wide parapet. Thirty-five yards away are the remains of a bomb proof. Col. James D. Radcliffe's North Carolina Eighth Regiment Volunteers, with one thousand men, were stationed to the south of Gonzales in support of his battery and defending the road parallel to the Broad River (today's State Road 462). The Beaufort Volunteer Artillery, with brothers Ralph and William Elliott Jr., was redeployed to Hardeeville. A week later, General Lee issued special furloughs for state legislators to attend the capitol session, and Ralph Elliott went to Columbia, where he also arranged to have the family silver plate deposited in the Branch Bank vault.[47]

Gonzales wrote on November 22 to Secretary of War Judah P. Benjamin, his nemesis during the 1851 filibuster trials, requesting that his nephew, William Elliott, slightly wounded at the Battle of Manassas, be reassigned to his staff. Artillery captain Thomas S. Rhett and Lt. John L. Boatwright were appointed by General Lee to duty with the Gonzales siege train on the twenty-fourth, although two days later, Rhett was detached to General Ripley in Charleston. The Gonzales defense preparations did not go unnoticed by his adversaries. A Charleston merchant, F. T. Sharratt, departed for Philadelphia on November 29, with information for U.S. Navy secretary Gideon Welles that "General Gonzales has a new 4-gun battery near Beaufort, a new style of gun, VIII-inch howitzers, 5 feet long, invented by General Ripley, which now lie, partly masked, back of Beaufort. They were sent secretly, and it is supposed that our [Union] troops are not aware of their existence." The intelligence was immediately forwarded to flag officer Samuel Francis Du Pont, commanding the South Atlantic Blockading Squadron, at Port Royal. Du Pont responded to Welles on December 4 that his blockaders had already "closed up North Edisto, Stono, and Bull's Bay, besides maintaining the existing force off Georgetown and doubling that off Charleston."[48]

During mid-December, Gonzales visited his family at Oak Lawn, where his son Brosio was recuperating from a recent illness. He collected from numerous citizens of Beaufort, St. Helena, and Charleston petitions for the Confederate War Department on his behalf, for the position of brigadier general on the South Carolina coast. Recommendations were also sent by General Ripley, Sen. Robert Barnwell from Beaufort, and one on December 28 from General Pemberton, Lee's second-in-command, saying that he considered Gonzales "well qualified to fill the position of Chief of Artillery and would be very glad of his services as such, as Brigadier General or in any other grade." General Lee had requested the previous month an additional brigadier general, preferably from South Carolina, but indicated that he "would be very glad to have" fellow Virginian Col. Henry Heth. President Davis had previously appointed Heth major general of the Trans-Mississippi Department, but an outcry from Missouri legislators and the press forced Heth to withdraw his nomination before congressional consideration. Heth had graduated last in his West Point class and was regarded by many as rash and incompetent. This was epitomized eighteen months later when Heth, going against orders and neglecting reconnaissance, prematurely touched off the Battle of Gettysburg, considered the turning point of the war, and lost a third of his division. The intransigent Davis promoted Heth to brigadier general on January 6, 1862, bypassing Gonzales for the third time.[49]

At Oak Lawn, Gonzales found Ralph Elliott, recently detached from the Beaufort Volunteer Artillery to serve as a local guide to Gen. "Shanks" Evans at Adams Run. Ralph then placed a two-week advertisement in the *Charleston Courier* indicating that he was "anxious to form a cavalry company, to be attached to the First Battalion South Carolina Volunteers, commanded by Lieutenant-Colonel John L. Black. . . . Term of service for the war. Men to furnish a good horse and saddle. Arms and equipment furnished." His anxiety was fueled by growing Union encroachments on the South Carolina coast and the occupation of the outer islands.[50]

Gonzales quickly returned to the battle lines when Northern troops in Beaufort, supported by five gunboats and artillery, crossed the Coosaw River on January 1, 1862, and attacked Port Royal Ferry. In the ensuing skirmish, siege train Lt. John L. Boatwright was wounded, as the sixteen-man detachment of the Palmetto Guard retreated with their eight-inch howitzer. They spiked and abandoned a twelve-pounder iron gun accidentally thrown from its carriage by a frightened mule. Union forces withdrew the same day after demolishing the redoubt. A week later, the *Courier* announced that "all letters and packages for members of the Palmetto Guard must be directed to 'General Gonzalez's Command, Pocotaligo, S.C.'" Gonzales was near the Bethel plantation of his brother-in-law Thomas R. S. Elliott, who had been transferred to the staff of Brig. Gen. Daniel Smith Donelson. Tom wrote his father that he was "fortunate in having a very Gentlemanly Regt. of Virginians, Commanded by Col. [William Edwin] Starke, quartered on my land. They give me no trouble & conduct themselves with the utmost propriety. The South Carolina & Tennessee Regts., are rascally fellows, they plunder hen roosts, & burn fencing at a great rate—if [John] Dunovant's Regt. are near you, I would advise you to keep a guard over your turkeys."[51]

A Union expedition temporarily occupied Edisto Island on January 22. A month earlier, General Lee had ordered the withdrawal of the guns Gonzales positioned there. Capt. Thomas Abram Huguenin, commanding the battery on North Edisto, had the cannons dismounted and transported by steamer to Sullivan's Island. The Confederates considered James Island the gateway to Charleston and were making "a desperate effort to prevent a landing" by constructing defense lines. At the same time, the chief engineer of the Union Expeditionary Corps on Hilton Head, Capt. Quincy A. Gillmore, submitted a plan to capture Charleston by way of James Island, bypassing forts Sumter and Moultrie. It included a naval bombardment of the batteries at Fort Palmetto and Fort Pickens, on Cole's and Battery Islands. This would secure command of the Stono River up to the first practical landing place on James Island. Maj. Gen. George Brinton McClellan, general in chief of the Union Army, instructed General Sherman that "the greatest moral effect would be produced by the reduction of Charleston and its defenses. There the rebellion had its birth; there the unnatural hatred of our Government is most intense; there is the center of the boasted power and courage of the rebels. To gain Fort Sumter and hold Charleston is a task well worthy of our greatest efforts and considerable sacrifices."[52]

Meanwhile, the Gonzales siege train, awaiting a Union incursion in the Pocotaligo region, on January 28 was assigned a contract physician, T. W. Hutson, the neighbor of Thomas Rhett Smith Elliott at Pocotaligo. Shortly thereafter, Gonzales was left without support when the siege train was detached from General Ripley's command, after the Palmetto Guard, the Charleston Light Dragoons and the Rutledge Mounted Riflemen volunteered for twelve months Confederate service. The Cuban, who had been serving without official rank or salary, went to Columbia to obtain another position. He stopped at Oak Lawn to visit his ailing wife and children. There he found William Elliott Jr., who was absent without leave for three days from Camp Elliott, near Red Bluff, after longing to see his family, "as well as to be in a house, and sleep in a bed." When Elliott returned, General Drayton, who meanwhile had evacuated his artillery position at Red Bluff, near the mouth of the New River, ordered his arrest and court martial. Confined at Camp Sturgeon, a mile from Purysburgh on a Savannah River bluff, Elliott enjoyed exemption "from drills, guard, roll call, and digging entrenchments." He was found not guilty after producing an earlier leave authorization from Drayton in case of urgency. Elliott claimed having returned home to prepare his tax and property returns.[53]

When Gonzales arrived in Columbia, he appealed to the state Executive Council for the rank of brigadier general. He paid a personal visit on February 16 to Executive Council member James Chesnut Jr., accompanied by former governor John Manning, who had been with him on

Beauregard's volunteer staff, and stressed the dangers facing Charleston and its railroad to Savannah. When Governor Pickens and the Executive Council met three days later, they approved the motion of Lt. Gov. William Wallace Harllee to urge, on their behalf, that President Davis appoint "General Gonzales to a position of such rank as will enable General Pemberton to assign him Chief of Artillery in his Military District." This was the only appointment request that the Executive Council made to the Confederate Government, which on the eighteenth had replaced the Provisional Government.[54]

Gonzales then joined, without pay, a South Carolina Military Commission headed by Harllee, to inspect coastal defenses and locate waterway obstruction points. The Cuban, accompanied by Charles Alston Jr., scouted the Georgetown area, which had been recently reconnoitered by Union gunboats. He then joined Harllee and other commissioners on the twenty-fourth at Laurel Hill Island on the Waccamaw River. Although the commission had been instructed by Chesnut to block the Pee Dee River further north, Gonzales advocated closing the channel at the Georgetown entrance. Harllee concurred that this suggestion "would be best as it saves this whole region." Andrew Johnstone, owner of the Annandale plantation near Georgetown, found deplorable "the Santee defenses—both forts on South Island demolished, as worthless (after much expenditure of labor and time). The remaining fort upon receiving the few pieces of ordnance from the dismantled fortifications, *caved in* but they were trying to rectify it. The river obstructions had floated away—and yet the people were leaving, expecting an immediate attack upon Georgetown."[55]

On March 5, General Lee was called to Richmond to act as military adviser to President Davis. John Pemberton, promoted to major general, became chief of the Department of South Carolina, Georgia, and Florida, and moved his headquarters from Pocotaligo to Charleston. He created a bureaucratic boondoggle by reorganizing Lee's district plan and concentrating his defense efforts on the harbor approach. In March, Pemberton withdrew the battery at Chapman's Fort, on the Ashepoo River, and the twenty guns defending Georgetown. He made a costly tactical mistake when, concurring with Lee's earlier opinion, he transferred the Lucas' Heavy Artillery Battalion and the guns in Fort Palmetto on Cole's

Island at Stono Inlet upstream to a newly built battery on the Stono River, which he considered "a more defensible position." This earthwork was gradually strengthened and named Fort Pemberton.[56]

By mid-March, seventeen thousand Union troops had consolidated their position on the entire Atlantic coast from North Edisto River to St. Augustine, Florida. The area, designated the Department of the South, was commanded by abolitionist major general David Hunter, who had succeeded Gen. Thomas W. Sherman. Repeated Union advancements made Gonzales anxious, especially after three months without an answer from the War Department to his petition for promotion to brigadier general. An article appeared in the *Charleston Courier* on March 14, with his assumed collaboration, saying, "No citizen has done more without office for South Carolina and the South, since the commencement of the present war, than A. J. Gonzales." It outlined his prior services to the state during the war and indicated that the loss at Port Royal of the guns he acquired in Richmond "was not the fault of any engineering or plan of Gen. Gonzales." The article complained that "men less qualified" had been commissioned, although generals and state authorities had urged an appointment for Gonzales as chief of ordnance on a division staff. Five days later, the *Mercury* praised the "zealous and unremitting" labors of Gonzales "in providing defences for the security of our seaboard; yet, while honors and military ranks have been showered upon others far less deserving, he has, thus far, gone unrewarded. We sincerely hope that a fitting commission will ere long be conferred upon this skillful and indefatigable officer." That weekend, the *Mercury* editorialized against "the false policy of President Davis in carrying on the war, and the weakness, indifference or incompetency of his administration." The newspaper objected to "continuing the same men in power, under whose auspices our disasters have occurred."[57]

Gonzales wrote on the twentieth from Columbia to President Davis, responding to his scathing letter of the previous October. Returning the terse salutary "Sir," Gonzales enclosed copies of two letters Davis had sent him as secretary of war, dated October 28 and November 20, 1854, regarding the Mississippi controversy over the Pierce administration's Cuba policy during the Congressional elections that year. Gonzales rectified his previous belief that it involved the Mississippi Senate race but nevertheless claimed that "the political relationship is obvious." He emphasized that his promotion request was due to the merits of his acts, as exemplified in all the supportive letters sent to the War Department by Pemberton, Ripley, Barnwell, the governor and council of South Carolina, and many notable citizens, and not in exchange for the president's gratitude. Gonzales asked Davis to "listen to those who know me best." The executive chief ignored the plea for the fourth time and noted on the edge of the letter that no "political relationship" was perceived, since the filibuster activities "for which he and others had been arraigned was not service to me personally."[58]

Gonzales visited James and Mary Chesnut the same day he wrote the letter. She was impressed by his "fine soldierly appearance in his soldier's clothes—and the likeness to Beauregard greater than ever." The Cuban expressed the "bitterness of soul" he felt in writing Davis, believing that had he been the president's West Point classmate, his brigadier general's request would have been granted, like the favorite treatment given to Lucius Northrop, James Heyward Trapier, Thomas Drayton, and Henry Heth. All of them proved to be ineffective in their commands. Gonzales said that he had worked for and earned a promotion, while the others who received it had not. Still, he would continue fighting for the Confederacy but at the same time "go on demanding justice from Jeff Davis" until he got his dues and "go on hitting Jeff Davis over the head" every chance he had. Mrs. Chesnut, who supported Davis but also considered him "greedy for military fame," replied, "I am afraid you will find it a hard head to crack." The "fiery" Gonzales responded "in his flowery Spanish way: 'Jeff Davis will be the sun—radiating all light, heat, and patronage. He will not be a moon reflecting public opinion. For he has the soul of a despot. And he delights to spite public opinion.'" The Elliott family also felt that Davis was being unfair with Gonzales, especially sister-in-law Emily. When South Carolina statesman James Louis Petigru visited Oak Lawn during the first week of May, Emily occasionally chastised the Confederate president for "neglecting the merits" of Gonzales.[59]

Meanwhile, Pemberton's chief of artillery and ordnance, Maj. Armistead L. Long, was promoted to colonel and assigned as secretary to Gen. Robert E. Lee. Gonzales then wrote from Charleston on May 8 to Confederate adjutant and inspector general Samuel Cooper, formally applying for the vacant position. Pemberton recommended Gonzales for chief of artillery of the Department of South Carolina, Georgia, and Florida,

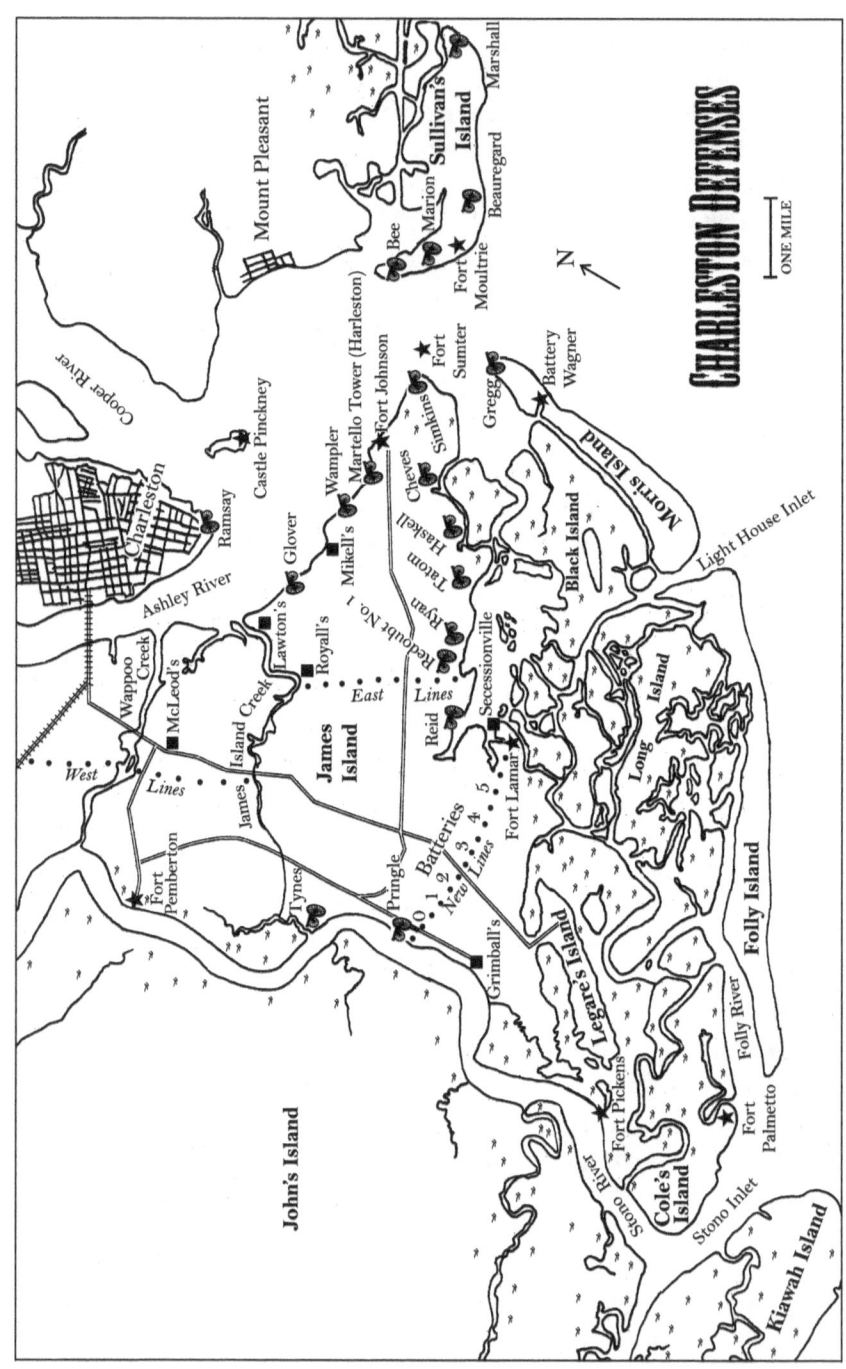

Charleston Defenses, 1861–65

with a rank of not less than lieutenant colonel. He wrote Cooper that "it is very important that an officer well acquainted with artillery in all its branches, should have the supervision and direction of this branch of the service within this Dept. of S.C. & Geo. I know of no one so well qualified to fill the position as Mr. A. J. Gonzales." This petition was endorsed by the South Carolina Executive Council and Governor Pickens, who credited Gonzales with carrying out his duties in the service of the State "with zeal, fidelity, & decisive military ability." Three weeks later, General Ripley sent another endorsement acknowledging that Gonzales had "rendered valuable and efficient service to the Confederate States" while in command of the siege train. Gonzales then petitioned General Cooper on May 28, enclosing the endorsements, "in the hope that, the President may find in them sufficient reason not to limit my rank to the lowest grade designated by Genl. Pemberton." He stressed that his services to the State and the Confederacy, from the time he commanded the siege train to the present, had been done without requesting a salary, incurring all expenses. This plea did not influence Davis to give Gonzales other than the minimum rank Pemberton had asked for. On June 8, General Orders No. 26 announced Gonzales as lieutenant colonel and chief of artillery for the Department of South Carolina, Georgia, and Florida. Sister-in-law Mary Johnstone then wrote to her mother, "I am so glad to hear Gonzie has at last got the position he so much wished. I only hope the means to do good with his Artillery will not be withheld."[60]

After fourteen months of military service, the Davis dispute kept Gonzales from the coveted rank of brigadier general, which had been denied four times. The position was twice awarded to men who proved less competent. Gonzales had been zealous in pursuing the war effort from the start. He proffered his services to General Beauregard and rushed to defend Morris Island during the attack on Fort Sumter. The Cuban worked arduously as inspector of South Carolina coastal defenses. He also procured at Richmond thirty heavy guns for the defense of his state. Gonzales was heralded as a pioneer in devising a siege train of heavy artillery. He commanded four howitzers and eighty men at a forward position for three months after the occupation of Port Royal. References from Generals Pemberton and Ripley, with those of Governor Pickens and the executive council, achieved his appointment as chief of artillery. Gonzales now looked forward to a brigadier general promotion through military merit on his quest toward winning the war.

## Six

# CONFEDERATE COLONEL

Gonzales was urgently needed at his new post as chief of artillery. The week that Pemberton recommended him, Union gunboats, encouraged by the evacuation of Cole's Island, met no resistance as they entered the Stono River up to Battery Island, where Fort Pickens was abandoned by Maj. Cleland Kinloch Huger and the Palmetto Guard. The Confederates destroyed all buildings and fortifications before obeying the order to retire. Since Pemberton had also dismantled the gun positions Gonzales designated for Georgetown, Union troops disembarked there on May 22, and sailed eight miles up the Waccamaw River to raid the J. Izard Middleton plantation. The same day, another Union excursion went up the Combahee River to the William Henry Heyward plantation. A week later, the 50th Regiment of Pennsylvania Volunteers had made a raid from Beaufort to destroy the railroad bridge at Pocotaligo, but were driven back by the Rutledge Mounted Riflemen and a light artillery battery led by Capt. Stephen Elliott Jr. Twenty Union vessels had sailed into the Stono Inlet on June 2. Union troops disembarked on the southwestern tip of James Island, on the Grimball plantation, five miles from Charleston, and captured three guns. Confederate artillery in the Tower Battery in Secessionville started shelling the new Yankee positions.[1]

The first task Gonzales faced as chief of artillery was to reinforce the Charleston defense lines. These were a series of detached artillery redans (V-shaped fortifications), placed between higher-profile infantry redoubts.

On the mainland, in St. Andrews Parish, the entrenchments stretched from Old Town Creek, on Brown's Plantation, for a mile-and-a-half to Wappoo Creek, crossing the Charleston and Savannah Railroad track and the Turnpike Road (today's Highway 17). On James Island, the lines were built two miles from the Stono River, to avoid gunboat fire, but this design flaw readily ceded half the island to the invaders. The West Lines, commanded by Maj. Cleland Kinloch Huger, ran a mile and a half from Wappoo Creek to the Holmes plantation on James Island Creek. The East Lines, commanded by Col. Thomas Gresham Lamar, South Carolina 1st Artillery Regiment, stretched from Redoubt No. 5 at Royall's House, the plantation residence of Crosky Royall, on a tributary of James Island Creek, to Redoubt No. 1, at the Reverend Stiles Mellichamp's plantation on the eastern shore behind Secessionville.[2]

After inspecting these defenses, Gonzales selected the most advantageous positions for deploying two forty-two-pounder carronades and two twenty-four-pounder navy guns stored in St. Andrews depot, with one hundred rounds of ammunition per gun. He then submitted to Pemberton a plan for establishing parallel railroad tracks behind the East Lines that would convey an extended siege platform over both lines. Each moveable carriage would carry "a formidable siege battery overlooking the embankment, to whatsoever portion of the Lines the enemy might attack. These guns could be separated or moved together and by giving the necessary turns to the Lines, might be made to take the enemy in flank as well as in front." The artillery platforms would be moved "with the rapidity of steam, or horse power, or when necessary by hand, with drag ropes." Gonzales also advocated establishing a similar rail plan for a *barbette* battery in Charleston to be moved from White Point Battery on East Bay Street to the other end of South Bay Street. He hoped that this plan would make up for the deficiency in artillery, but the "difficulties were so great, that nothing was done regarding it."[3]

The chief of artillery also mustered a reliable staff of officers and clerks and requested as his assistants artillery lieutenant Willis Wilkinson and former Siege Train subalterns, Pvt. Theodore Dehon Jervey of the Calhoun Guards and Lt. John L. Boatwright, Ripley's drill master. Soon thereafter, Maj. John G. Barnwell and Lt. George Upshur Mayo were transferred to duty in the Artillery Department. The headquarters clerks arranged the official papers, which were copied, classified, and indexed in a systematic manner for easy referral. Gonzales assigned officers to see that the artillery troops were properly armed and supplied with proper ammunition, or if needed be, correct the deficiencies. This entailed inspection tours and submitting reports to the chief of artillery on the condition of the guns, limbers, battery wagons, and traveling forges. On June 14, six days after assuming command, Gonzales was ordered by General Pemberton to assist Capt. William H. Echols, chief engineer for South Carolina, in building a redoubt on the Stono River, near Newtown Cut, containing "a forty-two-pounder rifled gun, *barbette,* four siege eight-inch howitzers, and two siege twenty-four-pounder rifled guns." Col. Thomas M. Wagner chose the position on a bluff on Dill's plantation. The next day, the commanding general instructed Gonzales to designate the points on the West Lines where field artillery ramps were to be constructed by Capt. William M. Ramsay, the Fort Pemberton engineer.[4]

As soon as these preparations began, some 2,800 federal troops led by Gen. Henry W. Benham prematurely attacked the Tower Battery near the James Island village of Secessionville before dawn on June 16. The battery, an unfinished M-shaped fortification commanded by Colonel Lamar, was a nine-foot-high earth-packed parapet protected by a six-foot-deep frontal entrenchment that occupied a one-hundred-yard neck midway on a two-thousand-yard-long peninsula, flanked by marshes. Its armament consisted of an eight-inch Columbiad that Gonzales had procured in Richmond eight months earlier, two rifled twenty-four-pounders, and two eighteen-pounders. Benham had been instructed six days earlier by Maj. Gen. David Hunter, commanding Department of the South, not to advance until largely re-enforced or upon receiving specific instructions from headquarters. Benham insisted on securing the position of his camps from enemy fire. He ordered a "reconnaissance" of the Secessionville fortification, which turned into a major battle.[5]

During two and one-half hours, five hundred defenders poured artillery and musketry volleys at their opponents. Four Union regiments advanced with supporting fire from two gun boats and ten artillery pieces, which fired over five hundred rounds. Three federal bayonet charges, two reaching the parapet, were repulsed, resulting in 683 Northern casualties and 204 for the Southerners. Major General Hunter soon withdrew all his forces from James Island and had Benham tried by court martial and demoted for violating his instructions. In honor of Colonel Lamar's heroic leadership, the Tower Battery was renamed Fort Lamar. After the battle, Confederate general "Shanks" Evans, commanding the Second Military District of South Carolina, praised the gallant conduct of the defenders and commended Capt. Ralph Emms Elliott, one of his volunteer aides-de-camp, for rendering "valuable services in transmitting orders under fire." Gonzales congratulated his brother-in-law and inspected the damage to the fortification.[6]

That same day, the *Charleston Courier* congratulated the Cuban on his recent appointment: "No citizen, native or adopted, has labored more zealously, efficiently, and disinterestedly for South Carolina since the opening of the war, than General A. J. Gonzales, as he is known to his friends." The article indicated that although Gonzales had "military studies, habits, and experience," he had "served thus far without adequate commission or reward, beyond the consciousness of duty, and the flattering testimonials of all under whose commands he has acted." The same journal announced eleven days later: "Lieutenant-Colonel A. J. Gonzales has submitted to the General Commanding this department, plans for the efficient use and application of *barbette* and siege guns, with special reference to the speedy change of batteries and the concentration of fire towards any required point. It is not proper that we should say anything of the details which have been for sometime well considered by the author who has devoted himself zealously and perseveringly to the matter."[7]

Three weeks later, on July 18, an emergency prompted Pemberton to temporarily assign Gonzales as chief of ordnance of the Department of South Carolina, Georgia, and Florida, in addition to his duties as chief of artillery. Gonzales assumed the responsibilities of Lt. Col. Thomas M. Wagner, killed that day due to the bursting of an experimentally rifled and banded thirty-two-pounder iron gun at Fort Moultrie. Gonzales rode his horse in the funeral cortege from the Wagner residence to St. Michael's Church. The procession included a military escort of regulars from Forts Sumter and Moultrie with arms reversed, a band playing a dirge, and the carriages with relatives and friends. It was the largest funeral in Charleston since that of Gen. Barnard Elliott Bee, attended by many distinguished civil and military personages, including Generals Pemberton and William Duncan Smith and Commodore Duncan N. Ingraham, commanding the naval station. Wagner was interred in his family vault in St. Michael's church yard.[8]

After the funeral, the chief of artillery and ordnance returned to his Charleston quarters, occupying four rented rooms, which cost the army $432 annually. His monthly salary was $185 and a ration of one cord of wood fuel. Commutation vouchers indicate that Gonzales did not have a slave and probably employed a private as a valet. In contrast, Capt. Ralph Elliott and his brother William, a lieutenant in Company C of the new South Carolina 2d Infantry Battalion Sharpshooters, were accompanied by their body servants in the field. The duties of the interim chief of ordnance included superintending the manufacture of gun carriages, the preparation and rifling of artillery, the transportation and mounting of guns, the building and equipping of some batteries, the manufacture and distribution of shot, shell, and ordnance stores, and increased bureaucratic paperwork. Gonzales had extra officers assigned to the Ordnance Department, who inspected all ordnance assigned to the troops to assure that it was sufficient and not defective. Their summary reports were reviewed by the chief of ordnance.[9]

One of the major problems Gonzales faced with his new duties was the scarcity of lead for making munitions. Since the previous November, the Ordnance Department had been advertising in the newspapers to buy scrap lead at fifteen cents a pound. They also asked patriotic Charlestonians "to contribute their lead window weights to the Confederate Government, for war purposes. . . . Those contributing will have the weights removed by careful workmen and iron weights substituted for them." The shortage was so desperate that lead was being taken from "the tops or casings of some inkstands" in school desks and some citizens donated the water pipes from their homes. Churches throughout the state contributed their steeple bells for conversion into artillery. Over one hundred discarded cannons, many from the Revolutionary War serving as street corner posts in Charleston, were recycled in a local foundry. Gonzales carried out his ordnance duties until August 16, when General Pemberton named Maj. J. J. Pope chief of ordnance. Five days earlier, Pemberton had recommended Gonzales to Secretary of War George W. Randolph for promotion to colonel, chief of artillery.[10]

The new rank took effect on August 29, with a $25 salary raise, totaling $210 monthly. Four days later, Hattie wrote from Oak Lawn to her husband in Charleston: "I want you to have your promotion announced in the papers. No one will know of it if you do not have it noticed." She thanked him for the new stockings included in a box of family shopping items and indicated that the children were quite anxious for his presence and to play wolf with him. Hattie was elated at the news that Beauregard had been reassigned to Charleston and wondered if Pemberton would remain.[11]

Pemberton's questionable defense preparations had prompted Governor Pickens to repeatedly ask President Davis for his removal. The *Charleston Courier* announced that General Beauregard, after recovering from an illness, had been assigned to the Department of South Carolina, Georgia, and Florida. Gonzales sent Beauregard a telegram on September 6, asking for his arrival date. The Creole replied that he had not received any orders yet, and cautioned Gonzales to "look out for the [Union naval] Monitors." Gonzales responded: "You can well imagine the delight with which I look forward to renewed service under you," and requested to be both his chief of artillery and inspector general, the latter being the last position he held under him. Gonzales stated that his "long devotion to the defence of this State" had made him "thoroughly conversant" with its personnel and materiel. He also informed Beauregard that the following day he was scheduled to sit in a Court of Inquiry at Adams Run. He had asked Pemberton to replace him with another officer, but his request was denied.[12]

The trial of Col. John Dunovant, South Carolina 1st Infantry Regiment Regulars, was ordered the previous month by Secretary of War Randolph on the charge of drunkenness upon duty. The detail for the court included Gonzales, Brig. Gen. Johnson Hagood, and Col. James M. Cullough. Dunovant had been tried and acquitted of the same charges the previous January in a court martial presided by Pemberton. The testimony presented the second time, including that of Lt. Mitchell King, compelled the court to find the accused guilty, but they did not consider the drunkenness "to the extent set forth in the specifications," as disqualifying him from performing his duties. In spite of this ruling, Jefferson Davis dismissed Dunovant from the service two months later. Gonzales and the two other court members then petitioned the president for executive clemency for Dunovant who, "if restored to the service will not again subject himself to a likely charge. . . . Col. Dunovant's high personal character, his general efficiency as an officer, and his services in Mexico, on the frontier and in this war appeal strongly on his behalf." Davis did not reverse his decision and instead referred the matter to the secretary of war, who upheld the dismissal.[13]

While serving on the court martial at Adams Run, Gonzales was able to stay with his family, which had prematurely returned to Oak Lawn from their Adams Run summer house, after twenty of their prime hands were caught trying to escape to the enemy. The plan was betrayed by the black driver and another slave. Gonzales found Capt. Ralph Elliott recuperating from an illness in his field tent at Adams Run. The pinelands village, with a telegraph station and railroad depot, was the headquarters of the Third Military District. It was encircled by camps of the South Carolina artillery, cavalry and infantry companies, and five private homes served as hospitals. The Gonzales boys—Brosio, age five, and Nanno, four—often went to district headquarters accompanied by servants, or were taken by their aunts to the Convalescent Hospital with fruits, flowers, or delicacies for the patients. Brosio recalled sixty years later how the infantrymen "sometimes let us 'pop' huge caps on the nipples of their long muzzle-loading muskets—the soldier holding the piece, while one of us tugged strenuously with both hands at the stubborn and unresponsive trigger." The Gonzales children also visited the artillery depot, being "set astride the bronze or brass or black-iron field pieces and, with the assurance of youth, discussed with the smiling artillerymen the relative efficiency of the long, slim Napoleons and the short, thick-lipped howitzers."[14]

The Adams Run District Headquarters and other department defenses were inspected by General Beauregard after he arrived in Charleston on September 13. The general, accompanied by his staff, Gonzales, and Pemberton, also visited the nearly finished Battery Wagner, an earthwork across the narrow neck of Morris Island, three-fourths of a mile from Cummings Point, named after Lt. Col. Thomas M. Wagner. Beauregard ordered strengthening the position by closing the gorge with an infantry parapet, adding three heavy guns to its sea flank, and including traverses and bomb-proofs. He then instructed erecting Battery Gregg at Cummings Point. The new department commander asked Gonzales on the nineteenth to request from Col. Josiah Gorgas, Confederate chief of ordnance in Richmond, the speedy delivery of heavy guns promised for the department, in lieu of a delayed order from the Noble Brothers and Company cannon foundry in Rome, Georgia, and that they "be 10-inch instead of 8-inch Columbiads, in view of the formidable character of the iron-clad ships preparing for the attack of Charleston." The next day, Gonzales accompanied Beauregard and his entourage on a special train from Charleston to Savannah. They spent the weekend under a torrential downpour inspecting the city defenses, totaling 241 guns. Upon returning to Charleston, Gonzales was dismayed to read in the *Mercury* that Thomas Elliott, the eldest son of Capt. Thomas R. S. Elliott, had died of yellow fever in Nassau, Bahamas, while engaged in blockade running.[15]

Beauregard assumed command of the department on September 24 and established his headquarters in a two-story house at 27 Meeting Street. Gonzales was located nearby at 22 Meeting Street. Pemberton was promoted to lieutenant general in charge of the Department of Mississippi and Eastern Louisiana, which included the Vicksburg fortifications. After his departure, Beauregard met on the twenty-ninth with the department chief officers, discussing the defense of Charleston, its harbor and forts, against an expected ironclad naval attack. Those present included Gonzales, Comdr. Duncan N. Ingraham and Capt. John R. Tucker of the Confederate navy station, Brigadier Generals Thomas Jordan and States Rights Gist, Col. and Insp. Gen. George W. Lay, and Engr. Capt. Francis D. Lee. Gonzales suggested sinking rubbish ships to block the Ashley River, placing obstructions under the plunging fire of Fort Pemberton, moving large-caliber guns to Lawton Plantation, and substituting heavy cannons in Fort Sumter, three on Morris Island and one on a gunboat, for lighter artillery. The chief of artillery had recently urgently requested a complement of ten-inch Columbiads from Chief of Ordnance Col. Josiah Gorgas in Richmond.[16]

The Cuban was ordered to reorganize the Charleston defense lines. His department staff consisted of five officers: In South Carolina, Capt. Willis Wilkinson was in charge of inspecting heavy artillery, and Capt. William J. Saunders of light guns. In Georgia, Maj. John G. Barnwell reviewed light and heavy artillery, assisted by Capt. H. Laurens Ingraham. Gonzales had recently obtained a promotion for headquarters staffer 2d Lt. John L. Boatwright, to first lieutenant. When Beauregard expressed a desire to reassign Boatwright, Gonzales requested at the end of the month the services of Maj. Cleland Kinloch Huger as his assistant. The chief of artillery initiated his new task on October 8, with a reconnaissance tour of the Stono River, from Wappoo Cut to John's Island Ferry, looking for vulnerable places. He rearranged the position of some guns at the mouth of Wappoo Creek and ordered enclosing the works at Church Flats, with construction of additional siege gun positions, to impede a land attack. Gonzales also ordered the construction of service magazines for the redoubts and redans on the East Lines on James Island. The next day, he attended with Beauregard a review of the James Island troops. Brother-in-law William Elliott, commanding one of the companies present, saw Gonzales "looking handsome and very military."[17]

A week later, Gonzales submitted to the commanding general the idea of building "one or two iron-clad Gun boats of but eight or nine feet draught and short enough to turn in the inland waters" around Charleston. The proposal was "heartily approved" by Beauregard and sent to the War Department. Secretary of War Randolph responded to Beauregard the following month that the Gonzales letter and his endorsement had been submitted to the Engineer Bureau chief. He, in turn, reported that such vessels would be of great value to the defense of Charleston, but that they could not be paid for from the Engineer Bureau's appropriation and referred the matter to Stephen Mallory, the Confederate navy secretary. Mallory claimed that the Navy Department was already constructing in Charleston all the boats for which it could find mechanics and materials. He referred the matter to the local authorities by saying that if there were parties in Charleston able and willing to build ironclads, the navy would immediately employ them.[18]

The urgent need for these vessels was apparent when on October 22, the new U.S. Army commander of the Department of the South, Maj. Gen. Ormsby M. Mitchel, sent an expedition led by Gen. John M. Brannan with 4,448 men and seven artillery pieces to destroy the Charleston and Savannah Railroad track at Coosawhatchie. The force disembarked from its transports at Mackays Point, a narrow land neck between the Tullifinny and Pocotaligo Rivers. The element of surprise was lost when one of the Union pilots ran a transport aground for almost four hours near the landing. This allowed Confederates to converge their forces on the Caston and Frampton plantations and destroy the Pocotaligo River bridge to impede the federal advance. Carolinian resistance was spearheaded by Capt. Stephen Elliott Jr. and the Beaufort Volunteer Artillery, the Rutledge Mounted Riflemen and the Charleston Light Dragoons. Fifty Union engineers reached the railroad track west of the Coosawhatchie River trestle. They cut the telegraph wire and removed two rails before retreating upon the arrival of the South Carolina 3rd Cavalry Regiment led by Col. Charles Jones Colcock. The expeditionary force returned that evening to its transports with 340 casualties, while Confederate losses amounted to 163. Capt. William Waight Elliott, ordnance officer of the Third Military District of South Carolina, recovered from the battlefield 68 artillery shells and 46 rifles and muskets left behind by the Northerners.[19]

As a result of this raid, Gonzales spent all of November inspecting and rearranging the artillery around Charleston and Savannah. The chief of artillery assigned a guard for the gun in the recently constructed battery near the New Bridge over the Ashley River. He examined in Fort Sumter a faulty crank and lever on a Columbiad carriage. Gonzales also tested a rifled thirty-two-pounder gun in Fort Johnson, an earthwork with guns *en barbette,* on the northeastern tip of James Island. His brother-in-law William Elliott, accompanied by his slave Frederick while stationed there, provided a cheerful welcome. After returning to Charleston, the department chief engineer, Maj. David B. Harris, gave Gonzales maps of the city defense lines, "with a statement showing the location of the guns as well as their descriptions and calibers." A similar sketch of the Savannah entrenchments and the forts and batteries on the Savannah River, were forwarded to the chief of artillery by Gen. Hugh Weedon Mercer, with an escort of twenty soldiers.[20]

General Mercer, commanding the District of Georgia from Savannah, had recently impressed the first contingent of chattels and free blacks into Confederate service, employed on city defenses. In November, Mercer despatched a loyal slave, David Johnson, to spy in the Union camp at Port Royal. Johnson proceeded to Pocotaligo, from where he made his way to Hilton Head, seeking to become the body servant of the commanding general or of another headquarters officer. He was instructed to purloin military documents, count the number of ironclads, gunboats, and light artillery pieces, obtain the names of all the generals and the regiments stationed there, and give clear warning of an impending attack. Three months later, Johnson received "a suitable reward for his services in the enemy's lines." Another slave, Diria, afterward "left the Yankees in disgust" and returned to his master William Habersham in Pocotaligo, providing "a great deal of information" that Thomas Elliott passed on to his brother-in-law Gonzales.[21]

The chief of artillery reorganized the Palmetto Guard Artillery Siege Train, commanded by Capt. George Lamb Buist, which had been converted into Company C of the South Carolina 18th Heavy Artillery Battalion. This unit, headed by Capt. Charles Alston Jr., was also known as the South Carolina Siege Train. Alston's own battery comprised Company A, and the former McQueen Light Artillery, led by Capt. M. B. Stanly, made up Company B. Gonzales reviewed Alston's company in the First Military District and filed a scathing inspection report on November 13, censuring the conduct of district commander general Roswell Sabine Ripley and the condition of the batteries. As a result, Beauregard ordered Ripley to have the battery inspected by a "competent artillery officer" and to "correct the condition of the company in question." He also reprimanded the chief of artillery because "the tone & temper shown in this communication are not those of an official paper intended for the files of this office." The Gonzales report, missing from official records, probably mentioned Ripley's alcoholism problems. Although Beauregard did not want another Dunovant case in his hands, two years later he relieved Ripley from duty due to his "unreliable habits" and a full investigation was made into his being "exited or influenced by liquor so as at any time to interfere with the proper discharge of his duties." Governor Pickens had earlier complained to President Davis that Ripley's "habit is to say extreme things even before junior officers, and this is well calculated to do great injury to General Lee's command." The contentious Ripley was "forever at odds with both his superiors and subordinates," and the Gonzales report terminated their friendship. Command of the reorganized South Carolina Siege Train then passed to Captain Alston. It contained Companies A, B, and C, each with their own battery of eight-inch howitzers, for a total of twelve guns. Company A included

G. W. Nott, a former discount clerk with the Bank of Louisiana in New Orleans, whom Gonzales knew from his filibuster days.[22]

The chief of artillery then inspected the western half of James Island, submitting his findings to Beauregard on November 16. The next day, he completed reports on the superfluous armament and ammunition at Fort Johnson, on the disposition of Lt. Thomas Davis Waties' Battery and the sea coast howitzer in Fort Moultrie, and on the progress of Capt. M. M. Gray in building submerged torpedo rafts. Beauregard, trying to ameliorate the discord between Gonzales and Ripley, ordered them on November 21 to conduct a joint inspection of the obstruction on the Stono River and of Fort Pemberton to see if its armament should be moved to a more advantageous position.[23]

After the tour, Gonzales applied on December 4 for the South Carolina battery of four eight-inch siege howitzers stationed at the entrance to the Charleston New Bridge. Consent was granted by the new governor, Milledge Luke Bonham, as long the weapons did not leave the state. The Horry Artillery organized the guns into a siege train. Gonzales subsequently recommended that Capt. George Lamb Buist be promoted to major and given command in Savannah of the Siege Train of Georgia, which Beauregard approved immediately. The chief of artillery also got consent for the Siege Train batteries to retain their full compliment of four officers. He supervised the mounting of a ten-inch Columbiad on a new platform and battery in Fort Johnson on December 8, and five days later ordered the deployment of a thirty-two-pounder rifle gun at White Point. These latest moves tightened the ring of fire around Charleston harbor against the expected Union naval ironclad assault.[24]

When Beauregard reorganized his staff on December 8, the chief of artillery was assigned the additional responsibility of department chief of ordnance, replacing Maj. J. J. Pope, relieved at his own request in order to join Gen. Hugh Weedon Mercer in Georgia. Gonzales initially faced some bureaucratic problems in fulfilling his new ordnance orders due to delays in the Charleston Arsenal, which was subordinate to the Ordnance Bureau. He had only two ordnance assistants, Captains Franklin B. Du Barry and L. Jaquelin Smith. Gonzales applied twice in December to have Siege Train artillerist G. W. Nott transferred as clerk to Captain Smith's office. He also had Beauregard assign a courier for the Central

Depot of Ordnance from Ripley's force. Another problem arose in mid-December, when Lt. Col. Olin M. Dantzler reported that his regiment had received new Enfield rifles three weeks earlier without the accompanying ammunition. Gonzales quickly filled the order, replying that the original requisition was not in his office and that he had delivered all Enfield cartridges to the corresponding ordnance officer of the First Military District. Beauregard admonished Gonzales for appearing "to misapprehend object of reference of within papers," by furnishing a report on the cause of tardiness while filling the requisition. The same day, Chief of Staff Thomas Jordan reproved Gonzales for not submitting an order to send a twelve-pounder Napoleon gun to Georgetown and a Prioleau gun for repair at the arsenal. As Gonzales had predicted to President Davis a year earlier, bureau work was cramping his energies and destroying his capability for military action.[25]

The new ordnance chief was also plagued with other requisition delays. He replied to the commanding officer on December 16 that the recent distribution of the Rodman gun powder sent by Lt. Col. George W. Rains from Augusta had been "filled to the extent of the capacity of the [Ordnance] Bureau." Beauregard, realizing that Gonzales was understaffed, responded, "Chief of Ordnance is wrong in supposing when papers are referred to him, that it is done to censure him." Gonzales apparently felt that he was being blamed for the inefficiency of Maj. Frederick L. Childs, commander of the Charleston Arsenal. To set the record straight, he ordered an inquiry into whether the Rodman gun powder had gone directly to the arsenal, and if so, what quantity had been forwarded to General Ripley; requested a copy of all arsenal requisitions since the beginning of the war to count how many had not been fulfilled; and asked for an abstract of all ordnance stores at the arsenal.[26]

Major Childs appealed to Col. Josiah Gorgas, the Confederate chief of ordnance in Richmond, claiming that the Gonzales request would be too time consuming, resulting in the neglect of his regular duties. Childs indicated that the requisition copies "occupy a space of about ten inches of closely packed papers," and admitted not processing many requests some of which were ultimately not wanted. He claimed that "to be required to give satisfactory evidence why a requisition made a year since has not been completely filled would be very difficult to do." Childs was concerned about wasting "two or three weeks of valuable time to making out a statement which when presented to a person unacquainted with all the facts might look like 'inefficiency' or something worse."²⁷

The Christmas season was marred for Gonzales and his family with the death from chronic bronchitis of his thirty-five-year-old sister-in-law Caroline Elliott. On December 22, Gonzales accompanied the Elliotts to the obsequies in Magnolia Cemetery on the outskirts of Charleston. Six weeks later, tragedy struck again. William Elliott, the seventy-four-year-old patriarch, got ill on a trip to Charleston and was taken to the Mills House. He was soon under the care of doctor Thomas L. Ogier, with son Ralph at his bedside. When word of his condition spread, more than twenty friends crowded into his room to wish him well. Gonzales was "very much worried at not being able to visit," being laid up with a foot infection. His boot had pinched his instep, which became irritated with the stocking, and the skin came off after it was treated with adhesive plaster. Ralph wrote to his mother at Oak Lawn on February 1, 1863, urging her to go to Charleston, as his father's condition had not improved, his stomach was "in a very irritated condition" and he had not retained nourishment for three days. When Mrs. Elliott arrived, her husband asked in a very weak voice for Hattie and her children. Elliott's death was diagnosed by Dr. Ogier as "Inflamation of the Intestines." He was interred on February 4 at Magnolia Cemetery, in the large plot that he had bought when Caroline died. His widow, alarmed by reports of an imminent Union invasion, purchased twelve days later the adjacent tract for one hundred dollars, for Gonzales and other family members in uniform.²⁸

At the beginning of 1863, Gonzales dedicated much effort to inspecting and organizing the artillery around Charleston harbor against a Union naval assault. Lincoln's secretary of the navy, Gideon Welles, had after the fall of New Orleans given top priority to the capture of Charleston, the "cradle of secession." Gideon informed Rear Admiral Samuel Du Pont on January 3, 1863, that he was sending him five monitors and the ironclad frigate *New Ironsides* to be used in obtaining the surrender of the city. A constant clamor by Northern newspapers, politicians, and preachers had influenced public opinion to believe that capturing Charleston had greater symbolic importance than taking Richmond. Fourteen months earlier, the Reverend George Hughes Hepworth had harangued from the pulpit at the Church of the Unity in Boston, "We want to see the city of Charleston, the home of treason, the hot bed of treachery, laid in ashes.

This is not revenge; this is retributive justice in its mildest aspect; and we want her ground ploughed up and sowed with salt, that no green thing may ever grow there." Union generals felt that a victory there would redeem their failure to relieve Fort Sumter in 1861. Gen. Alexander Schimmelfennig, who commanded Charleston after its capitulation, warned that "Charleston must be considered a place of arms. It contains a large arsenal, military foundries, etc., and has already furnished three ironclads to the enemy. It is our duty to destroy these resources." The city was an important commercial seaport, with six shipyards and numerous docks, constituting the main blockade-running lifeline to Europe. During the previous year, large quantities of cotton and other agricultural products had been exported from Charleston, while thousands of weapons and tons of munitions arrived there to maintain the Confederate war effort. Charleston was a railroad hub and contained the South Carolina Railroad workshops. Its fall would open the way to the southern heartland and the industrial centers in Columbia and Augusta.[29]

Charleston was also an important military target. The city had numerous foundries, workshops, and manufacturing facilities that contributed material to the military. The former United States Arsenal there had been seized and kept operational by the Confederates. Between March and November 1863, it "issued 16,000 heavy artillery projectiles, repaired nearly 10,000 small arms, fabricated three million small arms cartridges, and, toward the end on the year, turned out 2,000 metallic friction primers a day." The 329 arsenal employees, including many African Americans, also banded, rifled and repaired heavy guns and built carriages for them. The arsenal also cast nearly thirty twelve-pounder Napoleon guns.[30]

The work of Gonzales as chief of artillery was being hindered by his dual task as department chief of ordnance, which proved to be a bureaucratic imbroglio. This included keeping records of all requisitions on the Charleston Arsenal, and all invoices and receipts of the officers making them. Gonzales was soon at odds with Col. Josiah Gorgas, Confederate chief of ordnance in Richmond, over the quantity of ammunition urgently needed for the defense of Charleston. He had sent Gorgas two requisitions a few weeks earlier for ordnance stores that the Atlanta Arsenal refused to release unless approved by the Confederate chief of ordnance. Gonzales indicated that his department did not have Enfield cartridges, that the limited supply of lead had the Charleston Arsenal producing only thirty-eight projectiles daily for the rifled thirty-two- and forty-two-pounder guns, and that he was unable to meet requisitions from Florida. The Augusta Arsenal could not manufacture the shells for lack of casting flasks and other material, and the Macon Arsenal did not have the drawings needed to make them for heavy guns. Gonzales indicated that the situation was desperate, since the rifled guns in Forts Sumter and Moultrie had an average of less than fifty rounds each, and there was no ammunition for a number of twelve-pounder rifle guns belonging to the Siege Train and others deployed around the harbor.[31]

Gorgas complained to Beauregard after Gonzales's first requisition, but the general objected to the opinions expressed against his chief of artillery, and told him to mind his own business instead of meddling with his staff. Gorgas informed Gonzales on January 6, 1863, that the Atlanta Arsenal was "so fully occupied in supplying the army in Tennessee that I cannot call upon it to serve your department," but that he had ordered the Macon and Columbus arsenals to each send 50,000 Enfield cartridges to Charleston as soon as possible. Gorgas concluded by asking Gonzales to forward a list of his artillery needs, "limiting your requisition to, say, 150 rounds per gun." That same day, Beauregard wrote to Adj. and Insp. Gen. Samuel Cooper in Richmond that Gorgas was causing "needless and possibly prejudicial delay in the supply of ordnance stores required in this Department for an emergency of which I have been notified by the Secretary of War as probable, namely an early attack of the Enemy." Beauregard complained that his troops had on the average sixty-five rounds of ammunition per man, and that he wanted to increase it to at least one hundred. He denounced the Ordnance Bureau in Richmond as being "fraught with delay and obstruction and I have to invoke the attention of the War Department to it as a growing evil."[32]

While the Carolinians prepared for a naval attack against Charleston, Union gunboats boldly penetrated further into the lowcountry waterways. On January 30, the U.S. gunboat *Isaac P. Smith*, mounting one thirty-pounder Parrott gun and eight eight-inch heavy Columbiads, anchored on the Stono River between John's and James Islands. In expectation of a repeated incursion, the Confederates had constructed the previous night masked batteries at Legare's Point Place on John's Island and at Thomas Grimball's plantation on James Island. These positions were occupied by companies A and B of the Siege Train, commanded by Maj.

Charles Alston Jr., and three other artillery units. The Carolinians executed a crossfire against the *Isaac P. Smith,* which returned fire while attempting to escape. As the vessel tried to pass Legare's Point Place, it was disabled by three hits to her machinery. The gunboat dropped anchor and her crew of eleven officers and 108 men, including three African Americans, surrendered. The Union had twenty-five casualties while the Confederates lost one man. The vessel's large guns were removed and employed by Gonzales in the defense of Charleston and Savannah. The prize was repaired and renamed the *Stono,* and used as a harbor guard boat. Buoyed by this victory, that night the Confederate gunboats *Chicora* and *Palmetto State,* under Commodore Duncan N. Ingraham, attacked the Union blockading squadron outside Charleston harbor and damaged the steamer *Mercedita.* The federal fleet, although initially dispersed, resumed their positions hours later. Among those celebrating the recent victories with Gonzales was Callender Irvine Fayssoux, a veteran of the four López expeditions. He had recently arrived in Charleston from New Orleans, and was appointed to the staff of Gen. "Shanks" Evans.[33]

At the end of January, Gonzales applied for a promotion to "rank becoming his position & duties," which was approved by Beauregard and forwarded to the adjutant and inspector general's office. Two weeks later, growing supply problems and frustration with the Confederate Ordnance Department led Gonzales to ask to be relieved as chief of ordnance. Beauregard immediately denied his request because it could not be effected "without material injury to the public service." Gonzales continued to pursue his ordnance duties with zeal in spite of the supply difficulties. His work during February included forwarding guns and distributing the available artillery projectiles throughout his department. The Cuban assigned a battery of field guns, manned by a company of the Georgia 46th Infantry Regiment, to guard the Charleston side of the New Bridge. He also suggested that the dredge boat *Moultrie* be converted into a war vessel. Beauregard approved the idea and conveyed it to Commodore Ingraham.[34]

The chief of artillery and ordnance then had a third smooth-bore ten-inch Columbiad mounted at Castle Pinckney and recommended that artillery ammunition from General Ripley's district be supplied to newly installed twenty-four-pounders at Causton's Bluff, Georgia. Ripley rejected the proposal, claiming a shortage of less than fifty rounds for all twenty-four-pounders under his command, and stated that the projectiles should have been shipped with the guns. During the last half of February, Gonzales received four railroad shipping consignments. When the first one arrived on the nineteenth, Quartermaster Maj. Motte A. Pringle denied Gonzales's request for private transport wagons, as his supply was "too limited to permit one officer to have transportation for 1,500 lbs., or the allowance of 15 officers." It seems that Gonzales was helping his family and in-laws prepare to move from Oak Lawn to an upcountry refuge in the face of an imminent Union attack. Brother-in-law Lt. William Elliott Jr., recently assigned as district enrolling officer in Greenville, had expressed concern regarding their peril.[35]

Gonzales spent March preparing for an invasion by inspecting and reinforcing the artillery positions on the Ashepoo River, Combahee Ferry, Coosawhatchie, and Dawson's Bluff at Pocotaligo. Thomas R. S. Elliott wrote from Pocotaligo to his sister Emmie that the local fortifications and rifle pits were strong and that torpedoes (mines) had been placed in the mouths of the Pocotaligo and Coosawhatchie rivers and along Ferry Road and Mackays Point Road. He described the land mines as "ten inch Columbiad shells filled with powder, with a fuse or friction match so nicely adjusted that the slightest pressure of the foot will cause them to explode." That month, Gonzales received eight railroad shipments of ordnance supplies. To deal with this increased volume, he applied for the services of Lt. Joseph J. Legare, who was assigned as military store keeper of ordnance.[36]

The chief of artillery and ordnance issued a circular on the twenty-second for district ordnance officers to submit regular reports to him or face penalties. This prompted a protest from General Ripley, the First District commander, claiming that Gonzales lacked authority to promulgate penalties or to require inspections from district staff officers that were not deemed necessary by the district commander. Ripley, still resentful of the accusations Gonzales made against him four months earlier, complained to department headquarters that to implement "the proceeding indicated would be needlessly to multiply papers, and entail the necessity of employing a large additional force of Staff Officers & clerks." Chief of Staff Gen. Thomas Jordan later called Gonzales's attention to the "redundancy" of the paragraph in question in his circular and the "irregularity" with which the document was sent directly to the First District ordnance officer and not through the district commander. According to Capt. Stephen Elliott Jr., Ripley, Jordan, Col. Alfred Moore Rhett, and others, belonged to a clique "with whom it is dangerous to interfere."[37]

Two days later, Ripley jumped at another opportunity to criticize Gonzales to department headquarters, claiming that his district was deficient of artillery friction tubes and that "the Chief of Ordnance has 4 or 5,000 on hand which he declines to issue." Gonzales responded that he had not issued the 4,500 friction tubes in store due to the probabilities of an imminent attack on Charleston and because he awaited a requisition made three months earlier for 130,000 friction and primer tubes from the Richmond and Atlanta arsenals. Just two days earlier, he had received from Atlanta six thousand friction tubes and five thousand primer tubes. The complete shipment, except one thousand friction tubes, had already been ordered sent to Ripley, leaving in store for the rest of the department 2,500 friction tubes and 2,000 primer tubes. Gonzales added that "if the General Commanding so desire it, they will also be issued to Brig. Genl. Ripley." This requisition was never fulfilled, but Beauregard replied to Ripley that the course of action taken by Gonzales had his "entire approval."[38]

The chief of artillery and ordnance retaliated against Ripley in early April by reporting to department headquarters that the First District ordnance stores depot at Summerville, twenty-five miles northwest of Charleston, was left unguarded. Ripley responded that he was unaware what might be there and that headquarters never told him to assign a sentinel, which would "probably be as much hindrance as protection." Beauregard ordered Ripley to send a small company to Summerville to establish a garrison and guard the depot.[39]

The expected Union invasion force appeared on April 5, 1863, when the most powerful ironclad fleet ever assembled gathered outside Charleston harbor under Rear Adm. Samuel F. Du Pont. Maj. Gen. David Hunter commanded a large part of his force of twenty-three thousand men, which was on Folly, Seabrook, and Cole's Islands and in transports on the North Edisto River, expecting to enter Charleston behind the fleet. The next day, when the advance was postponed due to hazy weather, Gonzales asked Beauregard for the Siege Train, to mobilize its twelve howitzers of reserve artillery to impede the enemy troop advancement.

The chief of artillery reviewed the strategic locations of the sixty-four guns and five mortars positioned for the defense of the harbor entrance. Fort Sumter had thirty guns and three mortars. On Sullivan's Island, there were nineteen guns and two mortars in Fort Moultrie, three guns in Battery Beauregard, and six guns in Battery Bee, including the eight-inch Columbiad previously employed in the Tower Battery during the battle of Secessionville. Morris Island had two guns in Battery Gregg and four in Battery Wagner. Two companies of infantry were sent to those islands to reinforce their garrisons against a land attack.[40]

The Union plan was to run this gauntlet, without returning fire, until they were within close range of Fort Sumter. They would target the center embrasure to create a penetrable breach. After reducing Sumter, the fleet would attack the Morris Island batteries. The monitors, with two guns in a single revolving turret, got under way on the seventh at 1:15 P.M., a clear day with calm waters, and maneuvered into the main channel in the following order of battle: the *Weehawken*, pushing a raft fifty feet long and twenty-seven feet wide, with chains dangling twenty feet deep to detonate torpedoes; the *Passaic*; the *Montauk*; the *Patapsco*; the flagship, the armored steam frigate *New Ironsides*, equipped with sixteen guns; followed by the monitors *Catskill*; the *Nantucket*; the *Nahant*; and the armored steamer *Keokuk*, with two fixed turrets each containing one gun. The fleet carried a total of thirty-two guns and 1,200 men.[41]

The Confederates had marked the channel with colored buoys, to indicate the location of mines and exact ranges for their guns. The artillery action began at 3:00 P.M. and lasted two hours and twenty-five minutes, while Charlestonians crowded the rooftops and the waterfront to view the bombardment. The fleet, outnumbered in guns two to one, were easy targets. The *Weehawken* was hit fifty-three times and fired twenty-six shots. The mine-detonating raft was released when it hindered movement. The *Passaic*, struck thirty-five times, fired thirteen shells before its turret rails were immobilized by two direct hits. The *Montauk* fared better, firing twenty-seven times and being hit fourteen. The *Patapsco* had one of its guns disabled after the fifth round, and received forty-seven hits. The ironclads had difficulty maneuvering in the narrow channel and upon retreating two of them collided with the flagship. The *New Ironsides* only fired eight times at Fort Moultrie, got hit sixty-three times, and had to anchor twice to prevent from being forced ashore by the current. The

*Nantucket* got off fifteen shots and received fifty-one in return. The *Nahant,* firing three times before its turret jammed, was hit thirty-six times. The last vessel, the *Keokuk,* defiantly passed the other ironclads, but managed to fire only three times before being peppered by ninety shells. Nineteen projectiles tore through its armor just below the waterline, forcing its retreat at 4:40 P.M.[42]

The Confederates fired a total of 2,209 shells, making 520 hits on the Union vessels. The fleet managed to fire 142 times and struck Fort Sumter with thirty-four shells from a distance of nine hundred to fifteen hundred yards. Confederate casualties were five men wounded by masonry fragments in Fort Sumter, and one of the fifty slaves working on the fortification; one man killed at Fort Moultrie when the flag staff fell on him; and three men killed and five wounded at Battery Wagner when an ammunition chest accidentally exploded. Rear Admiral Du Pont called off the attack the next day due to heavy damage to three monitors and after the *Keokuk* sank in thirteen feet of water three-fourths of a mile off the south end of Morris Island. During low tide at nighttime, the Carolinians removed from the *Keokuk* the two eleven-inch Dahlgren guns, which Gonzales assigned to the defense of Charleston. The failure of the expedition would cost Du Pont his command of the South Atlantic Blockading Squadron. In his combat report to Richmond, commanding general Beauregard praised and thanked Colonel Gonzales and Chief Engineer Majors David B. Harris of his department and William H. Echols of South Carolina "for valuable services in their respective Departments." Beauregard concluded his report advising that Fort Sumter should be armed with additional heavy guns, on both tiers of casemates and the *barbette,* conformably to its original plan, to render it impregnable. Although this later proved to be impracticable, it was heartily approved by the Carolinians who had opposed Pemberton's disarmament plan.[43]

Six days after the battle, anticipating a renewed attack, Gonzales submitted to Beauregard a new series of plans for fortifications in Fort Sumter and Fort Johnson and rearranging the guns around the inner harbor. He recommended removing all of the thirty-two-pounder guns from Fort Sumter and placing fifteen of them on *barbette* carriages on the high hills of Morris Island. The only artillery to remain were the large Columbiads and the heavy rifle guns. To strengthen Fort Sumter's damaged masonry, Gonzales suggested filling up the casemates and the quarters of the south face with concrete, or part concrete and the other sand. In anticipation of another ironclad attack, Gonzales proposed casting thirteen-inch mortars, to be placed on Morris and Sullivan's Islands. Their high-arch trajectory could penetrate the thin armor on the ironclad's decks. The proposal to drastically reduce Fort Sumter's artillery smacked of heresy to most Carolinians, who regarded their beloved fortification as a bulwark of freedom against federal tyranny. Beauregard responded that "some of the ideas contained in this communication are good, others not so, & others again, although good, are impracticable at this time." Beauregard considered the reorganization of the defenses such a momentous subject that a board of engineers and artillery officers should be convened to decide on it and examine the views of the artillery chief.[44]

Gonzales published a note in the *Charleston Courier* on April 13, indicating that "it is the desire of the general commanding that in accordance with Paragraph I, General Orders No. 90, 1862, Adjutant and Inspector General's Office, Chiefs of Artillery in the several commands of this Army will make, before going into action, such disposition of the teams attached to the battery wagons and traveling forges as will render them available for the purpose of securing artillery captured on the battle field." This prompted Beauregard's chief of staff, Gen. Thomas Jordan, to reproach Gonzales against issuing certain orders "as before explained to you on several occasions, belong to the Department of orders of an Army and cannot properly emanate from a bureau such as yours. They are clearly of the nature of orders—and I must insist on your observance in this matter hereafter of the usages and regulations of the service." Nine days later, Gonzales published another announcement in the *Courier*, listing the six artillery and nineteen ordnance officers in the department. He also advertised for "three competent engineers, for service in Foreign Trade." The chief of artillery and ordnance probably planed to send these officers abroad on a mission to purchase engineering equipment, explosives or torpedoes. The Cuban then applied for the services of infantry lieutenant J. G. R. Gourdin, "for copying drawings of carriages of heavy guns," before dispatching a ten-inch Columbiad carriage to Savannah.[45]

Gonzales again inquired about the Siege Train on April 14, and was informed the next day that it was under the command of General Ripley.

In attempting to assign it to Gen. States Rights Gist, Gonzales circumvented the chief of staff and directly addressed Beauregard, who turned him down. This caused General Jordan to send Gonzales a scathing letter saying that "present circumstances are not favorable to such discussions, when the time and attention of these Head Quarters are absolutely required by more important considerations. When an officer of the general staff does not clearly understand an order from these Head Quarters it is his duty to carry it out in *Spirit* if he cannot in *letter* and when in doubt, he will call in the 'Chief of Staff' for a verbal explanation. The discussion of orders is unmilitary, tends to obstruct the public service & to increase the already overtaxed labors of the Adjutant General's office of the Department."[46]

Gonzales then tried to curry favor with Beauregard and Jordan by publishing an article in the *Charleston Courier* on April 25, which has all the markings of the Cuban's flowery prose, ending with a *mea culpa*. It detailed the services he rendered since the war started, and how he "succeeded where others would have done little or nothing," in obtaining thirty guns for Port Royal. The article credited the "sagacity and perseverance and determination" of Gonzales with procuring the seven-inch rifled and banded Brooke gun for Charleston. It stated that Gonzales was "thoroughly acquainted" with all aspects of artillery and ordnance, "as taught in the books," that he was "fertile in resources and prolific in suggestions," and "ever eagerly ready and willing to receive, apply and acknowledge the suggestions of others." The article concluded by saying that "the most prominent trait" in Gonzales's mind, "is a devoted love for his Chief and General, to whom in his person he bears a resemblance, which is often noticed, and which with the same uniform might confound any but the most discriminative eyes."[47]

Beauregard, who was very susceptible to flattery, was apparently swayed by this article to change his mind. Three days later, he ordered "the return of the Siege Train to the command of the Chief of Artillery." Gonzales requested on May 1 that the Siege Train be brought back to its original location at the race course in Charleston, but Beauregard later decided to leave it at Hampstead Mall in St. Andrews Parish south of the Ashley River in Ripley's First Military District. Beauregard fully supported Gonzales in the Ripley feud three weeks later when, citing General Orders No. 7, paragraph II, War Department, he ordered all field batteries not attached to brigades or assigned to defensive positions to be parked as reserve batteries under the command of the chief of artillery. Gonzales applied on May 2 for a seven-day leave of absence to spend with his family, which had remained at Oak Lawn in spite of earlier evacuation preparations. It was his first furlough request since the war began, and Beauregard immediately approved it.[48]

During this time, Gonzales posed for an oil portrait in his colonel's uniform. The unsigned work was apparently done by A. Grinevald, a Charleston artist and lithographer who had recently returned to the city after a year at the Confederate lithographic works in Columbia. Grinevald had created portraits of Beauregard, Robert E. Lee, the Charleston harbor fortifications, the bombardment of Fort Sumter, and the monitor attack. The Gonzales painting has Grinevald's distinctive detail style, and depicts the Cuban's light brownish-gray hair and three aging lines creasing his brow.[49]

Upon returning to duty, Gonzales prioritized the defense of the Combahee and Ashepoo rivers. For this task, he obtained as a courier St. Helena planter Pvt. Thomas B. Chaplin, Company B, South Carolina 5th Cavalry Regiment. The chief of artillery and ordnance hurried production of four Napoleon guns being cast in the Charleston Arsenal. Due to the lack of rifle guns in the ordnance depot, he suggested the transfer of Capt. Frederick C. Shulz's artillery battery to Combahee Neck. Gonzales consolidated the Siege Train and advised the promotion of officers in two of its companies. He applied for light artillery batteries to be sent from Virginia to strengthen the department defenses. During the advent of the fever season, the batteries on Red Bluff and the Ashepoo, Combahee, and Coosawhatchie Rivers had their guns removed and their garrisons were relocated to healthy areas. Gen. William Stephen Walker, commanding the Third Military District, sent some of the artillery to Savannah without informing Gonzales of their destination. At the end of May, the Cuban was relieved as department chief of ordnance, in compliance with his request three months earlier, while continuing as chief of artillery. The vacancy was assigned to Lt. Col. John R. Waddy, until then the department's assistant adjutant general.[50]

Although his bureaucratic obligations decreased after relinquishing the command of the ordnance department, Gonzales still had to micromanage a massive amount of paperwork generated by his artillery headquarters.

He was responsible for a dozen officers at headquarters, thirteen others assigned to the Siege Train, and numerous assistants. There were monthly employee pay vouchers to fill out, including one for forty-five dollars for Central Depot porter C. A. Coste and a ten-dollar voucher for Elias, the African American who served as the headquarters janitor. Gonzales also had to submit requisitions for ordnance stores, quartermaster supplies (including blank forms and four bottles of black ink), goods obtained from local merchants, and forage for three horses assigned to his office couriers, amounting to 1,116 pounds of corn and 976 pounds of fodder monthly. Chief of Staff Jordan once again admonished Gonzales in June, together with the department chief engineer, Lt. Col. David B. Harris, for having referred to Beauregard's office a bill for twenty coils of rope from John Fraser and Company. Jordan wrote: "Staff officers who use articles must take measures to ascertain the proper Government accountability & to secure private individuals against prejudice. This will hereafter be rigidly expected of them to the end that cases like this & of the iron purchased for Capt. [John] McCrady shall not have to be repeatedly referred to this office."[51]

The chief of artillery applied on June 2 to Col. Josiah Gorgas, Confederate chief of ordnance in Richmond, for "better guns for Light Artillery in the Department," and ten thousand McAvey fuse igniters. Gorgas returned the order with the admonition that all ordnance requisitions had to be sent by the new chief of ordnance. After the request was rerouted, Gorgas agreed to send a battery of Napoleon guns from Macon and indicated that some eighteen-pounder rifled guns were at Bermuda and would be ordered if needed in Beauregard's department. That same day, in an effort to recruit local German artillerists, Gonzales requested that Capt. William K. Bachman's company of light artillery be returned to South Carolina and exchanged with the Nelson Light Artillery of Virginia Volunteers. Although Gen. Robert E. Lee rejected the proposition, Beauregard insisted three weeks later that there were more than fifty Germans in Charleston with consular protection who could not be conscripted, but who would be influenced to join Bachman's mostly German unit by "an influential German citizen." The exchange was made later that summer. Gonzales spent nearly a week in mid-June visiting the Second Military District, and recommended the disposition and organization of certain guns in preparation for another Union offensive. Northern forces now controlled some 250 miles of coast, from Light House Inlet, South Carolina, to St. Augustine, Florida, and occupied nine islands. To oppose this encroaching foe, the Department of South Carolina, Georgia, and Florida troops amounted to 6,488 infantry, 6,046 cavalry, and 7,329 artillery.[52]

The Siege Train was again requested by General Ripley on June 22. Beauregard authorized it but ordered its return to Gonzales as soon as circumstances permitted its substitution for a light battery, "as it is desirable to keep the Siege Train together as habitually as practicable." That day, command of the Siege Train was given to Maj. Edward Manigault, a forty-six-year-old Charlestonian bachelor from a prominent French Huguenot colonial family. He and his brother Arthur had been officers in the Mexican War and after secession he became chief of ordnance of the state of South Carolina. Manigault had written to Secretary of War James Seddon the previous month, requesting command of the Siege Train after the resignation of Maj. Charles Alston Jr.[53]

A week later, Gonzales announced the promotions of Siege Train Company B lieutenants Samuel J. Wilson and J. B. W. Phillips. On July 4, he requested orders for the Siege Train, proposing two days later the appointments of Sergeants I. W. Girardeau and W. T. Logan, Company A, to senior and junior second lieutenants of Company C, stationed at the Artillery Cross Roads on James Island. Gonzales made the recommendations because Company C had recently rejected its elected officers and there was no prospect of filling the vacant positions. Beauregard authorized the chief of artillery to hold new elections for the positions and detail an officer to superintend and certify the process. If the sergeants suggested by Gonzales failed to win a majority vote, outside officers would then be assigned.[54]

The Cuban was ordered by Beauregard on July 9 to "hold the siege train in readiness to move at a moment's notice." That day, Charleston mayor Charles McBeth, notified by Beauregard of an imminent attack on the city, temporarily suspended all commercial transactions and requested the evacuation of women, children and noncombatants. The Elliott women and Gonzales children had left Oak Lawn that week to spent the summer at their Flat Rock, North Carolina, home. Ralph Elliott, stationed nearby, had a "new picket arrangement for guarding the negroes" on their plantation. Mayor McBeth also complied with

Beauregard's request that all free male blacks in Charleston, between the ages of eighteen and sixty years, be conscripted to work on the unfinished Morris Island defenses. To protect the First Military District, Beauregard transferred from Georgia captain John F. Wheaton's Georgia Chatham Artillery Battery, five mortars, and the rifled guns of the Georgia Siege Train at Savannah, along with a battery from the Third Military District. Governor Bonham, commander in chief of the state forces, relocated his headquarters from Columbia to Charleston on the tenth.[55]

At 5:00 A.M. on July 10, forty-seven Union guns, covertly installed the previous week on the north end of Little Folly Island, shelled Confederate positions on Morris Island. Under cover of the bombardment, two thousand federal infantry crossed Light House Inlet in small boats, drove back the defenders, captured eleven artillery pieces, and occupied the southern portion of Morris Island. The plan had been orchestrated by Brig. Gen. Quincy A. Gillmore, who had assumed command of the Department of the South from Maj. Gen. David Hunter the previous month. Gonzales was ordered to immediately inspect all the heavy batteries on James Island, starting with Fort Pemberton, and consult their commanding officers regarding their most pressing needs. The chief of artillery was also instructed to forward the Siege Train to James Island, where it could advantageously engage enemy gunboats.[56]

Gonzales ordered Major Manigault to move the Siege Train from the Charleston race course to William Wallace McLeod's plantation house, near Wappoo Creek, on James Island. Companies A and B were dispatched, while Company C, on the island for the previous fortnight, remained in position at the Artillery Cross Roads. The Siege Train crossed the Ashley Bridge at 9:00 A.M., arriving at their destination one hour later. When Gonzales reached McLeod's around 11:00 A.M., Manigault reported to him. They conferred with the island's commander, Col. Charles H. Simonton, South Carolina 25th Infantry Regiment, after which Gonzales deployed the Siege Train one mile east to James Island Creek. The chief of artillery then inspected the nearby West Lines and sent a report to the Ordnance Department detailing "the want of ammunition chests and arms, etc., for the troops."[57]

The next morning, July 11, federal forces advanced toward Battery Wagner, supported by heavy artillery fire, three monitors on the left flank, and several barges with howitzers in Light House Inlet. The attackers retreated after losing two officers and 95 soldiers killed, and six officers and 113 soldiers prisoners. Confederate casualties were six dead and six wounded. Six months earlier, Gonzales had recommended that fifty "shells on land with sensitive tubes" (also known as land mines or torpedoes) be placed in the approaches to the fortification, which was approved by Beauregard and referred to General Ripley for action. This defensive measure was delayed until the twentieth, when over sixty torpedoes were planted two hundred yards from Battery Wagner, across the width of the island. Beauregard then ordered mining the roads and approaches to the James Island defensive lines. Three types of mines, with plungers at ground level, were used: over thirty consisting of ten-gallon gunpowder kegs with conical extremities, about twenty employing twenty-four-pounder shells, and the rest were armed with fifteen-inch navy shells. Although Gonzales originated the idea of planting the torpedos at Battery Wagner, historians have not given him due credit.[58]

The chief of artillery was back in Charleston on the eleventh, when Beauregard gave him a letter to deliver to Capt. W. F. Nance, assistant adjutant general of the First Military District. It gave defensive instructions for James Island, which the Cuban was to augment with additional details. Beauregard instructed Gonzales to increase the armament in Battery Wagner with four twelve-pounder howitzers and two thirty-two-pounder carronades on siege carriages, and in Fort Moultrie with guns withdrawn from Fort Sumter. The commanding general also ordered the construction of a new gun position on Shell Point, later named Battery Simkins, to deploy two ten-inch Columbiads, a 6.40-inch Brooke gun, and three 10-inch mortars. Beauregard had eight mortars and a battery of rifled guns from the Georgia Siege Train at Savannah brought to the First Military District. Gonzales correctly estimated that they faced a "protracted siege" and on July 14 urged caution in the use of ammunition. He also recommended that President Davis be solicited to transfer the heavy guns on the State of Georgia floating battery in the Savannah River, from the Navy Department to the War Department, so that they could be used in the defense of Fort Sumter.[59]

Two days later, Gonzales requested the assignment of his relative, Maj. Stephen Elliott Jr., the chief of artillery for Gen. William Stephen Walker, as his assistant. That same day, due to the recent massive landing of enemy troops on Morris Island, Beauregard ordered that at least twenty guns on siege carriages reinforce the new batteries on the eastern shore of James Island. These artillery positions, facing Black Island, were later named Batteries Ryan, Tatom, Haskell, and Cheves. Union troops, "working like beavers on Morris Island," were being monitored by Lt. A. E. Morse in a reconnaissance balloon over Fort Johnson. On the seventeenth, Governor Bonham sent William Porcher Miles to Richmond, urgently requesting President Davis authorize three Brooke guns and all available ten-inch Columbiads be sent to Charleston. Bonham received a few weeks later two 9,000-pound 6.4-inch caliber Brooke rifled guns that Gonzales placed in Fort Johnson, a 20,000-pound triple-banded seven-inch Brooke gun from Navy Secretary Mallory, who replied that his Department had no other surplus guns, four ten-inch Columbiads, and four ten-inch mortars.[60]

Union troops launched a second assault on Battery Wagner with six thousand men on July 18, spearheaded by the 54th Massachusetts Infantry Regiment, the first regularly organized African American regiment to fight in the Civil War. Eleven hours before the bayonet charge, federal land and sea artillery threw over nine thousand shot and shells into the earthwork to obliterate resistance, and score an easy propaganda victory for the black troops. Union general Truman Seymour reported to his superior that Battery Wagner "was subjected to such a weight of artillery as had probably never before been turned upon a single point." The Confederates had intercepted a message ordering the attack at dusk, giving the 1,300 defenders time to seek shelter in the garrison bombproof and in scores of rice barrels buried in the rear sand hills. As a result, they had only four killed and fourteen wounded during the saturation bombardment. When the attackers advanced after the shelling ceased, they were hit with a "most destructive musketry fire" from the redeployed troops on the Battery Wagner parapet. According to Commanding Brig. Gen. Quincy A. Gillmore, the 54th Massachusetts Regiment "was soon thrown into a state of disorder, which reacted disadvantageously upon those which followed, and rendered it necessary to send in the supporting brigade." After three assaults and vicious close combat for nearly three hours, the Union forces retreated with more than 1,500 casualties, including 246 men killed, 890 wounded, and 391 captured. The African American regiment suffered the greatest loss, 272 men, including their commander, Col. Robert Shaw, killed while mounting the parapet. Confederate losses that day were 174 killed and wounded. Three days later, Gillmore urgently requested from Army Headquarters in Washington, eight to ten thousand seasoned troops.[61]

Expecting a third attack, Gonzales recommended to Beauregard on July 25 that sixty double-barrel shotguns in the Charleston Arsenal be sent to Battery Wagner. These weapons would be more effective than muskets in close combat. The commanding general referred the matter to General Ripley, who sent the weapons a month later. Beauregard then wrote Ripley a lengthy complaint of his leadership failures, including the lack of a proper defense of Morris Island, which prompted perpetual animosity between them. The next day, Gonzales arrived before 5:00 A.M. at Battery Haskell, on Legare's Point, James Island, where the Siege Train was stationed. He had breakfast with Maj. Edward Manigault, commanding the post. They examined the works, recently readied to receive four mortars and eleven siege guns. Manigault's report criticized the position as being unprotected and having a narrow field of fire. Gonzales concurred with the observations. Both officers then dined with Captains A. N. Toutant Beauregard and Stephen Proctor. The following morning, the Cuban sent General Ripley a report suggesting that rifled ammunition, in short supply, be obtained for Battery Haskell. He mentioned that the Charleston Arsenal also lacked projectiles for various calibers of Brooke and Blakely guns, while the thirty-two-pounder and forty-two-pounder artillery in Fort Sumter had no shells. To alleviate the shortage, the Confederates were recycling "a large quantity of old shot and shell" fired by the Federals on Morris Island.[62]

Gonzales recommended to the commanding general on the twenty-eighth that the Brooke and other heavy rifled guns at Fort Sumter and Batteries Simkins and Cheves, bearing upon Union troops on Morris Island, be immediately replaced by eight-inch and ten-inch Columbiads, and redeployed in positions facing the sea. The recently built Battery Cheves, between batteries Haskell and Simkins, had just received five 8-inch naval guns. The chief of artillery emphasized that rifled guns could not withstand repetitious fire, and were better suited for close range shooting at ironclads or firing at distant wooden vessels. The Columbiads, whose shell fire was more rapid and destructive, had greater effect on wrecking enemy land guns and annoying their working parties. Gonzales stressed: "They will also answer better for general purposes against land attacks, flank attacks, barges, etc." He also based his rearrangement decision on the scarcity of rifled ammunition and the greater availability of Columbiad shells. The Cuban recommended the construction of an enclosed work in Fort Johnson, "mounting heavy guns, to be removed from another point" that "could be made to bear upon the channel, upon Morris Island, and upon the land approach upon the lines." This enclosure was never built. General Ripley, whose grudging disagreements with Gonzales had become more frequent, indicated that the heavy guns should not to be taken from Fort Sumter too indiscriminately because of the moral effect it would have on the garrison. Ripley erroneously assumed that "should Morris Island fall, Sumter becomes the salient point of our defense and, as it must hold out and repulse the enemy, too great a reduction of its offensive armament is deemed unadvisable." Beauregard agreed with Gonzales and ordered on the twenty-eighth that the six eight-inch Columbiads in Fort Sumter be removed, sending one to Battery Simkins, one to Battery Haskell, two to Battery Cheves, and two to the battery east of Fort Moultrie. He had planned for over a year to mount a new heavy gun at Fort Johnson. A week later, the commanding general instructed Gonzales to consult with Chief Engr. Col. David B. Harris for preparing at Battery Haskell "two guns on Columbiad carriages and six siege guns in embrasures," to fire on Morris Island.[63]

The James Island defense lines were inspected on August 4 by their former commander, Gen. Johnson Hagood, now in charge of Morris Island, and his replacement, Brig. Gen. William Booth Taliaferro, who had commanded Morris Island until July 26. Hagood considered the island's armament "a good deal out of fix," and wanted it organized forthwith. Taliaferro reported that the East and West Lines, stretching for seven miles, were "very weak and altogether too long." Taliaferro requested for his First Sub-Division of the First Military District "a chief of artillery, a competent and experienced officer," to replace Lt. Col. Delaware Kemper, who was on sick leave. The next day, Beauregard issued Special Orders No. 152, assigning Gonzales "to special service as chief of artillery on James Island" under General Taliaferro. On the seventh, the department commander approved the recommendation of Col. Charles H. Simonton, abandoning the obsolete East and West Lines and replacing them with a new two and one-half mile defense line, denominated the New Lines, from Dill's plantation house on the Stono River east to

Fort Lamar at Secessionville. These consisted of six batteries, numbered zero through five, were about one-half mile apart, and connected with *crémaillère* lines. Gonzales continued as department chief of artillery, with the extra duties of inspecting and improving the Fort Pemberton armament and the James Island defensive system. His first task was determining the placement of two rifled and banded forty-two-pounder siege guns at either Battery Ryan, on Mellichamp's plantation, or Battery Haskell.[64]

After touring Fort Pemberton, Gonzales recommended, on August 8, removing the four thirty-two-pounder guns on its left flank battery, which interfered with a concentrated fire on the southeasterly mainland approach, and moving them to dominant positions on the other parapets. He also proposed a traverse and merlons "to protect the guns on east curtain from an enfilading fire from below on the Stono." To further strengthen the river approach, the chief of artillery suggested anchoring a hulk, above the obstruction, with a guard of marines and a squad of artillerists manning a forty-two-pounder carronade available from the Charleston Arsenal. General Taliaferro approved the recommendations, Ripley forwarded them to Beauregard without comment, and the commanding general referred it to Chief Eng. Col. Harris. Gonzales sent two other reports that day to General Taliaferro, who approved them, recommending moving the twenty-four-pounder smooth-bore gun at Battery Haskell to the bend on the West Lines, sending to the Charleston Arsenal an old English twelve-pounder rifled siege gun to be banded, and installing a sight on the twelve-pounder Napoleon gun on board the steamer *Juno*. The latter, together with an identical gun, were later issued by Gonzales to the Palmetto Light Artillery Battalion, Company G, commanded by Capt. William Lambert De Pass, which had defended Battery Wagner during the assault on the eighteenth.[65]

The next day, Gonzales proposed better protection for the guns in embrasures at Battery Wagner and resubmitted his idea, presented to Pemberton a year earlier, of establishing siege guns on railroad platforms behind the James Island lines. He urged the quick deployment of moveable siege guns on two sets of railroad tracks to be located behind an embankment at the rear of batteries Ryan, Tatom and Haskell. The guns, pulled with horses or by hand, would cover Light House Inlet, Black Island, Morris Island, and all the approaches to Secessionville. Gonzales indicated that "such rapid changes in the position of our guns would also baffle the enemy's artillerists." Beauregard concurred with Chief Engineer Harris's opinion not to alter the sandbag arrangement shielding the Battery Wagner guns. The railroad track suggestion was "theoretically good, but practically impracticable, with our present means. I would be well satisfied if a common, good dirt road could be made in rear of our defensive lines."[66]

The chief of artillery was ordered by Beauregard on August 10 to move his headquarters from Charleston to James Island, and to quickly put into fighting condition the New Lines, whose guns were only to be fired under special orders from Beauregard's headquarters or from General Ripley. The Cuban settled in Royall's House, the telegraph-equipped headquarters of the First Military District, commanded by General Taliaferro. Pvt. Richard Jacques, Gonzales's clerk, described the place as "a neat little house surrounded by a grove of beautiful oaks." The one-story dwelling had small boat access to Charleston Harbor through James Island Creek. It no longer exists, but the dozen large oaks are still located at the end of Targrave Road. Gonzales again applied the next day for the services of Maj. Stephen Elliott Jr., who was ordered to report to Beauregard for special service. On the morning of the fourteenth, the chief of artillery, accompanied by Engr. Capt. William M. Ramsay, visited Batteries Ryan and Haskell. Major Manigault informed him of the failure to bring out the 3.5-inch Blakely guns.[67]

Three days later, recently installed Union batteries on Morris Island, five monitors and the ironclad frigate *New Ironsides,* began a week-long bombardment on Fort Sumter, firing 5,643 shots which destroyed most of the parapet and ramparts of the gorge wall, forcing the defenders to remove all the *barbette* guns which had not been dismounted by enemy fire. The gorge wall "would have crumbled to its level," had it not been for Gonzales's recommendation the previous month to reinforce the casemates and officers' quarters with "wet cotton bales, filled in between with sand." Eight rooms on the lower tier and nine on the upper floor had been packed to the ceiling in sixteen days. Union colonel John W. Turner, chief of artillery for the Department of the South, reported to Brigadier General Gillmore after the bombardment, that the gorge wall stood as a "mass of ruins," up to the floor of the second tier of casemates. He concluded that it "would have long since been entirely cut away," had it not been "for the sand-bags with which the casemates were filled, and which have served to sustain the broken arches and masses of masonry." Col. Alfred Moore Rhett, commander of Fort Sumter, declared that the fortification was "unserviceable for offensive purposes," being "impracticable to either mount or use guns on any part of the parapet." Beauregard reported that the gorge wall was turned into "a mass of debris and rubbish on which the enemy's powerful artillery could make but little impression" and that "Fort Sumter must be held to the last extremity."[68]

The scarcity of artillery ammunition continued to frustrate Gonzales and hamper his department's military operations. Beauregard requested on August 18 that the Ordnance Bureau in Richmond "exert every effort" to provide him with shells for eight- and ten-inch Columbiads, ten-inch mortars and thirty-pounder rifled guns. Gonzales and his superior were anxiously awaiting "two Blakely guns, carriages, and 60 tons of shot" which arrived the next day at Wilmington, North Carolina, on the blockade runner *Gibraltar*. At noon on August 19, the chief of artillery visited Battery Gregg and ordered Commanding Lt. James R. Pringle to cease firing his two guns because the malfunctioning fuses were doing little damage to the enemy on Morris Island and a shattered carriage was about to dismount one of the cannons. Gonzales was waiting for five thousand fresh fuses recently sent by train from the Richmond Arsenal. That same day, the Cuban applied again for an increase in rank. He informed Secretary of War James Seddon that his commission as colonel of artillery dated over one year; that he was senior artillery officer commanding 157 heavy, siege and light guns; and that the duties of his office would be best rendered if he had the assistance of a staff. Beauregard endorsed the application, indicating that "Col. Gonzales is active, zealous and intelligent in the discharge of his duties."[69]

Six days later, Gonzales informed President Davis of his application for promotion, to which he was eligible, since he had nearly double the eighty guns needed to form an artillery brigadier general's command. He pointed out that he was in charge of all artillery on James Island and the mainland, comprising Fort Pemberton, Fort Johnson, and Fort Lamar, and all the lines armed for the defense of Charleston. Gonzales expected the number of artillery pieces he had to increase when the works on the harbor and the Stono River were completed and supplied with guns. He closed his letter emphasizing, "Had I not devoted myself to the defence of this State where my family reside, I would have gone to Vicksburg with Genl. Pemberton and I am satisfied that he would have applied for me, advancement in rank." Seddon quickly responded, "I have no Brigadier's Command to assign to, if I should appoint." Davis delayed more than three months in sending the Gonzales letter to the secretary of war, affirming that the appointment of an officer to command the artillery at Charleston was justified, but he was noncommittal toward Gonzales. The issue remained unresolved for another year.[70]

The chief of artillery relentlessly pursued his duties on James Island. He relocated his office on August 20 from Royall's House to the Red Top House, usually called "the house at Bennetts," near the mouth of James Island Creek on the Lawton plantation. The dwelling, destroyed at the end of the war, had three floors, including the cupola, grand piazzas facing the harbor, high ceilings, wide hallways, a richly decorated interior, and multiple fireplaces. Gonzales's clerk, Pvt. Richard Jacques, frequently gazed at Charleston with a spy glass from the cupola, hoping to see his fiancée, Tute, promenading on the Battery. He was "strongly tempted" to row across the harbor "to get a kiss" from her, but feared getting caught without a pass. Jacques assured Tute that there was "no chance on my going to Morris Island."[71]

The chief of artillery visited Fort Johnson on August 20, and that afternoon inspected Battery Haskell with General Taliaferro. The next day, Gonzales sent a report to Lt. Col. George W. Rains, commanding the Augusta Arsenal, indicating that experiments on James Island with the sensitive Girardey Percussion Fuse had proven advantageous over other fuses "for firing upon earthworks, into thickets, woods, or rubbish, as well as against craft." He requested a large quantity of ammunition with Girardey fuses, for his "two very fine batteries of Blakely shell, $3\frac{1}{2}$ inch guns." The available fuses had been mostly defective. Thirty-seven percussion shells fired from a Brooke gun in Battery Beauregard on Sullivan's Island had failed to explode when hitting Union positions on Morris Island. The "often-reported" problem of defective fuses was still plaguing Confederate artillery six months later. The following night, after Brigadier General Gillmore threatened to fire on Charleston if Fort Sumter and Morris Island were not surrendered, a Union eight-inch Parrott rifle dubbed the Swamp Angel commenced an indiscriminate bombardment of the city with two-hundred-pounder incendiary shells. The gun, located in the tidal marsh near Black Island, burst three days later on its 36th shot. The chief of artillery was ordered by Beauregard on August 25 to report the number and type of additional heavy guns needed at Battery Ryan and Redoubt No. 1, which had just been armed with an eight-inch navy shell gun from the captured Union vessel *Isaac P. Smith*.[72]

Gonzales returned at 4 P.M. to Battery Haskell, inspecting the unfavorable condition of five gun platforms. Heavy rains the previous night had left some of their planks floating in water. The chief of artillery also viewed the thirty-pounder Parrott gun which had burst three days earlier and wrote a report of his visit the next day. A week later, Gonzales recommended that one of the two new seven-hundred-pounder 12.75-inch Blakely guns that arrived on the blockade runner *Gibraltar*, be installed on Sullivan's Island and the other one at J. M. Mikell's plantation on James Island. The move was opposed by Maj. Gen. Jeremy F. Gilmer, recently appointed as second in command to Beauregard, due to the risk of exposing the Blakelys to Union Parrott guns on Morris Island. Beauregard endorsed the objections and sent both guns to Battery Ramsay at the foot of Meeting Street in Charleston, to cover the inner harbor with a range of up to a mile and a quarter. Weeks later, one of the Blakelys burst "due to a want of forethought" by General Ripley. Neither gun was ever fired at the enemy. Gilmer, formerly an engineering instructor at West Point and until recently the chief of the Bureau of Engineers in the Confederate War Department, was "generally regarded as the ablest military engineer in Confederate service." The major general, who had earlier suggested using sandbags laid to an eight-foot thickness to cover the guns in the lower tiers of Fort Sumter, had been outdone by Gonzales's quicker solution of cramming wet cotton bales and sand into the casemates. Thus began a strained relationship in which, for the next nine months before returning to Richmond, Gilmer frequently overruled Gonzales on defensive matters.[73]

During the previous fifteen months, Gonzales had been appointed chief of artillery of the Department of South Carolina, Georgia, and Florida, with the rank of lieutenant colonel, and quickly ascended to colonel for meritorious service. He was senior artillery officer, commanding 157 heavy, siege, and light guns, including the Siege Train. The Cuban had reorganized and reinforced the crucial defense lines on James Island and the artillery ring around Charleston Harbor, which held up against massive Union attacks. He contributed to the defense of Battery

Wagner by requesting the mining of its approaches and having the defenders armed with double-barreled shotguns for close combat. His innovative suggestion of filling the Fort Sumter gorge casemates with cotton bales and sand prevented its complete disintegration by enemy shelling. Gonzales had double the responsibilities of most officers. He had twice temporarily held the position of chief of ordnance for seven months. Gonzales also held the dual role of department chief of artillery and General Taliaferro's chief of artillery on James Island, as well as occupying a forward defensive position during combat.

His work was hampered by bureaucratic paperwork, delayed shipments, and a shortage of lead, shells, fuses and artillery. His zealousness prompted personality conflicts with Generals Ripley and Jordan, Colonel Gorgas, and Major General Gilmer. In contrast, Generals Pemberton, Beauregard, and Taliaferro, and Governors Pickens and Bonham held the Cuban in high esteem. His dedication to his adopted state, the return of General Beauregard to Charleston, the advancement of the invaders, and the proximity of his family, prompted him to remain in South Carolina rather than accompany General Pemberton to Mississippi. Meanwhile, he fretted over the continuing Union advancement on Charleston, now by land instead of by sea.

## *Seven*

## HONEY HILL

During the first week of September 1863, as Union forces were about to take Morris Island, Gonzales moved his James Island headquarters from the Red Top House to McLeod's plantation dwelling, about two miles from Charleston. This house, built in 1854 by William Wallace McLeod, still stands today. Department paperwork continued consuming much of the chief of artillery's time. He had to micromanage special requisitions for equipping his new office with furniture, blank forms and writing supplies. Each request had to be certified as being correct and that "the articles specified are absolutely requisite for the public service." His clerk, Private Jacques, informed his fiancée that the move had "entailed much labor" and kept him "extremely busy" for two more days. Jacques regarded the department orderlies, privates F. W. Stender and Thomas B. Chaplin, as "dead heads when anything other than the usual routine of business occurs, even then they are not of much use." After the move, Gonzales had Private L. L. Cohen replaced in his office with a Private McCoy, and complimented Jacques for being "the clerk that has principally done its business, that keeps its books, *and alone* is conversant with all matters relating to the office from the day of its permanent organization." Jacques longed for a two-day pass, to visit his fiancée, who had moved to Aiken.[1]

On September 4, Beauregard summoned Gonzales to a staff meeting where it was determined to hold positions on Morris Island as long as communications could be kept open. Enemy forces were using Drummond calcium lights to assist in maintaining a round-the-clock siege and bombardment of Battery Wagner. Union "engineers and black infantry were employed exclusively on fatigue duty," while "groping their way forward among the torpedoes" previously deployed under Gonzales's recommendation. Most of the digging was done at night under a constant fire from Confederate artillery and sharpshooters. The sappers marked their progress with an American flag. Brig. Gen. Quincy A. Gillmore reported to Army Headquarters in Washington that the area in front of Battery Wagner "is filled with formidable torpedoes" and that his sappers had encountered eight in their advanced trenches. The explosion of six torpedoes caused a dozen Union casualties. One African American corporal who touched off a mine in the darkness was blown twenty-five yards away, "entirely naked, with his arm resting on the plunger of another torpedo." When discovered the next day, it prompted "the absurd story that the enemy had tied him to the torpedo as a decoy." The mines caused "considerable trouble and anxiety" for the Union forces, who failed to explode them by having their marksmen shoot off the plungers. The torpedoes had also killed two Confederate artillerists who had been warned not to venture out in front of the parapet and had gravely wounded one soldier who was planting them. Two days later, Beauregard decided to evacuate the island after Col. Lawrence Massillon Keitt, commanding Battery Wagner, informed that the shelling had "shattered large parts of the parapet" and Union trenches had reached the moat. The Confederates safely evacuated all but forty-six men from Morris Island, leaving behind thirty-six spiked artillery pieces and much ammunition in Batteries Wagner and Gregg. Brigadier General Gillmore credited part of his Pyrrhic victory to "the great mistake" of the Confederate engineer in not building the fortification "2 miles farther south, near Light House Inlet." He told his superiors, "The city and harbor of Charleston are now completely covered by my guns." As a result of his success, Gillmore was promoted to major general of volunteers.[2]

Gonzales telegraphed Beauregard from Royall's House on September 7, requesting permission to mount the ten-inch Columbiad in Fort Pemberton at Battery Pringle, on Dill's plantation, overlooking the lower Stono River. He also asked for a thirty-two-pounder and a forty-two-pounder in Fort Pemberton to be *barbette*-mounted at Battery Pringle.

Three days later, Beauregard's Special Orders No. 179 assigned Gonzales and Chief Engineer David B. Harris to a board of officers, presided by Maj. Gen. Jeremy F. Gilmer, "to determine the armament of the batteries of the New Lines on James Island," from Battery Pringle to Secessionville. They would also inspect the old lines and the upper Stono River defenses to "determine which guns may be transferred from the present Lines to the New Works and the batteries on the Upper Stono." The board advised no further platform removals from Fort Pemberton, gave a statement on the sixteen guns in Fort Lamar at Secessionville, the twenty on the East Lines, the fifteen on the West Lines, and called for the rifling and banding of a dozen smooth-bore guns of various calibers. The report proposed arming the New Lines and Forts Lamar and Pemberton with fifty-one artillery pieces, and the four earthworks in St. Andrews Parish with a dozen guns. Beauregard then approved on September 14 the board's recommendation delaying the movement of guns to the New Lines until the positions were ready. He ordered Gonzales to chose the order in which the dozen guns to be modified would be withdrawn.[3]

Three days later, Gonzales wrote to Lt. Gen. John Pemberton, who had been paroled to Atlanta after he surrendered at Vicksburg two months earlier: "Your heroic Defence has had all my admiration enhanced by your silent endurance of so much obloquy. It has been my pleasant duty to pay a tribute to truth and justice by meeting the numerous attacks made on you among people you strove so hard and so unpopularly to defend. Fort Sumter, the dismantling of which you have been accused to have counseled, *is* utterly and completely *dismantled* & laid in ruins by the enemy, & still Charleston is not his and it is not his because of Wagner, which you constructed, as you did Battery Bee, the obstructions & the works of the inner harbor, now stronger because in a great measure, of the guns *taken from Sumter*, of which ten 10-inch Columbiads due to your personal endeavors in Richmond contributed largely to the repulse of the 7th of April, and are contributing now to hold the enemy at bay. Although a friend of Genl. Beauregard, whose merit I recognize, I am still more the friend of truth."[4]

The chief of artillery concentrated his efforts on containing any Union advances from the Stono River on James Island. To strengthen the New Lines, he recommended on September 18 the rifling and banding of two eighteen-pounder guns and acquisition of an additional twenty-five ammunition chests. Beauregard quickly approved this plan, but three days later he admonished Gonzales for allowing Maj. Edward Manigault, commanding the Siege Train and Batteries Haskell, Tatom, Ryan, and Redoubt No. 1, to use solid shot at a great elevation instead of shells against Union positions on Morris Island, due to the lack of heavy mortars in the department. Gonzales returned on the evening of September 26 to Battery Haskell to supervise the construction of a bombproof and to inspect its strengthened parapet and new traverses. Three days later, he was overruled by Major General Gilmer, with Beauregard's approval, after proposing further armament transfers from the Old Lines to the New Lines. The chief of artillery immediately responded by recommending "the appointment of a board to determine the character to be given certain works on James Island." Gilmer rejected the plan, with Beauregard's consent, who added that the works mentioned were "under process of construction by the Engineers, and are of the proper description."[5]

Gonzales solicited the same day the return to James Island of "6-pounder iron field guns or 12-pounder iron field howitzers" from Maj. James Davis Trezevant, who commanded the Charleston Arsenal. He also applied for an assistant, requesting either Lt. Col. Joseph A. Yates, in charge of the artillery at Fort Johnson and Batteries Simkins and Cheves, or Col. Stephen Elliott Jr., who had been placed in command of Fort Sumter three weeks earlier. Beauregard replied that neither officer, nor any other, could be spared to assist him. He therefore relieved Gonzales "from the duty of inspecting the heavy artillery of the Department (except in the Sub-District of James Island)" for as long as he remained chief of artillery to General Taliaferro.[6]

In September, the Union raids into the lowcountry caused "pecuniary loss" for the Elliott family at Oak Lawn. When Gonzales heard that Gen. Beverly H. Robertson and three cavalry regiments were being sent to Adams Run, he wrote to his wife that "the family had much better go to Oak Lawn and save it from destruction" because the plantation would then be better protected than ever. Emily agreed, suggesting to her mother that "the negroes down at Chehaw might be brought to Oak Lawn to farm and make us comfortable." Gonzales expected "to get a house in St. Andrews Parish safe from shelling" to be with his brood. The chief of artillery went to Charleston on October 3 and acquired a passport for

Emily and Annie Elliott, which he sent to them in Columbia, allowing his sisters-in-law to travel to Oak Lawn. By now, the Gonzales-Elliott family had already invested $49,300 in war bonds, including $2,000 by Gonzales and $5,050 by his wife.[7]

During the first week of October, Gonzales received from the Charleston Arsenal, as ordered by Beauregard, two forty-two-pounder carronades for placement on James Island. The chief of artillery advised transferring an eight-inch Columbiad from Battery Cheves to Battery Haskell, to be used as a shell gun, which received Beauregard's approval. He then suggested "looking to the better defense of John's Island," but was overruled by Major General Gilmer, who insisted that it was "thought to be sufficient." Beauregard concurred, telling Gonzales that "it is not contemplated or advisable to extend our present defensive lines, which are already much too long for our available forces." The Cuban was vindicated two months later, when Union forces assaulted John's Island and seized there two Confederate heavy howitzers. Gilmer and Chief Engineer Harris were also reticent to suggestions made by Gonzales of transferring to the New Lines four thirty-two-pounders to Batteries 2 and 4 and two guns in Battery Reid to Battery 5. They recommended that "the general change of armament be not made until the lines are reported ready by the Chief Engineer of the Department." When Gonzales suggested improving certain roads on James Island, Gilmer and Harris replied that it "has been already ordered, and is in the hands of the Engineers."[8]

The relocation of guns to the New Lines progressed on October 9, with the transfer of an eight-inch shell gun from Secessionville, a thirty-two-pounder from Redoubt No. 1, and three thirty-two-pounders from the water side of Fort Pemberton. Gonzales reported that every redan and redoubt on the East Lines was armed except the two at Royall's House, and that a few minor changes, including the rifling and banding of some guns, would result in "a complete double line of defense for the east of James Island, cutting off by a double line all approach to the harbor, or to the western side of the island." Major General Gilmer replied that "the question of a line of defense is one for future consideration." Three days later, Gonzales went to Major Manigault's headquarters at Legare's Point and spent the afternoon examining Batteries Haskell, Tatom, and Ryan. Beauregard opposed his recommendation to exchange the thirty-two-pounder rifled gun at Battery Haskell for a rifled forty-two-pounder. He ordered the latter to be placed in Battery Pringle, next to a ten-inch Columbiad.⁹

The New Lines were finished on October 20 by some four thousand slave laborers. The 2,531 picket troops assigned to this 4,060-yard entrenchments and earthworks were Brig. Gen. Alfred H. Colquitt's brigade and two of Brig. Gen. Johnson Hagood's regiments, in addition to the artillery crews. The next day, the batteries along the eastern shore of James Island from Secessionville to Fort Johnson were reassigned artillery as follows: Redoubt No. 1, three guns; Secessionville (rear), four guns; Battery Ryan (right), four howitzers; Battery Ryan (left), two twenty-four-pounders and two eight-inch siege howitzers; Battery Tatom, six guns of various calibers; Battery Haskell, five guns, three howitzers and two forty-two-pounder carronades; Battery Cheves, two eight-inch Columbiads; Brooke Gun Battery, one eight-inch shell gun; Battery Simkins, two eight-inch guns and three ten-inch seacoast mortars; Bay Batteries (Fort Johnson), four Columbiads, three ten-inch and one eight-inch; Martello Tower Battery (four hundred yards west of Fort Johnson), six guns; Battery Wampler, two ten-inch Columbiads; Battery Glover, three thirty-two-pounder rifled guns; and Battery No. 5, two twenty-four-pounder smooth-bores and two twenty-four-pounder howitzers. Gonzales was instrumental in assigning the position of all the guns.¹⁰

The progress of the new fortifications on James Island was reported on October 20 by Union Maj. Gen. Quincy A. Gillmore to Army Headquarters in Washington. He desired executing "the original project of getting possession of the inner harbor with the monitor fleet," but questioned "whether to attempt to enter with the present monitor force, or await the arrival of the new ones." That same day, Gonzales rushed to the artillery post at Grimball's plantation, under attack by the Union gunboat *Pawnee*, which "fired two shots on John's Island & two on James Island," before retreating. The chief of artillery devised a plan to attack the ship during its next weekly incursion. He suggested a partial unmasking of Battery Pringle, to allow using its long-range guns on the gunboat; placing four siege guns in separate sunken masked batteries at Grimball's; stationing a light battery across from Battery Island; deploying on John's Island "the Palmetto Guard, with their light battery of steel rifled Blakelys, and the section of the Georgia Siege Train, with their twenty- and ten-pounder Parrotts, all easily moved or withdrawn"; and placing torpedoes in the gunboat's path. The plan was endorsed by Maj. James Jonathan Lucas and General Hagood and received Beauregard's approval.[11]

Also on the twentieth, Gonzales's clerk Private Jacques wrote to his fiancée that another monitor had joined the blockading fleet and that the frigate *New Ironsides,* attacked by the torpedo steamer *David,* was not sunk by a torpedo exploding under its bow. He indicated that Confederate batteries on James Island had "kept up an almost incessant firing upon the enemy's works on Morris Island, which, from all appearances, are very nearly completed." Jacques described how "the once beautiful fields of this Island present now one mass of formidable batteries and the quiet yeoman with a willing heart prepares of the soil he has so often tilled, engines of destruction. . . . Citizens of all states from every part of the confederacy are assembled here to drive the vile invader back or to die in the attempt."[12]

The successful Union advances on the Atlantic coast prompted on October 22 the rearrangement of the Department of South Carolina, Georgia, and Florida. Gen. Henry Alexander Wise received command of the Sixth Military District, which extended from south of the Ashley River and west of Wappoo Cut to Rantoles Station and the Church Flats Battery on the Stono River and had its headquarters at Adams Run. He replaced Gen. "Shanks" Evans, awaiting a court martial for disobeying orders. The Seventh Military District, under General Taliaferro, comprised two sections of James Island. Brig. Gen. Johnson Hagood headed the eastern division, and Gen. Alfred H. Colquitt was in charge of the western division. This further multiplied the duties of Gonzales, who was now chief of artillery of the department and of the Sixth and Seventh Military Districts. Five days later, Union artillery on Morris Island commenced an intense two-week bombardment of Fort Sumter, reducing to rubble the northeast sea face next to the ruins of the gorge wall.[13]

Meanwhile, Gen. William Stephen Walker, commanding the Third Military District, requested the return of the guns removed during the start of the fever season five months earlier from the batteries at Red Bluff and on the Ashepoo, Combahee, and Coosawhatchie Rivers. Walker had only eighteen artillery pieces and 2,168 effectives on duty. Gonzales replied that he had not been informed of the destination of the missing guns and stressed that "this case out of many illustrates forcibly the necessity for the chief of artillery being informed of the movement of every gun of every description in the department, so that when called upon he may furnish to headquarters the desired information, or, in other cases, make recommendations from knowledge in his possession." He added that until the strength of St. Andrews Parish was fully ascertained, after the transfer of the guns from the old lines, as approved by the Board of Officers, it would be "impossible to determine what guns and how many might be spared from the defense of Charleston for that of the Third Military District." Gonzales recommended that the artillery removed from Walker's district be returned. Beauregard indicated that the guns could not be placed in position until a sufficient number of artillerists and support infantry, unavailable at that moment, were obtained. The transfer of the guns to St. Andrews Parish was authorized on October 28, and the chief engineer was instructed to prepare the necessary platforms and traverse circles. In late October, other Gonzales proposals were approved by Beauregard: An eight-inch Columbiad was mounted at Fort Pemberton, and the siege train was withdrawn from the James Island lines and encamped as a reserve artillery. The Cuban also received a request from artillery major George Upshur Mayo in Alabama to serve under him again. It was granted by the War Office a week later.[14]

The chief of artillery went to Battery Haskell on November 2, detaching from there 2d Lt. Thomas Morritt Hasell, to be his assistant and inspector of artillery. His recommendation to transfer the forty-two-pounder carronade at the Charleston Arsenal to Battery Haskell, was approved by Beauregard. Gonzales left at Haskell a damaged twenty-four-pounder, commanding a creek, with orders not to be fired "with a greater charge than 4 pounds of powder." The same day, the Gonzales family and his in-laws returned to Oak Lawn from Flat Rock and Greenville, South Carolina. When they stopped in Charleston on the way home, Gonzales received "permission to spend a few evenings in the city with his family." Two days later, President Davis arrived in Charleston. He gave a patriotic speech at City Hall and the following day toured the James Island defenses with Governor Bonham, Generals Beauregard and Hagood, Colonel Gonzales, and others. The entourage had lunch at Royall's House and arrived at 1:00 P.M. at the Legare's Point headquarters, near Battery Haskell, where Davis received a military salute from the

South Carolina 27th Infantry Regiment, the Georgia 54th Infantry Regiment, and the Siege Train's Company A. Gonzales, Davis, and the officers observed with spyglasses the enemy positions on Morris Island. The chief executive inspected the New Lines on his return to Charleston. Gonzales was unable to bid him farewell in the city due to "pressing duties on James Island."[15]

During the next four days, the chief of artillery, accompanied by Generals Taliaferro and Hagood, continuously supervised Battery Haskell. A large slave force was building bombproofs there and at Batteries Ryan and Tatom. The laborers altered two gun chambers, added a second mortar platform to Haskell, and thickened the Tatom parapet facing the shore. During the following week, Gonzales inspected Fort Sumter, James Island, St. Andrews Parish and Secessionville, making recommendations for upgrading their defenses. He observed that while Battery Haskell was approachable only by one creek and was defended by twelve guns, the eastern and northern portions of Secessionville, accessible by several creeks, had only two thirty-two-pounders for its protection. Gonzales predicted a probable attack on Secessionville and proposed constructing heavy works on the weak sides "with numerous positions for siege and field guns" and moving there the two ten-inch mortars at Battery Haskell. He suggested removing the few twenty-four-pounders left on the East Lines, which were now nearly dismantled after the New Lines were erected, cutting down its redans to accommodate Napoleon guns, and moving part of the siege train to Royall's House. Beauregard endorsed and forwarded the plans to Chief Engineers Major General Gilmer and Colonel Harris. They replied that the engineer department had been engaged "for some time" in increasing the waterfront defenses of Secessionville and had "no apprehension" of its capture by the enemy.[16]

In mid-November, Beauregard issued the new daily passwords (officers' surnames) and countersigns (city names) to be used in the First, Fifth, Sixth, and Seventh Military Districts for the rest of the month, when the districts were rearranged again. The password assigned for November 20 was "Gonzales," and the countersign, "Nashville." As winter approached, the commanding general was "more concerned about the want of shoes, clothing, and blankets for our troops than about their want of food." He also complained to Richmond of lacking ten-inch mortar shells, which had forced his ten mortars to "have been almost entirely silent nearly a month." Three days later, Beauregard asked Gonzales to meet him at noon at his headquarters on the southwest corner of Meeting and John Streets in Charleston. They apparently discussed the frequent disagreements between Gonzales and the chief engineers. General Ripley was also strongly complaining to Beauregard that week about the work of the Engineer Department, their "slow progress of the works on Sullivan's and James Islands," and the interference, "carelessness and inattention of engineer officers." On the twenty-fourth, the Cuban requested reassignment to the "inclusive performance" of his duties as department chief of artillery. He indicated that his temporary assignment, "the reorganization and completion of the armament of James Island and St. Andrews," had been accomplished. Gonzales claimed that "the duties of the Chief of Artillery of the Department do not seem well to harmonize with those of Chief of Artillery to officers subordinate to the Commanding General." He therefore "considered impossible conscientiously to serve, at the same time, three different authorities without giving umbrage to some one of them." His conflict with the Engineer Department diminished after Major General Gilmer was transferred to Savannah a few weeks later. Gonzales then applied for and received a ten-day leave of absence, after four months on James Island, to spend Thanksgiving with his family at Oak Lawn.[17]

The Elliott women and Gonzales children had returned to Oak Lawn after the first frost, and left their field hands safely at White Hall, Abbeville County. At the Charleston railroad station, they assigned two servants to guard the baggage trunks, but one disappeared, containing "forty odd pounds of coffee, loaf sugar, five pounds of green tea & all of our little comforts." Emily complained of "rascality & extortion in high places. Rascality, theft, squalor in low places seems to be the order of the Confederacy." The Elliotts found that the soldiers encamped on their plantation "were beginning to make themselves at home thinking the place nicely given up for the war, if we had not come down, our home would soon have been a desert." The family faced numerous tasks: "cotton, potatoes & rice to be harvested, & no hands to work." The house servants "all turned out con amore" to assist. The Elliotts advertised in the *Charleston Mercury* to sell their "six horse power engine" at Chehaw plantation. William Elliott Jr. had been reassigned on November 11 to Georgetown, as conscript officer, "to keep the District free of deserters."

He was accompanied by his slave Frederick, but the new position left him "much depressed." Ralph Elliott, now a private with the Charleston Light Dragoons, was stationed at Accabee plantation, six miles north of Charleston on the east bank of the Ashley River. He suffered from "the exposure on picket duty, & from being obligated to eat cold food, every alternate 24 hours . . . leathery beef & sour meal which compose the soldiers food." His guard duty comprised stopping "deserters & negroes trying to force their way across all night." Mrs. Elliott informed Ralph that she wanted him near Oak Lawn, on the staff of General Robertson. He suggested that she or Gonzales could persuade Robertson "that I would be serviceable to him." A month earlier, Ralph Elliott had sold some of his cattle to General Robertson to feed the troops.[18]

Gonzales was reassigned by Beauregard on December 3, under Special Orders No. 258, to his post as chief of artillery of the Department of South Carolina, Georgia, and Florida. A week later, he moved his office from James Island to a two-story house at 46 Rutledge Street (today 184 Rutledge Avenue), in upper Charleston. All military offices and hospitals had been relocated to that area, out of range of the federal guns on Morris Island which for the previous six months had dropped 472 Parrott shells on the civilian population, killing an aged female slave, two women, an eighty-three-year-old man, and a fireman. The chief of artillery's voucher for commutation of quarters reveals that he was renting five rooms for seventy-five dollars monthly. The form indicates that Gonzales did not have a servant and was allotted eight cords of wood for fuel. The chief of artillery soon recommended the addition of platforms for the Siege Train on various works at St. Andrews Parish. He also made reference concerning guns to be used in an expedition. Gonzales was ordered by Beauregard on the twenty-third to put in position the guns intended for the Second and Third Military Districts, upon their arrival at the railroad depot nearest their destination. The Cuban returned home for the Christmas holidays, accompanied by his brother-in-law, Private Ralph Elliott, who had been assigned as his guide for fifteen days.[19]

While at Oak Lawn, the chief of artillery effected his military duties by courier and telegraph. He had another dispute on military procedure with Col. Josiah Gorgas, when the Confederate chief of ordnance ordered Col. John R. Waddy, the department chief of ordnance, to inspect the district's light artillery. Waddy, knowing that this infringed on the duties of the chief of artillery, referred the matter to Beauregard. Gonzales wrote to the commanding officer on the twenty-eighth, saying, "I think it proper to inform Col. Gorgas that the inspection of ordnance in batteries pertains to the Chief of Artillery and to send him a copy of General Orders 95, series of 1862: that he may know how matters are regulated herein that respect." Gonzales also protested to Beauregard about a request made by Lt. Col. Charles Colcock Jones Jr., the chief of light artillery in Georgia, who applied directly to Gorgas "for guns (Napoleons) for batteries in Georgia which should have been sent to me and not to Richmond." The Cuban also told his former schoolmate that his family sent their regards and that Oak Lawn had been "cleaned up and made to look bright for your reception." Brother William was unable to join them, being sick in bed at his post in Georgetown with a severe attack of nascent tuberculosis. The party was also attended by Col. Stephen Elliott Jr., commanding Fort Sumter, who later wrote to his cousin Emily that it had been similar to those of the old Norsemen, whose "feasts of their 'heroes' was a prominent feature in their economy, and this custom I see you are disposed to promote." He praised Gonzales for several suggestions of "great value" that were adopted in the arrangements for the defense of Fort Sumter.[20]

Two days later, Gonzales left Hattie in bed with a severe cold and neuralgia, to inspect the artillery platforms being erected on the Combahee and Ashepoo Rivers, that protected the advance toward the Charleston and Savannah Railroad. He had a conflict with Chief Engineer Col. David B. Harris, which was settled when General Jordan sent a circular on December 31, 1863, ordering that "hereafter, when either the Chief Engineer or of Artillery shall deem it proper to suggest any changes or improvements which would require a reference to the Chief of the other branch, he will consult with him before making the recommendation to these Head Quarters." The next day, Jordan reprimanded Gonzales for circumventing the chain of command by not addressing his dispatches to the chief of staff, and instead directly addressing Beauregard.[21]

New Year's Day 1864, with "a cold spell of unusual severity" and windy overcast skies, found Gonzales at Camp Jackson, near Jacksonboro on the west bank of the Edisto River, four miles from Oak Lawn plantation. The South Carolina Washington Artillery Battery, commanded by Capt. George H. Walter, was located there. The chief of artillery believed that the unit, two Napoleons and two ten-pounder Parrott guns, along with the Marion Light Artillery Battery stationed in the district, offset with discipline their numerical inferiority. While there, Gonzales wrote to Gen. Henry Alexander Wise, the district commander at Adams Run, suggesting placing in position the siege guns assigned to the works at Pineberry and Willtown Bluff on the South Edisto River. Two days later, he recommended to Colonel Harris that a mountain howitzer at the Charleston Arsenal be sent to the camp of Capt. Frederick C. Schulz's South Carolina Palmetto Light Artillery Battalion, Company F, at Willtown Bluff, a move approved by Beauregard. Returning to Oak Lawn, Gonzales wrote a report on January 3, advising defense improvements for the Sixth, Second, and Third Military Districts. General Wise visited Oak Lawn shortly thereafter to personally thank the Elliott ladies for their attentions toward him. This contrasted with Ralph Elliott's earlier opinion of the general, whom he described to his mother as "a rough, jolly, whiskey drinking, hard fighting, hard swearing old cove, who no doubt will be less esteemed by persons of refinement."[22]

The chief of artillery was back in his Rutledge Avenue office in Charleston by January 9, under inclement weather that hampered all local operations. The artillery headquarters was out of range of the Union guns on Morris Island, which had devastated the southern half of the city, up to Wentworth Street. Gonzales recommended that heavy guns be transferred to Red Bluff and Burnetts and that light artillery be shifted to strengthen the defenses of the Sixth, Second and Third Military Districts. Beauregard forwarded an eight-inch Columbiad to Red Bluff, withheld moving a rifled thirty-two-pounder from Fort Pemberton, and ordered that the requested six-pounder and twelve-pounder guns be sent "to the Districts named when the positions designated for them shall have been selected and the artillerists to serve them shall have been found." That night, Union artillery intensified their bombardment of Charleston, firing over one thousand shells in one week. Three days later, Gonzales returned to the camp of the Washington Artillery Battery, under a "weary spell of rain, cold and endless mud," and filed a report related to "the works on the Ashepoo and Combahee and to the 4.62 rifle siege guns at Green Pond." On January 20, he visited the Chehaw River defense lines in the Second Military District and requested other cannons for the Siege Train.[23]

On Friday, February 5, Union guns ceased firing on Charleston for the weekend, and there was "a comparative lull in operations on both sides." Two days later, Gonzales sent from Charleston an unofficial letter to President Davis regarding a recently passed bill in the Confederate Congress that regulated the Army General Staff but ignored its artillery department. Gonzales presumed that the only form of promotion would be upon the order of the president if an officer had a proportionate number of guns under his command. He enclosed copies of six 1863 documents with orders, endorsements and recommendations, to indicate the importance of the work he had been doing as department chief of artillery, and highlighted some of the major points in two paragraphs. The Cuban also emphasized, "As we, of the Artillery, cannot grasp a ten-inch Columbiad, as my old compatriot [Don Quixote] of La Mancha grasped his lance, to sally forth in quest of an adventure and opportunities for renown, it may be reasonable to expect of the public at large that, if some notice be taken of our keeping the enemy at bay for months and years of continuous labor and exposure, envy shall not construe it otherwise than as our share of public honor and reward." He sent his best regards to Mrs. Varina Davis and asked "that you will say to her that, should she again visit this State, she should not forget that she has friends in the 'Low Country' in a pleasant spot, from which our guns have, so far, with God's help, kept aloof our surrounding foe."[24]

Gonzales rightly attributed responsibility to his artillery for neutralizing the Union attempts to occupy of Charleston. The effective work of the harbor guns during the ironclad attack made it too costly for the North to ever attempt to fight their way in again. Union major general Gillmore had discarded that idea in mid-December. The defensive artillery organized by Gonzales had established fields of fire that rendered virtually impossible the advancement of superior Union forces, turning the conflict into a war of attrition. As a result of this stalemate, the ironclad fleet, which had been prematurely ordered to the Gulf of Mexico in April 1863, remained in South Carolina for another year. This delayed the scheduled attacks against Wilmington, North Carolina, and Mobile, Alabama, until a new class of ironclads became operational in the summer of 1864. However, Davis was unmoved by the Gonzales letter and routed it to Gen. in Chief Braxton Bragg "for perusal." When it was returned to the president, he delayed answering for three months, as with his previous

Gonzales correspondence, and tersely replied, "This being an unofficial letter . . . it requires no action that this office can take."[25]

Gonzales was appointed by Beauregard on February 10 to preside over a board in Charleston, with Chief Engineer Col. David B. Harris and Chief Ordnance Officer Lt. Col. John R. Waddy, "to determine what guns both in and out of positions in the Districts contiguous to Charleston, may be made available for works to be and those already constructed in the State of South Carolina, reserving if required a 42-pounder rifled gun for the State of Georgia." A few days later, Beauregard reviewed the troops in Pocotaligo. Thomas R. S. Elliott and his family, returning to their Bethel plantation, had the pleasure of seeing the commanding general and "recognized him immediately from the striking resemblance to Gen. Gonzales." At the end of the month, Tom went to Charleston at the request of Gonzales, to participate in the "greatest sale" of the Exporting and Importing Company of South Carolina, presided by William C. Bee, the Elliott family cousin and factor. The enterprise advertised in the *Charleston Mercury* a variety of dry goods, shoes and shoe thread, groceries, and cotton cards that had just arrived on the blockade runner *Alice*. The chief of artillery was "very busy" and could not fill the order requested by the Oak Lawn family. Tom bought a bottle of vinegar for two dollars, a pound of coffee for seven dollars, brown sugar for two dollars, and a pair of ladies' shoes for fifty-five dollars. He also got some calico dresses for his daughter and a few other necessary articles.[26]

The Cuban spent the first week of March inspecting the Fourth Military District and the district of Georgia, while Beauregard made a three-week tour of northeast Florida defenses after the Battle of Olustee. Gonzales left his assistant inspector of artillery, Maj. George Upshur Mayo, in charge of the Department Artillery Headquarters, attended by two temporary orderlies, Privates J. H. Kendall and J. H. Happoldt. The chief of artillery spent the last half of the month in Pocotaligo, accompanied by brother-in-law Ralph Elliott, inspecting defenses and "recommending the assignment and arrangement of Light Artillery." Part of his duties consisted in presiding over a board of officers, with Lt. Col. Delaware Kemper and Maj. Andrew Burnet Rhett, that submitted a defense report. While in the Pocotaligo area, Gonzales and Ralph visited the Bethel plantation of Tom Elliott, accompanied by Tom's son, William, who was temporarily detached from his unit to serve as their guide. Upon returning to Charleston, Gonzales received a detailed narrative of the Sullivan's Island defenses, prepared by Mayo on the twenty-ninth. The island had 71 heavy and light guns and mortars of various calibers, served by 1,065 artillerists.[27]

The Sullivan's Island works constituted a continuous parapet from Battery Bee on the western extremity, through Battery Marion and Fort Moultrie, to Battery Beauregard at the harbor entrance. Along the south beach there were four two-gun batteries, numbered one through four, spaced five hundred yards apart, between Batteries Beauregard and Marshall. Major Mayo found all in good working order, except Battery Kinloch, "with a damp magazine, and the implements and tackling for working the gun much abused." The inspector criticized "the want of care which infantry guards at batteries have displayed since I have been upon inspection duty in this department, and beg leave again to suggest the assignment of a noncommissioned officer of artillery at all batteries not garrisoned." Gonzales approved the report and suggested to the commanding general transferring from Sullivan's to James Island the nine ten-inch sea-coast mortars "to cover a field of fire ranging from Black Island, through the Swamp Angel Battery and Wagner, to Battery Gregg." Beauregard instead ordered an additional two mortars for Sullivan's Island and reducing those on James Island from eight to six.[28]

During the first week of April, Gonzales returned to the Sixth Military District to inspect the artillery defenses and adapted a telescopic sight to a rifle cannon. He then asked the commanders of the First, Fifth and Seventh Districts to designate a rifle gun for a telescopic sight. On April 7, Gonzales attended a funeral mass held at St. Joseph's Church for General Beauregard's wife, who had passed away the previous month in federal-occupied New Orleans. Four days later, the chief of artillery informed the commanding general of the need of light artillery for the defenses of the coast and requested that "application be made to Richmond to supply deficiencies in heavy, siege and field guns." On the sixteenth, Secretary of War James Seddon requested a list detailing the number of guns in the Department of South Carolina, Georgia, and Florida. Gonzales drafted the report, listing 622 guns under his supervision: 472 in position and 150 in moveable batteries of light and siege artillery. He also asked Seddon for "a uniform rule of action and definition on the rights, privileges and authority of Chief of Heavy & Light Artillery of all armies and Departments."[29]

The War Department had recently ordered to the Virginia front about half of the South Carolina cavalry, consisting of three regiments and five companies, and one cavalry regiment and three companies from Georgia. It also reassigned Maj. Gen. Jeremy F. Gilmer from the command of Savannah defenses to his former duties as chief of the Bureau of Engineers in Richmond. He was replaced by Maj. Gen. Samuel Jones, a West Point graduate and instructor who was Beauregard's chief of artillery at the First Battle of Manassas. Beauregard complained to Gilmer that the War Department had not promoted any of his own officers in over a year in spite of his recommendations and was sending permanently their least esteemed officers to South Carolina. He then accepted an offer from Gen. in Chief Braxton Bragg to head the new Department of North Carolina and Southern Virginia against an expected Union offensive. After Beauregard left Charleston on April 20, he was substituted by Jones. Maj. Gen. Lafayette McLaws received control of the Savannah defenses. Stephen Elliott Jr., commanding Fort Sumter, was promoted to full colonel and given command of the Holcombe Legion, going to Virginia. A month later, Elliott became brigadier general and headed Evans's brigade after the capture of Brig. Gen. William Stephen Walker at Petersburg, Virginia. Ralph Elliott, who had been detached for service at artillery headquarters, informed his mother that he and Gonzales were staying in Charleston "to post Jones in artillery affairs for the present." Mrs. Elliott supplied them with food from Oak Lawn. William Elliott Jr., had been transferred that month as enrolling officer to the isolated Midlands town of Manning in Clarendon District. He regarded his work as "dull & monotonous," and resented having to enroll free blacks.[30]

The following month, Gonzales requested that Lt. Col. John J. Garnett, awaiting orders in Virginia, be assigned to light artillery duty in his department. The 150 light and siege moveable batteries that remained after the fall of Morris and John's Islands were commanded by Lieutenant Colonels Delaware Kemper and Charles Colcock Jones Jr. Kemper soon left for the war front in his native Virginia and Garnett was ordered to report to Gen. Robert E. Lee, engaged in the Spotsylvania campaign. Gonzales then inspected Capt. William Lambert De Pass' Palmetto Light

Artillery Battalion, Company G, composed of two Napoleon guns and two twelve-pounder howitzers, requested quinine for the artillerists to ward off fevers during the upcoming summer season, and filed a report and requisition on the department's artillery needs. Maj. Gen. Samuel Jones forwarded the request for additional guns to Colonel Gorgas in Richmond. The Cuban went to the Adams Run district headquarters on at least two occasions, visiting his family at their summer home in the village. He was delighted to once again see Hattie, who was four months pregnant, their children, and his in-laws. Upon returning to Charleston, Gonzales found that all of his assistants and clerks except Private Richard E. Jacques had been detached by Major General Jones for a new infantry battalion on James Island. Jones had also included his own clerks and those of other staff officers to confront a May 22 Union landing on James Island. Another two thousand troops from the department had just been ordered by General in Chief Bragg to the Virginia front. Gonzales had to now transact the entire business of the Artillery Department with only one clerk, and requested four days later that Jacques "be not ordered away from the city on military duty."[31]

Federal forces in South Carolina were also reassigned during May. Major General Gillmore, ordered to Virginia, was replaced by Maj. Gen. John G. Foster, the former chief engineer of the Charleston Harbor fortifications, who incremented coastal raids. Lt. Gen. Ulysses S. Grant had ordered the demonstrations around Charleston to keep Confederate forces concentrated there and prevent their reenforcing the Virginia front. Charleston was being impacted daily by about thirty Parrott rifle shells and Fort Sumter was subjected to a slow mortar fire from federal artillery on Morris Island. A recent Union attack had destroyed the two-gun battery at Pineberry on the Pon Pon River. In early June, Yankee batteries on Light House Inlet and Long Island fired over seven hundred shells at Secessionville, demolishing the cabins, shanties and other camp buildings. Gonzales spent that week inspecting the defenses of the Combahee River and the Second and Sixth Military Districts, making artillery recommendations and requesting that "orders be issued to ascertain if defects affecting important points on James Island have been remedied."[32]

The Civil War again affected the Gonzales and Elliott families in June. Thirty-year-old Ralph E. Elliott, captain of the Palmetto Guard, South Carolina 2d Infantry Regiment and brother of Brig. Gen. Stephen Elliott

Jr., died on June 6 while defending Richmond. Then, on June 15, Gonzales and his brother-in-law Ralph Elliott received in Charleston the "most horrible and distressing news" that their brother-in-law, sixty-year-old Andrew Johnstone, had been murdered by Confederate deserters in his *Beaumont* estate at Flat Rock five days earlier. This shocking incident was reported during the next ten days in five articles in the *Charleston Courier* and once in the *Mercury*. Johnstone had just finished dinner with his family when a servant informed him that there were six soldiers at the back door asking for food. He invited them to the dining room table, but when the men showed nervousness and behaved suspiciously, Johnstone told his fifteen-year-old son Elliott to unlock the gun cabinet and secure the door leading to the ladies' chambers. After eating, the guests drew their guns on their host attempting to take him prisoner. As Johnstone went for his pocket .22-caliber pistol, he was shot in the abdomen but managed to empty the weapon at the fleeing assailants. Elliott responded by firing a double-barreled shotgun, hitting four of the attackers and mortally wounding two. Johnstone was carried by his servants to bed, where he passed away forty-five minutes later in the company of his grieving family. He was interred in the St. John-in-the-Wilderness Episcopal Church graveyard. A week earlier, Major General Jones had ordered a secret expedition of forty soldiers with hound dogs to exterminate the ruffians operating in that mountainous region.[33]

Gonzales, unable to attend the Johnstone funeral, was busily arranging the disposition of light batteries on the coast. On June 28, he was ordered by Major General Jones to renew his earlier requisitions on the Ordnance Department in Richmond "for 12-pounder Napoleons and 10-pounder Parrotts." Two days later, the chief of artillery prepared a list of guns for which positions were to be made ready in the Second Military District. The war of attrition continued between Confederate artillery on James Island and Union guns on Morris Island. Charleston endured shelling "thrown at random, at any and all hours, day and night," at points remote from each other. Major General Jones notified his Union counterpart that the bombardment had "never suspended for an instant, the labor on or in any military or naval work, factory foundry, arsenal, or depot of supplies," and the only persons killed were an octogenarian man, women, and children. During the first weekend of July 1864, the Northerners launched failed assaults on White Point on North Edisto,

Fort Johnson, and the New Lines on James Island, were defeated at Burden causeway on John's Island, and built a double line of entrenchments on Legare's Island, some three thousand yards from Battery No. 2 on the New Lines. The temporary occupation of a part of John's Island allowed them to enfilade the Confederate works across the Stono River on James Island. Throughout the eighth, six federal naval vessels kept an "incessant cannonading" on Battery Pringle, commanded by Maj. Ormsby Blanding, South Carolina 1st Artillery Regiment, turning it into a "slaughter-pen" and causing a "good deal of damage." All the heavy guns, except a smooth-bore, were disabled. Two days later, a Union attack on Batteries Haskell and Simkins was repulsed.[34]

Gonzales spent the first fortnight of July strengthening Battery Pringle, the New Lines (by then called the West Siege Lines), and the James Island defenses. Their disabled guns prompted Major General Jones on July 11 to inquire from Col. Josiah Gorgas about the status of the artillery requisition that Gonzales had made two months earlier. The commanding general asked the Confederate chief of ordnance for his particular attention and prompt reply. Ten days later, Gonzales accompanied Jones and his entire staff to St. Paul's Church, for the funeral of Capt. John C. Mitchel, who had been mortally wounded while commanding Fort Sumter. That same day, Jones issued Special Orders No. 184, instructing Gonzales to travel to Richmond and expedite the delayed artillery order. Maj. George Lamb Buist, commander of the Georgia Siege Train, was temporarily assigned to cover the duties of the chief of artillery. Before departing, Gonzales requested the banding and rifling of the eight-inch Columbiad "Old Pattern" at Battery Tynes, on Dill's plantation, northwest of Battery Pringle, and to substitute it with a ten-inch Columbiad. After arriving in the Confederate capital, Gonzales visited his nephew William Elliott, recuperating from a hernia operation. The chief of artillery requested that Elliott, who had a six-month certificate of unfitness for cavalry service, be detached from Capt. William L. Trenholm's Rutledge Mounted Riflemen, during that time and reassigned to duties in his Charleston office, where he had been "left but one clerk and an orderly." The transfer was quickly approved.[35]

While in Richmond, Gonzales wrote on August 13 to Adj. and Insp. Gen. Samuel Cooper in reference to his promotion application of the previous year. The Cuban indicated that for two years he had been senior colonel and department chief of artillery and commanded over 620 guns. He indicated that lesser ranking artillery officers in his department had been promoted based on the strength of the number of guns under their control, and requested advancement on the same principle. Among the recommendation letters sent by his superior officers, Beauregard affirmed that during the nineteen months he headed the Department of South Carolina, Georgia, and Florida, Gonzales "displayed great zeal, energy and intelligence, in the discharge of his laborious duties, thereby entitling himself, in my opinion, to promotion." Another from John Pemberton to Secretary Seddon highly praised Gonzales for his experience, command and great responsibility, and asked his promotion to the rank of brigadier general. Pemberton, who had capitulated with his twenty thousand men at Vicksburg in 1863 and had recently resigned his lieutenant general's commission, was still held in high esteem by President Davis's, who gave him the rank of lieutenant colonel in the artillery corps. Beauregard's recommendation was insignificant, since a recent entry in Col. Josiah Gorgas's diary indicated that Davis was "strongly prejudiced against Beauregard." Although the Cuban's hopes to achieve a generalship were dashed for the fifth time, he did not lose enthusiasm for the Confederate cause.[36]

The chief of artillery remained in Richmond until mid-September, dealing with the obstinate Colonel Gorgas and the government bureaucracy to fulfill the order for twenty mortars placed four months earlier. Gorgas, who that summer had been treating a facial neuralgia with opium and quinine, omitted from his diary Gonzales's two-month visit. Gen. Samuel Jones had to further prod Gorgas on September 2 with a lengthy telegram, requesting a large supply of projectiles for rifled artillery. The fall of Atlanta that day to U.S. general William T. Sherman opened the way for Union troops to advance through central Georgia and South Carolina. Gonzales quickly returned to Charleston, discussing the situation on September 19 with a board of artillery officers convened on Sullivan's Island. That morning, Confederate mortars retaliated "severely" against Union guns that were firing shells into Charleston every fifteen minutes, after Brig. Gen. Rufus Saxton, a radical abolitionist, had declared to Major General Foster his intent to destroy the city.[37]

The following week, Gonzales spent three days with General Beauregard inspecting the harbor and city batteries and those on Sullivan's

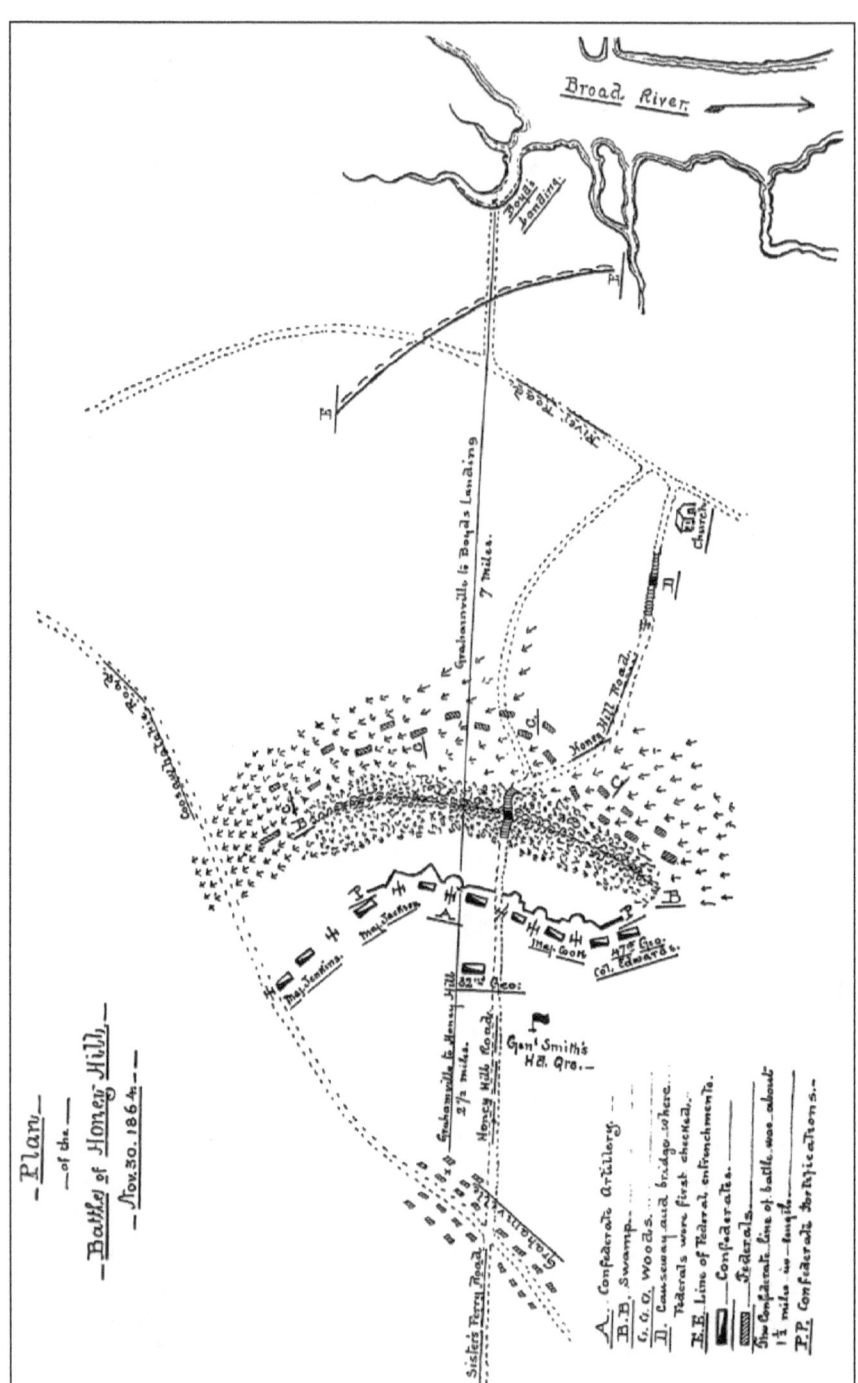

Plan of the Battle of Honey Hill, November 30, 1864. Duke University Library

Island. Beauregard had just returned on the twenty-fifth from a successful campaign in Virginia. President Davis then instructed Beauregard to proceed to Augusta, Georgia, where on October 2 he received command of a new department called the Military Division of the West, headquartered in Jacksonville, Alabama, extending from Georgia to the Mississippi River. Three days later, command of the Department of South Carolina, Georgia, and Florida passed to Lt. Gen. William J. Hardee, who appointed Gonzales as his chief of artillery. The staff included Maj. Cleland Kinloch Huger as chief of ordnance and Medical Director Thomas L. Ogier, the physician who had attended the dying William Elliott. Gonzales kept his headquarters at 46 Rutledge, where on October 10 he permitted Joseph Quash, a free black, to operate a barbershop out of one room. Quash published a card in the *Charleston Mercury,* informing his patrons of his new location.[38]

The chief of artillery spent the last half of October occupied with inspecting department defenses, while a yellow fever epidemic proved fatal to over twenty people daily in Charleston. He provided ten six-pounder iron guns for the flanks of a new bastion earthwork on John's Island, and "two twelve-pounder iron howitzers to replace 24-pounders moved from Secessionville." Four other iron field pieces were sent to Battery Marshall for flank defense and a rifled and banded thirty-two-pounder gun at Fort Pemberton was exchanged for an eight-inch smoothbore Columbiad at Battery Bee. On the seventeenth, Hardee ordered the rearrangement of the seven military districts in the state into five subdistricts under the new District of South Carolina. The next day, Gonzales wrote to Colonel Gorgas, chief of ordnance in Richmond, requesting the twenty eight- and ten-inch mortars previously ordered for the department. He indicated the urgency to "silence Battery Gregg" if Fort Sumter was subjected to another serious bombardment, and to impede a possible naval incursion into Charleston Harbor. Gonzales believed that an additional thirty mortars were needed to protect Battery Marshall, John's, Sullivan's, and Long Islands, and for Whitmarsh Island, near Savannah. In the midst of this activity, Gonzales received the news that Hattie had given birth on October 22 to their first daughter, Gertrude Ruffini Gonzales. The widowed Mary Johnstone wrote to Mrs. Ann Elliott that "a daughter will be an especial trial to you all. Gertrude is too harsh a name. I think Anita or Annita much prettier." Two of the Cuban's aunts were named Gertrudis and Anita. Ruffini was his mother's maiden name. The following day, the chief of artillery ordered that a 4.62-inch rifled and banded siege gun at Green Pond be placed in position on the Combahee River. He also recommended preparing two gun positions on the Charleston New Bridge.[39]

In early November, Gonzales turned his attention to "the precarious condition of the defence of the 4th Military Sub-District of South Carolina." Facing "a protracted and desolating war," on the tenth he submitted to General Hardee his 1861 plan for transporting his artillery Siege Train on fifty platform railroad cars. They would be pulled by two locomotives that he had already procured from Robert L. Singletary, the president of the Charleston and Savannah Railroad. The proposed flying column, similar to the one used by the French to conquer Algeria, would be composed of two thousand men and two artillery batteries stationed at Green Pond, a central point between Charleston and Savannah. He suggested using "the two best companies in this State," the Washington Artillery Battery and Capt. Henry Middleton Stuart's Beaufort Volunteer Artillery. Guns and ammunition would always be kept on the cars and special ramps would facilitate unloading at any point. From Green Pond, Gonzales indicated, the flying artillery could reach by train within thirty minutes to two hours any one of what he deemed the four probable landing points for a large enemy force: "White Point on North Edisto, Field's Point on the Combahee, Mackays Point on Graham's Neck, and Boyd's Landing on the Broad River."[40]

As winter approached, Lt. Gen. Ulysses S. Grant ordered ending the artillery fire on Fort Sumter, unless it prevented establishing new batteries there, and limited the firing on Charleston "to an occasional shell," to "economize ordnance stores." Brig. Gen. John P. Hatch, commanding Morris Island, concurred that further battering of Fort Sumter was "an idle waste of material." Meanwhile, in mid-November, U.S. general William T. Sherman began his devastating march to the sea from Atlanta to Savannah with 60,958 troops. He ordered Maj. Gen. John G. Foster, commanding the Department of the South, to send troops inland to cut the Charleston and Savannah Railroad around December 1. This would impede Confederate reinforcements and supplies reaching Savannah from South Carolina and would eliminate the only escape route open to Hardee's army if they decided to retreat. Foster asked Brig. Gen. John P.

Hatch, commanding Morris Island, to pick the place to strike, and Grahamville was chosen. At 1:00 A.M. on November 27, Hardee received in Savannah an erroneous report that Union cavalry had crossed the Savannah River into South Carolina twenty miles below Augusta. He immediately wired Maj. Gen. Samuel Jones, the recently appointed District of South Carolina commander in Charleston, "Send Gonzales to me for a few days," and urgently asked for entrenching tools and any available batteries of light guns. The Cuban departed for Savannah by train the next day, after submitting a report to Hardee for the reorganization of light artillery in the department, with several recommendations for promotion.[41]

On the night of the twenty-eighth, Major General Foster departed Hilton Head with his entire available command, consisting of two brigades totaling five thousand infantry, cavalry, and artillery, plus five hundred sailors and marines. The first brigade, led by Brig. Gen. Edward E. Potter, included the 56th, 127th, 144th, and 157th New York, 25th Ohio, and the 32d and 35th U.S. Colored Troops. The second brigade, commanded by Col. Alfred S. Hartwell, consisted of the 26th and 102d U.S. Colored Troops and the 54th and 55th Massachusetts Infantry Regiments. Lt. Col. William Ames commanded Batteries B and F, 3d New York Artillery, and Battery A, 3d Rhode Island Artillery. Comdr. George Henry Preble headed the naval brigade and Capt. George P. Hurlbut was in charge of four companies of the 4th Massachusetts Cavalry Regiment. The expedition headed for Boyd's Landing twenty miles up the Broad River to cut the Charleston and Savannah Railroad at Grahamville. Gonzales had predicted weeks earlier that this would be one of the four places used for an enemy disembarkment. The troop transports, protected by six gunboats, started sailing toward the Broad River, but were soon engulfed in a dense fog. Most vessels dropped anchor, but two or three that continued either grounded or mistakenly went up the Chechesee River. When the first federal troops reached Boyd's Landing at 11:00 A.M., the lone Confederate sentry hurried to the Grahamville headquarters of Col. Charles Jones Colcock, commanding the Third Military District since June. The only defenders available were a portion of one squadron of the South Carolina 3d Cavalry Regiment. The assistant adjutant general, Lt. E. W. Fraser, sent a messenger to Colcock, who the previous day had gone sixty miles to Matthew's Bluff to superintend the construction of breastworks to obstruct Sherman's army. Fraser also telegraphed Lieutenant General Hardee at Savannah, Major General Jones at Charleston, and Maj. John Jenkins at Pocotaligo.[42]

The unopposed disembarkment was led by the Marine brigade under Commander Preble. Three battalions, dragging eight howitzers by hand, advanced two miles west on Boyd's Avenue, presently Salisburg Road, to where it ends at the road now called Highway S.C. 462. Preble had been ordered by Brigadier General Hatch, who commanded the land force, to wait for the army after encountering the first crossroads. Since this was a "T" intersection, not the four-way crossroads depicted on Hatch's crude map, Preble was unsure of his position. Leaving his force behind hastily erected artillery abatis on opposing sides of the road, Preble, with his adjutant and fifteen men, proceeded two miles north, on what is now Highway S.C. 462, in search of the crossroads. They were abruptly halted by a twelve-pounder canister blast from behind a Confederate picket line at Bee's Creek, manned by Capt. H. C. Raysor and Company E, South Carolina 2d Cavalry Regiment. After a briefly exchanging rifle shots, Preble's squad retreated to the intersection, leaving behind a wounded comrade. Preble regrouped his force, which now included the recently arrived 32d U.S. Colored Troops Regiment, and at 4:00 P.M. advanced two miles up the same road, and began encamping. Just then, Brigadier General Potter arrived and told Preble that Grahamville was in the opposite direction. The three naval brigades then pulled their artillery back to the "T" intersection, where they were ordered to rest and spend the night.[43]

Brigadier General Hatch arrived on the scene and decided to head south immediately on present-day Highway S.C. 462 with General Potter's First Brigade and the men from the 4th Massachusetts Cavalry Regiment. Major General Foster had returned to Port Royal at 4:00 P.M, while the debarkation proceeded. Other troop transports arrived that night and the following day. One mile south of the "T" intersection, the Union vanguard encountered on their right, the Grahamville Road (today Highway 278), leading five miles west to Grahamville. At the southwest corner was Bolan's Church, founded in 1860, now the First Euhaw Baptist Church, built on the same site in 1971. Their guide, a runaway former slave, purported to recognize the area and led the column four miles further south, away from Grahamville and into a marsh. The guide was described by a *New York Times* correspondent as "either faithless or ignorant." Hatch, determining that they were heading in the wrong direction, ordered a countermarch to Bolan's Church. The exhausted troops, who had been awake the previous night on the transports, bivouacked on the church grounds at 2:00 A.M. If the advance had not been led astray, they would have reached Grahamville with little or no opposition.[44]

After the initial skirmish was reported to Maj. John Jenkins, he arrived in Grahamville at 4:00 P.M., wiring Lieutenant General Hardee in Savannah and Major General Jones in Charleston: "Ten Gunboats, with transports and barges at Boyd's. Landing troops near Grahamville. Four Gunboats coming up Broad River to Mackays Point, which is the approach to Pocotaligo and Coosawhatchie. Reinforcements needed." Jenkins then went to Coosawhatchie, ordered two guns from Kanapaux's artillery to Grahamville and another force to Bee's Creek. Hardee asked Jones to send the Georgia 32d Infantry Regiment from Charleston to Pocotaligo. He had Gonzales answer Jenkins that he had just ordered Maj. Gen. Gustavus Woodson Smith and his First Division of Georgia Militia, riding on two trains from Macon to Savannah, to be routed to Grahamville, and would arrive there the next morning. Smith's forces had been falling back from Atlanta, skirmishing with Major General Sherman's advance, which had defeated them at the battle of Griswoldville on the twenty-second. Gonzales also told Jenkins that local troops should check and delay the enemy's advance to prevent their occupying the railroad before Smith's arrival. He also urged them to have horses ready for the general and his staff. Jones replied from Charleston that Jenkins should expect a reinforcement regiment at the Grahamville depot at 8:00 A.M. the next day. One thousand South Carolina reserves, led by Gen. James Chesnut Jr., were sent from Augusta to report to Hardee.[45]

Colonel Colcock was notified by courier of the federal advance at 6:00 P.M. on November 29. Returning to Grahamville, he left a message in Robertsville to postpone his wedding, scheduled for the following day. He arrived at his district headquarters at sunrise the next day, after an all-night equestrian ride. The telegraph operator gave Colcock the messages received, the last one indicating that General Smith and half of his division, the State Line Brigade, composed of the Athens and Augusta reserve battalions and the First Brigade of Georgia Militia, had passed Hardeeville by train on their way to Savannah and were soon expected at Grahamville. The exhausted Georgia Militia had arrived in Savannah at 2:00 A.M. on the thirtieth. Upon receiving the order to proceed immediately to South Carolina, Smith went to General Hardee's headquarters to protest the illegality of sending the Georgia Militia to another state. Hardee and Gonzales, who had known Smith since the 1848 filibuster events, convinced Smith that if Sherman's troops destroyed the Charleston and Savannah Railroad, they would cut off the only evacuation route available to the Confederate army in Savannah. The Cuban was ordered by Hardee immediately to accompany Smith to Grahamville.[46]

The vanguard of Georgia reinforcements arrived in Grahamville by train about 7:20 A.M. on Wednesday, November 30. Gonzales introduced Smith to Colonel Colcock, his adjutant general captain Louis DeSaussure, and Major Jenkins. Colcock, the founder of the Charleston and Savannah Railroad, had owned three local plantations. Smith had him select a defensive position while awaiting the second train with the remainder of his 1,500 troops. Colcock chose the Honey Hill earthworks, built in 1861 by Col. Thomas Lanier Clingman. The fortifications are located on both sides of Highway 278, 2.7 miles northwest from Bolan's Church. They consist today of a 25-yard-long, 2-foot-deep entrenchment 50 yards south of the road, which terminates in a dense swamp. The main bastion, on a 25-foot escarpment, is located 100 yards north of the highway, in a wooded area, behind a small stream. It is a semicircular earthwork, presently three feet high and three hundred yards in perimeter, with a 100-yard rear opening. Contemporary maps placed it at a bend on the Grahamville Road, which has since been rerouted. Stretching northward from the earthwork, there is now a 30-inch-high rifle pit, curving northwest for miles, to the road from Bee's Creek to Grahamville, which is today's Highway S.C. 13.[47]

Smith later wrote in his battle report: "Colonel Colcock, the commander of the district and next officer in rank upon the field to myself, was assigned to the immediate executive command of the main line; Colonel Gonzales was placed in charge of the artillery, and Major Jenkins of all the cavalry; Captain DeSaussure, who was thoroughly acquainted with the whole country, remained near me." Gonzales, who three years earlier spent over a month in the area supervising the construction of defenses, directed the deployment of the guns. Due to its defensive characteristic, the artillery was assigned the principal role, while the troops were placed in subordinate protective positions. Gonzales made sure that the batteries were not encumbered by musketry defense within their limits, as the proper place for riflemen was on the flanks, in sufficient numbers to prevent the artillery being overrun. He verified that the five guns were spread out enough so that a successful enemy charge would not capture them all at once. He checked that the cannons had a good sweep over the approach route, covering the ground intervening between them. Limbers and caissons were positioned to provide easy access to ammunition. The Cuban reviewed the quality and quantity of available shells, gunpowder and friction primers. He inspected the bronze twelve-pounder Napoleons for flaws and scarcely perceptible cracks, any weakness in the vents, or warped trunnions, and the proper function of the carriage machinery. Any malfunction, exacerbated by continuous firing, could cause major damage and inoperability. The chief of artillery assured that no other troops were positioned behind the guns, where they would be exposed to the enemy return-fire. He then gave a complete report to General Smith at his headquarters one-half mile behind the lines.[48]

A courier arrived after 8:00 A.M. saying that a federal column had just departed Bolan's Church moving toward them. Colcock and his staff sallied to meet the challenge. Accompanying them were Company K of the South Carolina 3d Cavalry Regiment, under Capt. William B. Peeples, and a twelve-pounder Napoleon from Capt. J. T. Kanapaux's South Carolina Lafayette Artillery Battery, commanded by Lt. C. J. Zealy. The gun was placed on a slight elevation, behind a bend on the Grahamville Road, covering the causeway over Euhaw Creek, one and one-half miles from Honey Hill near the entrance of today's Good Hope Plantation. Peeples's cavalry dismounted and deployed on the right as skirmishers, across a large field with tall broom grass, intersected with numerous canals. The left flank contained an impenetrable swamp.[49]

Before leaving Bolan's Church, General Hatch sent two mountain howitzers back to hold the "T" intersection, with four companies of the 54th Massachusetts Infantry Regiment. This force soon repulsed a Confederate attack emanating from the Bee's Creek picket line. General Potter advanced on the Grahamville Road with four regiments, the 127th, 144th, and 157th New York and the 25th Ohio, without scouts or flankers. While crossing the Euhaw Creek causeway at 9:15 A.M., the first shell from Zealy's gun cut a nine-casualty swath through their lines.

Before the panic subdued, the Confederates fired three more shells, the last one killing thirteen men. Their opponents responded with two guns of Battery B, 3d New York Artillery, under Lt. Edward A. Wildt. The bluecoats formed a line of battle and marched across the field to flank the Confederate position. Colcock ordered staff officer Capt. William Waight Elliott, the cousin of Brig. Gen. Stephen Elliott Jr., to burn the broom grass. A favorable wind fanned the flames across the field, forcing a hasty Union retreat to the road, with many soldiers dropping their haversacks, blankets and superfluous baggage. Zealy retreated to another point, where he was joined by one gun from Earle's Battery, Furman Light Artillery, commanded by Lt. E. H. Graham, and an artillery duel ensued between opposing sides.[50]

The Union advance had been delayed one hour, but the Confederates needed more time for the Georgia reinforcements to arrive. At 10:00 A.M., a courier informed Colcock that the Second Georgia Brigade had arrived by train and General Smith was leading them to the Honey Hill entrenchments. Colcock then ordered his advance to withdrawal behind the earthworks. After retreating half way, Zealy was ordered to make a stand at another bend on the road. Further shelling mortally wounded Lieutenant Wildt as he sighted one of his guns. Zealy finally pulled his Napoleon gun back to the Honey Hill redoubt. Some 1,400 Confederate troops, with seven light artillery pieces, held the lines on both sides of the Grahamville Road. The Beaufort Volunteer Artillery, commanded by Capt. Henry Middleton Stuart, had five guns posted, there was one gun from Kanapaux's Lafayette Artillery Battery and another from Earle's Battery.[51]

Colcock informed General Smith around 11:00 A.M. of the latest events and deferred to him control of the battlefield. Smith, declining command of South Carolina troops on their soil, left Colcock in charge, while he withdrew to his headquarters in the rear. As the regrouped Union forces advanced on the Honey Hill defenses, Confederate artillery hit them at six hundred yards with "a terrible volley of spherical case." The first shell blew up two ammunition caissons of Mesereau's Battery, 3d New York Artillery, throwing a gun into a ditch.[52]

The regrouped Union forces had difficulty attacking the Confederate position behind a small stream bounded by a marsh with thick undergrowth, which became a troublesome obstacle to surpass under enemy fire. Repeated federal charges "with vigor and determination" were driven back by the artillery and musketry. Capt. Henry Middleton Stuart later recalled that the defending artillery fired double-headed canister at the rate of five times per minute. A reporter for the Confederate *Savannah Republican* wrote: "The gallant Col. Gonzales was an active participant in the fight, and might have been seen everywhere along the line posting the guns, and encouraging the troops." The Cuban galloped on horseback behind the entrenchments, scanning the battlefield with his field glasses, assuring that all seven artillery pieces were coordinating their fire. He reviewed enemy damage to the artillery positions, monitored the ammunition on hand, and verified that reserve artillerists were relieving the gun crews at proper intervals and replacing the wounded men. The Savannah correspondent also described how "the artillery was served with great accuracy, and we doubt if any battle field of the war presents such havoc among the trees and shrubbery. Immense pines and other growth were cut short off or torn into shreds."[53]

During five hours, "from eleven in the forenoon to four o'clock in the afternoon," the attackers made two direct charges and an abortive third try before attempting an advance and flank movement on the right through the swamp, but were driven back by the Georgia 47th Infantry Regiment. Five companies of the 55th Massachusetts, led by Col. Alfred Hartwell, advanced in a double column. The African Americans moved against the left flank rifle pits occupied by Company B, South Carolina 3d Cavalry Regiment, but withdrew after sustaining over one hundred casualties. Gen. Beverly H. Robertson arrived from Charleston late in the day with the Georgia 32d Infantry Regiment, an artillery battery, and a cavalry company "in time to render most effective aid." Brigadier General Hatch finally ordered a Union retreat to Bolan's Church at 7:20 P.M. They left "the swamp and road literally strewn with their dead," many "horribly mutilated by shells." A number of bluecoat cadavers had stacked up in front of the earthworks, others were "floating in the water," and six African Americans were "piled one on top of the other" in a roadside ditch. A Confederate who later viewed them said that the black troops had their pockets "full of money" and superstitiously adorned their caps with a sprig of the herb Life Everlasting. In Bolan's Church, the pews were removed to create a makeshift hospital for the wounded. According to Major General Foster, they were unable to win because "the unfavorable nature of the ground" allowed them to employ only one section of artillery. Federal losses were 88 killed, 623 wounded, and 43 missing, including about a dozen prisoners. Confederate casualties were described by General Smith as eight killed and forty-two wounded. General Hardee arrived at Grahamville on November 30 and directed Smith to turn over the command to Brigadier General Robertson upon his arrival to relieve the Georgia Militia, which returned to Savannah.[54]

On December 5, Gonzales joined Maj. Gen. Samuel Jones at Pocotaligo, where they established temporary headquarters. Jones ordered Brig. Gen. James Chesnut Jr., to keep the Georgia 47th Infantry Regiment and a section of artillery on a train at Coosawhatchie, to quickly move to any threatened point along the Charleston and Savannah Railroad. The next day, Major General Foster left one Union regiment and a section of artillery at Boyd's Neck, and sent over four thousand troops on a dozen barges to Gregory's Point at James Gregory's plantation on Tullifinny Creek. The force advanced on the road to Coosawhatchie, until checked at the Coosawhatchie River bridge by a battalion of the Georgia 5th Infantry Regiment. The bluecoats fell back to an entrenched position behind a dense woods three-quarters of a mile from the railroad. The federals attacked again on the ninth with thirty-pounder Parrotts, firing a heavy barrage on the railroad between Tullifinny trestle and Coosawhatchie. They advanced to within two hundred yards of the Confederate battery near the train track, commanded by Brig. Gen. Beverly H. Robertson. After a four-hour skirmish, the invaders retreated one mile back to their camp, from where they continued to shell the railroad. The bombardment disabled a locomotive, killed an engineer and a fireman, and temporarily cut the rails in three places.[55]

Five days later, Foster met with Major General Sherman at Fort McAllister on the western outskirts of Savannah and expressed his desire to "wipe out" the city. Maj. Gen. Henry W. Halleck, Army Headquarters chief of staff, wrote to Sherman, hoping for the destruction of Charleston. Sherman informed Lieutenant General Grant that he planned to march his 60,958 troops to Columbia, South Carolina, thence to Raleigh, North Carolina, and then join their armies in Virginia. Sherman wanted to "punish South Carolina as she deserves," and desired to "have this army turned loose on South Carolina to devastate that State, in the manner we have done in Georgia." Facing this onslaught were ten thousand

Confederate troops in Savannah and 5,500 at Pocotaligo. At the Pocotaligo headquarters of Major General Jones, Gonzales received a telegram from Beauregard on December 15 ordering his immediate presence in Charleston to discuss the arrangement of his requested armament. Beauregard had arrived in the city eight days earlier from Montgomery, Alabama, after President Davis extended his Military Division of the West "to embrace the Department of South Carolina, Georgia and Florida." Beauregard then went to Savannah on the eighteenth.[56]

On the night of December 20, Hardee evacuated his forces from Savannah to Hardeeville, crossing the Savannah River over a hastily constructed pontoon bridge. Maj. Gen. Gustavus Smith and his command were sent to Augusta. The Confederates scuttled their gunboats, destroyed the ironclad ram *Savannah*, burned the navy yard, spiked 167 artillery pieces, broke the wheels of gun carriages, and dumped tons of gunpowder and ammunition in the water. During the siege of Savannah, Gonzales's railroad "flying artillery" idea had been put to the test: "Sherman's men marveled at the Confederate battery that had been mounted on a railroad car; it was moved from one front to another with celerity, firing accurately into exposed groups of men and generally harassing the Union army." Hardee gets credit from his biographer for this Gonzales idea.[57]

After the fall of Savannah, Beauregard decided that a new South Carolina defense line of 2,500 troops would be established at the Combahee River, with a reserve force of 1,000 between them and the second line at the Ashepoo River. The third defensive line was the Edisto River and the last would be the Ashley River. On December 20, Beauregard wrote a memorandum of orders to be issued by Hardee, who was at Hardeeville, that also instructed Gonzales to assign light artillery units "to the most appropriate positions for the defense of the 4th Sub District and the line of the Combahee from Salkehatchie [Railroad] Bridge to the coast." This became official two days later, when under Special Field Orders No. 17, the chief of artillery was instructed by Hardee to carry them out with 14 of the 49 light artillery pieces he received at Pocotaligo from Savannah. These were four Georgia light artillery batteries, Daniell's, Guerard's and Maxwell's, commanded by Capt. J. A. Maxwell, and Capt. Cornelius R. Hanleiter's unit. The remaining salvaged guns were distributed to the troops at Hardeeville, New River, Honey Hill, Coosawhatchie, to cavalry general Joseph Wheeler, and to the militia accompanying Smith's division.⁵⁸

The new defense line was commanded by Maj. Gen. Lafayette McLaws, relieving Maj. Gen. Samuel Jones at Pocotaligo, who returned to Charleston and resumed command of his district. Beauregard ordered the retreating cavalry to apply a scorched earth policy before the enemy by driving off "all cattle, sheep and hogs, not necessary for its consumption, and impress and send to Charleston, to be turned over to the Chief Engineer, all negros [sic] capable of bearing arms." He also ordered the destruction of every mill, boat and building that could be used by the Union forces for military purposes, and all surplus rice, corn and provisions beyond what was necessary for their troops, the owners, their families and slaves. The next day, Beauregard told Hardee to "impress all negro men, teams, wagons, and supplies *not* removed by owners in section of Country about to be abandoned to Enemy—destroy all Bridges and Trestles behind you." Hardee soon had the countryside stripped up to the Combahee River defense line.⁵⁹

Beauregard assumed that Union troops were headed for Charleston, but Sherman wrote to his superiors on Christmas Eve that he would move on Columbia, ignoring the "mere desolated wreck" of Charleston. Sherman decided that in spite of the historical and political significance attached to the city, he hardly had time for siege operations. He regarded Columbia "as bad as Charleston" and was not going to "spare the public buildings there." Sherman stated, "The truth is the whole army is burning with an insatiable desire to wreak vengeance upon South Carolina. I almost tremble at her fate, but feel that she deserves all that seems in store for her." Three days later, Beauregard directed Hardee to "silently and cautiously" prepare the evacuation of Charleston. Light artillery batteries were to be organized into battalions of three batteries each, with one battalion assigned to each division. Gonzales would command all others held in reserve. Beauregard showed great faith in the military capabilities of the chief of artillery when he instructed that these newly organized battalions, although controlled by the division commanders, would make all their returns and reports to Gonzales, who would also be privy to all other correspondence related to their management. When not on the field, all the batteries would be commanded by the chief of artillery.⁶⁰

On December 31, Major General Jones received charge of the District of Florida, and Beauregard relinquished his local command to reorganize in Tupelo, Mississippi, the remnants of the Army of Tennessee, until then led by Lt. Gen. John Bell Hood. Beauregard instructed Hardee to apply to Charleston the same principle he used in Savannah: "Defend it as long as compatible with the safety of your forces." He indicated that while the fall of Charleston would be a "terrible blow to the Confederacy," more fatal to the cause would be "the loss of its brave garrison." Prior to evacuation, all cotton was to be removed and any remaining destroyed. Depots of provisions and forage were to be established at Columbia and Florence for the retreating troops. Two days later, Sherman wrote Grant that he would use part of his sixty thousand troops to make feints toward Charleston on the right and Augusta on the left, while his army headed for Columbia. Sherman told his officers that South Carolina should be punished for having started the war, and the more private property they destroyed, the better. Sherman's right flank, led by Maj. Gen. Oliver O. Howard moved from Beaufort against Pocotaligo on January 13, 1865, while the Union navy shelled the Confederate positions near Charleston seaboard.[61]

Five days later, Gonzales wrote from Charleston to his mother-in-law at Oak Lawn, urging that it was "a question of very few days" before Charleston would be evacuated and that there was not a minute to lose "for things material or immaterial." The plantation lay forty miles from Pocotaligo, on the road to Charleston. Gonzales indicated that the South Carolina Railroad would be running the next evening and that everything to be vacated from Oak Lawn "shall go *at once*" on that route. He asked that Hattie, the children, and their nurse join him immediately with whatever bedding, indispensable crockery, and utensils they could carry, and that the rest of the family should follow as soon as possible. Gonzales asked Mrs. Elliott to accompany Hattie, if she was not needed at home, because at Charleston she would be "on the way to any point and in case of a miracle like that of the parting of the Red Sea might go back to Oak Lawn." The sixty-two-year-old matriarch had been reluctant to relocate until her cattle and cotton were moved to the countryside. Gonzales admonished his family: "You are all behind time in every thing that has had to be done, and I fear that unless someone gives you all a military order to do so, at the imminent risk of his head & reputation from a feminine courtmartial you will not be moved in time."[62]

Gonzales suggested that Mrs. Elliott send her chief clerk to negotiate the sale of her cotton for sterling with W. T. J. O. Woodward, the agent of the Southern Express Company and various blockade running enterprises in Charleston. Woodward had been involved with Gonzales in selling and delivering Maynard rifles in 1860 through the Adams Express Company. During the war, Woodward had been the express agent in Charleston, a position that exempted him from military duty. He had advertised in a Nassau, Bahamas newspaper for the shipping of goods to Charleston on blockade runners. Woodward was indebted to Gonzales, who had recently secured the protection of all his wagons, teams, wagoners and a pass to Cheraw. He offered to send the Elliott cotton to Nassau at the lower rate of two bales for one, instead of the usual three to one exchange fee, and asked that the offer be kept secret. Small private vessels had successfully run the blockade of Charleston harbor. There had been thirty-eight runs in 1864, and, although the fall of Savannah and Wilmington had added more Union warships on the coast of Charleston, four other vessels managed to enter the port in January. Gonzales concluded the letter to his mother-in-law saying that he was "overwhelmed with labor . . . in such a crisis" and had great "anxiety at seeing you still at Oak Lawn with uncertainties before you & so much yet undone." Mrs. Elliott made a "desperate" sale of her cotton to Benjamin Mantoue, of the firm Guerin, Mantoue, and Company, in Charleston, who agreed to pay her two-thirds of its value. The twenty-seven-year-old Frenchman had migrated there a decade earlier. The transaction was witnessed by Gonzales, Mrs. Elliott, and a Mr. Levy.[63]

Eleven days later, only Emmie and the house servants were at Oak Lawn, making departure preparations. She had used the railroad to keep her family in Charleston supplied with fish, fowls, potatoes, milk, and wood for cooking and heating. Gonzales's orderly retrieved the goods from the depot in his wagon. Emmie wrote to her mother on January 29: "T'is a splendid morning, frost over everything & cannon firing in the direction of White Point, but they sound like ours." The Salkehatchie defense line protecting White Point, on the North Edisto River, fell after Sherman's right flank reached Pocotaligo on February 1. Gonzales then decided to remove his family from Charleston to Cheraw. He was unable to secure shelter for them at the house of S. S. Solomons, the engineer and superintendent of the Northeastern Rail Road, and on February 10 wrote to Scottish-born state militia colonel Allan Macfarlan, a planter and president of the Cheraw and Coalfields Rail Road Company. Macfarlan and Ralph Elliott had been friends while serving in the State House of Representatives in 1860. Gonzales, slightly acquainted with Macfarlan, informed him that the Elliott women and his children would leave in a day or two for Cheraw and asked him to procure for them a house in that city or in a nearby farm. Their house servants would accompany the wagons with provisions, bedding, and some furniture.[64]

That same day, as the Confederacy was rapidly collapsing all around them, former Maj. Gen. John Pemberton, recently appointed inspector general of artillery in Charleston, nominated Gonzales for promotion. In a telegram to Adj. and Insp. Gen. Samuel Cooper in Richmond, Pemberton indicated that he had 106 guns organized and equipped as mounted artillery, which lacked a general or field officer to serve them. He requested that a brigadier general be at once appointed to their command, and highly recommended Gonzales to the position. Hardee endorsed the petition, adding, "Col. Gonzales is Chief of Artillery in this Dept. & his long experience, his thorough & practical knowledge of Artillery, & his great industry and zeal, fully entitle him to the position of Brig. General of Artillery."[65]

The request was forwarded to Secretary of War Seddon who, in turn, submitted it to Jefferson Davis. The Confederate president responded that an inquiry should be made as to the brigadier generals "fit for artillery command and who are available for this position," and with Machiavellian cunning suggested that Brig. Gen. Josiah Gorgas, chief of ordnance of the Confederate States, make the choice. Gorgas, who previously had various disputes with Gonzales over ordnance supplies and artillery requisitions, recommended Brig. Gen. Francis A. Shoup for the job. Shoup had fought at Shiloh as Hardee's chief of artillery, surrendered with Pemberton at Vicksburg, and after being exchanged was chief of staff to Lt. Gen. John Bell Hood, recently relieved of command of the Army of Tennessee. While Beauregard had delegated great responsibility on his chief of artillery six weeks earlier, Gorgas resentfully claimed, "Col. Gonzales is, in my opinion, not fitted for the position, and in this opinion I am joined by Genl. Gilmer, Chief Engineer, who had better opportunities than myself to observe Col. Gonzales professionally." For the sixth time in four years, Gonzales was turned down for the rank of brigadier general, but he was not about to abandon the Confederacy at its worst moment. In contrast, some Confederate generals had resigned because others had been promoted over them, including Milledge Bonham and Gustavus Woodson Smith.[66]

Beauregard's Army of Tennessee was ordered to South Carolina on February 11. He returned to Charleston three days later at the request of Hardee, who was told by President Davis to hold the city. That same day, while Sherman marched north to Columbia, Beauregard ordered the evacuation of Charleston so that Hardee's troops would not be cut off from the rear, as occurred to the Continental Army during the Revolutionary War. In April 1780, British troops had lain siege to Charleston from the north, forcing Gen. Benjamin Lincoln to surrender 5,500 men. On February 17, Charleston residents began to carry out Beauregard's order to systematically destroy in Charleston anything valuable to the enemy. Huge piles of cotton in the public squares and thousands of bushels of rice were set ablaze, along with the wharves and the bridge over the Ashley River. The steamers and all vessels on the stocks were burnt, the gunboats *Palmetto State, Chicora* and *Charleston* were burnt at their moorings, and the Blakely gun on the Charleston Battery was blown to pieces. The fall of Charleston was a psychological blow, signifying the end of the war for many. South Carolina governor Andrew Gordon Magrath called the military withdrawal the "death march of the Confederation."[67]

Gonzales accompanied Hardee's general staff out of Charleston, taking the artillery on the only available railroad going north. Hardee had two divisions totaling ten thousand men, commanded by Generals McLaws and Taliaferro. Gen. Stephen Elliott Jr., heading one of the three brigades under Taliaferro, joined Gonzales on the retreat. These forces began regrouping on the nineteenth at St. Stephen, forty miles north of Charleston, before crossing the Santee River. Beauregard telegraphed Hardee that day from Ridgeway, ordering him to Cheraw by rail and from there to march to Charlotte, North Carolina. Hardee's adjutant, Lt. Col. Thomas B. Roy, noted a "great many desertions from the Command on the march from Charleston. Some art[illery] companies almost disbanded by desertion." Most of the deserters worried about their families being in Sherman's path of destruction.[68]

Union troops from Beaufort who fought at Honey Hill under General Hatch had been pushing north toward Charleston along the coastal road. Company G of the 157th New York Regiment camped at the Elliotts' Oak Lawn plantation on February 20 and were informed that Charleston had been taken. They set up their tents on the front lawn and along the avenue of oaks, destroying the fences to build campfires. The bluecoats looted whatever they could carry off. The next day, orders were given to burn the English-style mansion and outer buildings before the column moved out. One two-story slave cabin was left intact after much pleading from its occupant. Along the ninety-mile road from Savannah to Charleston, only two plantation homes were left standing.[69]

That same day, Gonzales arrived at Cheraw with Hardee's artillery, slowly followed by the rest of their forces. Hardee and his staff, including Gonzales, were invited to supper and entertained at the home of a Colonel Woodward. On February 25, the last of the half dozen evacuation trains, with Generals Wilmot G. DeSaussure and Beverly H. Robertson, crossed the Santee River at midnight, destroying the bridge after they passed over it. Gen. Joseph E. Johnston arrived in Charlotte, North Carolina, on the twenty-fifth. Gen. in Chief Robert E. Lee had commanded Johnston to assume command of the Army of Tennessee from Beauregard and all forces of the Department of South Carolina, Georgia, and Florida from Hardee, making both generals his subordinates.[70]

After leaving Charleston, the Gonzales family and the Elliott ladies had found refuge in Springville, in a house provided by forty-four-year-old Maj. Albertus Chambers Spain, which belonged to the Gibson family of Darlington. Three miles north of Darlington on the road to Cheraw, Springville was an unincorporated village that served as a summer resort. Mrs. Elliott, her daughters Anne, Emily, and Hattie, and the four Gonzales children settled into a two-story Central Hall Forms residence. They would remain there for more than a year. The author located this abandoned house in 1995, on the north side of Lide Springs Road (State Road 29), one mile east of Society Hill Road (State Road 133), near the corner of Willowtree Road. Its gabled roof was partially collapsed and all of the symmetrical windows lacked glass panes. A small porch led to a central doorway, opening into a hallway with a missing circular staircase. Bilateral front parlors, measuring fifteen by seventeen feet, with twelve-foot-high ceilings, contained brick fireplaces and the remnants of the once elegant wainscot and plaster work. These rooms led to smaller rear chambers. The second floor had four rooms of an identical pattern.[71]

Springville was occupied by three federal regiments on the night of March 3. They encamped at Major Spain's home, evicting him the next day. Spain described them as being "like bees going and coming in every direction." The Yankees looted and destroyed a number of residences in the area, including those of John Dargan, John L. Hart, and the widow Harriet McIver, who were neighbors of the Gonzales family. Three separate Union cavalry raiding parties descended on the Oak Lawn refugees. According to Emily, "they robbed our negroes but got from us very little. We met them on the piazza & so kept them from entering the house—a circumstance we are a little proud of."[72]

On March 4, General Johnston moved his headquarters to Fayetteville, North Carolina, where he planned to regroup the retreating Carolinians, who arrived there five days later. He immediately appointed Gonzales his acting chief of artillery, ordered his guns reorganized at Hillsborough, and transferred his headquarters to Raleigh on the tenth. That evening, Hardee's Corps arrived five miles south of Averasboro, where six days later they briefly halted Sherman's advance. The battle of Averasboro served as a delay tactic that allowed Gonzales to advance the artillery to Smithfield, where Johnston had established his headquarters on the fifteenth. The cannons were boarded there on railroad cars for their transfer to Hillsborough via Raleigh. Gonzales was in Raleigh on the twentieth, when he received General Johnston's orders to temporarily send all the artillery without horses to Greensboro, seventy-five miles west, until the animals arrived in Raleigh.[73]

On March 23, the invincible Union army reached the railroad junction at Goldsboro, twenty-four miles southeast of Smithfield, after cutting a sixty-mile-wide swath across the Carolinas. Sherman then went to the coast and took a ship to Virginia, to consult on final strategy with General Grant and President Lincoln. Sherman wanted to crush Lee in Virginia, but Grant did not need any help, and Lincoln was not specific on the terms of surrender that Sherman should demand. Sherman returned to Goldsboro on the thirtieth, organizing his enlarged force of ninety thousand to move on Raleigh against twelve thousand Confederate troops.[74]

The following day, Gonzales, who had taken the artillery to Hillsborough, was ordered to send by railroad four Napoleon guns with horses to Gen. Braxton Bragg at Salisbury to thwart the advance of Gen. George Stoneman's raiders. Another battery was to be forwarded to Beauregard at Greensboro and a third should be "ready to move at a moment's notice." Beauregard warned Gonzales that the raiders could threaten his position and advised him to be ready to move his artillery back to Raleigh or Smithfield. Two days later, Beauregard notified Gonzales to send three fully-equipped batteries with good field officers to the commanding officer at Danville.[75]

Gonzales remained at Hillsborough until Johnston capitulated to Sherman near Durham, North Carolina, on April 26. Confederate officers were allowed to keep their sidearms and their horses after signing a parole. Gonzales was paroled at Greensboro on April 30, as chief of artillery of Hardee's Corps, General Johnston's Army, with his eight staff members. The Cuban was released with a group that included Generals Beauregard, Matthew Calbraith Butler, Louis Hebert and Thomas L. Logan, and Admiral Raphael Semmes, who as commander of the CSS *Alabama* from 1862 to 1864 had captured and sunk fifty-five ships, including the U.S. warship *Hatteras*. Gonzales returned to his adopted state, probably with Beauregard, who continued on to Louisiana. South Carolina was in the midst of social and economic upheaval, with more than 18,000 of its Confederate soldiers dead, Charleston and Columbia destroyed, lowcountry plantation rice fields flooded and their homes burnt, and 400,000 slaves emancipated.[76]

Gonzales had played a crucial role in the defense of South Carolina throughout the war. His artillery department, with limited resources, had impeded repeated, massive Union onslaughts during more than three years. The Cuban had performed multiple tasks in addition to his regular duties, including being chief of ordnance for seven months and chief of artillery for General Taliaferro for another four months. The former filibuster led the artillery at the battle of Honey Hill, which defeated a federal attack four times greater than the defending force. It was his finest moment in combat and his actions, commended by General Smith, were exemplary. Gonzales's military record would have merited promotion to brigadier general in almost any other army. President Davis's spiteful character hindered not only Gonzales, but also all the officers serving under Beauregard who were denied promotions during the last two years of the war. The conflict had a devastating effect on the Gonzales and Elliott families. They had lost their property, most of their wealth, and were surviving as refugees in Springville. Gonzales reunited with them after a three-month absence, burdened by the weight of unemployment, poverty, and an uncertain future.

**HANC**

# PRO SUA

## IN LITTERARIO CERTAMINE

AD IURIS CIVILIS

BACCALAUREI GRADUM OBEUNDUM

**ESPEDIENDA VICE**

*Filii familias tutores esse possunt.*

DIE 2 MENSIS *Mai* ANNI A SALUTE MDCCCXXXIX.
In Regiæ ac Pontificiæ Universitatis Habanensis Musæo.

**SUSTINEBIT**

*Dom. Ambrosius Josephus Gonzalez.*

**DEDUCTAM**

*Ex lege 22 tit. 1 lib. 26. D. D.*

**MANE HORA SOLITA.**

*In Typographio Litterario. Regens Dominicus Patiño.*

Ambrosio José Gonzales's Havana University diploma, May 2, 1839. Courtesy of Elliott McMaster, New Mexico.

*Cuban Independence general Ambrosio José Gonzales. Courtesy of the late Dr. Ambrose G. Hampton, Columbia, S.C.*

*Narciso López (1797–1851). Courtesy of the Library of Congress, Washington, D.C.*

*John Anthony Quitman (1798–1858). Courtesy of the Library of Congress, Washington, D.C.*

Filibuster monument, Shively Park, Louisville, Ky. "As a tribute to the valiant Kentuckians who fought for the liberation of Cuba in 1850." Author's collection.

*Harriett Rutledge Elliott Gonzales (ca. 1869). Courtesy of the late Dr. Ambrose G. Hampton, Columbia, S.C.*

*The oak allée at Oak Lawn Plantation, Colleton County, S.C. View from the front porch toward the main gate. Author's collection.*

*Front view of The Anchorage, home of William Elliott (1788–1863), Beaufort, S.C. Author's collection.*

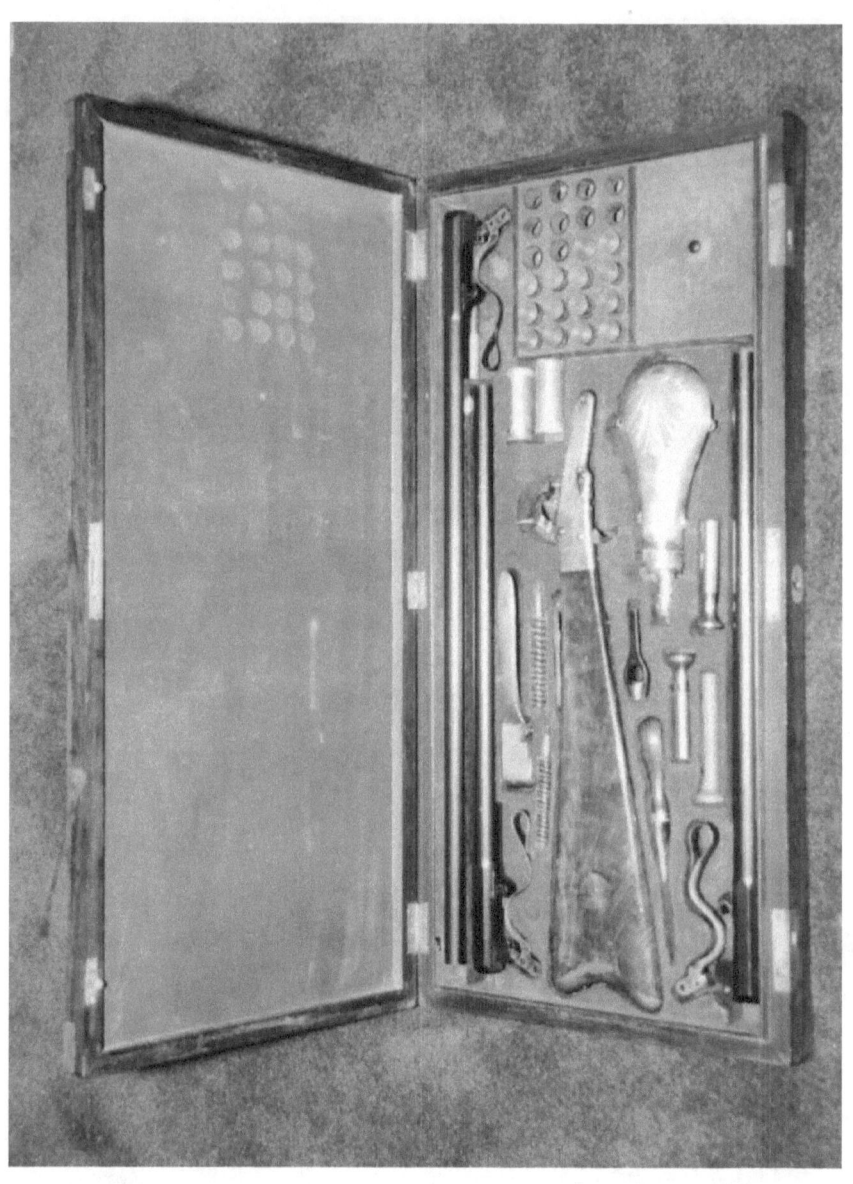

A disassembled Maynard rifle, with its accoutrements and case. In 1860, Ambrosio José Gonzales was a sales representative for the Maynard Arms Company of Washington, D.C. Author's collection.

*Confederate colonel Ambrosio José Gonzales. Courtesy of the late Dr. Ambrose G. Hampton, Columbia, S.C.*

Confederate general Pierre Gustave Toutant Beauregard (1818–93). Courtesy of the Library of Congress, Washington, D.C.

I ...A. J. Gonzales... solemnly swear that I have carefully read the Amnesty Proclamation issued by ANDREW JOHNSON, President of the United States of America, on May 29th, 1865, and that I am not excepted from the benefits of that Proclamation by any one of the fourteen exceptions therein made, except

A. J. Gonzales

Sworn to and subscribed before me at Darlington, S. C. this 28" day of July... 1865.

B B Munny
Lt. Col. & Pro. Marshall
Mil. D. E. S. Co.

NO. 2

## United States of America.

I, A. J. Gonzales of the County of Darlington, State of South Carolina, do solemnly swear, in the presence of Almighty God, that I will henceforth faithfully support and defend the Constitution of the United States, and the Union of the States thereunder, and that I will, in like manner, abide by and faithfully support all laws and proclamations which have been made during the existing rebellion with reference to the Emancipation of slaves—So HELP ME GOD.

A. J. Gonzales

Sworn to and subscribed before me at Darlington S.C. this 28 day of July... 1865.

B B Munny
Lt. Col. & Pro. Marshall
Mil. D. E. S. Co.

The above named has dark complexion, gray hair, and hazel eyes; is 5 feet 8 inches high, aged 45 years; by profession a Planter.

(The original oaths will be transmitted, by the officer administering them, to the Department of State, through intermediate channels, and a certified copy thereof will be furnished to the applicant for pardon, to accompany his petition.)

Oath of allegiance to the Union signed by Ambrosio José Gonzales, July 28, 1865. Courtesy of the National Archives and Records Administration, Washington, D.C.

On June 24, 1866, Gonzales, Woodward & Company was looted during a race riot in Charleston, S.C. From Harper's Weekly, August 4, 1866.

*After the Civil War, Ambrosio José Gonzales operated a sawmill manufactured by George Page & Company of Baltimore, Md. Woods's Baltimore City Directory, 1871.*

In 1872, Ambrosio José Gonzales resided in this boardinghouse, at 333 N. Charles Street, Baltimore, Md. Author's collection.

*Three "ghostly" faces surround the aged Ambrosio José Gonzales in this rare example of spirit photography. A woman's face, purportedly his deceased wife, Hattie, is apparent over his left shoulder. Gonzales, a cigar smoker, keeps a visible one in his coat pocket. Courtesy of the late Dr. Ambrose G. Hampton, Columbia, S.C.*

*Narciso Gener Gonzales wearing his Cuban Liberation Army uniform. Courtesy of the late Dr. Ambrose G. Hampton, Columbia, S.C.*

*William Elliott Gonzales (ca. 1913). Courtesy of the Library of Congress, Washington, D.C.*

*Ambrose E. Gonzales (1857–1926), author of* The Black Border: Gullah Stories of the Carolina Coast *and other books on South Carolina's African American folklore. Courtesy of Mrs. Joanne Hampton, Columbia, S.C.*

Gertrude Gonzales. Courtesy of Mrs. Joanne Hampton, Columbia, S.C.

*Ana Rosa Gonzales. Courtesy of the late Dr. Ambrose G. Hampton, Columbia, S.C.*

Home for Incurables (ward), Bronx, N.Y.

*Dr. Ambrose Gonzales Hampton erected a monument at the gravesite of his great-grandfather Ambrosio José Gonzales, Woodlawn Cemetery, Bronx, N.Y. Author's collection.*

## Eight

# RECONSTRUCTION RETAILER

After Gonzales was paroled on May 1, 1865, he joined his family in Springville, South Carolina. He had kept in touch with Hattie by private messenger after the evacuation of Charleston, but the letters took a month to reach her. Ralph Elliott also joined his relatives after his cavalry unit disbanded at Chester without surrendering. The Oak Lawn clan was reunited with the arrival of brother William, suffering from tuberculosis. Emily described how they were "very comfortably located & find the good people of Darlington & Springville abundantly kind. They supply all of our wants & we have not felt any of the horrors of refugeedom." Even so, Annie started planting vegetables on a "very poor piece of land." Emily lamented that "we can not bring ourselves to believe that our four year struggle has been in vain, that our heroes have lived & died in vain & that we are to be a despised & conquered people." Yet all was not grief, as the Gonzales family attended that summer the wedding reception of Mary Gray Crockett at Col. Allan Macfarlan's home in Cheraw. Gonzales sang at the nuptial festivity and the bride recalled almost sixty years later how he had been a "splendid performer on the piano and gave us some fine selections."[1]

Mary Barnwell Johnstone, the eldest Elliott sister, had been living in a rented cottage in Greenville, South Carolina, with her six children since the previous year, after her husband was murdered by Confederate renegades. She was "mortified" that her mother and sisters "should be inconvenienced by the desertion of their well cared for servants," since she had felt "quite attached" to some of them. A few freedmen remained with Mrs. Elliott, including Dick, Jacob, John and his wife Chloe and their children. Chloe had been Caroline Elliott's servant. Mary still had her German governess, Miss Hinckel, and at least four former slaves attending her family.[2]

In June, Ralph Elliott went to Charleston to arrange housing; he was the first of his family to return. He found that the rents were going to be "exceedingly high" during the winter and suggested that his relatives stay in Springville. Ralph was using their servant Dick as a courier because the mail routes were deficient. He relayed rumors that Oak Lawn had been destroyed, but his sister Mary refused to believe it, especially after learning that their brother Tom's plantation house at Bethel was unburnt. Ralph advised his relatives to "try to be patient & hopeful, & bear the misfortune, poverty, & degradation which has been put upon us, with fortitude." Gonzales sent Ralph his ornate saddle to sell in Charleston, but it was returned when no buyers appeared.[3]

Gonzales then suggested that the family emigrate to Cuba, where his relatives and friends would provide support while they rebuilt their lives. Emily wrote for advice on June 28 to her cousin, Bishop Stephen Elliott, refuging in Augusta, Georgia. He replied that he had "no faith in emigration" to uncivilized countries where they would "only find greater misery." Talk of going to Brazil or Cuba, where slavery existed, was "mere nonsense" since they were "very expensive countries and have no use for the kind of talent we should carry there." He recommended that the Elliotts "remain in Carolina among your own people & kindred" and claimed that they would soon be able to recover their lands. Bishop Elliott also advised that if they had to dispose of their silver plate to survive, it was best to sell it in New York, where it would bring a higher price than in the South.[4]

The Elliott properties had been forfeited under the Confiscation Act of 1862, which authorized the U.S. Treasury Department to seize the 195 plantations on the Sea Islands for nonpayment of taxes and placed them on the auction block. General Sherman's Sea Island Circular of January 18, 1865, had reserved for the settlement of freedmen the islands from Charleston to Port Royal, and the adjacent lands thirty miles inland.

According to native abolitionist Dr. William H. Brisbane, white Northerners were "gobbling up the lands at the expense of the freedmen," forty thousand of whom had been moved to the area. African Americans purchased only three thousand acres of the eighty thousand auctioned off. A Boston joint-stock company bought eight thousand acres for a trifling seven thousand dollars. Four months later, president Andrew Johnson granted amnesty, with some exceptions, to former Confederates. Johnson claimed to be abiding by the Constitution when restoring all property rights, except slaves, to the Southerners. Lands not sold by the government could be recovered if the owners paid the tax, took the oath of allegiance, and received a presidential pardon. In July, Mrs. Elliott wrote for legal advice to her Charleston attorney Richard De Treville, a West Point graduate, former lieutenant governor of South Carolina, Secession Convention member, and colonel of the 17th South Carolina Volunteers.[5]

The lure of recovering their lands and restoring their civil rights prompted Gonzales and some of the Elliotts to take the presidential amnesty oath. Gonzales swore before the provost marshal at Darlington on July 28 to "henceforth faithfully support and defend the Constitution of the United States, and the Union of the States thereunder," and to "abide by and faithfully support all laws and proclamations which have been made during the existing rebellion with reference to the Emancipation of slaves." The next day, Mrs. Elliott took the same oath, followed by her daughter Anne a week later. Ralph later claimed that he was willing to take the oath of allegiance, on account of his mother, if it necessarily contributed to recovering their homes. He would "die rather than take it on my own." He never took the oath and remained an inveterate Confederate the rest of his life. By the fall of 1865, more than a hundred pardons a day were being granted, for which pardon brokers and attorneys were charging fees ranging up to five hundred dollars.[6]

Leaving his family in Springville, Gonzales was back in Charleston by August 2, visiting a family friend, the Reverend Dr. John B. Bachman, pastor of the English Lutheran Church and former associate of naturalist John James Audubon. The city was recuperating from the extensive damage of shelling, fire, and prolonged looting. Flocks of buzzards roosted on rooftops and chimneys. Rural freedmen had created squatter camps along the East Bay wharves and coal docks and had occupied many of the abandoned residences. They had migrated to Charleston in droves, seeking work and rations from the Bureau of Refugees, Freedmen, and Abandoned Lands. Most of the former merchants had returned to the city and were "making arrangements for resuming business." Gonzales was almost forty-seven years old when he explored the possibilities of starting a business in Charleston with merchandise and funds advanced by his wealthy relatives and friends in Cuba. He returned to the island in August, under a political amnesty that had been granted by the Spanish Crown in 1856.[7]

Hattie, pregnant with their fifth child, stayed at Springville with her children, mother, sisters, and invalid brother William. Ralph went to visit Mary Johnstone at Greenville and then proceeded to Flat Rock, North Carolina, to inspect their *Farniente* estate. It was "all in decent order"; two-thirds of their china was robbed, but they still retained a box of valuables in the Abbeville bank. In early September, the Freedmen's Bureau published an order calling for applications for the restoration of captured or abandoned property. The Elliott family then heard a rumor from their servant John that the silver plate and china that had been buried at Oak Lawn had been stolen. He also stated that the freedmen expected to return to the plantation in the fall and claim the land for themselves. To prevent this, the following month the family had attorney De Treville file a petition on their behalf for the restoration of their plantations on the mainland. Ralph found portions of the document "hard to stomach."[8]

A second petition was presented in November after De Treville met with abolitionist general Rufus Saxton, assistant commissioner of the Bureau of Refugees, Freedmen, and Abandoned Lands, to discuss the legal recovery procedure and had an interview with the land commissioner regarding the St. Helena properties. He was informed that the plantations Shell Point, Grove, and Ellis on Port Royal Island and Myrtle Bank on Hilton Head Island had been sold by the government and that the Beaufort house was being advertised to be auctioned for taxes on December 6. When Gonzales returned from Cuba that month, he was instrumental in helping De Treville obtain from Saxton an order on December 9 restoring to Mrs. Elliott the lands of Oak Lawn, Social Hall, The Bluff, and Middle Place after she relinquished "all claims against the U.S. Government for damages." De Treville later billed Mrs. Elliott one thousand dollars in legal fees.[9]

Three days later, the *Charleston Courier* announced that the former "popular energetic" superintendent of the Southern Express Company, W. T. J. O. Woodward, had returned to the city, and that "whatever business he undertakes is sure to be successful, for he knows no such word as fail." Gonzales brought Woodward and Peter J. Esnard into his business plans, which accelerated after City Hall announced on December 29 the renewal of applications for liquor licenses. The *Courier* advertised on January 1, 1866, the formation of the partnership Gonzales, Woodward and Company, "a general commission, shipping and factorage business in this city." It was located at 73 East Bay Street, an antebellum liquor and cigar store run by Capt. J. Gadsden King, first commander of the South Carolina Marion Light Artillery Battery. Today, the former business is the Adam Tunno House at 89 East Bay Street. The three-story brick house, built around 1780, had "a store front with square columns, and a large gateway with square columns and a balustrade, on the south portion of the property." Double shop doors led to store space on the ground floor. There were offices on the second floor and residential quarters above it. The rooms on the first two floors contained "paneled wainscoting, architrave molding in door and window surrounds, six-paneled doors, wooden cornices and interior shutters." The upper residential area had "plain mantels, plain-beaded flush board chair rails, simple door and window surrounds and board and batten closet doors with strap hinges." The fireplaces on each floor were encased in half vaults of English bond brickwork. The backyard led to Bedon's Alley. The storefront was removed during restoration in the 1930s.[10]

Gonzales, Woodward and Company advertised Cuban products of sugar, molasses, rum, coffee, and cigars, and it also bought and sold cotton and other southern produce. A week later, the credit agent for the firm R. G. Dun and Company gave the firm a "B" rating, with a credit risk, if any, of not over four hundred dollars. His company report described Gonzales as "a Cuban & is said to have means, expects to control consignments from that quarter." Woodward is mentioned as "formerly Agt. for the Adams' Express for yrs, is indus[trious] & atten[tive] to bus[iness] means." Esnard was a sixty-seven-year-old New Yorker who moved to South Carolina in the 1830s, where he raised a family and had been one of four partners in the commission merchant firm Hall and Company before the war. He appeared on the credit report as "an old citizen, don't kn[ow] his means." Three days after the firm was established, Gonzales traveled to Havana on the steamship *Isabella* to promote his new mercantile endeavor.[11]

While he was gone, Ralph Elliott visited Oak Lawn accompanied by planters William B. Means, Nathaniel Heyward, the Rhetts, and a Yankee guard, who toured the area. Many freedmen had returned to the large lowcountry estates due to local familiarity, kinship, and economic necessity. Military squads were going through the plantations forcing squatters who did not contract with the owners to leave. Ralph's former slaves Dick and Jacob had recently visited Oak Lawn and confirmed that all of the buried treasures were stolen. The Elliott summer house in Adams Run was found in a dilapidated condition, with some of the doors missing and all the outbuildings destroyed. Ralph began contracting the former bondmen to start a crop at Oak Lawn, and among the first to go back were Jacob and his wife. Tom Elliott and his son William returned to their Bethel plantation in Pocotaligo, where they began planting cotton and erected a sawmill with ninety black and white laborers.[12]

While Gonzales was in Cuba, he received a letter from his mother-in-law in response to an earlier proposition he had made to her to acquire Social Hall and The Bluff plantations for seventeen thousand dollars. In order for Mrs. Elliott to acquire legal title to all her lands or to obtain a mortgage, her husband's will first had to be executed, which included the settlements of Hattie's ten-thousand-dollar marriage bond and Mary's seven-thousand-dollar marriage bond. Gonzales agreed that, in exchange for his ownership of Social Hall and The Bluff, he would forfeit Hattie's bond and the legacy of six thousand dollars left to her in the will, and he would pay in installments the remaining one thousand dollars to Mary's trustees. He considered that planting there was "out of the question this year" and had "very little hope that it will be made successful for a long time." Gonzales had "no potential use" for Social Hall, except for its timber. While in Matanzas, he tried procuring a two-thousand-dollar, two-year loan for Mrs. Elliott to start planting at Oak Lawn. The new entrepreneur also forwarded to his company a shipment of molasses and cigars by schooner.[13]

Gonzales left Cuba on March 10 with three Chinese coolies from Matanzas, whom he offered wages of fifteen dollars per month and rations. The Chinese labor used in Cuba since 1847 had proven "docile, industrious, frugal, temperate, hardened to rural labor," reliable, cheap, and submissive. Gonzales was innovative by introducing Asian workers to South Carolina. Six months later, Chinese Commissioner A. H. Yue traveled from San Francisco to the southern seaboard states, "offering to furnish field hands." Mrs. Elliott commented three years later that "Southerners are taking hold of the idea of having Chinese labourers." The coolies would be accustomed to the arduous conditions of low-country rice cultivation. Upon returning to Charleston, Gonzales moved his family into a rented house on Society Street. Their cow Sallie had a shed in the backyard. The family was vaccinated against smallpox, which had swept through the city the previous year with devastating results. Charleston was still rebuilding from the ruins of war, with laborers engaged in "removing rubbish, pulling down walls and chimneys, and arranging the bricks in squares." The stench from filthy streets, stables and garbage-laden scavenger carts prompted the *Courier* to demand a city ordinance against such nuisance. Attorney De Treville described the city as "no place for a native of So. Carolina to live in. Of every twenty persons you meet in the street, nineteen are Yankees or negroes." Mary Johnstone considered Charleston "the least desirable place possible (next to Beaufort) to live in now."[14]

After Ralph failed to plant at Oak Lawn, his brother William, briefly recovered from a tuberculosis bout, decided to try it. Arriving there at dusk in early March, he gazed up the avenue of oaks: "I could almost believe myself in a dream for there stood a dwelling house at the other end, a kitchen and out house & from the last a cheerful fire seemed to bid me welcome. Still there was a strangeness in its appearance. T'was not the house I had formerly known. It required a much nearer approach to find that the still erect walls of house and kitchen were but the skeleton that once encased the spirit (forever fled) of the Hospitality & Refinement of bygone days." He found that the oaks and all the younger trees had not been cut or damaged except those near the mansion, "blackened as the hearts of the villains who fired the house." William saw that holes in the garden contained fragments of the Bohemian glass and china sets that had been unearthed. Some of the missing green and gold china was later seen in the home of a former slave and other pieces were being sold on Edisto Island. The kitchen walls stood well, although the floor and the ceiling were burnt. William reported the summer house still standing, but with damaged floor and walls, and concluded, "Indeed the old place is still beautiful in its decay, and worth the trouble of repair." He moved into the two-story slave cabin formerly occupied by servant Chloe and her husband. William decided against raising a crop at Oak Lawn, fearing that cattle would destroy the unfenced fields, and instead hired eleven freedmen to plant one hundred acres at The Bluff.[15]

Many former slaves in the lowcountry obstinately rejected the new labor contracts due to distrusting their late masters and fear that the new arrangement might jeopardize their chance for "forty acres and a mule." On the Elliott lands, freedmen had already planted their own patches and were living on rations provided by the Freedmen's Bureau, but they finally signed the contracts under coercion from an army lieutenant. According to Gen. Robert K. Scott, the assistant commissioner for the Freedmen's Bureau in South Carolina, most workers refused to fulfill the terms of their contracts. He therefore issued two months later General Order No. 9, due to the increase of "theft, drunkenness, and vagrancy" among the freedmen. It ordered those who refused labor contracts to be expelled from the plantations by Bureau agents. Squads of the 35th U.S. Colored Troops Regiment, who fought at the battle of Honey Hill, enforced the evictions. Any freedmen violating their contract terms and wandering off the land were to be conscripted into public road work.[16]

The presidential proclamation of April 2, 1866, declared that the rebellion was over and that South Carolina and the former Confederate states could have civilian authority to enforce their own laws. A few days later, Emily and Annie left Springville and moved into the "dilapidated negro house" at Oak Lawn. William went to live in a tent at The Bluff to closely supervise the work force. The women started making plans to plant tobacco and corn at Oak Lawn and to grow rice in the pond. They were surprised to find their former slave Jacob very glad to see them and "talking all the time." Jacob expected to resume his previous servant duties, as he "confesses he can not do much himself with the hoe." Emily wrote her mother that the freedmen "nearly all carry guns" and were "civil & quietly respectful. We are better waited upon than we have been since the fall of the Confederacy, with cheerfulness & alacrity, showing the advantage of being land owners, if nothing else."[17]

Mrs. Elliott was staying in Charleston with Hattie, who was in her ninth month of pregnancy. Gonzales got "very vexed" when his mother-in-law took over making the boys beds and spanked Gertrude. He sent one of his coolies, Isidore, to assist at Oak Lawn. When the train stopped at the main gate, conductor William Crovatt bellowed to the Elliott ladies: "I've brought you a gentleman who does not speak English." At first impression, Emily found Isidore "rather wild," with "fine boots & bandy legs," and his speech was "not very intelligible." The antebellum socialites felt "strangely rather, domiciled with negros & a china man." After twenty-four hours, it was evident that Isidore objected to doing the field work expected of him. When Emily showed him the tobacco seed for planting, "the only notice he took was to say that he did not smoke." Instead, he proposed "cutting down all of the Laurels & large trees to sell in Charleston to make 'too much money.'" When he left Cuba, Isidore had expected either to supervise field hands or to have congenial employment in Charleston. He demanded to talk to Gonzales, departed for Adams Run, and returned in the evening. Emily had wanted him to "take a hoe or spade & show the negroes how to work," especially those who had reverted to the idleness manifested during bondage. According to Reconstruction historian Joel Williamson, African Americans "labored less arduously in freedom than they had in slavery. To many whites, the slowdown seemed a stoppage." He added that the freedmen "had to be instructed in the necessity of constant and assiduous labor." Emily concluded that with Isidore, "so ends, another attempt to better our fortunes & we must confess, great disappointment." Emily, who was busy hoeing and raking, bristled at the "tea caddy" and "heathen asiatic . . . only fit for the cholera," sending him back to Charleston the next day.[18]

By contrast, Hattie was enjoying her Chinese servants who "bowed & smiled & looked so happy & clean & so much like servants that I was quite pleased with them. They speak only broken Spanish." She regretted that they would soon leave her for other employment in Charleston. Many Carolinians complained that black servants, some being former field hands without refined training, had a rapid turnover and when irritated would resign without notice. Cooks were habitually wasteful and unclean. Hattie delivered her fifth child, Benigno Gener Gonzales (later christened William Elliott Gonzales) on April 24, 1866. He was named after Matanzas planter Benigno Gener Junco, a personal friend with whom Gonzales was doing business. It was a difficult birth, with Hattie remaining unwell for more than a week, causing great anxiety to the family. The baby was later described by his grandmother as "an uncommonly fine child—intelligent & good. His mouth may improve as his nose has & then he will be handsome."[19]

Meanwhile, the Elliotts were divided over the future of their Chehaw River lands, Social Hall, and The Bluff, which had gone up in value due to the growing Northern demand for yellow pine timber. New steam circular sawmills were being erected on plantations in the proximity of Charleston. Commission merchants were advertising to negotiate prices with lumbermen and timber cutters, and the Charleston City Railway Company was submitting public proposals for acquiring more than eleven thousand pieces of lumber. Emily advocated establishing a sawmill at Social Hall to exploit the timber lands. She expressed that Social Hall should go to Hattie and her children, but insisted that Gonzales pay sixteen or seventeen thousand dollars for it. Emily complained that Ralph was unsuccessful and indicated that had she "been the fortunate owner of a pair of pantaloons, the debts of the Estate would have been paid by this & the lands would remain in hand for the legacies." The Elliott sisters contracted J. Z. Johnson and four lumbermen to cut cross ties at Oak Lawn before departing to Flat Rock for the summer. Hattie, her mother, and the children joined them to avoid the prevalent summer fevers.[20]

In May, Gonzales applied at the Spanish consulate in Charleston for a new passport to travel to Cuba, trying to expand business contacts. Gonzales, Woodward and Company had steadily grown during its first five months. They started paying a city tax of $11.86 in January on sales merchandise, which the following month quintupled to $55.17, indicating a large growth in revenue. Their first merchandise consignment arrived from Cheraw on January 28. The company received a cargo of molasses and cigars from Matanzas on February 1; cigars, tobacco, fruit and other goods from Havana on the twelfth; merchandise from Baltimore on the twenty-fourth; and a railroad shipment two days later. Goods arrived from New York on March 5, and there were five railroad consignments that month. In April, the shipments increased to two from New York, on the thirteenth and the nineteenth, and there were six railroad consignments. In May, two more merchandise shipments arrived from New York, on the twenty-third and the thirty-first, and there were five railroad consignments. Federal tax records for May 1866 indicate that Gonzales, Woodward and Company was a wholesale liquor dealer and commercial broker, with a tax of seventy dollars, indicating a steady increase in business during the previous five months. The firm received three merchandise shipments from New York on June 11, 20, and 25, along with six railroad consignments. Gonzales had traveled to New York City on June 12 on the steamship *Quaker City* to promote his business ventures.[21]

That month, Gonzales, Woodward and Company apparently became the target of looters. On Sunday, June 24 at 5:00 P.M., a rock-throwing incident between black and white children at the Battery escalated into a fight between adults of both races. A *Harper's* artist on the scene sketched what he described as a race riot, depicting an officer's futile effort to halt the melee. The police dispersed the crowd and arrested about half a dozen ringleaders. Hours later, a group of African Americans who eluded the authorities instigated others in the immediate vicinity. Soon, a mob of two or three hundred "directed by some eight or ten negro soldiers" rampaged up East Bay Street and turned left on Tradd Street. The Gonzales wholesale liquor business at 73 East Bay Street was located three doors north of Tradd, close to the path of the rioters. The liquor trade was a hated symbol for African Americans, who were prohibited by the Black Code of September 1865 from making or selling alcoholic beverages. Richard M. Brantford, a white man who tried to flee the mob, was knocked down on Tradd Street by a shower of brickbats. The victim was kicked and beaten to death, abetted by an agitator shouting "kill the rebel son of a bitch." Although news accounts of the damage done by the rioters is superficial, it appears that the prosperous "Gonzales, Woodward & Co." suffered heavy looting. During July, the business only received one railroad consignment on the twenty-fifth, in contrast to the nine shipments the previous month. After the riot, the firm disappeared from state and federal tax records.[22]

Gonzales then turned his energies toward establishing a sawmill at Social Hall. The previous month he had bought one acre of land from John Raven Mathews for one hundred dollars. The land was on Chapman's Bluff on the west bank of the Ashepoo River, near Social Hall plantation. It was the site of the May 24, 1864, Battle of Chapman's Fort, where Confederate artillery had sunk the Union navy transport *Boston*. The landing had docking facilities to handle lumber shipments. Gonzales then traveled to Baltimore, where on July 17 he bought a steam-powered circular sawmill from George Page and Company with a promissory note for $6,726.63 to be paid in eight months to the

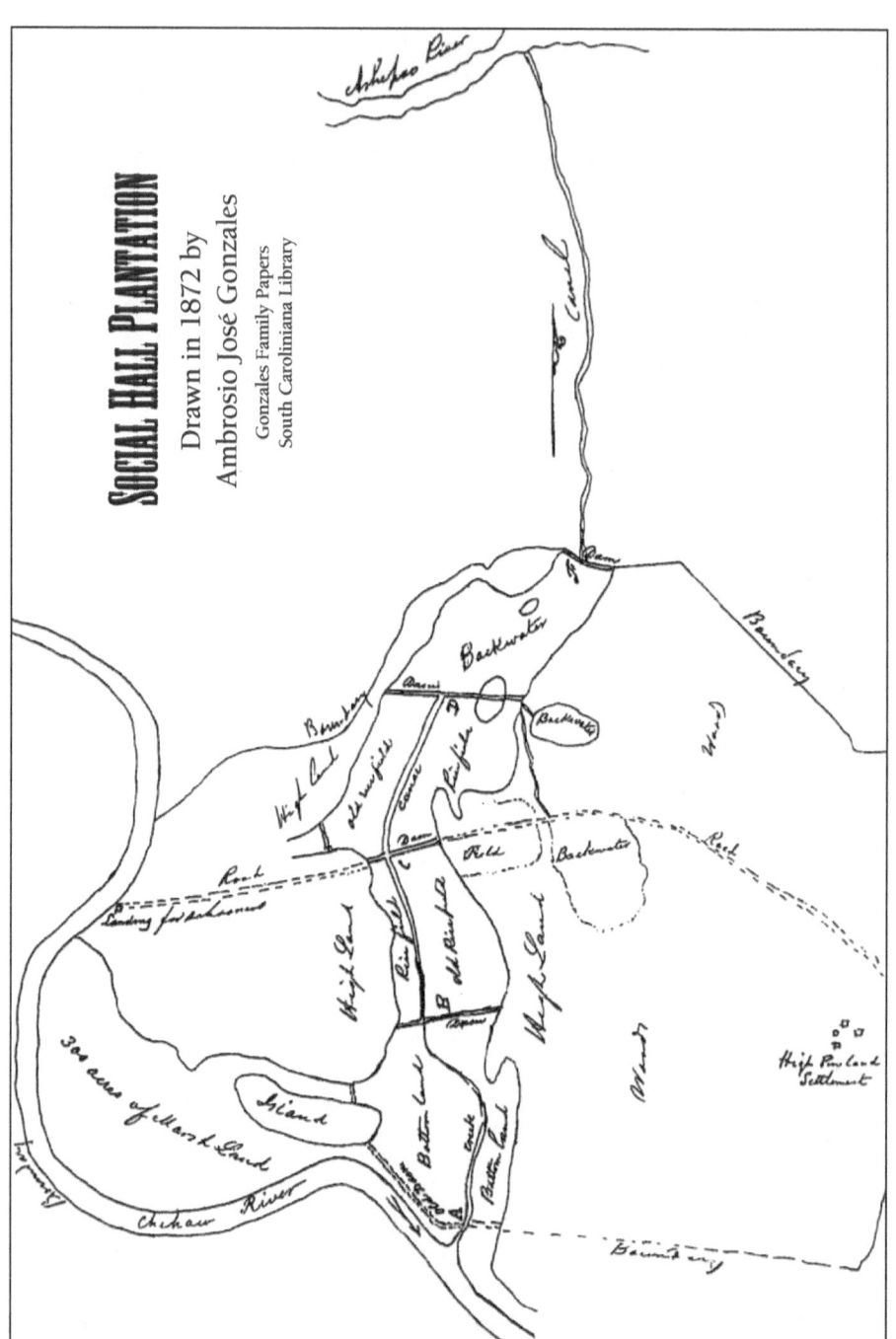

Social Hall Plantation, 1872

Commercial and Farmer's National Bank. The machinery arrived in Charleston by schooner from Baltimore on July 30. It consisted of "one Portable Boiler and Steam Engine of 25 horse power, one 3rd class Saw Mill with upper saw, 2 pair timber wheels, grist mill etc." The equipment was transported to Social Hall, where Ralph assisted Charleston engineer John F. Caha in assembling it. Unfortunately, they placed the mill in a bad location distant from navigable waters, built a shoddy water supply well dangerously close to the engine, and left the machinery exposed to the weather. Ralph, William, and one of the Chinese coolies had moved into tents at Social Hall, where laborers cut timber to build six log cabins. They subsisted on terrapins, bacon, and four grams of quinine daily to avoid fevers. When William's tuberculosis left him bedridden, he moved to his brother Tom's Bethel plantation on the nineteenth. Soon after arriving, William viewed from his bed the accidental explosion of Tom's steam sawmill, killing three workers.[23]

Gonzales, Woodward and Company received its last railroad shipment on August 18. By the end of the month, they vacated the premises at 73 East Bay and the locale was advertised for rent in the *Courier* on September 4. Partner Peter J. Esnard established his own commission merchant enterprise at 18 Boyce's N. Wharf. W. T. J. O. Woodward had started working after the riot as general superintendent of the National Express and Transportation Company at 157 Meeting Street. He went bankrupt in seven months and his business property was auctioned off. Woodward apparently departed the city, being later suspended from Masonic Washington Lodge No. 5 for nonpayment of dues. Gonzales left Charleston and moved to Social Hall in late September to oversee his sawmill operation.[24]

As part of the agreement to purchase Social Hall plantation, Gonzales signed on October 31 a bond for $6,175, with interest, as collateral to the trustees of the marriage bond of Mary Johnstone, freeing the Elliott estate of the debt. He had previously signed a ten-thousand-dollar note in May to Ralph Elliott, as trustee of Hattie's marriage bond, relieving that amount from the estate. The Social Hall deed of sale required the signature of Mrs. Elliott and her seven offspring, the heirs of their father's estate. Ralph, who was an employee of Gonzales at the Social Hall sawmill, exceedingly regretted having signed, and advised his brother Tom to confer with attorney De Treville on the matter "before giving your sanction to the transfer." Ralph wanted to sell Social Hall for cash to Northerners, in spite of the objections of the majority of the family, who wanted Hattie and her children to have it. He felt that Gonzales might not be able to pay off Mary's bond, making the Elliott estate liable for that debt. Ralph quoted De Treville as saying that if Gonzales lost Social Hall by defaulting on Mary's bond, he would then have the right to claim his ten-thousand-dollar marriage bond from the Elliott estate. In spite of this opposition from Ralph, Tom signed the deed giving Gonzales title to Social Hall plantation. Prior to finalizing the transaction, Gonzales and Jacob rode to Social Hall to verify the land marks, but the freedman was doubtful about the property limits.[25]

After acquiring Social Hall, Gonzales advertised in the *Courier* on November 10 the sale of "twenty bags of superior Coffee, warranted Cuba grown, over two years old, and to have been fifteen months in the pod before being shelled." The shipment had just been released by Customs after its July 2 seizure when it arrived from Matanzas on the British schooner *Aid*. It was part of a larger invoice, consigned to F. P. Salas and Company from the firm Da Costa and C. P. Madan, which also contained smuggled brandy casks and cigar boxes packed in seven sugar hogsheads and four kegs of rum marked as California wine. In a landmark admiralty case, the federal court ruled in favor of exempting from forfeiture Gonzales's goods and upheld the confiscation of the entire Salas invoice. In November, Hattie and the children joined Gonzales in a Charleston boardinghouse. He was back in Matanzas by the end of the month, with free passage from his friends Mordecai and Company in Baltimore, the agents for the two-mast propeller steamers *Cuba* and *Liberty*, which frequently traveled to Baltimore, Charleston, Havana, and New Orleans. Gonzales spent the next two months shuttling between Matanzas and Havana, seeking contracts to sell yellow pine timber to five railroad enterprises, including the Sabanilla Rail Road Company, directed by his friend Benigno Gener Junco. Spanish colonial bureaucratic transactions were very slow. Gonzales had to make personal appointments and give individual presentations and bribes to each member of the various committees involved in the negotiations.[26]

Part of his business difficulties stemmed from the bad luck of arriving in Cuba at a time when the government bank was in its biggest economic crisis in a decade. Gonzales had offered his friend Cándido Ruiz a partnership in planting at Social Hall, but Ruiz reneged at the last moment due to financial difficulties. His contract proposal to the Sabanilla Rail Road was rejected by the board of directors in December because he asked for a two-thousand-dollar advance for resuming sawing operations. The lack of business progress delayed a Christmas homecoming. Gonzales wrote Hattie about his predicament, saying that "it is hard to go back empty handed after all the trouble and the still greater suffering I experience owing to my anxiety about yourself & the children such as I have not had before." Gonzales later received promises for future orders from the managers of three railroad companies and hoped to obtain an order from his friend and former separatist conspirator Miguel Aldama Alfonso, owner of the Havana Railway Company. The Cubans were interested in yellow pine timber, which rarely rots, for building railroad cars and cross ties. The high-quality hardwood trees, native to the southeastern United States, did not grow on the island.[27]

Gonzales planned to return home with at least one lumber contract from Aldama, who gave him a four-hundred-pound "box of refined sugar, a hogshead of brown sugar, and a large bag of black beans." He was taking back two guardian dogs and Luis, his aunt Lola's young Chinese coachman, on loan free of charge for one year, who could "cook, sew, and do almost anything." In his letters to Hattie, Gonzales expressed great suffering and anxiety over being absent from his family for so long. He told his wife: "I am anxious to go to our little home and cease this roving life for since you have been with me and satisfied, I have loved you and cherished you so much the more and I know that we can be happy in a home of our own, if I can only pay my debts." Gonzales instructed Hattie to obtain from merchant Edward H. Lafitte all their family needs and that if they were too uncomfortable at the boardinghouse where he left them, to move to the Carolina House until he arrived on January 22, 1867. Mrs. Elliott, living with Hattie, described her as still looking "remarkably young, has her father's buoyant spirits, but the unassisted care of five children is telling upon her strength." The Elliott matriarch preferred the country life for her daughter and grandchildren.[28]

The Elliotts were also encountering emotional and financial difficulties. The crop from their Chehaw River lands was "less than hoped"; after paying expenses, they earned less than three hundred dollars' profit. Emmy and Annie returned to Oak Lawn in January, taking six-year-old nephew Alfonso, called Fonsy, with them. They asked their mother to "promise not to cry when you see the old place but to behave like the mother of the Gracchi!!" Mrs. Elliott was feeling "sorry & much shocked" at hearing of the death from cholera of former family servant John at Hilton Head. She regretted how "we were never grateful of those times that are now gone for ever." The family was overcome with grief when thirty-six-year-old William Elliott Jr., died of tuberculosis on January 21 at Tom's residence and was buried in the Stoney Creek Church Yard in Pocotaligo near the graves of Tom's children.[29]

After Gonzales returned from Cuba without funds, Ralph went to New York and Boston to obtain loans for himself and for Charleston merchants, expecting to make a five-thousand-dollar commission. He carried a letter of introduction from his sister Anne to William Amory, a family friend who had stayed at Oak Lawn in May 1861. The unreconstructed rebel was dour and uncomfortable among the aristocratic Yankees who refused to loan him money on the little security he had to offer. The Amorys gave Ralph a "small sum very hastily collected among several friends" who kindly remembered the Elliotts. Ralph departed Boston so abruptly that other family friends were unable to give him over one thousand dollars in donations that were later remitted to Oak Lawn. Four months later, Amory sent Annie Elliott two hundred dollars for a stone marker and a cast iron fence for her father's grave in Magnolia Cemetery. The money was instead spent on the land tax for Oak Lawn, which had doubled.[30]

Ralph returned home on the steamship *Quaker City* from New York on February 12. He befriended passenger George H. Hoppock, a thirty-two-year-old New York carpetbagger who lived in the Mills House and had run a wholesale grocery and commission merchant business on East Bay Street since 1865. Ralph later boasted how he "soon made him succumb to my eloquence," and as a result two weeks later Mrs. Elliott signed a risky two-year, ten-thousand-dollar mortgage on Oak Lawn to Hoppock. In exchange, Ralph agreed to cultivate cotton and rice at Oak Lawn and The Bluff and deliver the produce to Hoppock, who after discounting "the usual and customary factorage charges and commissions," would equally divide the profits between them. Ralph's previous failures can be attributed in part to his apparent bouts with alcoholism. Soon after he signed the deal with Hoppock, he wrote to his nephew William:

"I am sober & industrious, & feel the better for it." Ralph contracted thirteen hands and got seven mules to plant one hundred fifty acres of short cotton and twenty acres of corn. William received twelve hundred dollars of the mortgage to raise a crop at Tom Elliott's plantation. Tom had rebuilt his sawmill and sent lumber to Oak Lawn for housing renewal. The Elliotts also tried raising money by having an inexpensive edition of their father's *Carolina Sports* published in Great Britain that year.[31]

The Congressional Reconstruction Act of March 2, 1867, started a process that gave the military control over the governments of South Carolina and other former Confederate states. Two days later, the Gonzales clan, minus Alfonso who remained at Oak Lawn, departed Charleston by railroad for Social Hall plantation. The property is presently the Ashepoo plantation on state Highway 26 ten miles south of Green Pond. It had 1,450 acres of woodland and marsh, with 100 acres under cultivation. Its western boundary was the Chehaw River, and the southern and eastern property lines were contiguous with John D. Warren's farm. To the north were the plantations of Alexander Robert Chisolm and Haskell S. Rhett. The area abounded in "excellent yellow-pine, oak, hickory, cedar, etc.," and the wildlife included "deer, wild turkeys, squirrels, hares and in winter wild ducks." Social Hall was valued in the 1868 Colleton County auditors tax return at $1,970, a far cry from the $16,175 sale cost, indicating how much South Carolina land prices had been declining since the war. Its residents included Luis, the Chinese coachman, and Irish domestic servants Margaret Fludd and Roseanna Roach.[32]

The plantation was centrally traversed by a road from the main gate in the east to the Chehaw River landing dock on the west. Social Hall Creek and its canal extension flowed from north to south, pouring into the Chehaw River. The junction between this waterway and the road divided four rice fields. The creek and canal had four wooden dams, destroyed during the war, for regulating the tide water flow on the rice fields. The one-hundred-acre crop field, partitioned by the road, was at the center of the plantation. Near the entrance, on the highest land elevation, in a cluster of pines thirty-eight feet above sea level, were six log cabins with weather boarding. The parental quarters contained their bedroom furniture and a crib for Gertrude and Benigno. Hattie described how "an aged dimity, well darned & starched is hung between the sleeping & eating room—& is pushed aside when desirable." The room had a small cooking stove that also provided heat during the winter. Outside, under an awning, a safe stood on one side and a large white table on the other. Ten-year-old "Brosio" and eight-year-old Narciso, called Narto or Nanno, slept in the cabin occupied by twenty-seven-year-old Margaret Fludd, "on a cot & mattress, an old white curtain on a rod, divides their apartment, which is kept clean & is naturally ventilated." The abodes contained trunks for storing clothing and linen. The kitchen cabin had a stove with a waffle iron and coffee toaster, pots, dishes and utensils. A storeroom served as a commissariat for sawmill hands. The residential area had three enclosed lots: a vegetable garden by the main cabin, an oat field next to the storeroom, and the third contained the stable and cow house. Hattie made a constant effort to keep the place tidy, but the cabins were "so low to the ground & so impossible to keep clean. The poultry, pig & even the pony 'roam at their rise' & come into the shanty when ever they plaise [sic]. I don't object to clean poverty but I do rebel against dirt & dirty we must be as long as we are in such a low building." Other animals at Social Hall included two dogs, a rooster, some mules and horses. The latter were fed on fresh marsh, which was cut daily at the sawmill and transported in the wagon to the settlement.[33]

Five days after arriving, Hattie felt ebullient and optimistic, marveling at the fine complexion of her children, who had never looked better, were nourished, and in excellent health. She was "too glad to get the boys out of Charleston. The boys there are so profane & vulgar." The Elliotts sent from Oak Lawn baskets with reading material, correspondence, flowers, food, and clothing for the children. Hattie delighted in picking wild flowers with two-and-one-half-year-old Gertrude, called Tula or Tulita, whom she described as "a darling little foreign thing, very dark, & graceful, & a natural coquette," whose father "idolizes her." When Margaret admonished Tula that her father would be told of her misbehavior, she replied, "You can tell him, Papa is not going to 'Spunish' his only little daughter." Hattie felt partial to one-year-old Benigno, called Minie or Mino, and hoped that God would grant him as her last male child. Hattie was breastfeeding both children and wrote her sister Annie how "Tula considers that B. deprives her of a part of her nourishment. She says that *one* belongs to her & teases me very much at times. B. bellows whenever he sees me & Margaret says will never be good until he is weaned but I will nurse him until he is fourteen months old."[34]

The two eldest sons were being educated at home. They were "reading the history of France, studying that of England, spelling with meaning, Geography, Spanish, writing." The boys also read a biography of Alexander the Great, *The Life of Stonewall Jackson* (which the boys did not find interesting), and their grandfather's *Carolina Sports*. Gonzales personally instructed them in arithmetic and French, subjects that Hattie claimed she was "a poor hand at." The kids spoke fluent Spanish with their father and Luis, and they helped tend the vegetable garden. Hattie proudly wrote to Oak Lawn how "my boys bring the cow from pasture, cook for the dogs, feed them, go on errands & rake the yard, besides learning their lessons & having plenty of time to play." Sallie the cow provided the family with seven quarts of milk daily, clabber curds, and butter. Their diet was supplemented with poultry, fish, terrapins, crabs, mullets, shrimps, ducks, pork, and venison. The family sometimes savored roast lamb or a soup containing fox or squirrel meat with curry and rice. During lean times they ate crows, which according to Hattie, their "flavor is almost identical with the rice bird & they afford a much larger mouthful." Gonzales held family prayers at least four times daily, which Narciso considered tedious.[35]

One of the first tasks Gonzales completed was building a shed over the sawmill, which for seven months had been left exposed to the corroding ocean mist and the elements. The steam engine also operated a gristmill on Saturdays, and the toll for its use helped to ration the workers. A dock was built on the Chehaw River for schooners and rafts to transport the timber. A small rail line connected the landing with the sawmill. Hattie told her sister that "the decline in lumber has been terrible to all engaged in sawing, if it were not for the orders from Cuba, the prices of which were fortunately fixed, milling would be ruinous." To fulfill his Cuban contracts, Gonzales found that it was "cheaper to buy timber at the present prices, than to haul it from this pine land." The former school teacher, unaccustomed to rudimentary farm work, accidentally sliced his finger to the bone with a fodder-cutting machine. He was treated with home remedies, although his wife insisted that he see Dr. Henry E. Bissell in Ballowville, to whom the family recurred during emergencies.[36]

Social Hall received occasional visits from the neighboring Rhett family. Hattie, still harboring the Elliotts' resentment toward them, objected having "to see people whom I don't respect" and was "much disturbed at their proximity." The Elliott-Rhett antebellum property-line feud was rekindled when Haskell S. Rhett proposed building on a spot that he claimed was his land, and Gonzales considered it part of Social Hall. Rhett alleged to have recently had his place surveyed, to which Hattie informed her mother, "If it was done during the war I have no doubt he took what he wanted." Hattie was going to keep her eyes "very wide open on all transactions with him." She later felt relieved upon hearing that her close neighbors were going to be Nathaniel Heyward and Alfred and Marie Rhett instead of Haskell. Col. Alfred Rhett, brother of secessionist *Charleston Mercury* editor Robert Barnwell Rhett Jr., had led the South Carolina 1st Artillery Regiment and commanded Fort Sumter during the April 1863 ironclad attack.[37]

Former slaves stopped by Social Hall to pay their respects, including three females who worked at the neighboring Rhett plantation. Freedman Quash George gave the Gonzales family "a fine present of eggs." Hattie commented that "the negroes are ragged & look *so* hungry that I can't help feeding them—encouraged, they beg extensively." In response to difficulties that the Elliotts were having with their contracted field hands, Hattie complained that some freedmen agreed to work on certain days and then not appear. Others frequently arrived at work late and periodically vanished during task hours. According to African American historian Alrutheus Ambush Taylor, when blacks "worked for white people for wages they found that by two or three days' work they could procure money enough to support them in idleness the next week." He indicated that Colleton County laborers received "from $6 a month to $120 and $150 a year." To add to the family's problems, the Social Hall sawmill stopped working in early May. The shoddily-built water supply well was caving in, threatening the engine. Gonzales bought a new water pump and hired Charleston machinist Archibald McLeash to mend the leaking boiler. He lost "a valuable order for lumber from Edisto" due to a rumor that his mill had "blown up." The lumber shipments to the Cuban railroads were "too long delayed."[38]

When the sawmill became operational after two weeks, only a few local orders were finished, and it "was done so badly that it had to be done over again." Hattie complained that the nine white sawmill workers were fed three times a day, "received high pay & did nothing. The freedmen now employed there seem really to have more conscience." She lamented to her mother in mid-May, "if we only could get a cargo off for Cuba I should breath [sic] freely. . . . Well, if we fail here entirely we can still find a home & something to do in Cuba." Hattie expressed in the same letter anti-Union and antiannexationist feelings, saying that she hoped Cuba would be saved "from the clutches of the detested eagle!" Her previous desire for Cuban annexation was now dampened by the recent passage of the Fourteenth Amendment by the United States Congress. It had disenfranchised thousands of former Confederates and the Southern states still under military rule were required to draft new constitutions before rejoining the Union.[39]

Gonzales was frequently absent from home seeking sawmill contracts and doing business transactions. He went on horseback or wagon to Green Pond, Salkehatchie, Walterboro, and adjacent areas, or took the four-hour train ride into Charleston. While in the city, he occupied his old quarters at Mrs. Sheen's home at her invitation, thereby saving the expense of a hotel. Hattie heard "little but Irish" all day. Her husband was "so often away & when he is at home—at the mill & farm all day—& when he returns he finds me using some very unenglish expressions" adapted from Margaret and Roseanna. Hattie considered them "excellent servants, but they lack the refinement of language so striking in our former [house] slaves." In the evenings, Hattie read to her husband all the political articles she had saved from the publications sent by the Elliotts. When the patriarch was absent, Hattie had one of her sons sleep with her, ready to blow a horn in case of an emergency, because she could not do it. When the horses and mules "broke through their fence & took to promenading on the piazza," the blaring horn summoned Luis and the overseer to the rescue. Hattie was very grateful to Luis, whom she considered invaluable, honest, and watchful: "Sleeps in the store house & guards it with 'macheta' [sic] & pistol, weighs rations, goes to the R.R., works at the mill, paints, has made a nice boat in the last two days—cuts wood & draws water for the cook—& gives us delightfully cooked dinners when we feel self indulgent & is besides a safe & kind companion for the boys who are perfectly devoted to him."[40]

Gonzales hired an Irish overseer, "Old" Simmons, who did not stay long. He "didn't like to get up early, & feed the horses," and "could not direct others & was so afraid of negroes that he would not stay by his garden to guard it at night." The garden contained corn, peas, turnips, cabbages, leeks, squashes, snap beans, tomatoes, okra, potatoes, plums and watermelons. It became the frequent pilfering target of freedmen and raccoons. Gonzales then employed an armed African American to protect the patch. Even so, when the blackberry bushes were picked clean, the watchman purported that the crows "had eaten them all." Other sentinels proved just as unreliable, making Hattie exclaim, "The man he left in charge I know is a thief—but I wont worry & am grateful when they allow me a dinner from the garden." The habits and attitudes of slavery still needed considerable modification. Poor crops, low wages, and high prices increased larceny. To solve the problem, a ferocious mastiff named Leon was kept chained near the storeroom during the day and allowed to roam freely at night.[41]

The Elliotts left Oak Lawn on June 1 to reside in their rebuilt house in Adams Run until the fall, sending seven-year-old Alfonso to Social Hall on the train with his father. While staying with his grandmother and aunts, Fonsy had been troublesome, dedicating "little attention to his book" and being "indignant at the idea of cleaning the cows house." His early abhorrence of work and study became perennial traits. Hattie decided to bring Alfonso home after Mrs. Elliott complained that he "needs control" and was "racing about in the sun & won't come in when he is called." When Fonsy departed, Mrs. Elliott and Anne "almost cried," but Emily laughed, an act that the child considered "very reprehensible." The mischievous Alfonso, who enjoyed being with mules, may have gotten kicked by one. His body had an "ugly" scar and prior to returning home, his mother assured him, "None of our mules kick they are all good mules."[42]

Alfonso's joyous homecoming was marred by the news that the largest commission merchant firm in Charleston, John Fraser and Company, with offices in New York and Liverpool, had gone bankrupt. The United States government had brought suit against the firm and its partners, making them pay import duty plus interest on all goods arriving on blockade runners during the Civil War. The confiscation "terribly depressed" the Charleston business community, including many of Gonzales's creditors, who started calling in their loans and canceling further credits. The failure affected the merchant firms of E. H. Lafitte and Company and Willis and Chisolm, to whom Gonzales owed $3,700. Hattie informed Emily that as a result of the John Fraser and Company bankruptcy, "our good friend Lafitte does not know if he owns a dollar." Gonzales was also indebted to merchants Thomas Bonnel, $200 for a note; West and Jones, $125 for a produce box; W. Mathisson and Company, $25 for clothing; and druggists Charles F. Panknin and Raoul and Lynah, $45 for medicine and quinine. Hardware debts to keep the sawmill operational amounted to $246 with the Charleston firms of Cameron and Barkley, Adams and Damon, John Toumey, J. E. Adger and Comapny, Cleland Kinloch Huger, and Bissell Brothers. Gonzales also owed $255 in grocery bills to Klinck and Wickenberg, R. A. Pennal, and Laury and Alexander. The great scarcity of money made corn "a cash article." It was consumed boiled or parched, and was used for making corn bread, corn meal or whiskey. At Social Hall, hired hands would "only work for corn," fearing greenbacks might become worthless like Confederate money. Gonzales was obligated to buy corn to also feed his family and horses, paying extra for drayage, railroad, and carting expenses.[43]

The financial crisis prompted Gonzales's debtors to default on him. The purchaser of their Charleston furniture had only partly paid for it. The family had not heard from the agent who was selling their silver in Baltimore. A gentleman owing them $35 for lumber begged to be let off after he could not raise the money in Charleston. Gonzales had not received payment for the work done by his blacksmith and grist mill. Hattie felt bitter when the Seabrook plantation manager obtained only twenty dollars worth of lumber from them while sending a major order of several thousand dollars to someone else. She called neighbor Alfred Rhett "mean" for not purchasing floor boards from their sawmill. By mid-June, Gonzales had not filled his Cuba orders and there was "little timber to be bought now, the low prices have discouraged the cutters & raftsmen." A month later, Hattie lamented that "Chisolm, Colcock & others who bought lumber 'for cash' can't pay for it. Men who were to furnish timber have had fever & could not cut it. Mill not working, hands to be paid & family to be fed! Nothing sure but debts. Poor Gonzie, he does not sleep at night, & is up at day light & is looking very badly. I trust something may 'turn up' for him soon."[44]

Neighboring sawyers were also having difficulties. The sawmill of William Simmons exploded after its well caved in. A Mr. Glover had "stopped sawing for want of orders" and desired to subcontract with Gonzales for the Cuba orders. The spring crop throughout the state had been severely damaged by insects and bad weather. Ralph Elliott wrote to his brother Tom that it had "utterly destroyed our prospects, and we have to sell mules, to eke out provisions until the corn ripens. I have no hope of being able to do more than pay expenses this year, & fear that even may not be accomplished. We are a doomed people."[45]

The summer season brought legions of mosquitos, sandflies, and other bugs to Social Hall, which were "piquing the poor children unmercifully" and had "Brosio murmuring about the [Biblical] plagues of Egypt." The family used a "portable smoke" to fight the insects and slept under mosquito nets. Poisonous reptiles crawled about and Hattie reveled at having "killed a young rattle snake with my own hands." The mettlesome Alfonso kept the Irish servants and his brothers "in a constant row" and had his cot moved into his parents' room. The boys would cool off with "a cold bath in the piazza which they enjoy the more perhaps from having to draw the water for it themselves." Hattie shaved Tula's head to prevent lice, and quipped how she "resembles Mr. Petigru at present. She is very merry & not at all unhappy at her appearance." The Social Hall inhabitants began their daily ration of quinine on July 1 to avoid fevers, which affected some of their sawmill hands and the neighboring Rhett children. Even baby Minie received a half-gram daily dose mixed with clabber. Hattie discovered that making quinine pills for a large household consumed much time. Her other tedious labors included darning and the weekly washing of over sixty articles of clothing.[46]

When Leon died that summer, he was replaced by two "very efficient" and "very much dreaded" guardian dogs, Marengo and Chino. The Gonzales family appreciated these animals, "for we have lost nothing while our neighbors have met with serious losses," of stolen mules, puppies, ducks, and vegetables. Luis accompanied Gonzales one night to guard the corn field from depredation by bears and freedmen. Gonzales also hoped to kill a deer, but was unsuccessful. Luis later said that he would not undertake another such venture, not even "for $1,000, the sandflies were so dreadful!" Hattie indicated that "undisciplined blacks . . . have taken *all* of our corn." The young freedman in custody of the field alleged that "Mr. Rhetts mules ate it but the white man says he sees where it has been plucked." She also complained of having no "pecans, for the negroes have robbed us of nearly all of those at Social Hall." Mrs. Elliott blamed the freedmen and the crows for taking all the corn planted at Oak Lawn. She accused former slave Robert Kinsey of stealing her turkeys.[47]

The Gonzales family was able to survive when aunt Lola in Cuba sent Gonzales money in mid-September which allowed him to buy corn, bacon, and supplies and to feed the sawmill hands. Their family diet was so deficient, that according to Hattie, "rice is a great luxury with us, we never have it but on rare occasions when the grist gives out." When rice was available, it was given to the baby with milk for breakfast. Gonzales then hired W. W. Stubbs, a Northeastern Rail Road sawyer from Mount Holly, to run the sawmill. Hattie wrote to her mother on October 2: "You know that the mill has been idle for months. We have now an admirable manager & *so* far an honest man. He has removed the mill to the very bluff (where it should always have been) saving a great deal in the future thereby." She also pleaded for the Elliotts to plant or hire out the adjacent Bluff plantation, which had become "a harbour for vagrants who steal where they can." The Charleston correspondent of the *New York Times* noticed a general "feeling of despondency" among the planters, due to the increased need "for capital and the unprecedented scarcity of ready money," the dismal future culture of cotton, and the "stagnation which has existed on the business streets during the Summer."[48]

Prospects for a crop at Oak Lawn were "unpromising" that fall. Ralph was only able to raise two bales of green seed cotton, which sold for fifty-five dollars each after prices had fallen by half. Sea Island planters Seabrook, Jenkins, and others, had completely failed. It was the third year of poor crops for the state, due to droughts, storms, and a caterpillar plague. Tom Elliott bemoaned: "I am financially Ruined my Crop will yield me nothing." Hattie noticed that there was "no money & scarcely any business in Charleston!" When Gonzales went to Charleston during the first days of October in search of milling contracts, his wife hoped that "this last visit will determine if our business is to go down entirely, or *go on* for a while,—success I do not dream of—The expenses of the mill are too enormous & labour too high to permit it. The lost crops, of course diminish the demands for lumbers, planters, don't need to build houses when they have nothing to put in them."[49]

In mid-October, Hattie received three letters in one day from her husband in that had been delayed in transit. He expressed "great anxiety at his prolonged absence & in great suspense kept waiting day after day upon the Gentleman who expects money & who has partly promised to lend it for carrying on the milling." Hattie informed her mother, "We are so much in debt that failing to get this money—will be complete ruin—but I am perfectly prepared for the worst. I am most anxious to leave this country for ever. The only attraction here is yourselves & although so near the difficulties of meeting are innumerable." The Elliott matriarch replied, "Hattie, tis a consoling idea that every one, the most acute Merchant & Planter, has failed. We have worked hard & must submit to our doom." Emily expounded their situation: "Money is scarcer than ever, & we hear of nothing but failures & losses. We are up to our ears in debt for the crop, won't think of paying & even had we made a crop the prices forbid profit, $9.00 tax on every bag of cotton."[50]

Hattie wrote back to her husband that the family was well and asked him to remain "until the business can be settled one way or the other." Gonzales responded on October 18, thanking her with all his heart for the very comforting letter. He planned on being home in a week, as he had better hopes than previously of making a business arrangement. The Cuban sent his wife by the Southern Express Company some postage stamps, "63 lbs. of bacon, some nice crackers for the family & 1 oz. genuine ham." He asked her to have Stubbs, the sawmill foreman, buy and grind corn for the family and the laborers. In late October, the sawmill was handling a new order from the Seabrook plantation. The Gonzales family was "put to great inconvenience & suffering by the failure of N. Heywards son to fulfil *his* engagement of paying for his lumber in corn." Hattie complained that "no one now seems to care about breaking their engagements." The Elliotts had been anxious to see the Gonzales children for months, but the railroad trip had to be postponed as "there was literally no money to buy their passage." The pesky Alfonso was described by his mother as "a lazy little scamp & is never energetic but when he [is] riding or driving a mule." Stubbs was so impressed, that he told his timber haulers that Fonsy "could beat them all at driving." Benigno was teething, had learned to walk, and delighted in being drawn about in a cart that Luis built for him.[51]

While in Charleston, the forty-nine-year-old Cuban was bedridden during the last three days of October with "awful" back pains due to a cold and "a great disorder of the liver." He was prescribed "tonics and two pills" by family physician Thomas L. Ogier. The city was "very unhealthy," with "fevers, congestive, & membranous croup," resulting in some deaths. Gonzales acquired from grocer R. A. Pennal some provisions on credit and sent them to Social Hall by express. He enclosed a note to his wife saying, "The 5 gallon whiskey you can have transferred to smaller vessels & have it dealt out to mill hands. The five gallon keg of molasses likewise. The one gallon demijohn of whiskey & 1 gallon demijohn of syrup are for *home* use. The self raising flour is in a box for Roseanna." According to Hattie, "Bacon, molasses & whiskey is what the freed people cared for. They have passed the stage for fancy hats & Jewelry at least in these regions. T'is very amusing to see Gonzie trading with them, several times he was about to cheap himself badly when Brosio & self came to the rescue—he is an excellent shopper however & has made some wonderful bargains lately."[52]

Gonzales returned home on November 5, worried about his unsuccessful commercial dealings. He went hunting with the Rhett brothers several times, to get "meat for the family," failing to get a deer, "although the woods are full of them." The Cuban found that one of his mules had been stolen when it was turned out to pasture to avoid starvation. Sallie the cow was pregnant and unproductive, and Brosio was being sent to purchase milk from a freedman at twenty cents a quart. The two servants had not been paid in months and, according to Hattie, were restless due to "sandflies, fear of fever, absence from the Priest (what a mountain they will have to confess) & want of company." Roseanna soon gave a month's notice that she would return to Ireland at the end of November.[53]

When the fever season ended in early November, the Gonzales family was relieved that their children had overcome slight fevers, thanks to their daily dose of quinine. Tom Elliott, his six-year-old son Apsley, and his daughter-in-law Belle were afflicted with congestive fever in October, resulting in the death of the last two. The neighboring Rhett family had become ill and some African Americans in Walterboro had died of bilious fever. Gonzales and his wife imbibed a mixture of gin and whiskey as a medicinal tonic. Hattie had helped nurse the sick field hands and wrote to her sister Annie: "You would be amused to see what a doctress I am considered by the nigs & poor whites—& t'is lamentable to find what ignorance prevails among these people. They die for the want of proper medicine." With winter approaching, the Elliott women were soon sewing warm clothes for the Gonzales children, who were also fitted out with "Paddie" jackets by Margaret.[54]

Gonzales, after spending a week in Social Hall, returned to Charleston on November 12 to collect cash "for lumber bought at the mill." His wife described the situation as "hard work to get this money for the factors are rarely in funds for their patrons & it takes a deal of coaxing and teasing to induce them to pay up." Gonzales also learned that George Page and Company, whose promissory note for the sawmill had been due in April, had sued him in the Federal District Court in Charleston. The plaintiff, represented by the firm of F. Rutledge and H. E. Young, was asking for double indemnity, a total of $13,460. Gonzales hired thirty-six-year-old defense attorney Michael Patrick O'Connor, a Beaufort native and former State representative. During the war, O'Connor had organized the Jasper Greens and was later a first lieutenant in the Lafayette Artillery Battery. He had also been the conduit of relief funds donated by Cuban planters for the Confederacy. In 1865, O'Connor had taken his family to live for two months in Cuba at his father-in-law's Delta plantation in Sagua la Grande. Gonzales returned home a few days later, stopping to visit the Elliotts at Oak Lawn.[55]

Hattie envisioned a bleak future for her family: "We are trying our best to make the mill pay but it has done little more so far—I have to ration the hands. T'is awfully expensive, all the time requiring something which costs money & does not return any. The present manager is very capable, hardworking & is altogether a businessman so I trust something may be done to make the speculation not altogether so forlorn a failure as it now appears to be." The mounting vicissitudes wore down Gonzales's natural optimism by mid-November. Hattie noticed that "he has not been so hopeful for some time past." She had also lost her enthusiasm for Social Hall, saying, "One would think that 'Sodom' was the veritable place of this spot." Hattie called it a "great place this, for insects of every description, & they 'never say die' fleas 'skeeters' sand flies, house flies, cockroaches, crickets, grasshoppers & moths of every kind—a splendid situation this for a naturalist." She had stopped teaching the boys because of "too much to do & too little time to do it in. Our servants do very little: Margaret only minds Benigno, & Roseanna confines herself to her very little cooking & washing . . . I wait more upon my cook that she does upon me."⁵⁶

Gonzales hired a white carpenter from Whippy Swamp named Fields, who claimed to have been "a good soldier," to build a clay chimney onto the cabin. Fields's wife agreed to do the wash and her sister, "a colored gal," would cook and do chores. After the Fields clan occupied a sharecropper's shack at Social Hall, the carpenter alleged that he and his children were sick, and delayed his work on the chimney. The Gonzales family provided medicine and soup for the children and sent Fields his meals, in addition to his rations of coffee, flour, tobacco, and whiskey. The carpenter spent weeks without finishing his work before absconding. The chimney was completed on November 29 by a freedman from the sawmill, who also built an addition to the main cabin.⁵⁷

Gonzales returned to Charleston, summoned before the federal district court on December 2, regarding the pending lawsuit. Since the family letters never mentions this affair, Gonzales may not have initially informed Hattie about it to avoid her further grief. She only told her mother that her husband had gone to Charleston to collect "the paltry sums due him by a 'big planter,'" and to acquire supplies for his family. They had "not had a drop of milk for nearly a fortnight & have been out of sugar & butter too." Gonzales grieved at seeing his eighteen-month son "so thin & pale," and Hattie regretted having stopped breastfeeding him. She indicated that Benigno "cries dolefully after corn bread. We give him coffee & gin & water, very weak." Gonzales was invited to stay in Charleston at the residence of railroad conductor William Crovatt, who lived "in comfort & style" at 83 Cannon Street. Crovatt "gave him a party, at which were present very respectable people." Gonzales heard that after New Year's, many King Street stores were closing and large numbers of clerks were being dismissed.⁵⁸

The Cuban was back home by December 9 with corn rations and very "cheerful for he had the prospects of forming a lucrative contract North—Money to be paid as cargoes left the mill." His glee turned into despair the next morning, when Stubbs and his employees told him that the sawmill boilers, which had been in disrepair for a long time, had broken down. The workers then departed to spend Christmas with their families. Stubbs remained a few days, traveling to neighboring islands to look at boilers for sale. When a deal could not be arranged, Stubbs returned to his Mount Holly home with a note for $140 in back wages. Other calamities followed that week, with the loss of a fine mule and Gonzales's buggy breaking down when Luis drove it to Green Pond.[59]

In mid-December, fifty-three-year-old Charleston merchant Edward H. Lafitte stayed at Social Hall with his nephew for a few days, enjoying a hunting invitation from Gonzales. Hattie described him as "a perfect gentleman & has been most kind & considerate of us," but she was embarrassed by "the horrors of our present condition." She borrowed silverware from her mother for the visit, making her melancholic as she recalled "the good old times." Lafitte, who was the family's biggest creditor, was no longer able to assist them, as he had "been ruined through endorsing for the Trenholms whom he believed to be good pay." The Gonzales household was experiencing their worst Christmas season ever, and lack of money prohibited them from traveling to Oak Lawn to be with the Elliotts.[60]

The Cuban made another business trip to Charleston during the first days of January 1868. The city was abuzz with the arrival of 124 delegates elected to the state constitutional convention. After Gonzales returned to Social Hall, he visited grocer "Large Trees" Hutchinson, bartering provisions of corn, bacon, sugar, and molasses for a lumber debt. The economic depression, the money shortage, the mounting debts, and the difficulty of repairing his sawmill, made Gonzales decide "after enduring a summer of anxiety" to "give up all his plans" at Social Hall and find something more reliable in Cuba. In February, the family abandoned their farm for the long-awaited reunion with the Elliotts at Oak Lawn. Their remaining domestic servant, Margaret Fludd, owed $135 in back wages, went to work in Charleston for the family of lumber merchant E. H. Sherman.[61]

The Elliotts had renovated their slave-cabin home with two additional chambers in the loft, each with three large windows. One room was for the matriarch and the other was used by the Gonzales family during their visit. Some of the children slept with their grandmother and aunts. The kitchen house had been rebuilt and brother Tom was also staying at Oak Lawn. The family reunion was an ecstatic event, but Elliotts were shocked to see the emaciated twenty-eight-year-old Hattie. A year of exposure in the pinelands had taken its toll on the Southern belle. Her weight had dropped under ninety-six pounds, she had three ridges across her forehead and her hands were "as hard as bricks." She was accustomed to sleeping anywhere and did not mind "sleeping on the floor." Still, Hattie arrogantly claimed, "I am more aristocratic now, than I ever was & the poorer I am, the more proud I am of my good blood—perhaps in *heaven* two classes might live together on the same footing, *religion* making ladies & gentlemen of all, but on earth certainly not." Her sister Mary considered Hattie "a heroine to have stood those fevers so long uncomplainingly."[62]

In March 1868, when Luis was scheduled to return to Cuba, the entire Gonzales family accompanied him. That month, Jefferson Davis stopped in Havana on his way from New Orleans to Baltimore. He received a telegraphed invitation from an acquaintance, Edward Sánchez, to visit his home in Matanzas. After Davis accepted, Sánchez quickly gathered a welcoming committee that included his cousin Gonzales, Hattie, and former Confederate Gen. Birkett Davenport Fry, living on the island since 1865. When Davis arrived in Matanzas, he shook hands with the former colonel, whom he had denied a general's rank on six occasions, and they quickly forgot their differences. Hattie wrote on March 20 to a friend in Greenville, South Carolina, that when she saw Davis, "I was so happy, I took his hands in mine and we instinctively pressed them in long silence." Davis rode in a carriage with Gonzales, Hattie, Fry, and Sánchez, to the latter's residence, where they dined together. Davis was introduced to the Gonzales children. He took three-year-old Gertrude in his arms and spoke of his own family. They spent the evening reminiscing and then accompanied the former Confederate president to the Matanzas Lyceum, where he was made an honorary member. Hattie found Davis to have "a look of melancholy, still he was entertaining and amusing at times. Joked about many incidents during the war." Gonzales "enjoyed his visit greatly."[63]

The Gonzales family recovered their health while guests of wealthy relatives and friends. The patriarch arranged to settle permanently in Matanzas, before returning with his kindred to South Carolina, after receiving passports on June 17, 1868. The following month, the state was readmitted into the Union after ratifying the Fourteenth Amendment. The Gonzales family settled in Charleston, but their movements are sketchy during the next six months. The federal lawsuit for nonpayment of the sawmill was pending. The machinery was removed from Social

Hall and returned to Baltimore in August. Appraisers from the Baltimore Steam Boiler Works "found the whole in a broken and damaged condition," worth only $1,250. By mid-August, Brosio, Fonsy and Mino were staying with the Elliotts in Adams Run, while Gonzales, Hattie, Narto and Tula were living in an 1820 two-story boardinghouse run by a Prussian couple, fifty-nine-year-old grocer John Scharlock and his forty-nine-year-old wife Amanda. It was located at 75 Cannon Street, a few doors from the home railroad conductor William Crovatt. Amanda was apprenticing two African American girls, eight-year-old Rebecca Chisolm and her five-year-old sister Anna. Tula liked Becky "very much & is playing with her all the time." A week later, Gonzales obtained a visa from the Spanish Consulate in Charleston to travel to Matanzas with his family, but he failed to use it because the legal claim against him had not been settled.[64]

In early October, the Gonzales family once more quartered with the Elliotts, and continued making plans to return to Cuba. Hattie sent a warm cloak to her niece in Baltimore, where her sister Mary Johnstone had moved with her children that month to work as a matron in the recently founded Edgeworth School for Young Ladies. Mary was "much pleased at the idea of seeing Hattie and her children on their way to Cuba." The crop at Oak Lawn, planted by Tom and Ralph with hired hands, was again a failure. Gonzales and most of the children returned to Charleston first, where Hattie and Benigno joined them on October 31. The family visited a traveling circus the next day, except Alfonso, who was feverish and remained with Mrs. Scharlock.[65]

Gonzales wrote on December 2 to General Beauregard in New Orleans, asking to intercede with the owners of Perkins and Company for reduced rates or free steamer passages to Havana. Beauregard replied eight days later that it violated company rules, adding, "I regret to hear of your disappointments, but you have this to console you, if it be a consolation, that few, very few of our late associates have met with any success since the War." Beauregard wished Gonzales better luck in Cuba but warned him against involvement in the independence revolution which had started two months earlier: "Your past experience will give you, at once, an important position there, should you desire to take a part in the coming struggle. You should consider well, however, whether Cuba is not happier under Spanish rule, than it will be under the complications which may ensue, should she separate from the mother country." The first Cuban War of Independence was declared by planter Carlos Manuel de Céspedes and a group of Masonic conspirators on October 10, 1868. Céspedes freed his thirty slaves, who joined the insurgency, and issued a proclamation calling for abolition and independence. The struggle was initially contained in the eastern provinces of Oriente and Camagüey.[66]

Beauregard later wrote to his former chief of staff, Gen. Thomas Jordan, in New York, enclosing a letter of recommendation Jordan needed to be hired by the exiled Cuban Junta to organize their rebel army on the island. He informed Jordan that Gonzales had returned to Cuba and could be of service to him there. Beauregard stated that he opposed filibustering and classified Narciso López and William Walker as no more than "Military Carpetbaggers." Beauregard's letter, his recent contempt for filibusters, and his animosity toward Jefferson Davis disappointed Gonzales and resulted in a lengthy estrangement between them. No further correspondence between them has been found, and Beauregard later omitted his former chief of artillery from his Civil War reminiscences.[67]

As the verdict of the sawmill lawsuit against Gonzales drew closer, he filed for bankruptcy on December 16 before Charleston federal district court judge George S. Bryan. Attorney Michael Patrick O'Connor again represented him. Gonzales gave a statement of "creditors whose Claims are Unsecured," attesting that during 1866 and 1867, he owed $5,088 to Charleston merchants, of which $1,900 was due to Edward H. Lafitte and $1,800 to Willis and Chisolm. Twenty other debts were under $200, including $20 he borrowed from Joseph Purcell, the livery stable keeper at the Mills House. The smallest debt, for $10, was due to druggists Raoul and Lynah for medicine. He also declared that he owed George Page and Company of Baltimore $7,000 for the ruined sawmill, the $6,175 bond of Mary Johnstone, and his $10,000 marriage bond, which he signed over to Ralph Elliott as trustee. The total debts of his bankruptcy petition amounted to $28,263. The ownership of Social Hall reverted to the Elliott estate since the two bonds Gonzales assumed for it, totaling $16,175, were unpaid. The Reconstruction state constitution had exempted the property of married women from their husband's debts. Gonzales surrendered to the court register his entire estate: a deed for one acre of land at Chapman's Bluff, bought in June 1866, worth $100. The "Personal Property" inventory form filed with the court indicated that Gonzales owned "no Household Goods, 2 suits of clothes." That same day, Gonzales was adjudged bankrupt and was granted a certificate of protection. Four days later, the jury in the civil lawsuit brought in a verdict against Gonzales for $6,300.64 on behalf of plaintiff George Page and Company, and an additional $31.25 in court costs.[68]

In spite of the bankruptcy and the freezing weather, the Gonzales family had a "pleasant" Christmas Day. Hattie indicated that the children "were very happy in their own way—having sold some old clothes we could give them a little money which they spent in the most foolish way for such sensible boys—Trumpets, horns, accordions, organs, & other noisy toys—fire crackers of course." Mrs. Amanda Scharlock, their boardinghouse owner, supplied the family with a turkey meal, fruit, and pound cake. A Mrs. Huger gave the children a big bundle of candy and a little cup and saucer for Tula. She also received a beautiful "doll, cradle & real little mattress," which delighted her.[69]

The Spanish consulate in Charleston expedited visas on December 29 for the Gonzales family to travel to Havana. The consul noted on the passport: "This gentleman, I know for certain that, in accordance with acquired information, after the dismissal of the confederate army, to which he belonged, is dedicated exclusively to his agricultural interests." Two days later, the *Courier* published the Gonzales bankruptcy notice, calling on his creditors to prove their debts before the federal court register on January 11, 1869. Gonzales avoided that embarrassment. On New Year's Day, the family started bidding farewell to their Charlestonian friends and relatives. Cousin Ebet Burnet found that "they all looked bright & well, the children all seemed delighted with the idea of travelling, steamboats, etc." Hattie seemed "very bold, and determined to be cheerful. . . . She had a bright color, and looked just as pretty as ever." Upon boarding the steamer for Savannah that night, on the first leg of their circuitous journey to Havana, the sky was very dark and looked "so much like rain." It appeared to be an omen of the misfortune that awaited them before the year ended.[70]

The previous three and one-half years had been very trying for the Gonzales family. The stress, insecurity, anxiety, debts, despair, and poverty, prompted by Reconstruction, defined Gonzales's character. His religious faith, expressed through frequent family prayers, supplied spiritual comfort during times of crisis. Hattie, the consoling and loving wife, provided affection, understanding, and moral support. Gonzales proved very forgiving toward a despondent and powerless Jefferson Davis, forever ending their wartime bitterness with a sincere handshake. The Cuban, an honest businessman, was incapable of exploiting anyone, even the freedmen, in his transactions. His commercial failures were the result of fateful circumstances that affected many other Carolinians, including crop losses due to theft and inclement weather. His increasingly prosperous Charleston commission business was destroyed by race rioters. The economic depression of 1867 thwarted his lumber venture and drove him deeper into debt. Forever an optimist, the Cuban always expected something to turn up. He now viewed the return to his homeland as providing that forlorn opportunity.

## Nine

# FAMILY FEUD

The Gonzales family arrived in Savannah by steamer on Saturday, January 2, 1869, and stayed for a week with their thirty-eight-year-old cousin Leila Elliott Habersham. The Friday night before departing, forty-two-year-old banker Wallace Cumming and his wife provided a "delightful" carriage tour of Savannah and gave Tula "a tremendous wax doll." The next day, the Carolinians boarded another steamer for Fernandina, Florida, arriving at dawn on Sunday, January 10, and quartered in the elegant Virginia House. Two days later, they left on David Levy Yulee's railroad to the Gulf Coast port of Cedar Keys. The town, serving seven steamer lines, had "three hotels, three restaurants and bars, and several mercantile establishments." The Gonzales family lodged in a modest hotel and enjoyed "the salt air, fish & oysters" while awaiting the steamer *Alliance* that carried them to Cuba on the fourteenth.[1]

The vessel stopped for twenty-four hours at Key West, where the family visited Stephen Mallory, the former Confederate secretary of the navy who had assisted Gonzales after the Cárdenas expedition. Mallory drove them around town, and "was charmed with the boys & much pleased to find he was Alfonso's godfather!" The continuing voyage to Havana "was very rough [and] the ship rocked terribly" as it traversed the Gulf Stream, making some of the kids a little seasick. When the *Alliance* sailed into Havana harbor on January 19, the passengers' passports were taken ashore before being allowed to disembark. Capt. Gen. Domingo Dulce immediately sought an interview with Gonzales and refused "permission to land to all the passengers until he had seen him." The former revolutionary and Confederate chief of artillery assured Dulce that he was there "to make a support for his family" and not to succor the new independence movement. Gonzales was permitted to stay in Cuba and the transient steamer passengers disembarked for the day. Dulce's moderate attitude toward the rebels prompted some hard-line Spanish officers to force his resignation four months later. When the *Alliance* returned to the United States, a rumor spread that Gonzales had been incarcerated in the Morro Castle, alarming the Elliotts and their friends. They asked Captain Reid of the *Liberty* to find Hattie and the children and take them back to Charleston. She later blamed "Yankees" annoyed with the passport delay for making up the story.[2]

The Gonzales family were greeted by Luis, their Chinese former servant, who drove them by coach to Aunt Lola's house, which Hattie found "handsome & one of the best located in the city." She described Lola as "a very sweet & handsome old lady—said to be over seventy, looks about 48, she is delighted with the children & is as kind as possible." Lola, who never had children, supported a large family of grand nephews and nieces. Gonzales's stepmother, sisters and half brothers were also "charmed" with the kids. The Carolinians went for evening drives on the Paseo Isabela Segunda, which Hattie considered "one of the most beautiful places in Havana . . . . The wealth & magnificence of this city is very great, & I am somewhat bewildered by the life & noise which seems to be as great as in New York." The summer climate was "delightful" and the nights were cool. Hattie, Brosio, and Alfonso were afflicted with slight fevers but quickly recovered.[3]

A month after arriving, Hattie found the situation "very sad," with numbers of political exiles leaving the island. Brosio's godfather, Miguel Aldama Alfonso, had recently gone into exile in New York with his family after their Havana mansion was ransacked by loyalist militia who suspected him of supporting the revolutionaries. José Tomás Gener, the son of Brosio's second godfather, Benigno Gener Junco, had been incarcerated in the La Cabaña fortress, after being captured "in a schooner which was bringing arms from Nassau." The seventeen-year-old son of another friend, Antonio Guiteras Font, had been "banished & condemned to two years hard labor." Gonzales "was under constant surveillance by Spanish spies." Ten-year-old Brosio never forgot those "dark, sinister faces" watching even the children's movements. The Spanish *Voluntarios* militia paraded daily with martial bands, passing in front of Lola's residence, which was "a great delight to the children." The kids were "acquiring Spanish rapidly." Nine-year-old Alfonso quickly adopted "Cuban ways & cookery." The family consumed fried bananas for breakfast, chewed on oranges and sugarcane chunks, and dinner consisted partially of rice, hominy, and boiled bananas.[4]

Gonzales pursued employment in Havana to keep his family safe there during the summer fever season. His newspaper teaching advertisement and the influence of friends failed to procure adequate work in the capital because of the situation created by what Hattie called the "unfortunate revolution." Gonzales traveled to Matanzas in mid-April and obtained two teaching positions. When he returned to Havana, there were great festivities, with bright lights, music, and fireworks, honoring the arrival of the Catalonia Volunteers. During the first week of May, Gonzales moved his family to Matanzas, staying a few days at the home of his high school friend José Manuel Ximeno. Hattie and Tula wore new dresses provided by aunt Lola. Mrs. María Dolores Ximeno soon had a tailor outfit the boys in white linen suits and elegant Panama hats. Alfonso became "very fond" of two-year-old Dolores "Lola" María Ximeno, who later became a distinguished socialite, acclaimed for her seven-hundred-page memoirs. The Gonzales family attended the Ximenos' Sunday dinners, a traditional social gathering for the city's aristocracy.[5]

Matanzas is located in the quarter-mile delta between the Yumurí and San Juan Rivers. A *New York Times* correspondent described the metropolis in 1868 as "comparatively a modern city" and one of the urban areas with the most "natural beauty" on the island. "The houses are all of stone—nearly all of a single story; the streets are all narrow; the churches are all massive; every fifth house is a *café*, or a *bodega*, or a *tabaqueria*." The correspondent marveled at seeing the speeding two-wheeled *volante* carriages and the street vendors loudly hawking an assortment of goods from chocolate to lottery tickets. Across the San Juan River from Matanzas, connected by an iron hanging bridge, was the flourishing little town of Pueblo Nuevo, described by a local resident as "semi-cosmopolitan" due to the many American, British, German, and French people living there.[6]

The Gonzales family settled in mid-May into a Pueblo Nuevo hotel occupied by three other American families. The building was "built upon solid rock some seven feet from the ground," making them "avoid the dust & gain the breeze." Servants frequently washed the dark brown tile floors with large sponges. The Carolinians rented two comfortable adjoining rooms with a bath. Hattie found the view of the nearby seashore and hills picturesque. She wrote to Oak Lawn: "The windows are enormous. My room opens into that of the children's so we have a draught—by day & night. Our window is on the street which is very gray in the evenings but theirs is nearest the sun & opens on the yard." They slept on "very large cots with all the fixings of bedsteads, head boards, posts & pretty muslin curtains are looped up in the day & serve as mosquito bars at night." Cuban friends provided clothes and books in English. A Dr. Cartera offered free services for the pregnant Hattie and the children, two of whom had colds and fever. Gonzales was teaching English and French seven hours a day in two preparatory schools: the *Instituto de Matanzas*, established by the Government in 1864; and *La Empresa*, founded in 1840 by separatist José Antonio Echeverría. Conspirators Cirilo Villaverde and Ramón de Palma had taught there, and after 1852 its headmasters had included Eusebio and Antonio Guiteras Font. The latter had just resigned his post the previous month, which was assumed by José Delmonte. The school had over 160 students, including about 100 boarders, some of whom had become local leaders of the 1868 independence movement. Gonzales replaced Prof. Manuel de la Vega, recently killed by a *Voluntario* militiaman in reprisal for the defilement of the statue of Ferdinand VII in the Plaza de Armas of Matanzas. His companion, Prof. Rafael Oliva, was gravely wounded. Hattie was very pleased that her husband had "so quiet an occupation . . . I have seen more civil war here, than in the four years at the South." Although many of their acquaintances spoke English, Hattie's only complaint to her mother was that "my head can not well carry two languages." Mail between Matanzas and Oak Lawn took two weeks to a month, with Hattie writing to her family care of George H. Hoppock in Charleston.[7]

The Elliotts were quickly notified of the birth on May 21 of Ana Rosa Gonzales, described as pretty and blond like her mother. Named in honor of Gonzales's aunt, she became "a never ending source of delight" to all. Mrs. Ximeno provided a layette with "linen cambries & embroideries, beautiful shoes, lace caps, bibs & embroidered diapers." Hattie engaged as nursemaid an English-speaking black slave, Flora Barckley, and asked Mrs. Elliott to keep the news "strictly private." Two months later, she informed her mother, "I have not seen the children look so well for two years & I am fat & feel strong & well." They were "living luxuriously," compared to Southern standards. In reply to Mrs. Elliott being "miserable at reading of the virulent cases, direct from Matanzas," of yellow fever, Hattie reassured her not to fret, that dying from it was "almost unheard of" in Pueblo Nuevo. In another letter, Hattie told her mother, "I am sorry that you 'feel certain that I will fall a victim to the fever,'" but that she could not remove to the country without her husband, on whose salary the family depended. Gonzales was making about $250 monthly.[8]

On a typical weekday, Gonzales and the boys would leave the house at 5:30 A.M. The children attended school in Matanzas from 6:00 to 10:00 A.M. and were taught in Spanish the subjects of geography, grammar, history, reading, writing, arithmetic, and catechism. The mischievous Alfonso distinguished himself "by the quantity of ink he manages to get on his pants." He was "made uneasy by the 'cross eyes' of his teacher," who was "looking at *him* when he appears to be looking at another boy." Gonzales gave his sons French lessons when he returned home after 4:00 P.M. The family had dinner at 5:30 P.M. and later habitually strolled to the seashore. Gonzales remained under surveillance by the Spanish authorities, who at the Havana post office opened the family letters. Mrs. Elliott had sent a newspaper clipping informing that Beauregard's former chief of staff, Gen. Thomas Jordan, had landed in Oriente province on May 11, 1869, with the *Perrit* expedition, accompanied by over one hundred men with a large quantity of arms and ammunition. Upon receiving the torn envelope, Hattie was "glad" that the censors had read the news about Jordan, and "appreciated it very much, every thing here is very quiet & if he has had any success the Spaniards don't seem to know it."[9]

Pueblo Nuevo celebrated their patron saint feast of St. John on Sunday, July 4. The Gonzales family viewed the afternoon religious procession from their hotel windows. That evening, they rode to the central plaza in Matanzas, had ice cream at Aunt Lola's *Louvre* café, heard good music, and visited friends. The boys had a two-week school summer vacation in August, along with Gonzales, who retained his private lessons.

One enthusiastic pupil traveled thirty miles once a week for tutoring. The Gonzales children received reading material, including Charles Dickens's novel *Nicholas Nickleby* from the Elliotts, who called them "the little Cubans." In mid-August, Hattie promised her worried mother to keep her "informed every fortnight of our welfare, at least during the yellow fever season." Brosio also reassured his grandmother that "if 'Yellow Jack' had intended to visit us he would have done so long ago."[10]

The Elliotts were having their third consecutive bad planting season at Oak Lawn. The brown rust disease had damaged the green seed cotton and they were not going to "make a bushel of rice, & very little corn, & but a few potatoes." Emily had been making marmalade and orange preserves for shipment to Northern purchasers. Flowers from their garden were sold as far away as Savannah. Mrs. Elliott exclaimed, "The expenses incurred in the cultivation of this place renders it impossible to pay Hoppock his advances and I think that I shall insist upon his foreclosing the mortgage this year that I may determine on some other mode of life for my daughters & self for we have [no me]ans of support & cannot pay the Taxes." The Constitution of 1868 had raised annual property taxes on Oak Lawn from thirteen to ninety dollars, and on the Chehaw River lands from twelve to thirty-three dollars. The Republican legislators pursued property confiscation through heavy taxation and abusive assessments for state, county, municipal, and school tariffs. Corrupt legislators committed state bond frauds and stole more than one hundred thousand dollars annually from the school and poll taxes.[11]

The Elliotts were also afflicted on August 31 by the tuberculosis death of nineteen-year-old Annie Johnstone. Her brother Elliott accompanied her remains from Baltimore to Flat Rock, North Carolina, where she was laid to rest near her father's grave at St. John in the Wilderness church yard. Tragedy soon struck again. Tula, Alfonso, and Benigno contracted "burning fever," but began recovering during the first week of September. The family was then visited by Gonzales's relative, thirty-five-year-old Mrs. Cecilia Michel Poujaud, the sister-in-law of thirty-six-year-old Charleston merchant F. P. Salas. She found Hattie "so lively & amiable, & such cheerful resignation to her circumstances." Hattie told her of how she had become a good Christian, and then dwelt at length on her fear of yellow fever. Mrs. Poujaud replied, "Did I not tell you that those accustomed to Charleston & its neighborhood would not get yellow fever."

Twenty-nine-year-old Hattie had been nursing her baby and "a Yankee woman's *twins,* who lived in the same hotel," weakening her immune system. Two weeks later, the Carolinians went to the Matanzas bathing house. Stagnant pools nearby served as breeding grounds for the *Aedes aegypti* mosquitoes that transmitted yellow fever to Hattie and Narciso. She was sick for two days before developing the black vomit symptom. Recognizing her affliction, "she fell into a convulsion from fright." Yellow fever went through various stages, including retained excretions that caused swollen abdomen, uraemic poisoning, and coma; changed skin color to saffron or bronze; fluctuated body temperature from dry and hot to cold and clammy moisture; along with blood secretions which oozed from the mouth, eyes, ears, and uterus. Life ended with delirium and convulsions during the death struggle. Gonzales knelt by her bedside, bewildered, immersed in prayer, dabbing her brow, and holding her hand until she last exhaled on September 17, when "with loving and trembling hands he closed the sapphire eyes." When the physician returned, surprised to find her dead, he encountered Gonzales standing before the corpse, exclaiming: "There you see her, born in affluence, reared in affluence, & died in misery, without feeling the transition." Eleven-year-old Narciso also had the black vomit but managed to survive. Gonzales was devastated by the tragedy. Mrs. Poujaud stated, "Poor man! His grief is most touching!" Among those attending the funeral was the American consul in Havana, Henry C. Hall, a captain with the 8th Connecticut Volunteers during the Civil War. Mrs. Poujaud later informed Mrs. Elliott that she thought that Hattie's remains were "placed in a family vault" by one of Gonzales's wealthy friends.[12]

Another setback soon followed after Hattie's death. Gonzales became unemployed when the separatist *La Empresa* was closed in October by the colonial government for being "a nest of vipers," and principal Delmonte went into exile. Gonzales left Narciso and Alfonso on November 8 in the care of his friend Agustín Dalcour at the Monticello plantation in Canimar, halfway between Matanzas and Cárdenas. He decided to take the rest of the children to Oak Lawn and return in April to retrieve the boys and Hattie's remains. Three days later, Gonzales received a passport from the governor of Matanzas allowing him to embark for the United States from the port of Havana with Ambrosio, Gertrudis, Benigno, and Ana. The Gonzales family spent one night with their aunt Lola in

Havana and departed the next day on the steamer *Savannah*. They returned to Charleston by the same route they had taken to Cuba. Gonzales and the children arrived in Savannah on the steamer *City Point* on the morning of November 18. They stayed in the Marshall House, owned by former filibuster conspirator Alonzo B. Luce, who according to an R. G. Dun and Company report was doing a fair trade and keeping good credit. They spent two days with the Habersham family before departing on November 21 for Charleston on the steamer *Dictator*. In Charleston they stayed for two days with Mary Manigault, who expressed that her "greatest grief has been for these dear little expatriated babies, who to us are sacred legacies." Gonzales then accompanied his brood by train to Oak Lawn.[13]

The refurbished slave cabin at Oak Lawn, occupied by Mrs. Elliott and her offspring Annie, Emily, Ralph, and Tom and visiting seventeen-year-old niece Emma Elliott Johnstone, became overcrowded with Gonzales and his four children. A personal animosity soon festered between the pugnacious Emily and Gonzales that would have everlasting dire consequences for the entire family. It appears that forty-year-old Emily, who had been in love with the Cuban before he wed Hattie, expected the bereaved widower to propose matrimony. This was not an uncommon practice in the South, which allowed for orphans to have a stepmother who already loved and accepted them. Georgia filibuster activist David Bailey, for example, had married his sister-in-law two years after his wife's death. But Gonzales, who grieved for Hattie the rest of his life, never remarried. Twenty-five years later, he movingly told friends, "For me, she is not dead, she is in my heart." Emily, who blamed the Civil War for the lack of eligible bachelors, assumed that Gonzales was her last hope for marriage and motherhood. She was very conscious that her biological clock was expiring and feared becoming an old maid. For the next three years, in the letters to her Gonzales nephews, she sometimes signed off as "Old Aunt Emmy," "Old Girl," and "Ancient Maiden Emmy." It was a social condition then mocked in the *Charleston Daily News,* which claimed that "it is a 'reproach' to be an Old Maid! That is the universal cry."[14]

Emily sought revenge against Gonzales for spurning her by attacking his weakest emotional point: she manipulated his children against him and did not relent until her death twenty years later. Emily started badgering Gonzales about his responsibility to educate his children. In mid-January 1870, Gonzales stayed with the Hugers in Charleston while looking for employment. He boarded Brosio in the Orphan Home and School (today the Porter-Gaud School) directed by the Reverend Anthony Toomer Porter of the Holy Communion Episcopal Church, the former chaplain of the Washington Light Infantry. It was "a classical school for the children of parents in straitened circumstances." Another one of Emily's tactics was to turn the rest of the family and their friends against Gonzales. Mary Johnstone was among the first to join ranks with her sister. She had encountered Gonzales in Charleston that January before returning to Baltimore and informed Emily, "I quite hesitated about addressing him before the Pinckneys, he looked so rowdy attired in his usual hunting costume great coat, slouched hat & that hideous old scarf wound around his delicate throat. . . . I do dislike that man more than I can express, and I am afraid more than I ought too." A month later, Mary Johnstone was writing to Oak Lawn, advising them to "keep on terms with your undoubtedly selfish guest. At least he has done nothing dishonorable, and if his temper is rough and soured he was dear Hattie's husband and as the father of your little darlings he has the authority to remove them from your care, and may do so if you do not appease his vanity." The Elliotts had hired a nursemaid for six dollars a month to attend to the baby.[15]

Twelve-year-old Brosio was writing in mid-February from his Charleston school to Oak Lawn that he was homesick and had chill and fever. Emily started sending comforting letters, emphasizing his father's absence and assuring him to write "for anything you want." She instructed him to "burn our Letters" to prevent Gonzales from discovering her duplicity. After only ten days in school, Gonzales sent Brosio back to Oak Lawn to recover, then left for Savannah in early March to find employment. He returned two weeks later to the Scharlock family boardinghouse in Charleston. While riding on the train from Savannah, sitting in the back seat, "some one threw a brick in the car which broke his head." Emily then attempted indisposing the two boys in Cuba against their father and tried making them feel anxious to return home. She wrote to Narciso and Alfonso on March 9, pretending to be Mrs. Elliott, addressing the letter to "My Dear Little Grandsons" and signing it "GdMama." This missive was written in Emily's distinct manuscript, which contrasts sharply with that of Mrs. Elliott. She began by saying, "Your father has never given me your address but I feel that I must write to you & so will again trouble Mrs. Poujaud to send you my letter." Emily indicated that the family was "spoiling" Brosio since his return. She described his life at Oak Lawn as exciting and adventurous, "playing out constantly" in the flowers, riding with the younger kids on a railroad handcar that the workmen lent him, building a bateau to navigate the pond, and shooting "as many as twenty-five robins in one day." The phony grandmother stressed that Brosio would be happier "if he had his little brothers with him." In an apparent attempt to agitate Alfonso, she said that his favorite mule "Bill Arp had grown so old that Uncle Ralph exchanged him for a young Mare."[16]

The children had no desire to return to the privations of Oak Lawn, as life on a Cuban sugar plantation was more rewarding. Alfonso had become "nearly twice as stout," they both felt "as strong as bulls," and "the place is healthy and shady." The brothers had frequently visited Matanzas to see the sugar mill grinding during the harvest season. Narciso was "learning a little Astronomy," which the indolent Alfonso quit after the second lesson. Eleven-year-old Narciso developed a talent for writing very vivid and detailed accounts, which later made him a renowned journalist. He described to his father how "yesterday a large hawk flew by, pounced upon a chicken and carried it off about a hundred yards and then let it drop as we were running after him and shouting."[17]

Gonzales was still a guest at the Marshall House in Savannah. He wrote to Ralph on April 20, enclosing twelve dollars to pay for the baby's nursemaid, and requested that Ambrosio and Tula be sent to him with Dr. Pinckney, whom he paid for their passage. Gonzales indicated that "I feel the want of the company of my children and as I cannot have them all, send for those I can properly take care of and entertain. I intend taking them back before the sickly season sets in." Mrs. Elliott replied that Tula was still unwell and did not have decent traveling clothes, having outgrown her Cuban dresses. She failed to send Brosio, stressing that he only had three years left to get an education, before working for his own support. Mrs. Elliott told Gonzales, "If you have not the means be candid with me I beg and let me either advance the money that you may send him to school or allow me to pay for him."[18]

Returning to his Charleston boardinghouse in mid-May, Gonzales replied to his mother-in-law that he had waited for Robert E. Lee's visit to Savannah in early April, to arrange Brosio's enrollment in Washington College in Virginia. Lee kindly replied that his son had "neither the age nor the preparation" to enter his institution. Gonzales then placed Brosio in the Charleston Home School and paid his board with cousin Ebet Burnet. He also accompanied his eldest son to a public Masonic Memorial Celebration and to the German Schutzenfest Festival. The Cuban then spent two weeks with his children at Oak Lawn before returning to Savannah. During that time, he went hunting with Ralph and Tom Elliott to procure food for the family. The Elliott ladies were worried that he would take the children with him to Savannah or Cuba. Mrs. Elliott explained to Brosio, "The want of money & the necessity for working to attain some, makes the difficulty of your father's life." Annie assured Brosio, "Papa arrived safely in spite of the Buggy's wheel nearly giving him an upset." By contrast, Emily resented seeing Gonzales back at Oak Lawn, would not speak to him, and relentlessly excoriated him in her letters to Brosio: "*My child* [author's emphasis] can it be possible that Daddy Noster has given you no pocket money at all? depend upon it—he was afraid of seeing me in Savannah. I would have been many wet blankets upon his festivities." She also advised, "Forget the Tyrant as much as you can & be light hearted," and insisted that he not "talk of our folks to anybody & destroy our letters."[19]

Two days later, Emily was at it again, telling Brosio that his father planned returning to Cuba in the summer to bring back twenty coolies to work at Social Hall during the winter. She advised him, "We don't think there is any danger of his taking or sending you there & you know if he does—what to do." Emily insisted, "Let us all try to forget Somebody," and again sent Brosio money for "treating yourself to ice creams & other things." In replying to the earlier letter, Brosio said that his father did not leave him any money but advised him "to value one cent given by a [father's] hand more than a thousand dollars given by an imbicil [sic] woman who did not Produce me." Emily then escalated her invectiveness: "His meanness my dear boy is marvelous, if he was only poor it would be one thing. You could not complain if he had nothing to give you, but when he spends freely on himself & denies you requisites, t'is another. Because he got his feet wet in hunting, he bought a dollar bottle of whiskey from [Henry] Barnwell's [store in Adams Run] he said to rub with. Then he smoked daily an average of five segars. . . . You won't get much of the proceeds I fancy." Brosio spent a sad and lonely thirteenth birthday on May 29 and wrote to his aunt Emily that he had not seen his father nor was expecting a gift from him. Two days later, she informed him that Gonzales was going to Charleston that day and would leave for a teaching job in Savannah on June 3. Emily warned Brosio that his father "has not asked & we have not mentioned that you have been writing to us."[20]

In Charleston, Gonzales transacted business commissions on behalf of Tom Elliott, and bought him and Ralph colorful neckties. He took Brosio along, gave him a pair of high top boots, and provided some diversion. When the delighted boy returned to his boarding at Ebet Burnet's home, he immediately mailed Emily a full account. The jealous Emily wrote a sarcastic letter to her nephew, ridiculing his father and warning that "those boots will lame you." She indicated that during the time that Gonzales spent at Oak Lawn, she "didn't say a word to him, more than I could help, & never do. I did not trust myself to speak, for I should have insulted him. The mean ungrateful contemtible [sic] creature!" Brosio replied that his father had complained about the treatment he received during his visit. Gonzales commented that the Elliott women neglected Tom and Ralph, would not sew for them, and that when he was ill with terrible stomach pains, Emily did not respond to his pleas and Annie mocked him. He told Brosio that his grandmother and aunt Annie were constantly quarreling with Emily. Brosio wrote to Emily that his father had made these statements to all his friends, and that "every body says what a pity Gen. Gonzales has such ungrateful Sisters & Mother in law." Brosio pointed out that his dad was "very careful not to say how he mistreated me at Oak Lawn. . . . I have the black & blue mark to this day when he boxed me on my shoulder and my arm hurts here when ever I hold it up." Brosio asked Emily not to forward his Maynard rifle to his father in Savannah, who had expressed a desire to use it.[21]

The Elliotts were moving to Adams Run in mid-June, but Gonzales prohibited Brosio from going there during the holidays due to of the possibility of fevers and to steer him away from Emily's machinations. She answered Brosio, "As to its being unhealthy for you in Adams Run, that comes beautifully doesn't it? from one who lost you your Mother in that death hole Matanzas! . . . May God frustrate his plans." She insisted that Brosio disregard his father's wishes by visiting Adams Run for a weekend in late June, and he complied. Emily called Gonzales one of "the devils own children" and told Brosio, "T'is very very hard to bear all that you have to tell of that man. One so false to every duty must misinterpret & falsify others to keep up his own self esteem." Emily failed to recognize her own duplicitous behavior. Brosio was playing both sides against each other to reap maximum benefits. He pleaded with Emily, "I write to you with my last stamp & I have not a cent in the world." She replied, "We will not let you be pennyless, but you must be prudent or we shall run aground. $10.00 besides the two or three you took with you in a month is being a little fast."[22]

Gonzales had returned to the Marshall House in Savannah while eking out a living as a teacher. The fifty-one-year-old proud pauper told the federal census enumerator on June 9 that he was a "retired planter" and that his personal estate was "$10,000." This was the value of his worthless dowry, which saved him from the embarrassment of admitting bankruptcy and poverty in the presence of his friends. Five days later, Brosio sent him a one-paragraph letter claiming that he did not have time to write because he was "so busy going to school." Gonzales answered that it was "a sample of your indifference and disregard" and demanded a substantial letter every Saturday detailing his weekly activities. He asked about the children, of whom he had received no news.[23]

In contrast, Gonzales was getting detailed lengthy missives from Narciso in Cuba, who was fascinated with encountering a variety of colorful fish and wildlife. He told his father that he had caught dozens of fire beetles and loved "to see them flitting through the air at night like golden stars & sometimes read from their light when it is dark." The letter made Gonzales "much pleased as it was so well written & spelt," and he proudly showed it to many of his friends. Emily continued to drive a wedge between Brosio and his father, telling him to lie about being sick and having to go to his aunts and grandmother in Adams Run for care. She also wrote, but crossed out, "That you were lame from the boots," after the new leather had initially galled Brosio's feet. Emily advised her nephew, "If he writes to scold you for not obeying him & writing on a certain day you can say with truth that you had no stamp." She enclosed $1.50 for

Brosio to pay his railroad fare to Adams Run and to "buy some candy if you fancy, or lemons but *not* newspapers for Uncles."[24]

When the teenager returned to Adams Run, Tom Elliott wrote to Gonzales warning that "you must be on your guard on talking to Brosey, he tells everything that should not be mentioned & makes mischief. He has been so accustomed to hear you abused by his aunts that he thinks it all right & proper & forgets that you are his Father. The fact is he has been drilled into speaking of you in the most disrespectful manner. I think it my duty to tell you this and I think if you could manage to keep him as far as possible from the evil influence of Ann Elliott, it would be much better for him." After Brosio's first visit, Tom started getting abused by his mother and sisters, who in a note to Ralph called him a viper. Tom deduced that comments he made to Gonzales were repeated to Brosio, who conveyed them home. He had referred to Annie as "the Duchess" and had called a Mrs. Marshall "more of a Sister than these at Home" because she had sewn garments for him. Tom recommended against letting Brosio "spend his vacation in Adams Run where he would run a great risk of getting fever, & have also bad advice given to him." He told his brother-in-law, "It has always been my disposition General, to defend the injured & take up for the wronged & I will not & cannot up hold my nearest Relatives in their abuse of you." Tom, abandoned by all of his offspring except one son after his wife left him three years earlier, told Gonzales that he hoped that "by another year you will be able to have all of your children under your own care, in order that they may have that wholesome discipline which they all need." Thereafter Gonzales kept a strained and sporadic correspondence with Mrs. Elliott regarding his children, but he never again wrote to Emily after these incidents.[25]

Gonzales also worried that Alfonso had never written to him from Cuba, and wondered if he too was under Emily's influence. He kept asking Narciso that Alfonso write him. Narciso, exasperated with his indolent brother and desirous of pleasing his father, finally wrote the letter himself on July 25, and signed Alfonso's name to it. The missive has the same manuscript and style of Narciso, who was covering up for his illiterate ten-year-old brother. It described how during the daytime the boys climbed tamarind trees and shook down the fruit. One night, they helped a laborer pour gunpowder and sulphur into the ground holes of *bibijaguas,* large red ants that were devastating the foliage: "We then set fire to the powder it catches & bursts killing all of the ants that are inside." A month later, Narciso was more direct with Emily: "You cant get a letter from Alfonso for he can neither compose or write one." There are no surviving Alfonso letters in the family collections.[26]

Narciso and Alfonso were taken in early August by Teodoro Dalcour to the Reunión Deseada plantation owned by his brother Agustín, who had just left for Baltimore with his family. The children stayed there intermittently for two months, "had a splendid time of it, romping playing & reading." Aunt Emily continued trying to entice them home, saying, "We are sorry to hear that you find it so hot. . . . We have not suffered from heat here, having had more breezes than usual, and the nights are always cool." This account contrasted with Brosio's description of the weather four days later. He told his father that there had been a recent hot spell during which his grandmother had bought a hogshead of ice, "which has been a great comfort to us." Narciso asked Emily if Oak Lawn remained mortgaged, and she affirmed, "There is no chance of its ever paying for itself. I do not think that Mr. Hoppock will push the mortgage directly, but he has lost very heavily by planting." She held hope that "when the Democrats get back into power, our lands are to be restored, & you may yet I trust be able to go to college as the grand sons of your grandfather should." Ironically, although her father and uncle were Harvard graduates, none of her brothers had a college degree. Emily's egging prompted the studious Narciso to tell his father, "I am longing to go to school with Ambrosio." Gonzales, who planned to retrieve his two sons in April, wrote to Dalcour in August lamenting his "strained position," but said nothing of his return to Cuba.[27]

The Elliotts were visited at Adams Run in August by the federal census enumerator. The four Gonzales children living with them were unexplainably omitted. Ralph was the only family member listed as having his right to vote denied, for not taking the loyalty oath. He and Tom were described as planters, a profession they had consistently failed at during the previous five years. They were unemployed and the family was surviving with the piecemeal sale of their silver plate and valuables, the charity of Northern and European friends who occasionally sent money, and the proceeds from Emily's preserves shipped to Northern cities. In spite of their reduced circumstances, the Elliotts hired a cook, Julia, and a nursemaid, their former slave Chloe, to mind sixteen-month-old Ana Rosa, who was "very languid and fretful from cutting her eye teeth." Four-year-old Benigno had become "much more mischievous" than Alfonso, "meddling with every thing he can lay his hands upon, digging holes all over the yard, breaking up & destroying," and was considered "a veteran doll-smasher."[28]

Gonzales had overextended his lodging at the Marshall House after nearly four months. By the end of September, he was receiving his mail care of banker Wallace Cumming, while Narciso commented, "I don't know what makes your friend behave so. They cannot be true friends who would forsake you now that you are in trouble." Gonzales informed Brosio that he expected Narciso and Alfonso to return in November. He had received a "terrible" letter from Matanzas complaining about their behavior and asking that he take them back immediately. Gonzales felt broken hearted and disappointed that the sons he "had *educated* to behave well from their infancy should be complained of in the terms of that letter." According to Brosio, his brothers were "much in the company of corrupt Chinese and Negroes" and had "out staid their welcome." Gonzales asked Dalcour to send the children to Charleston by way of Baltimore in the care of Mordecai and Company, the steamship agents who gave him free passage, and to notify him of their travel arrangements. On October 3, he left Savannah for Beaufort, where he was unable to obtain employment. He then continued on to Charleston, arriving on the steamer *Pilot Boy* on October 13.[29]

Gonzales received in Charleston a letter of introduction from his former creditor, Edward H. Lafitte, for the firm H. M. and W. Le Count in New York City. He hoped to become a commission agent in the South for large northern companies. The Cuban visited his children at Adams Run before departing at the end of October. His reception, to judge from the increased abominations expressed by Emily, must have been very unpleasant. Gonzales asked Brosio not to write to him in New York "for fear of hearing of illness of the children, which I could not remedy and the thought of which, added to my other anxieties, would incapacitate me from the purpose, which for them (you, of course, included) I set out to accomplish." Gonzales knew that Emily was still coaching Brosio's letters, which had been a constant source of anguish and despair. The six terse letters Brosio sent his father during the previous two months all mentioned, except one, that baby Anita was greatly suffering from teething, "which gives her fever." Gonzales planned "to return in a few weeks" to Charleston.[30]

His arrival in Manhattan, with all of his worldly possessions in a steamer trunk and very little money, coincided with a record bitter winter. Gonzales sought out the Cuban exile community that had fled political persecution and the ravages of the war of independence being fought on the island. The 1870 federal census indicated that there were 1,207 Cubans among New York City's 942,292 residents. More than half of them were concentrated in the fifteenth through the eighteenth wards of lower Manhattan. Some exiles, like Miguel Aldama Alfonso, were wealthy merchants, but many others worked as cigar makers in their own homes or in small shops. The intellectuals like Gonzales were mostly employed as teachers, translators, or editors. A few émigré revolutionary newspapers flourished, some circulating one year and vanishing the next. Soon after Gonzales arrived, a classified advertisement appeared for one week in the *New York Tribune* under the "Teachers" section: "an experienced 'Tutor' in the Classics, Modern Languages and English Branches, wishes an engagement in a private family or school. Music, Piano and Organ taught if desired. Address Tutor, care Unger & Keen, 23 Maiden-lane." Gonzales probably placed this announcement, like others that he had published after arriving in Havana in 1869 and in Savannah earlier that year.[31]

The former Cuban aristocrat moved into the Maltby House at 21 Great Jones Street, on the southeast corner of Lafayette Place in Manhattan's Bowery section. It was a five-story building on a twenty-five-by-one-hundred-foot lot, managed by G. H. Costar, with a real estate value of $20,100. The proprietor was George A. McCleskey, the Savannah filibuster conspirator who hid Gonzales and López in 1851. Maltby House was located in the fifteenth ward of the city, between the Jewish Quarter on the east and Little Italy on the west, an area crowded with tenement houses that was rapidly deteriorating after the end of the Civil War. The Bowery contained theaters, cheap saloons, billiard halls, ten-pin alleys, brothels, pawnshops, rat-infested stale-beer dives, phony auction rooms, sensational dime museums, and German lager beer gardens. The area attracted gamblers, prostitutes, pickpockets, and other criminal elements, like the tough Bowery Boys, intent on breaking into its stores, shops, factories, and warehouses. Foreigners flocked there in search of cheap lodging.[32]

Gonzales sought spiritual refuge in St. Bartholomew's Episcopal Church, located across from the Maltby House. The church, a colonial structure with a tall steeple, had been consecrated in 1836. Gonzales became a salaried member of the St. Bartholomew's choir. He found moral strength in the sermons of fifty-five-year-old Reverend Samuel Cooke, who powerfully preached human salvation through the atonement of Christ and the punishment of sins in an eternal hell of fire and brimstone. Gonzales was facing the lowest point in his life, with the loss of his beloved wife, economic hardships, the separation from his children, and the raging family feud spearheaded by Emily Elliott.[33]

His life took another downturn after Narciso and Alfonso arrived at Oak Lawn on December 6, 1870. Three weeks earlier, Narciso had notified his grandmother that they would be returning home on November 27 via Baltimore. Mary Johnstone later lamented missing the opportunity of seeing her nephews and told her mother that, had they traveled on the *Liberty*, Captain Reid "would have known where to look for me." Dalcour had sent them on a different steamer line and failed to notify Gonzales or the Elliotts of the new arrangements. Four months earlier, Gonzales had asked Narciso if Dalcour and a Mr. Artús were receiving his letters and the reply was affirmative, but that "I really don't know why they don't answer your letters."[34]

When Gonzales failed to write to Oak Lawn by the end of November, Mrs. Elliott asked her daughter Mary Johnstone for news about him. She replied negatively, "But if he succeeds in getting anything for You—we can afford to forgive him for being so disagreeable I think." Mary also asked if George H. Hoppock's recent cardiac death had affected a foreclosure on the Oak Lawn mortgage. Six weeks later, Brosio obtained from Mr. Lafitte his father's forwarding address in New York. Mrs. Elliott wanted to transfer Hattie's remains to the family plot in Magnolia Cemetery, but she did not write to Gonzales in New York to arrange it, nor did she inform him that his sons had returned from Cuba. Instead, in January 1871, she asked Mrs. Cecilia Michel Poujaud to assist her in having Hattie brought home and made arrangements with her nephew, the

Reverend Charles Cotesworth Pinckney, to fix the lot and acquire a headstone. Mary Johnstone was still optimistic in February that Gonzales's "flitting about will result in something splendid. We can all forgive his selfishness, if He would strike out in the money making line."[35]

Gonzales wrote a month later to Brosio from New York, saying that he had "succeeded in obtaining the Agency for the Southern States of three large concerns, in this city, out of the commissions of which, I have every reason to expect, when I get full underway, ample means for the support of myself & children." He claimed being unable to pay for his children's expenses during the previous five months because he was barely surviving. Gonzales had moved to cheaper quarters at 39 West Washington Square and had spoken with Brosio's godparents, Miguel Aldama Alfonso and Hilaria Fonts Palma, who agreed to finance his education. This was the traditional role of a Hispanic godparent, or *compadre,* who was committed to the child's upbringing. Gonzales assured his eldest son that he had secured his education for eighteen months "in as good a school as can be found." He stated that he would later provide for Narciso and Alfonso to go to boarding school. There were no public schools for whites in Colleton County. Gonzales indicated that he was going to return to Charleston in mid-April "to commence my labors and to earn wherewith *to support you all.*" He assumed that Narciso and Alfonso were still in Cuba, according to a "very old" letter he received from Mr. Dalcour's partner, which had been forwarded from Charleston. Gonzales had earlier inquired with Mordecai and Company in Baltimore if his sons had returned from Cuba, but they replied that their names did not appear on the *Liberty* passenger lists. He told Brosio that he would ask Dalcour to send the boys via Florida so that they would arrive in Charleston in April, when he would be there. Gonzales asked his son to respond upon receiving the letter, but he never got an answer nor did he learn that Narciso and Alfonso were at Oak Lawn. Emily Elliott, who by then was a surrogate mother to the Gonzales children, probably dissuaded Brosio from replying.[36]

The Cuban then wrote on April 7 to John D. Warren, owner of the Ashepoo plantation, bordering Social Hall. The sixty-seven-year-old Warren lived in Walterboro, had fifty thousand dollars in real estate holdings, and owned personal property worth ten thousand dollars. Gonzales indicated that he had not been notified by the Colleton County treasurer on the payment of taxes for Social Hall or their due date and had requested a payment extension. He implored Warren to pay the property tax, which would be refunded upon his return to Charleston in a month. Gonzales informed Warren that he had a fair prospect for securing an independent position. He later changed travel plans at the last moment after getting a "tempting offer" of several thousand dollars if he sold a patent. There were also propositions to sell first mortgage railroad bonds and real estate which, had he succeeded, could have netted "fifteen or twenty thousand dollars." Gonzales apparently had some success. In the Gonzales Family Papers at the South Caroliniana Library, there is an unendorsed check dated June 3, 1871, for $250, on the National City Bank, 52 Wall Street, New York, signed by Rafael Carrasco, and made to "the Bearer or A. J. Gonzales." There is a two-cent Internal Revenue stamp on its upper left corner, canceled by an oval rubber stamp marked "Rafael Carrasco—New York." Gonzales probably sent this check to Oak Lawn to support his children and the Elliotts never cashed it. Since nearly all of the correspondence relating to the family feud was destroyed, it is speculative that the needy Elliotts rejected this "tainted" money. Nine months later, Mrs. Elliott refused to accept a smaller remittance from Gonzales. The Elliott ladies could thus boast in vain that Gonzales had not contributed one cent toward his children's welfare.[37]

That summer, when Gonzales did not receive further news about Narciso or Alfonso either from Cuba or from Oak Lawn, he visited Carlos Poujaud, who had just arrived from Matanzas, at the New York Hotel. Gonzales, accompanied by forty-nine-year-old Charleston commission merchant John B. Lafitte, queried Poujaud about Narciso and Alfonso, being told that "they were well at Mr. Dalcour's near Matanzas." Mrs. Aldama, who was arranging Brosio's schooling, suddenly died. The Gonzales projects again started unraveling when his hopes of economic success turned into delays "from week to week and from month to month," making "too long & too sad a story." Plans for planting at Social Hall had to be postponed after the South Carolina lowcountry was hit by a cyclone in late August, dropping fifty-three inches of rain in thirty-six hours. While the Elliotts derided the elderly Gonzales for his failures, they had overwhelming sympathy and understanding for the unemployed

Ralph and Tom Elliott and their twenty-three-year-old nephew Elliott Johnstone, who had been in Savannah for months fruitlessly searching for a job.[38]

Gonzales told Brosio that he spent the rest of the year with a "daily expectation of selling one thing or another and of being able to give a surprise to your grandmother sending her not only enough for the education & wants of all my children but for herself & family." He participated that fall in a Cuban celebration in Manhattan commemorating the October 10 independence uprising. There were raffles and games to raise rebel funds, and Gonzales won a tricolor Cuban cockade, which he later sent Gertrude. Three weeks later, he wrote to G. Decourt, the French consul at Baltimore, about the possibility of colonizing Social Hall with citizens of Alsace-Lorraine. The formerly French region had been recently ceded to Prussia as a result of the Franco-Prussian War. Decourt replied in French that he liked the idea but that the consulate was unable to help, so he referred him to the president of the Alsace-Lorraine Society of New York. Gonzales was emulating a Beaufort County planters' association that imported immigrant workers under the condition that labor disputes would not be settled before courts with predominantly African American judges and juries. Many civil proceedings during Republican rule, employing corrupt and uneducated justices and jurors, were flagrantly biased against the white employers. Foreign immigration was also promoted by Redeemer politicians and journalists as a source of white votes. Lowcountry fevers, poor wages, and racial tension drove away many prospective immigrants.[39]

Gonzales moved to Baltimore in December 1871, securing a five-month choir contract at St. Paul's Episcopal Church and a job teaching French at the Eclectic Institute until June. The school was operated by fifty-year-old Mrs. Letitia Tyler Semple, the fourth of eight children of president John Tyler. Her husband, James A. Semple, had served as paymaster in the United States and Confederate navies. Mrs. Semple had been a volunteer nurse to Confederate soldiers hospitalized in Williamsburg, Virginia after the Battle of Manassas. Her institute, enrolling about twenty young ladies, was located in the three- and four-story houses that today are 29 and 31 East Mount Vernon Place, owned by the Peabody Institute since 1962. Fannie and Emma Elliott Johnstone were interned there after arriving in Baltimore in October 1868. The school had very strict rules: girls received demerits for not wearing thick veils after sundown or for not speaking French at the dinner table.[40]

St. Paul's Church, on the southeast corner of North Charles and Saratoga Streets, was the largest and most spacious in the city, with a seating capacity of seventeen hundred. The structure was rebuilt in 1856 in an ancient basilica style after a devastating fire. Its high-ceiling flat-roof nave, supported by interior arches with lateral aisles, still stands today. The rector at St. Paul's was the English-born forty-one-year-old Reverend Dr. John Sebastian Bach Hodges, who shared Gonzales's passion for music. Both had a mutual friend in Charleston, the Reverend Anthony Toomer Porter. The church took pride in the beauty of its services and Hodges improved them by enhancing the choir. Most contemporary American Protestant churches had choirs of men and women dressed in mufti who were screened from the congregation. St. Paul's had a female organist and another lady led the choir. It was composed of at least ten young women and about four men. Gonzales received twenty-five dollars monthly, which was equal to the salary of a rural parish minister. During the Lenten season, a society columnist praised the choir and "noticed the dignified General G— a distinguished foreigner."[41]

Gonzales boarded at a modest 1815 federal-style four-story brick rooming house in what is now 333 North Charles Street, one block north of St. Paul's. The Cuban became acquainted with twenty-seven-year-old Danish composer Asger Hamerik, a friend of Rev. Hodges, who lived nearby. Hamerik was the musical director of the Peabody Academy of Music, at Mount Vernon Place near the Eclectic Institute. Gonzales was soon introduced into the Baltimore *haut monde*. In January he participated in the soiree of the Allston Association, which presented a program of vocal, choral, and chamber music selections from Mendelssohn and Schubert. The activity included an exhibition of "a fine collection of paintings, principally from the easels of the best known New York artists, and which were secured expressly for the occasion." A Beaufort lady spending that winter in Baltimore later told Brosio that she had heard his father sing with a splendid voice at a concert. Gonzales also contacted the Cuban exile community in Baltimore. His friend Agustín J. Dalcour and half a dozen Cubans were members of Masonic Centre Lodge No. 108, which met in the newly constructed Grand Lodge of Maryland, next door to St. Paul's. Gonzales undoubtedly attended some Masonic meetings.

Another point of émigré contact was the Cuban Cigar Store, operated by Moynello, Garmendia, and Castaneda in the post office rotunda, where Gonzales must have acquired a supply of cigars and heard Cuban news.[42]

After arriving in Baltimore, Gonzales called on his sister-in-law, the widow Mary Johnstone, a matron at the Edgeworth School for Young Ladies. Gertrude and Ana Rosa Gonzales would later be educated there. The finishing school, located one block west from where Gonzales worked, today is the Tiffany-Fisher House on 8 West Mount Vernon Place, described as "one of the treasures of Greek Revival architecture in America." The three-story building has a Doric portico and its dentiled cornice and gallery conceal two attic dormers. The academy was founded in 1865 by thirty-five-year-old Miss Sarah Agnes Kummer, who in 1870 was joined by thirty-two-year-old Lila Vance Fisher of Wheeling, West Virginia, a Confederate widow known as Mrs. Hubert P. Lefebvre. The school grew to have about one hundred students, half of them day students and half boarders, many of them Southerners. The curriculum at the Edgeworth School included foreign languages, diction, literature, history, drawing and painting, fine arts, musical instruction in piano, singing and harmonizing, and scientific studies.[43]

Gonzales appeared as Mary Johnstone was visiting Oak Lawn for the holidays. He called a second time to see her daughter Emma, but was not admitted. When Mrs. Johnstone returned after New Year, Emma thought to have heard him singing in the St. Paul's choir. Mary, subsequent to speaking with Gonzales in January, wrote to her mother, "He looks old, but was gentlemanly but tiresome, & with a wretched cold. I was simply polite of course not cordial." Gonzales inquired about his children with great interest and asked if Miguel Aldama Alfonso had arranged for Brosio's education. Mary told him that Narciso and Alfonso had joined Brosio at Oak Lawn the previous year. Gonzales responded that "no one had condescended to write him about the arrival of his sons from Cuba." He stated that "*he* was taught to love & reverence his father, and so surely as there is a God in Heaven, just retribution would fall upon those who instilled other than filial ideas." Gonzales then wrote an apologetic letter to Ralph Elliott, authorizing him to sell his Spencer rifle and ammunition left at Stalling's grocery in Charleston "for the relief of my children." The weapon was worth thirty dollars. The Elliotts regarded the offer as too little, too late. They refused to touch the rifle; instead, Emily asked fourteen-year-old Brosio if he wanted it.[44]

Brosio had recently boarded in the Island Home School in Beaufort, run by fifty-three-year-old Mrs. Catherine A. Hamilton, his grandmother's cousin. His teacher, thirty-year-old Robert B. Fuller, lived in the Hamilton residence with his mother, wife, and four children. Mary Johnstone informed Brosio that she had tried to arouse Gonzales's interest in educating his sons. She added that his father was sincere in his efforts, but that his aid would not be immediate, advising her nephew not to wait "for what is too uncertain." In contrast, Emily again began upbraiding Gonzales in her letters to Brosio, with exaggerations and lies: "Mary enclosed a piece from a newspaper mentioning him as one of the singers at an entertainment & puffing him, composed by himself or instigated by him at any rate. He is trying the musical dodge once more. Mary seems to be utterly disgusted, says he nearly made her sick but now she knows him & she thinks we need not fear any interference about the children." Emily insinuated to Brosio that he should not expect assistance from his godfather Aldama, who had not helped even Gonzales. She was right this time. A prominent Cuban revolutionary described Aldama as "a proud, overbearing and ignorant aristocrat all his life."[45]

Gonzales sent Charleston merchant William C. Bee twenty-five dollars on February 25, 1872, requesting that he forward the amount to his mother-in-law for the support of his children. Bee wrote to Mrs. Elliott that he would either express the money to her or pay anyone she desired. The Oak Lawn matriarch ordered Bee to return the check to Gonzales, so "that he will use the same for his own necessities, as the children are doing well and requiring no relief." Whereas Mrs. Elliott was civil in her rejection, her daughter Emily was the only family member actively inciting Brosio against his father. She gloated when telling her nephew that the check had been returned, "so if he meant to be great & magnanimous in sending his *first* earnings, he will have the wind taken out of his sails." Emily sent Brosio eleven dollars and wrote that Narciso would join him at school in April. Brosio then warned his brother that in Beaufort he would have to control his temper and "not behave as you do at Oak Lawn." There, he told him, he would encounter "a great many low & bad Boys who will provoke you in every way," and he suggested following his example of ignoring them. Brosio said, "It is a very hard thing to keep my hands off them some times I never go into the streets but I hear 'Gonzales! Gonzales!' calling me Spanyard & trying to provoke me in every way." When Narciso arrived at the Island Home School in April, the Elliotts paid forty dollars monthly for their boarding, plus three dollars for laundry. Mrs. Hamilton, a religious zealot, insisted that the boys attend church services twice on Sunday and also Sunday School, to which Brosio replied that "they had both graduated."[46]

Emily notified the boys on May 24 that the Reverend Charles Cotesworth Pinckney had visited Oak Lawn to christen the children. Ana Rosa was given her mother's name, Harriet Rutledge Elliott and Benigno was renamed William Elliott, but his brothers soon started calling him Plug, after the "Plug-Ugly" rowdies. Emily savored what she considered a victory over their "lamented father" when she told her nephews, "What will Pater say when he 'resumes his full relations as a Parent'? he need never know any thing about it I guess." She signed the letter as "Ancient Maiden Emmy," realizing that she was beyond her childbearing years. Five days later, Gonzales sent an apologetic letter to Mrs. Elliott from Baltimore. He had just received two of Hattie's trunks forwarded from Savannah, enclosed one of her letters, and tried to make peace with the Elliotts. Gonzales explained that he had spent all his spare time during the previous winter and spring studying agriculture. He had planned to return to Social Hall that spring to plant in partnership with José Ramón Cucalón. Gonzales had drafted a lengthy unsigned contract, committing Cucalón to invest $4,500 in Social Hall for cropping during five years. In return, Gonzales would repay the full amount in three years at seven percent annual interest. He needed $1,500 to repair the dams and canals, $600 to buy four mules, $400 for fertilizer, and $2,000 for a wagon, a cart, forage, plows, harrows, wheelbarrows, spades, and other implements. When Cucalón failed him, Gonzales sought another investor for the fall planting season. He informed his mother-in-law that he hoped "to do better for my children in the future, than I have been ennobled to do latterly. My love and devotion to them has been and is as intense as ever, and my feelings to yourself and family are what I hope is Christian like and what becomes one not unmindful of the kindness experienced in the happy days gone by." The letter prompted Emily to again roil fifteen-year-old Brosio against his father. He replied, "I do not think you need give yourselves any anxiety about Gonzales & rest assured he shall never make a swineherd out of me[,] it is now more than two years since he has given me even a *counterfit* cent & I do not think or consider that he has any authority over me."⁴⁷

In June, Gonzales presented to composer Asger Hamerik a plan for colonizing Social Hall with "thirty to fifty" immigrants from Denmark or Norway for five years. The sharecroppers could keep everything they cultivated on as much land as they wanted, and in exchange Gonzales wanted two days a week labor from the men and one day of housework from the women. The deal collapsed after Brosio wrote to his father a month later that a storm had converted the plantation into "a fine lake" without communication to the mainland, and that all the buildings had disappeared.⁴⁸

The Cuban attended a social function on June 13 at the residence of Mrs. Mary Armistead Byrd Wyman in Baltimore. Her wealthy husband, Samuel Gerrish Wyman, was a contributor to the Orphan Home and School that Brosio had attended in Charleston. Gonzales was accompanied by forty-six-year-old Miss Martha Custis ("Markie") Williams, first cousin of Mary Custis Lee and distant cousin of her husband, Robert E. Lee. Her father had been an officer killed at the Battle of Monterrey in 1847. Markie became General Lee's "confidant" during a correspondence exchange that lasted twenty-six years until his death. Mary Johnstone, present with her daughter Emma and Mrs. Lefebvre, described the woman as being "ugly" and "poor." She greeted Gonzales and "heard him sing the 'Marseillaise' very well, tho' his voice is broken, it made us sad to hear him. Emmy says he made an impression, and Mrs. Wyman gave him a cordial invitation to her Receptions—Monday Evgs—so his perseverance has had its reward." Emily immediately wrote to Brosio and Nanno in Beaufort, promising to send money soon and exaggerating her sister's account: "A Miss Williams took him there. Mary addressed him as *Col. Gonzales,* Think! how disgusted he must have been. . . . So he has succeeded in getting the entry into that fashionable—*rich* house—but as Mrs. W. leaves for Massachusetts now & G. is to become planter in the fall it won't profit him much perhaps. Miss Williams may intend to marry him. I don't know who or what she is—an old tabby probably—in want of a beau." Emily, who was reflecting her own frustrations in that last statement, persisted four days later, "I would have thought that you both would have liked to be out of Dads way in the fall," and called their father "that antique fossil pressure, Old G."⁴⁹

In mid-July, Gonzales sent to Brosio in Adams Run a valise with family mementos, books, and clothing for all the children, "two little dolls for Tulita and Anita," a Cuban tricolor cockade for the eldest girl, and letter paper, envelopes, and stamps for Brosio and Nanno. He explained his misfortunes to Brosio, assured that he loved his children, and that he had "postponed writing from the very excess of interest in you all." Brosio thanked his father for the gifts and updated him on the children. The baby had been christened with their mother's name, and "Gertrude has developed into a bookworm." Twelve-year-old "Alfonso is at last beginning to read after nearly two years of hard work on the part of his Aunt."⁵⁰

That month, Gonzales advertised in Baltimore for summer pupils, which Mary Johnstone hoped would "enable him to pay for his board," but he was unsuccessful. He spent the last week of July at the farm of Aniceto García-Menocal Martín in Warrenton, Virginia. The thirty-six-year-old Matanzas native had a civil engineering degree from Rensselaer Polytechnic Institute in Troy, New York and had been assistant chief engineer of the Havana water works from 1863 to 1870. García-Menocal had been a New York City Department of Public Works engineer until 1872, when he became a U.S. Navy Department engineer. Gonzales later stayed a few weeks with his former commander, Gen. John Pemberton, who lived on a two-hundred-acre farm four miles from Warrenton with his wife and teenage sons. He wrote to Brosio how the Pembertons "have been very kind to me, poor as they are, and I have been saved not only the heat in Baltimore but the expense of living, a part of the summer."⁵¹

Gonzales was back in Baltimore by August 28. Mrs. Semple had earlier offered him another teaching contract for September, but he instead returned to New York City, using the Maltby House as a mail drop. He informed Narciso and Brosio that he was trying to "find a Northern farmer who will go shares with me in farming at Social Hall and thus help me gradually to reclaim the place in order that it may become, hereafter, of some value to yourselves." The teenagers, who were under Emily's influence all summer, never answered his letter. Mrs. Elliott then sent Brosio back to school after seeing a newspaper advertisement for a Mr. Clark's school near Winchester, Virginia. She wrote of her plight to seventy-three-year-old Washington philanthropist William Wilson

Corcoran, a family friend, who sent two hundred dollars to pay for one year of her grandson's education. In mid-September, Brosio went via Baltimore to Mr. Clark's school, which was "a stone house on a stony hill & among the blue ridge mountains about 13 miles from Winchester," in Buffalo Marsh, Frederick County. He found the eight students there to all "smoke & drink" and to be "the most disgusting rowdy & unmanerly set of boys that I have ever seen . . . they go about the country robbing orchards[,] shooting peoples hogs & turkeys & getting into all manner of scrapes." Brosio had difficulty adjusting, similar to his Beaufort hazing experiences, because "they provoke me in all manner of ways & I can get no redress as they don't mind beating[;] their favorite amusement at present is to tie a string to my toe at night carry it out of the window & pull me out of bed & I never can find who it is as the whole School denies it." Soon after arriving, Brosio changed his name to H. H. Elliott. Emily asked him why, but his response letter has been mutilated. In a subsequent missive, Brosio indicated, "Folks up here talk a great deal about Old G. they have seen him lately in N.Y. & Balt. but I have changed my name & will only answer to the name of Elliott." Two months later, he went back to signing his letters as Brosio.[52]

After returning to New York in September, Gonzales spent "months of suffering and anxiety" attempting to obtain employment. In November, he mailed Gertrude a monthly subscription to a Philadelphia penny newspaper and tried to open a correspondence with his two eldest sons, who continued to snub him. According to Emily, Gertrude ignored the publication, knowing that it came from her father. Gonzales finally secured steady work on Wall Street in January 1873. He sent Mrs. Elliott fifty dollars through William C. Bee, "to do something towards the welfare of my children, with a fair prospect of doing better in the next few months." Gonzales indicated that he received his mail at the Maltby House, "although boarding at a cheaper place," and that he was indebted to his aunt Lola "for not starving heretofore." After consulting with her daughters, Mrs. Elliott asked Bee to inform Gonzales "that her health does not permit her to correspond with him." The Elliott women agreed that any further remittances should be accrued by Bee "until it amounts to enough for the education of one of his children." Emily did not want Gonzales "to say that he assists to support us or supplies us 'with luxuries'!" She spitefully suggested to Brosio that if he did not wish to continue corresponding with his father, to "date your letter Buffalo Marsh & enclose it to Aunt Mamie to forward—it will then have the Baltimore Post Mark & puzzle him, for he won't be able to 'spot' you, but you are old enough now to shape your own course, with regard to the man with rich friends."[53]

In contrast, Mamie Johnstone wrote to her mother, "I am sincerely glad that Gonzales has at last found occupation, his patrons are rich, and he may be able to get help for his children from them. Are You going to answer his letter? I wish he could get nice plans in New York for Brosio & Nanno and send You help for the younger children." Although Gonzales had managed to obtain work, Tom and Ralph Elliott had "not been able to do so," but the Elliott ladies never complained about them in their correspondence. Nanno indicated to Brosio that Tom's "work consisted of reading, and lately of writing all day to his 'prodigal' family." Emily's manipulation of the children against their father was evident when in February Narciso informed Brosio: "Emmy says I must tell you that if you write to Gonzales that you must not tell him anything about us or what we are doing etcet." Brosio's decision to stop writing his progenitor gave "great satisfaction" to Emily, while Nanno volunteered to consume his father's "epistles." Emily was taking advantage of Brosio's emotional weakness after six months in Virginia. He was being pressured by other students "to tell them their lessons & do their sums for them." Brosio wrote home in March, "I have suffered more since I have been here from different causes & from trials which a stranger & a poor friendless boy is subject to than I have since I left Cuba." Emily replied with further paternal vituperation, "Nothing more has been heard of Old G. (as Hattie begins to call him)." She had read in a newspaper that A. B. Stockwell, "the brother of one of his choice bevy of scholars," had a bad reputation on Wall Street. "So Old G. has probably got into the very heart of shoddy. I wish he may rest there. We have no use for him." In spite of the overwhelming scorn she heaped on Gonzales, the two eldest sons never forgot their Cuban heritage. Fifteen-year-old Nanno wrote a bilingual letter to Brosio that summer, using Spanish sentences to describe secretly how he accidentally killed a neighbor's dog.[54]

When Gonzales failed to surface by late June, Mary Johnstone told her mother, "Is it possible that no more has been heard from their father? He must have hoped too soon for those agencies, and the schooling for

Brosio—poor Man—of how little avail his talents are, unless his getting himself cared for by strangers is a proof to the contrary, but he always was a model sponge." In the same letter she expressed pity for her son Elliott, living off relatives in Savannah, "disappointed, at the difficulty in getting work—thousands are in the same predicament—even those who have friends to help them, which my boy has not." But Gonzales was not the freeloader depicted by the Elliotts. His *savoir faire* among the elite New York and Baltimore *beaux mondes* offered the opportunity to wed a wealthy widow and solve his financial problems. But the loving memory of Hattie, whose picture he carried in a locket, prevented him from remarrying. The Elliotts had incurred heavy losses that year, prompting Mrs. Elliott to again write to banker William Wilson Corcoran for financial assistance toward Nanno's schooling. Before returning home in September, Brosio dissuaded Narciso from studying at his school, which he called "a perfect *hell*."[55]

In September 1873, a financial panic closed thirty-seven banks and brokerage firms in New York City, and the Stock Exchange shut down for ten days. Before the year ended, eighty-nine railroads and over five thousand major commercial enterprises went bankrupt. The Panic of 1873 created a depression that persisted the next three years of the Grant administration. In New York City, long bread lines appeared and the unemployed crowded the streets. Wages were cut for those still working. Charleston merchant William C. Bee described the situation to Ralph Elliott as exceeding "any that I have witnessed in an experience of 45 years." Gonzales lost his Wall Street job and returned to living in a flophouse. The opportunity to provide for his children had vanished again.[56]

Narciso arrived on November 3 at Saint Timothy's Home School for Boys, in Herndon, Virginia, twenty-seven miles from Washington, D.C., after a steamer trip to Baltimore and riding two railroad lines. Gonzales had recommended it to Mary Johnstone two years earlier for Brosio. The two-story school house, on the northeast corner of Grace and Vine Streets, was established in May 1871. The former cheese factory was renamed Rawson Lodge and was affiliated with Saint Timothy's Episcopal Church. The newcomer described the headmaster, thirty-five-year-old David Sanford L. Johnson, as "religious in the extreme" and "very kind & attentive to his boys." His wife was "fair, fat & forty." Narciso complained that Johnson "commenced calling me Francisco when I arrived, but as it seemed a labor to him to do so, I told him my middle name was Elliott, so he and all here have adopted it as my cognomen." There were only five other students, between the ages of eleven and eighteen, who were "well behaved & aren't up to anything worse than moving the hands of the clock a half an hour or so when Mr. J. leaves the room." Narciso shared a bedroom with three students and told Brosio after arriving, "Fortunately the Pater is not known here, t'is a source of thankfulness."[57]

That month, Gonzales wrote to president Ulysses S. Grant regarding the *Virginius* affair. Grant had favored, in his State of the Union address in 1870, the annexation of Cuba to abolish slavery there and end the "exterminating" insurrection. The steamer *Virginius,* a gun runner flying the American flag, attempted disembarking in Cuba a contingent of revolutionaries and munitions. A Spanish cruiser gave chase for twelve hours, capturing the vessel in Jamaican waters on October 31 after it had jettisoned its cargo. The *Virginius* was escorted to Santiago de Cuba, where fifty-three of its 155 passengers and crew, including its American captain Joseph Fry and some U.S. and British nationals, were executed between November 4 and 11. The intervention of a British Navy frigate, whose captain threatened to bombard the city if the massacre continued, saved the remaining prisoners from death. Various rallies were held in New York denouncing the *Virginius* massacre. Hilario Cisneros was appointed chairman of the meeting held at Masonic Hall and other participants included former López followers Juan Manuel Macías and Ramón Arnao Alfonso. Gonzales, who closely followed the *Virginius* situation, probably participated in these acts to stay informed and express his outrage. The Grant administration sent the Spanish government an ultimatum on November 14, demanding the return of the *Virginius* and its survivors, a military salute to the American flag, reparations, and punishment for those who captured the vessel and ordered the prisoners executed. Spain accepted the demands and released the *Virginius* and the detainees on December 10.[58]

The Gonzales letter is not among the Grant Presidential Papers or in the Department of State Miscellaneous Letters. We can surmise its content because Gonzales again wrote Grant two years later, repeating "in substance" his previous message. The missive contained a brief autobiography and named attorney Caleb Cushing as a personal reference. Gonzales indicated that "although my sympathies have always been with the Cubans, I have taken no part in Cuban movements since the time of Lopez expecting to act in the future as a citizen of the United States." He asked Grant that he "be considered an instrument at hand for the Government of the United States," in regard to what he hoped "may prove the last scene" in the history of the Cuban movement he had pioneered. Gonzales believed that he could assist in the transition of Cuban annexation to the United States.[59]

The 1874 New Year found the Gonzales family scattered over the eastern seaboard. Sixteen-year-old Brosio was working in Grahamville, assisting his cousin Arthur H. Elliott, a twenty-two-year-old railroad agent, who began teaching him telegraphy. Narciso was at Saint Timothy's Home School for Boys in Herndon, Virginia, developing a passion for journalism after editing the student newspaper. The other children remained at Oak Lawn, where thirteen-year-old Alfonso was twice daily milking the family cow and nursing a persistent fever. Their uncle Ralph had obtained a low-paying position with J. L. Owen, "a brother mason" and lumber manufacturer, at the Altman Station depot of the Port Royal Railroad, overseeing laborers. He received seven dollars for the first seventy-five days of work. Ralph had repeatedly failed to raise a profitable crop at Oak Lawn, forfeiting in 1869 the ten-thousand-dollar mortgage to George H. Hoppock. After Hoppock's death, his administrator, Howell Hoppock, established foreclosure proceedings, which were ordered by a Colleton County court on October 10, 1872. That year, the *Beaufort Republican* published in one issue fourteen columns of farm auctions for delinquent taxes in Beaufort and Colleton counties. During the next two years, more than 720,000 acres of land were forfeited for taxes in South Carolina, while the state debt soared beyond $25 million. That amount included a $450,000 expense for "public printing" during 1873, used largely for political graft. Oak Lawn was advertised for auction by the county sheriff on January 5, 1874. Three months later, 2,000 real estate properties were forfeited in Charleston during one week for nonpayment of taxes. Benjamin Sauls, a St. Paul's Parish cattle driver, was outbid for the purchase of Oak Lawn by Charleston merchant William C. Bee, an Elliott cousin, who through a Mr. Bissell, paid the $700 in costs and disbursements. Sauls, a private during the war, had been found guilty by a court martial of being absent without leave, but had received a light sentence when deemed to be "quite an ignorant young man." Five years later, Emily and Anne Elliott obtained clear title to Oak Lawn and in 1886 they mortgaged The Bluff plantation to Sauls.[60]

Social Hall plantation, owned by Gonzales, was also on the verge of foreclosure. Gonzales's movements during the first nine months of 1874 are shrouded in mystery. He appears to have been traveling in and out of New York City, seeking better job opportunities. Mrs. Elliott was reporting to her son Ralph in February: "I was mistaken in Gonzales's whereabouts. He is still in New York." The fifty-five-year-old unemployed teacher, alienated from his family, had lost contact with South Carolina for over a year. In 1872, a Mr. Wichman had paid the seventy-dollar tax on Social Hall for one year in exchange for farming privileges, but on May 18, 1874, the plantation went on the Colleton County Treasurer's auction block for $83.88 in tax arrears, after being advertised in the *Walterboro News*. Neighbor John D. Warren, who had paid the tax for Gonzales in 1871, bought the property. Warren had already acquired in the same manner the adjacent plantations of Alexander Robert Chisolm and Haskell S. Rhett. He feared that the forfeited land would be divided by the South Carolina Land Commission into small farms and parceled out to African Americans. Warren wrote to Gonzales at the Maltby House on June 6, explaining that he had "been obliged in self defense on several occasions to buy more land in order to protect what I have." He promised to "be perfectly willing to restore it to you on equitable terms," upon the Cuban becoming economically secure. Warren's letter was unclaimed at Maltby House and returned to sender. Ralph Elliott then failed to raise $100 to redeem Social Hall. Mrs. Elliott received in June $150 for renting her *Farniente* estate in Flat Rock to J. B. Pryatt, but instead of buying back Social Hall, the money was used to redeem the Elliott's Myrtle Bank plantation on Hilton Head for $318.18 in back taxes and penalties. The following year, the government returned their Beaufort lot, taxed for seventy-five cents. The family tried to sell both places, but still had them six years later.[61]

Fall 1874 brought an upswing in the fortunes of Gonzales and his two eldest sons. Narciso returned to Oak Lawn and then joined his brother in Grahamville to learn telegraphy. The following year he became the telegraph operator at the Varnville depot of the Port Royal Railroad. Brosio had been promoted to railroad agent, replacing his cousin Arthur, who had left for New York. Gonzales began teaching modern languages, moved to 785 Broadway, and for the first time appeared in the New York City directory. His new job and good luck lasted only a few months: he moved again in February 1875 and was omitted from the subsequent directory. Former Union general Charles Cleveland Dodge wrote the Cuban that St. Bartholomew's congregation was "exceedingly sorry" to lose his good services. The church had relocated from the Bowery to a new building at Madison Avenue and Forty-fourth Street two years earlier. Dodge, a partner in the firm Phelps Dodge and Company, told Gonzales, "You have become so rapidly familiar with the peculiarities of our music—& have such a large, powerful voice, under good control & read so well, that it will be very difficult for us to replace the same." He offered Gonzales his recommendation to "any of our Protestant Churches." Our subject becomes difficult to track until late 1875, when he was appointed by the New York City Board of Education to teach English in the public schools "to a very large class of Cuban exiles."[62]

Gonzales was back in the Bowery by January 1876, at 44 Great Jones Street, near the Maltby House. That month, the United States had failed to rally the European nations to force Spain to end the Cuban insurrection. There were public fears that the problem would lead to war between Spain and the United States. When Spain proposed political reforms for the island, Gonzales again wrote to President Grant on January 8, 1876, offering his services as a diplomat, but was turned down. The Elliott ladies asked Narciso of his opinion on "war with Spain." He replied that it was not fancied by the people or the press, "but Grant is just fool or knave enough to start a row, & then for a while, Uncle Sam's ships will collapse into the jaws of the 'bloody Spaniard.'" In May, black agricultural laborers in Colleton County held a strike that "threatened the destruction of the crops and property of the planters." The unreconstructed Ralph Elliott started taking Brosio and Narciso to the Grahamville Democratic Rifle Club meetings. Nearly three hundred of these clubs, with some 14,350 members, were organized throughout the state under former Confederate General James Conner. Their uniform, the symbolic red shirt, was a rite of passage for teenagers who had been too young for the Civil War. They relied on "force without violence" to sponsor Democratic political rallies, gaining the support of "hundreds of Upcountry freedmen." The Red Shirts sponsored Wade Hampton's candidacy for governor in the 1876 elections, for which the Elliotts held "great Democratic rallies at Oak Lawn."[63]

The Oak Lawn family received "a great shock" when fifty-seven-year-old Thomas Rhett Smith Elliott died on July 14 of "Congestion of the Brain" in the Balls plantation stable, after recuperating from a stroke the previous month. He was interred in the Hutchinson burial ground at Stoney Creek Cemetery, near Yemassee, by his cousins and friends, who were unable to procure the services of a clergyman. Tom had been the only Elliott not involved in the family feud against Gonzales, who did not learn of his brother-in-law's death until seven months later.[64]

Gonzales remained in New York City, where Hilario Cisneros, one of the vice presidents of the Cuban Junta, placed an advertisement in *La Verdad*, indicating that "the nocturnal classes for the Cuban emigres" would start on October 9 in Grammar School No. 25 on Fifth Street, between First and Second Avenues. The "distinguished professor" Gen. Ambrosio Gonzales would be in charge of the English class, which was open to people of all nationalities. Gonzales's address book indicates that after arriving in New York he sought employment at various academic agencies, including Miss Herre's Educational Bureau on Fifth Avenue, Miss Young's Agency in Union Square, and Mr. Pinckney's Agency, among others. Gonzales was hired on one occasion as an interpreter for the court testimony of a Spaniard accused of seduction. He resumed contact with a number of prominent Cuban exiles, including Miguel Aldama Alfonso, José Manuel Mestre, J. M. Ceballos, Benigno Rico, Carlos Martí, Ricardo Acosta, and former political prisoner Joaquín Delgado del Valle. Gonzales also visited old Confederate comrades Gen. Roger A. Pryor and artillery Col. Charles Colcock Jones Jr. Pryor, formerly Beauregard's aide-de-camp, was a prominent lawyer and a delegate to the 1876 Democratic National Convention. The Cuban also met with former Union general Alexander Webb, president of the College of the City of New York, but was unsuccessful in getting a teaching job.[65]

Two weeks before Christmas 1876, Gonzales wrote to Brosio, care of merchant William C. Bee, and enclosed a one-hundred-dollar certified check. He had not left the Bowery tenement, and told Bee he was "still contending with adversity." The merchant forwarded the letter and the check to Emily Elliott, offering "assistance in obtaining the money for it."

The following month, Gonzales again wrote to Bee inquiring about his family. Bee responded that he had obeyed his instructions but that Emily had not answered. He informed Gonzales that one of his sons was an agent for the Charleston and Savannah Railroad, as he had seen "'Gonzales' signed to notes of shipments therefrom, but with regard to your other children I am without any information whatever." Gonzales afterward wrote an eight-page letter to Brosio that is missing from the family correspondence. Brosio's reply indicates that Miss Martha Custis Williams had notified Gonzales from Baltimore that Gertrude was attending the Edgeworth School. Gonzales had asked Miss Williams to visit Gertrude on his behalf, but Mary Johnstone retorted that the twelve-year-old was unwilling to receive her. When Miss Williams replied that Gonzales was corresponding with Brosio, the two women got into an argument. Mary later wrote to Emily, "She made Me very angry, by stating that *whenever* Gen. G. had sent down money, it was through Mr. W. C. Bee I told her *no* money was expected of him. So She will not come again I am sure." Mary informed Miss Williams that Gertrude "has a proud spirit, unlike her father who seems to have been willing to bemoan to others his ills and sufferings." Gertrude, whose signature was her initials G. E. G., later wrote to Emily that she had dropped her surname and was only using Gertrude Elliott. She also appears to have been influenced to write a negative letter to Miss Williams, telling Emily, "No other G on to my name I have not heard any thing more about the Old man or Miss W. That letter was a hard pill to swallow I expect." Gertrude's regard of Emily as a surrogate mother is evident when she ended her letters with "Your own loving child" or "Your loving baby." No letters from Emily to Gertrude, probably containing disparaging remarks about her father, survive.[66]

Brosio answered his father on February 21, 1877, mentioning everyone at Oak Lawn except Emily. He relayed having just received a telegram notifying that his grandmother, who had been ill since the winter, had just gotten worse, and he was quickly departing for Oak Lawn. The letter was signed "Ambrosio," instead of Ambrose Elliott Gonzales, the name he used since 1872. The seventy-four-year-old Mrs. Elliott died two days later of "nervous [sic] disability." She was buried on the twenty-fourth in the Magnolia Cemetery family plot, next to her husband and daughter Caroline. Two weeks later, Mary Johnstone again asked her friend Lena L. Cary in New York City for news regarding Gonzales but received a negative reply. Brosio received from his father a very sentimental condolence on the death of Mrs. Elliott. He responded with interest about his father's welfare and plans, informing that at Social Hall "the rice fields at high tide look like a large river, and are covered with marsh, five or six feet high. Looking across the water from 'Proud Hill,' you can hardly see the pecan trees for the growth of pine saplings, 'tis a perfect wilderness." Gonzales replied that he had been sick and was distressed knowing that Miss Williams was prohibited from visiting Gertrude. Narciso also renewed communication with his father. In April, Gonzales told his sons that he was leaving New York. Brosio suggested that he get employment in the south but considered that it was "next to impossible."[67]

That month, the Republican government had been unseated in South Carolina, and former Confederate general Wade Hampton, a conservative Democrat, assumed the governorship. The Gonzales and Elliott families identified with the new political leadership, sometimes referred to as "Bourbons" or "Redeemers" by historians. These were the former aristocrats, most of whom had been raised on plantations and were college educated in their native state prior to 1860. During the Civil War, the conservative leaders had fought as officers for the preservation of their civilization, which represented a glorious and sacred endeavor during their political campaign. Their speeches always contained an unrepentant nostalgic eulogy of their past heritage, in contrast to the Union fanatics who waved the "bloody shirt." The new Democratic leadership included former Gonzales associates Matthew Calbraith Butler, Johnson Hagood, William Wallace Harllee, Michael Patrick O'Connor, James Simons, Charles H. Simonton, and cousin William Elliott, a Beaufort attorney.[68]

Gonzales moved to Washington, D.C., in June and unsuccessfully applied for a choir position in various Episcopalian churches. He sent Brosio a trunk with family mementos and agricultural books. Brosio asked his father to "come further South before Winter sets in" and said that he "would like exceedingly to hear you sing again." Brosio empathized with his father's predicament by indicating, "Most people [are] living from hand to mouth & I see men of good family and education working at $15.00 per month & some with families to support at that!" Narciso then sent his father on July 1 his "most sincere wishes for your success in Washington." Gonzales failed to obtain employment in the nation's capital and spent the summer in Quincy, Florida. Brosio had asked his father to visit him, but it did not occur. A gap in the family correspondence obscures the situation.[69]

Quincy, twenty-two miles northeast of Tallahassee, had been a resort for Northern invalids since the 1850s. In mid-September 1877, the *Savannah Morning News* reported that "the people of Quincy have been enjoying some musical treats in the singing of Professor Gonzalez, a celebrated baritone of New York." Elliott Johnstone, living in Savannah, immediately sent the article to his aunts at Oak Lawn, asking, "Can this be the songster of whom we know?" A month later, Gertrude was inquiring from Baltimore about Brosio, asking her aunt Emily, "Does the Old Man still write to him? You know Miss Martha Williams has married a Col. [Samuel Powhatan] Carter from Virginia & gone off with him. I'm awfully glad now she can't bother me any more." Correspondence between Brosio and his father ceased in June 1877 and did not resume until after they met four years later. Emily continued to track her brother-in-law's movements through friends and relatives, manipulating the information to widen the schism with his children. A cousin reported to her in November that Gonzales had been on the same train from the spa at Sweet Springs, Virginia, to Washington, D.C., "where he gives music lessons." Mary Johnstone informed Emily in late April 1878 that a Mrs. Hopkins of Savannah had told her of having met General Gonzales at a reception in New York.[70]

The previous decade had been the worst period in Gonzales's life. His beloved Hattie had tragically perished, and his bitter and revengeful sister-in-law Emily had purposely and methodically alienated his six children from him. The Cuban insurrection, skyrocketing Reconstruction taxes, unemployment, and the Wall Street crash of 1873 had ruined Gonzales's economic fortunes. As he approached his sixtieth birthday, the terrible toll of the years showed on Gonzales. The attempts of this Democratic sympathizer to obtain a government position under the Republican Grant administration proved futile. The former wealthy aristocrat had lost his plantation and was reduced to living in Bowery flophouses, but he had survived due to his strong religious faith and the charity of family and friends. A perpetual optimist, he still hoped that something would turn up for him.

## Ten

# PARALYTIC PATRIOT

The estrangement between Gonzales and his children continued until the end of 1878. Nineteen-year-old Narciso had moved to Savannah the previous November, working as the night telegraph operator for the Atlantic and Gulf Railroad. He lived on the third floor of a boardinghouse owned by the widow Sarah A. Falligant. It is today the Jesse Mount House, a bed-and-breakfast inn, at 209 West Jones Street. Narciso worked eighteen hours daily for sixty dollars monthly, energizing himself with coffee. Mosquitoes forced him to "wear gloves all night to protect my hands, and my skin looks like a pepper box already." In May 1878, Narciso transferred to cheaper lodgings run by a Mrs. Gibbes. Brosio remained at his post in Grahamville, Gertrude was at school in Baltimore, and the rest of the children were at Oak Lawn under Emily's tutelage. Ralph Elliott, called "Seventeen" due to the record number of cocktails he consumed on one occasion, was being considered for the state legislature. He told his sister Annie that the "local idiots" were going to nominate him for the vacancy of Richard H. Humbert in Darlington County, and asked her opinion, preferring that she advise against it.[1]

Thirteen years after the Civil War ended, the Elliotts were still selling their silver heirlooms to survive. The money helped provide for a family excursion to Tryon, North Carolina, ten miles from Flat Rock, that summer. Narciso moved in August to the Keller boardinghouse in Valdosta, Georgia, working as a telegraph operator for the same railroad company.

He later lived in a hotel operated by Charlie Stuart, whose daughter was the postmistress. Narciso took the lower-paying job "on account of the incessant wear and strain" of the night shift on his health. He went to Tryon at the end of the month to spend two weeks with his aunts, Minnie, Hattie and Gertrude, who was visiting from Baltimore. Narciso later joined Brosio on Election Day in November, both wearing the first Hampton red shirts ever seen in predominantly black Beaufort, while campaigning with the Grahamville Democratic Rifle Club.[2]

Two days after Christmas 1878, Gonzales wrote to Brosio from New York, indicating that he was going to pursue better fortune in Cuba and would be stopping at the Charleston Hotel in early January. Brosio, who had resigned as telegraphist in Grahamville, was living at Oak Lawn and planting there and at The Bluff. The sixty-year-old Gonzales, disappointed when his son did not respond, wrote to him again from Charleston on January 6, 1879. He said that he "had so much to say in relation to the past and my purpose in going to Cuba that being matter for which no number of letters would have been sufficient." The patriarch wanted to give his son a few books and "some little relics of your mother's which coming from her, your sisters would like to have." Another gift was a recent oil portrait of himself that a friend had painted gratis. Gonzales planned to go "to Oak Lawn for a few hours to see the family and my three children." He wrote a second letter to Brosio that same day that was personally delivered by Doctor Henry E. Bissell. Gonzales reiterated, "I am *most anxious* to see you and . . . my stay here is quite limited. I have come at great inconvenience and pecuniary difficulty this way instead of going direct to Cuba from New York, in order to see yourself & brothers and sisters."[3]

Gonzales was not welcomed at Oak Lawn and his attempted visit made Emily very angry. Mary Johnstone later advised her, "Let Your *mind* have some rest, after your late excitement. Please Emmie don't worry over Gonzales, he is not going to claim his children and if You are too violent in Your condemnation of him to them it may lessen their gratitude to You, and weaken Your influence." Mary stated that Gonzales, whose "temper may have improved," had "neglected" his children because of what she considered his selfish nature, "finding his children sheltered, and having all confidence in the devotion of the Aunts & dear GdMother. He has done as he says, just supported himself, and his little trips to the Springs, to promote his health were considered as a good investment of spare funds." Mary begged Emily, "For pity's sake though, don't dwell on what may happen, life is not long enough for such continued friction of ones nerves. Gertrude does not speak on the subject, much, or little sly amusement at the Padre's expense occasionally is all she indulges in. I have no doubt if she is ever called upon to choose between You & him, which will be her choice & that ought to comfort you."[4]

While in Charleston, Gonzales tried to meet with John D. Warren, who had bought Social Hall five years earlier for tax arrears. Warren arrived at the hotel too late to see Gonzales, so he instead wrote to Brosio, offering to sell back the plantation at a loss for three hundred dollars, and promising not to dispose of it without his approval. Before leaving Charleston, Gonzales had written to Narciso at his new telegraphist job in Valdosta, Georgia, requesting a visit. Meanwhile, Emily had admonished her nephew to leave Gonzales alone, but Narciso replied to his father, "although a nine year's separation is not conducive to affection, and my feelings toward yourself have under gone wide fluctuations in that time, I still feel, and do not care to repress, a great interest in whatever betides you, and wish you much success in all your undertakings. I would be glad to hear from you as often as you feel disposed to write me." They agreed to meet on January 17 at the railroad junction in DuPont, Georgia, and jointly proceeded to Jacksonville, Florida. Before taking the steamer to Cuba, Gonzales had a studio photograph made with his son. During this interlude, Gonzales apparently explained to his twenty-year-old son the reasons behind Emily's scorn and his rejection of her marital intentions. As a result, Narciso would later respond sarcastically every time Emily derided his father.[5]

When Narciso returned to Valdosta, he wrote to Emily about the reunion, suggesting a suspension of judgment against his father until he could present the explanations given him. Emily did not reply, and Brosio warned Narciso that advocacy on behalf of their father would wean him from the family. Mary Johnstone commiserated with Emily, "I was relieved to find from Your letter that your bete noir had departed and trust that his position in Havana will be sufficiently lucrative & comfortable to keep him there. You must try and dismiss the subject. I do not imagine he will annoy You again—and his indolence will be Your security." Narciso visited Oak Lawn for ten days during March after both sides agreed not to mention Gonzales. The domineering Emily found a postcard that he had with a Spanish phrase, sending the words to Gertrude, asking that her Spanish teacher translate it. Emily said that it was from a female that Narciso was corresponding with. Gertrude sent back the meaning, "Nothing shall pent thee from me," and called it "a remarkable thing to send on a postal" by a young lady to a man. The note was obviously from Gonzales to Brosio or Nanno.[6]

When Narciso wrote to Gertrude promoting a reconciliation with their dad, the fourteen-year-old girl replied "rather bitterly" against her brother and father. Narciso later informed Gonzales that "I find that there is no possibility of a change of sentiment toward you on the part of the ladies, Uncle Ralph, or the children, and it is useless to expect one." Brosio could not correspond with his father without being ostracized by the family, which would result in "the overturning of all his investments and plans for their benefit." The eldest son had abandoned the priority of cotton planting on the family lands and had diversified into oats, corn, rice, cattle, sheep, and hogs. Nanno reiterated to his father, "It will be many years I am afraid before they will think as I do. I cannot give any letters or messages to the children from you, against the wishes of the household. It would be a breach of the truce now existing and the children would not alter their belief even were I to do so." In closing, Nanno suggested that any future financial savings accrued by his father be used toward "the payments of what little debts you owe in this country. It will have a good effect." Those liabilities included redeeming Social Hall.[7]

After arriving in Havana, Gonzales quickly found employment with the Havana railroad, where his stepbrother Ignacio worked at the wayside station as a locomotive engineer, and where his stepbrother Próspero had been an engineer mechanic. He lived with his stepmother and stepbrothers at 24 Calle de la Gloria in the Jesús María neighborhood. Gonzales used Ignacio's post office box to correspond with Narciso, sending him Spanish mastery books and Havana newspapers, which renewed his son's interest in Cuban politics. Narciso replied to the questions about the family that "the children know really less of you than of any other relative, and you therefore cannot expect them to be attached to you, or to have faith in you. Children are attached to those who are near them and kind to them. The little they remember of you is your strictness."

Nanno apologetically indicated, "I think you have injured yourself in this country by speaking of yourself in frank opinion," prompting his detractors to call him a conceited braggart. Nanno stated that he would be proud to see his dad overcome this and change his ways so "as to take the weapon from the hands of your critics."[8]

The telegraphist was feeling "wretched and worn out" from overwork at Valdosta, describing it as "the veriest drudgery from 7:30 A.M. to 10 P.M.," cooped up "in a maze of scribbling, corrections, and reports, no companionship, no rest, no peace, I don't think I can endure it very much longer." He was so anxious to leave that he suggested to the editors of the Charleston *News and Courier* to go as a correspondent to Memphis during a raging yellow fever epidemic. Narciso was immune to the disease after surviving it in Cuba. Bartholomew Rochefort Riordan replied that "they would not expose any employee, whether acclimated or not, to such a risk," but gave him "an opportunity of joining their staff" later. Narciso's misery lasted eight more months before he permanently abandoned telegraphy.[9]

Meanwhile, after Gonzales had worked one year in Havana, he was able to liquidate his debts and sent Theodore Dehon Jervey, the head of William C. Bee and Company, a check for eighty dollars, asking that it be given to the Reverend Charles Cotesworth Pinckney, the Elliott cousin who had performed his wedding ceremony. It was to pay for thirteen-year-old Willie's tuition for three months at Anthony Toomer Porter's school in Charleston, now called Porter-Gaud School, where Brosio had first enrolled. The gesture set off another round of the family feud that further alienated both sides for nine months. When Pinckney informed Emily of the plans, she proposed sending Willie the following year to the preparatory department of the University of the South, in Sewanee, Tennessee. She asked that Narciso administer the funds, along with the two hundred dollars Gonzales promised to send, but Pinckney refused to release the money without prior authorization from the donor. Gonzales had also offered Narciso money so that he could leave Valdosta and start a sheep herding enterprise at Oak Lawn. His son replied with gratitude, "But I could not accept assistance from you when I feel that *I* should tender it to *you*. I rather prefer to work my way out by myself if I can endure it, and certainly your surplus can be put to a better purpose."[10]

Meanwhile, Pinckney informed Gonzales of Emily's refusal, to which Gonzales replied that the minister should immediately put Willie at Porter's institute himself. Narciso then wrote to his father that the preacher "has been mistaken in your intentions, or that you have made a serious blunder. None of us have asked for any assistance for Willie or any of us: we feel able, when the time comes, to give Willie a better education than either of us elder ones has received." He also stated, "I have been on your side in the past, and would like to continue there, but if you begin by parading a small assistance to one of the children, and at the same time, ignore, slight, and possibly insult the people who have fed us all, I have little hope of your retrieving yourself in their, or any one else's estimations.... One thing is certain, Willie will not go to any school that is not acceptable to the family nor will he go except when sent by them." This response "highly incensed" Gonzales, whose reply, in Narciso's words,"vindicated himself in a dramatic manner." Nanno sent the missive to Emily, indicating, "I write to soothe him, but I am afraid he will not be reconciled, and that I will be 'cut off with a shilling.' Under the circumstances, I must decline handling his money for the present."[11]

Narciso finally left Valdosta on June 9, visited Oak Lawn for a day, then proceeded to Greenville, South Carolina, to start working as a *Greenville News* reporter. The federal census enumerated on June 18, 1880, shows that at Oak Lawn, forty-six-year-old Ralph Elliott and nephews Brosio and Alfonso, also called Bory, were dedicated to "Raising Cattle." Emily, Annie and the children were listed as having a "Home" occupation. The column naming the birthplace of the children's father was dittoed as South Carolina and eleven-year-old Hattie, born in Matanzas, appears as a Carolinian by birth. Whoever provided the information was trying to obliterate the children's Cuban heritage. By contrast, Narciso in Valdosta correctly identified his father's homeland in the federal census. Also living at Oak Lawn were former slaves Clytus Wilson, a thirty-eight-year-old laborer, and his thirty-five-year-old homemaker wife, Chloe, who had been Caroline Elliott's servant. That summer, Ralph and the two oldest boys stayed on the plantation, while the Elliott ladies and the Gonzales children went to Flat Rock, where they were joined by Gertrude, vacationing from school in Baltimore. After two months in Greenville, Narciso arrived on August 5 in Columbia, South Carolina, to be the

Charleston *News and Courier* reporter in the state capital, covering politics and local items. He visited Oak Lawn in mid-October to report on the Democratic rally at Walterboro, accompanying Ralph and Brosio "with some Red shirts from this section." They supported the successful gubernatorial candidacy of former General Johnson Hagood. Brosio planted cotton that fall at The Bluff, but protested being cheated out of his "whole year's work" by the freedmen sharecroppers he dealt with. The eldest Gonzales son, like his father had done at Social Hall, treated his employees honestly, but his generous nature was also abused.[12]

Gonzales remained in Cuba, where in January 1881 he moved to Matanzas to start teaching. He wrote to Narciso that "although some of my oldest friends are away and others, like the Ximenos, are utterly ruined, I still feel more at home in my native town than I did in Havana." His son, in turn, seemed to be taunting Emily when he sent her a copy of the letter, saying, "This will give you a topic for conversation, and so I'll leave you." Emily immediately asked Gertrude if Nanno had resumed mentioning her father, to which she replied, "I heard from Nanno twice last week, he told me nothing about the Old Man." Two months later, Narciso informed his brother Willie that Emily "is not corresponding with me now." William Elliott Gonzales had finally left home to be a cadet at King's Mountain Military School in Yorkville, South Carolina. Narciso, in spite of his journalistic duties, kept a profuse correspondence with his family. He complained to Brosio, "If the ancients had hated writing descriptive letters as much as you do, we wouldn't have any history." Nanno expressed his father's antiroyalist and revolutionary ideology when commenting on the assassination of Alexander II in Russia: "I can't help sympathizing with the Nihilists, not only because I don't believe in Tsars, but because I admire their perseverance."[13]

The redemption of Social Hall was constantly on Narciso's mind during the spring of 1881. Promoting his father's plan of a decade earlier, he told Brosio in March, "I will not cease my exertions until I get a colony settled down there." Their twenty-one-year-old brother Alfonso could not be counted on for these plans, as his idleness had now extended into saloons. Narciso presented his idea to U.S. senator Matthew Calbraith Butler, a former Confederate major general, who was considering taking "a load of immigrants" from New York to the South Carolina lowcountry. Butler also advocated black emigration from the state. The promotion of foreign agricultural workers had been dismal due to low wages and overwhelming black labor competition. The 1880 South Carolina census indicated that of 198,147 farmers in the state, only 122 were aliens. Butler was willing to invest in cattle and sheep raising, and Nanno impressed him with the possibilities of success at The Bluff and Social Hall. Narciso instructed Brosio, "If you work the people up to the right pitch on the political, moral, and agricultural effect of an influx of sturdy white men in their midst, and can furnish data on which to make a good show of advantage to the Germans, I will do the rest." He then informed Willie, "We are bringing German farmers into the State in small parties to work on shares with the planters. We hope to get thousands after awhile, and show the negroes that they are not to rule us again in politics or labor. I'm going to try to settle a colony of them at Chehaw or home next Fall. They are brought by the State and distributed all about." The recently created South Carolina office of Superintendent of Immigration, headed by Dr. E. M. Boykin, was recruiting new immigrants at Castle Garden, New York. The costly scheme was abandoned the following year after bringing only 534 Europeans into the state.[14]

The colonization project for Social Hall and The Bluff collapsed at the end of August when a hurricane devastated the crops on more than forty-seven plantations in the South Carolina lowcountry from the Combahee River to Charleston. Ralph acknowledged extensive damage to his field and garden fences, "and now caterpillars are sweeping the cotton, and chills & fever prevailing to an unheard of extent." In consequence, Brosio, who had been temporarily working as a telegraphist in Yemassee with the Port Royal and Augusta Railway Company, decided to seek better fortune. He obtained at Charleston free passage for Manhattan on the new steamship *City of Columbia*, with "two dollars & my old gold pen as my assets." Arriving on September 5, he went directly to the Western Union Telegraph Company building, was tested and approved "ordinary first class" operator, and ordered for duty the next morning at a starting rate of seventy dollars monthly. That night, Brosio probably understood what it had been like for his father to arrive a decade earlier in New York under similar precarious circumstances.[15]

Gonzales left Havana on the steamship *British Empire*, arriving in New York City on June 22, 1881. He proceeded to the health resort at Saratoga Springs, New York, from where he wrote to Narciso. His son replied in mid-September, "There is no marked improvement in the condition of any of us and the outlook is not promising." Narciso indicated that Brosio was working in the Western Union office in lower Manhattan. Although Narciso was very affectionate toward his father and had been sending him the weekly *News and Courier,* he concealed the renewed contact when he wrote two days later to his aunt Emily, who was spending the summer in Flat Rock, North Carolina. Instead, on the eighteenth, he told his uncle Ralph at Oak Lawn, "The old man has reappeared—this time at Saratoga. Col. McMaster who has just returned from that resort, met him there visiting the Misses DeLeon. Occupation unknown, but appearance 'youthful.' That's all."[16]

Brosio was living frugally with three other telegraphists in a "very nice furnished room" on East Twentieth Street in Manhattan. He walked three and one-half miles to work daily, considering it "good exercise" to elbow his way through the Broadway crowd, but sometimes took the omnibus. Brosio viewed New Yorkers as disrespectful to women and "mannerless in the extreme. If you tell a man good morning here he looks as much astonished as if you were to knock him down." He worked in a telegraph operating room 250 feet wide by 70 in length, "in the 7th story of the tallest building in town," with "a lovely view of the two rivers, the bay & ten miles of solid city." There were 300 men and 130 women operators in the office, under the "deafening" clatter of "700 sets of instruments going at once." Brosio worked 13-hour days, including many Sundays, to make $100 monthly. He was sending money to his uncle Ralph, still struggling with alcoholism, to help with stock-raising expenditures at Oak Lawn. Sometimes telegraphic work was very demanding, averaging "a message a minute for an hour or two." The pace grew frenzied after the assassination of president James A. Garfield, when Brosio had to send hundreds of condolence messages to his widow. During his spare moments, the Carolinian reposed in Castle Garden, contemplating the many vessels traversing the harbor, and the teeming arrival of immigrants "loaded with every kind of baggage & chattering & vociferating extravagantly." He frequently corresponded with his family and had direct chats with Ralph on Sunday mornings via the telegraph operator at Adams Run.[17]

Narciso also relocated on October 5 when the *News and Courier* assigned him as their Washington correspondent for one hundred dollars monthly. He lived in a furnished room at 623 Pennsylvania Avenue

NW. The reporter found the capital "a remarkably handsome city, full of life and activity entirely stimulated by government pap." He attended the Senate sessions, "raking up general news and gossip," assisted by Democrats Matthew Calbraith Butler and Thomas Francis Bayard. Butler invited Narciso along on a Congressional Commission steamer that went to the Yorktown centennial ceremonies. The twenty-three-year-old newsman described to Ralph Elliott going with "a mob of Senators and Governors and their attendant females swigging Roederer and Heidsieck [champagne] like water for four mortal days." Narciso was also sending his unemployed uncle money for "petty expenses."[18]

In early October, Gonzales paid Brosio a visit at the Western Union office. The details of this encounter are lost in a missing page from Brosio's letter to Emily. Other accounts from the aunt to her nephews dealing with this situation have been excised from the family correspondence. Their responses to her show that she felt bitter and "despondent" over the renewed paternal contact. Emily immediately sent Nanno a four-page diatribe for being the conduit of the reunion. He replied with a highly sarcastic letter, indicating that "you refer so feelingly to my paternal progenitor, reciting nice things about him which I have never heard before and commending Brosie for his filial joy at seeing him." Narciso said that he could not accept her "liberal praise for bringing the twain together," purporting to have written to his father in Matanzas on September 13, to provide an annual "bulletin of the condition of his descendants," without anticipating his speedy return. This was obviously a ruse, since Narciso had been sending his father the *News and Courier* and informed his uncle five days later that Gonzales had surfaced in Saratoga Springs. Narciso flatly told his aunt to stop writing about such conjectures in the future and to concentrate on "so much home and so little Gen. Gonzales." A week later, Narciso wrote an affectionate letter to his father, wishing him "every success" and saying that due to "financial stringency" he could not yet visit New York. He expressed gladness at knowing that Brosio and his father had renewed their acquaintance after meeting several times and conversing at length. From then on, the two eldest sons never again lost contact with Gonzales and tried to unite the other siblings with him, in spite of Emily's continued reprobation for another decade.[19]

Lack of funds kept Narciso from joining his father and brother in New York during the Christmas holidays. He was saturnine when approached by Robert Smalls on the night of December 23 in front of Willard's Hotel. Smalls, a former South Carolina Republican representative, became a Union hero two decades earlier when, as a mulatto slave and Charleston harbor pilot, he commandeered the Confederate steamer *The Planter* out to the federal fleet. The obese forty-two-year-old Smalls upbraided the young reporter for a "falsehood" written about him. Narciso described what happened next: "There were a lot of Republicans around. Smalls had a heavy stick, and I had nothing. I plugged him in the eye. He cuffed me, bear-fashion, on the side of the face twice. I renewed my blows, but failing to make an impression. I doubled him up with a kick in the expansive region 'below the belt.' A friend of his took his arm and led him away. . . . I was unscathed. If I had had a pistol I would certainly have shot him, and wanted to go for him afterwards but Mr. [Samuel] Dibble and Gen. Butler advised otherwise." Narciso was later surprised by the praise he received in various newspapers, "for the fellows, Democratic and Republican, love a reporter to whip a man who infringes journalistic license." Before the year ended, his sister Gertrude visited from Baltimore. They went sightseeing and climbed to the top of the Washington Monument, since at the time women were prohibited from riding the elevator. Narciso, who like his father smoked cigars, made a New Year's resolution to quit, but resumed his habit within a month.[20]

In mid-January 1882, Gonzales received a letter from his niece Irene Espinosa in Havana, informing him that his wealthy aunt Lola had died on New Year's Day. He returned to the island after learning that Lola had bequeathed him ten thousand dollars. Emily did not waste the opportunity to rankle his younger children against him, and Gertrude responded, "So the Old Man has gone to Cuba! How I wish he would stay there!" Emily then sent a "rather hard" twelve-page letter to Narciso that is missing from the family collections. He sternly and sarcastically replied, "Speaking of 'el Padre.' I didn't like to open the subject, knowing that you always consider his scarf a red rag, as it were, but I will now say that he isn't 'rich' exactly, as you intimate. $10,000 isn't much at interest—and I suppose he will put it at interest and leave it to you when he makes his will. He will not be able to live on it alone, and must, I suppose, continue his lessons. I congratulate him, as you do, and I consider that it required some judgement to get $8,000 out of old Lola with a hungry horde so near her. I wouldn't be surprised if the old gentleman were to send you an Easter card with a pretty angel on it. Wouldn't that be nice!" Emily blasted him back with "considerable defiance" in a letter which was also sanitized from the family papers. He responded, "Do you know that $10,000 at 5 per cent (which is all that a safe investment North will bring) gives $500 a year and that living in my pinched style with only meat and drink and no luxury or recreation costs over $600? Then observe the impossibility of the Paternal's living comfortably Northward on that income, without work. If he don't work, he can't make a good support so of course he will not be idle. Of course, also, $10,000 would count for a small fortune on a farm in our locality, but I tell you it isn't much at the North. And as you say he expects to live forever, he will not spend the principal. Be easy. He is *not* a nabob. But I hope he will receive all the benefit it can give him. He has had a pretty hard time, whether deservedly or not, and he is an old man." Before Lola's will was settled, a letter from Brosio, which has not been found, prompted Gonzales to rush back to New York without having time to bid farewell to his niece. He left Havana on the steamship *Niagara* on April 13, reaching Manhattan four days later.[21]

After receiving Lola's bequest, Gonzales generously rewarded his children. He gave Brosio a two-thousand-dollar loan, plus six hundred dollars to purchase Social Hall from John D. Warren, six hundred dollars to build "a one storied cottage with four or six large rooms with a wide piazza or Verandah" to replace the moldy, roach-infested slave cabin the family had inhabited for sixteen years, and another amount for Hattie's schooling in Baltimore. The Cuban then boarded the steamship *Labrador*, signed the register as "Gen. A. J. Gonzales," and departed New York on May 16 for Le Havre, France. He visited Paris to obtain medical attention for nasal problems that affected his singing. Emily Elliott, who for more than a dozen years had instilled in the children that their stingy father was living luxuriously while they wallowed in misery, wanted Brosio to reject the money given him. He defiantly responded, "I don't think tis necessary to discuss the Old mans character or his object in leaving this money in my hands. The present object is Hatties schooling which *I intend that she shall have*." Thirteen-year-old Hattie was later sent to the Edgeworth School in Baltimore. In October, Brosio departed for a better

Western Union position in New Orleans. While visiting his aunt Mary in Baltimore, he discovered that Emily had been excoriating him in her letters to Mary after he accepted his father's funds.[22]

Narciso had returned to Charleston two months earlier, during the congressional recess. He boarded in the home of Mrs. Lawton, on 13 Coming Street, where two other reporters resided. The building is today a dormitory of the College of Charleston. Narciso resumed his duties with the *News and Courier,* which consisted of "reading some 500 Southern daily and 75 State weekly exchanges a week, and culling therefrom the State news, political gossip, crop notes, marriages, deaths &c. &c." He spent six to eight hours daily doing this, plus rewriting letters, heading dispatches, writing from dictation, concocting condensations of miscellaneous clipped articles, revising proofs, and occasionally drafting an editorial.[23]

Narciso received in the mail the day after New Year a copy of the *American Register* of London, England, dated December 16, 1882, sent by his father from Paris, mentioning that Gonzales was recuperating his health. He forwarded the news clipping to his aunt Emily in a vain attempt to elicit sympathy for his ailing sixty-four-year-old father. The Gonzales address book reveals that he was in contact with a Dr. Crane, a staffer at the *American Register*, and in Paris he consulted physicians Charles Faunel and L. Real, who prescribed medication for his nasal affliction. He also visited revolutionary compatriots sixty-four-year-old José Güell Renté, Enrique Piñeyro Barri, Laureano Angulo and José Bueno Blanco, and he saw Judah P. Benjamin, the former Confederate secretary of state and his nemesis during the 1851 filibuster trials. Gonzales then went to London, stayed at the Hotel Mathis in Coventry, was treated for catarrh by two physicians, and bought clothes from two tailor shops. He also visited Plymouth and Liverpool, departing from the latter on the steamship *Bothnia*, which docked in Manhattan on July 31, 1883.[24]

Twelve days later, Brosio arrived in New York City with his seventeen-year-old brother Willie. They moved into a gaslight room at 35 West Sixteenth Street, two blocks from Frobisher's College, where Willie enrolled for three weekly lessons to cure his stammering. Brosio worked at Western Union and part-time in the Stock Exchange, eighteen hours a day, seven days a week, for $140 monthly. He was afflicted with a partial facial paralysis, possibly Bell's palsy, lasting three months. Brosio was unable to move the left side of his face, his neck, or close his left eye, which he had to bandage when trying to sleep. He could not smile or whistle and had difficulty eating. His condition improved after taking strychnine and giving himself daily shocks with a galvanic battery loaned by a physician. Gonzales found his son in that distressing condition when he returned to Manhattan in early September after visiting a resort. He met Willie for the first time in thirteen years, but the teenager, who did not recall his father, still harbored the resentment inculcated by Emily. Willie wrote to her: "The old man is here and visible every day or two, not doing anything of course. . . . About a week ago I went to see Rip Van Winkle, which was splendid—with & *at the invitation & expense of The Paternal*." Gonzales also took his son to a large Democratic rally, where they "heard speeches from some of the more important men in the state."[25]

Emily, obsessed with hatred, kept pumping her nephew for information on Gonzales, to which he replied, "You ask me what the Old Man aims to do; I really do not know, but think it is to 'lie low' for the present, and 'bob up serenely' every now and then in different parts of the country." Willie knew that his father planned to go to Charleston soon, but he did not mention it to his aunt. The following month, Gonzales was told by his stepbrother Ignacio that their aunt Juana González had died and left him a share of the family estate. It amounted to one fourth ownership of a tenement house at 7 Revillagigedo Street in Havana, worth 5,500 gold pieces. Gonzales sent Ignacio a proxy to act on his behalf and to forward his portion in a letter of credit, care of the banking firm Brown Brothers in New York, but the settlement took over a year. Gonzales gave $300 to Brosio for Alfonso to buy cattle and another sum for his daughters.[26]

Willie returned to Charleston by rail in early October, to begin classes at the South Carolina Military Academy, the prestigious Citadel. He was part of forty-five new cadets in a class of 160 students. One of his professors, former Confederate artillery officer Delaware Kemper, had served with his father. The school surgeon, Francis L. Parker, and the quartermaster, Lt. William W. White, also spoke kindly about the former chief of artillery and asked Willie about him. The youngest Gonzales son started overcoming his animosity toward his dad and sent him an affectionate letter describing his Citadel life, of breaking his left arm soon after arriving, and how his stammering renewed upon reciting before a large class. Emily was losing to maturity another ally in her fourteen-year battle to alienate the children from their father.[27]

The family feud exacerbated on November 20 when Gonzales arrived in Charleston on the steamship *City of Columbia* and checked into the Charleston Hotel. He used William C. Bee and Company as a mail drop and did translations for a South American newspaper. When Emily heard of his proximity, she resumed her vitriolic campaign against him. Hattie was writing from Baltimore to Willie that she supposed "Emmie is worried at the Old Man's being in Charleston," and believed that there was no danger of her father paying her aunt a call. Willie reassured Emily that Gonzales would not visit her, although he was anxious to see Alfonso. He tried appeasing his aunt, saying, "About my share of his fortune, he offered me some pocket money, but, altho' I had a perfect right to it, I declined." Gonzales apparently met with Alfonso, who was breeding two race horses named Zulu and Comet, because Emily soon wrote to Brosio that his father was going into the "hippodrome business." Brosio replied that it sounded "preposterous" and reassured her that Gonzales would not interfere in Trudie's life, since "it rests with her to renew the acquaintance or not as she chooses." Emmie then sent Willie on a "mission" to get further information from Gonzales regarding his plans. He reported back, "I do not think Gertrude would be disturbed by the Paternal, think that has been checked." Willie also indicated that Alfonso had received money from Brosio, instead of his father, for a trip to Florida to recuperate his health. By late January 1884, Willie was still informing Emily that he had "heard nothing of the Old Man." A few months later, he dropped out of the Citadel.[28]

Brosio visited South Carolina by rail during the first fortnight in February and presumably met with his father before returning to New York. Gonzales later departed by steamer for Fernandina, Florida, where, according to his address book, he met with former senator David Levy Yulee. He then traveled by railroad to Jacksonville, lodging at the boardinghouse of Mrs. N. L. Ward, on the southwest corner of Forsyth and Julia Streets. The Cuban went to Tallahassee on the Jacksonville, Pensacola, and Mobile Railroad. In the state capital he stayed in the boardinghouse of Mrs. Vining. The former aristocrat was introduced to a number of prominent people, including Gov. William Bloxham, former governor David Walker, Supreme Court chief justice Edwin M. Randall, Secretary of State John Crawford, and the editors of the *Tallahassee Floridian*, Robert B. Hilton and James W. Dorr. Gonzales proceeded to Pensacola, where he visited U.S. Navy physician Daniel Guiteras Gener, son of Antonio Guiteras Font, and his wife Laura Peoli. He also called on Angela Mallory, the widow of Stephen Mallory, and her sons. Gonzales probably reminisced about the Mallory home in Key West, his haven and shelter after being wounded at Cárdenas thirty-four years earlier. The Cuban left Pensacola in March on a steamer for New Orleans.[29]

Gonzales stayed in the Crescent City with former siege train artillerist G. W. Nott. His address book indicates that he tried one afternoon to locate General Beauregard "at 'La Variete' Club under the Grand Opera House, Canal St." Gonzales also visited Col. Alfred Roman, Beauregard's former inspector general, who was a criminal court judge. Roman had just published a massive two-volume work, *The Military Operations of General Beauregard in the War between the States, 1861–1865*, in which the role of Gonzales is minimal. The Cuban also met with cotton merchant A. H. May, a juror in the first filibuster trial of 1851 who had voted for acquittal.[30]

The memories rekindled in New Orleans, where he had not been for more than thirty years, prompted the former filibuster leader in late March and early April to publish two articles in *The Times-Democrat* at the request of its editor describing his role in the López expeditions. In the last article, Gonzales revealed for the first time that Sen. John C. Calhoun had told him in 1849 that Cuba was destined to be a part of the Union. He blamed "the hostility of England and France" for the failure of Cuban annexation during the Pierce administration and concluded that the island in 1884 was "of more vital importance than it was then." Gonzales sought employment as professor of Spanish at the University of Louisiana, presided by Charlestonian Randell Hunt, which was in the process of becoming Tulane University. He presented numerous letters of recommendation from various active and former politicians, including William Porcher Miles, and a petition signed by Charleston mayor William A. Courtney, banker Harmon Hendricks DeLeon, merchant Theodore Dehon Jervey, former Confederate generals Wilmot Gibbes DeSaussure, Benjamin H. Rutledge and Rudolph Siegling, attorney Cleland Kinloch Huger, state senator George Lamb Buist and *News and Courier* editor Francis Warrington Dawson. The first four petitioners were Freemasons. Fate played against Gonzales once more, as Hunt was replaced as rector and the Spanish professorship went to J. A. Fernández.[31]

The Cuban spent the summer in New Orleans, bought two Havana Lottery tickets on June 2, and was fitted with new eyeglasses. When Democratic presidential candidate Grover Cleveland, the governor of New York, won the November 1884 election, Gonzales wrote to his stepbrother Ignacio in Havana that "the election of a Democratic President has been for me, as for the rest of the country, a very happy event." The return of a Democrat to the White House after twenty-four years was regarded by many Americans as a significant step toward sectional reconciliation, while African Americans feared that the Democrats would "reestablish slavery." Gonzales urgently asked for his share of aunt Juana's inheritance to sustain him while he sought patronage in Washington, D.C. He traveled via Charleston, where he published an anonymous article in the *News and Courier* on January 21, 1885, titled "The Cause of Mr. Davis Is the Cause of Every Ex-Confederate." Gonzales mailed it to the former Confederate president and signed "with highest regards." The article defended Davis in a controversy that began two months earlier, when Gen. William T. Sherman publicly claimed that he had a letter proving that the Civil War had been a conspiracy by Davis to take over the United States. When Davis demanded that the purported document be produced, Sherman said that it had been destroyed in the Great Chicago Fire of 1871. The dispute extended to the U.S. Senate on January 12 when Connecticut senator Joseph R. Hawley, a former Union army general who led a brigade on Morris Island during the siege of Charleston, presented a resolution favoring Sherman's version of Davis's treason, equating him with Benedict Arnold and Aaron Burr.[32]

In his defense of Davis, Gonzales pointed out that Sherman could not be trusted. In Sherman's memoirs he admitted that he had lied about Wade Hampton to diminish his influence. Gonzales also maintained that "there are a good many original Republicans, no doubt, who would be unwilling to go as far as Senator Hawley goes." He continued: "If Mr. Davis was a traitor, then every Confederate soldier, and every person who gave, in any way, aid and comfort to the Confederate cause was likewise a traitor. . . . No ex-Confederate, therefore, can afford to be silent when the foul name of traitor is applied to Mr. Davis." He went on to say that if Sherman, Hawley, and others were not satisfied, "they can bring Mr. Davis to trial for treason. The lapse of time will not prevent it. No Statute of Limitations applies to such a case as this. The ex-President of the Southern Confederacy has never been pardoned. Try him, then, and see whether there is any Court in the United States that will adjudge him to be a traitor! It could not be accomplished twenty years ago. The Government abandoned the prosecution because it saw that Mr. Davis must be acquitted." Gonzales indicated that no other Confederate officer or politician had been branded a traitor and asked, "Is it to be assumed that Mr. Davis alone was responsible for the secession of the Southern States, and for the gigantic war which was caused by the invasion of the South by the Northern armies?" His article questioned the timing of the renewed controversy, on the eve of the first inauguration of a Democratic president in nearly thirty years. Cleveland, who hired an immigrant substitute to comply with his draft obligation during the Civil War, was determined to reconcile lingering sectional animosities that impeded national economic progress.[33]

While in South Carolina, Gonzales obtained many recommendations from prominent citizens in Charleston and Columbia and the endorsement of the entire South Carolina delegation in the House of Representatives for a diplomatic mission to Latin America. Narciso, whose journalistic pen was praised by some people and despised by others, wrote on his father's behalf to all of the South Carolina legislators in Congress. Gov. Hugh Smith Thompson recommended Gonzales to South Carolina chief justice William D. Simpson and Senators Wade Hampton and Matthew Calbraith Butler. The latter had been paroled with Gonzales at Greensboro. The senators later personally promised Gonzales to use their influence in his favor. In February, Gonzales moved to 1326 L Street NW in Washington, D.C. Narciso later joined him when covering the inaugural ceremonies for the *News and Courier.* Over a quarter of a million people, including many Confederate veterans, flocked to the capital to see Grover Cleveland inaugurated as president on March 4. The military parades on Pennsylvania Avenue had Northern regiments with African American veterans, while Fitzhugh Lee and the Richmond State Guards headed the Southern militia troops from various states. Upon returning to Columbia, Nanno wrote to Oak Lawn, "The *padre* is comfortably located in Washington waiting for a boost from S[outh] C[arolina].

Major [George Lamb] Buist is engineering the boost. I understand that Capt. [Francis] Dawson has promised his influence which is valuable and which he has refused to give for any candidate for office in the State. The old gentleman gave me a curious insight into his philosophy before I left him. He says that he believes in the Darwinian theory of 'the survival of the fittest' and is determined to prove himself 'the fittest' by 'surviving.' He said that he had been joked about resembling 'Micawber,' but that he felt encouraged by the comparison as something 'turned up' for Micawber at last."[34]

Narciso's literary allusion paints a vivid portrait of his father's character. Wilkins Micawber, a character in the Charles Dickens's novel *David Copperfield*, always maintained his optimism in the face of constantly poverty and disappointment. He pawned his family's household goods to survive yet was temporarily sent to debtors' prison. Similarly, the Gonzales family had lived in abject poverty at Social Hall, sold their furnishings, and subsequently declared bankruptcy. Micawber's long-suffering wife, like Hattie, supported her husband through all his hardships while they frequently moved with their children to different cities seeking better fortune. When Micawber's corn business became unprofitable, like the Cuban's, he advertised in the newspapers for a new situation. He pronounced grandiloquent speeches and had a penchant for writing long, flowery letters, traits common to Gonzales. Micawber's implicit faith that "something will turn up" resulted in a loan that allowed him to start a new life in Australia, where he became a successful magistrate. Likewise, Gonzales received an inheritance and hoped that his long-sought diplomatic position would result in a rewarding future.

Gonzales was among the thousands of office seekers who descended upon Washington, D.C., aggressively pursuing a political appointment in the various branches of government. He hoped that the Southern recommendations he had gathered with Narciso's help would make a personal impact on Cleveland. In early April, Narciso established in Columbia a branch office of the *News and Courier*, with his brother Willie as assistant and his uncle Ralph as bookkeeper and distribution manager. When Emily asked about his father, Nanno tersely replied, "Gen. G. hasn't got his mission yet. I haven't heard from him." Gonzales wrote on May 23 to former South Carolina representative William Porcher Miles, Beauregard's former volunteer aide-de-camp, and owner of Miles Planting and

Manufacturing Company, the second-largest sugar producer in America. The Cuban mentioned all the political recommendations he received from South Carolina, plus one from New York Supreme Court justice Theodoric R. Westbrook, a former congressman who knew him intimately during the Pierce administration. He asked Miles to write to Secretary of State Thomas Francis Bayard, indicating that he had been recommended before the war for the same Latin American mission by most Southern congressional delegations. Gonzales indicated that "this precedent is invaluable to me not only as determining the action of this administration, but as a justification before a Republican Senate if the nomination be sent in." He believed that his diplomatic appointment to Latin America would be beneficial to the United States, as his being "an assimilated member of the Spanish race" would serve "as a token of friendship & to draw closer commercial & political relations," which was "far more obvious" at that time than ever before. His son Brosio did not share the same optimism or racial perspective. Upon being laid off from his telegraphist job in New York City a few weeks earlier, he wrote to his sister Gertrude: "I could sing in a Bowery museum for $7 a week[;] and last, but not least, the field of Hurdygurdyism is always open to the Latin races." Brosio returned to South Carolina and sold *News and Courier* subscriptions in the lowcountry for a fifty-dollar monthly salary.[35]

The day after writing to Miles, Gonzales petitioned Secretary Bayard to withdraw from the files of the Department of State the letters of recommendation sent on his behalf by Southern legislators during the 1850s for a diplomatic post in Latin America. They were returned to him the following week. Gonzales used these documents to try to obtain other political endorsements, and afterward refiled them in their place of origin. Cleveland, emphasizing isolationism in foreign affairs in his inaugural address, paid scant attention to Latin America and limited his sporadic interest there to the Caribbean Basin. The presidential policies lacked the economic expansionism or annexationism envisioned by Gonzales. Cleveland and Bayard had charted a peaceful foreign policy void of jingoism and imperialism. In mid-June, the president announced that after July 1, he would "decline to receive personal visits from office seekers or their friends."[36]

Gonzales spent the summer of 1885 at the fashionable four-story Warren Green Hotel in Warrenton, Virginia, a resort frequented by Washington politicians, whom he lobbied for his coveted position. While there, Gonzales sent copies of his two articles published the previous year in the New Orleans *Times-Democrat* to José Ignacio Rodríguez, a fifty-four-year-old Cuban-American attorney residing in Washington, D.C. since 1870, who later wrote a history of the Cuban annexation movement. Returning to the capital in the fall, Gonzales established residence in the Mc Pherson House, 1423 I Street, an abode for civil servants, and patiently awaited a governmental response to his petition. Emily Elliott, by now an old maid, retained her bitter resentment against Gonzales after sixteen years. When making her last will and testament in November 1885, she left all her property to her sister Anne, "but after her death it is my will that the same shall be divided between the children of my Sister Harriet R. Gonzales, who shall then be living—their father to have no claim part or parcel in the same." It was a spiteful exercise in futility. Anne died in 1916 at the age of ninety-four, outliving Gonzales and three of his six children.[37]

When Gonzales did not hear in late January 1886 about his desired diplomatic post, he wrote to Cleveland's private secretary, Daniel Scott Lamont, reminding him of his recommendations for a mission to Latin America. The desired diplomatic employment failed to turn up by May 12, when Gonzales again wrote to President Cleveland. Lamont replied in three days, with a one-sentence acknowledgment, saying that Cleveland had "directed its reference for the information of the Secretary of State." Two weeks later, Gonzales gave up all claim to his ten-thousand-dollar marriage bond and had Ralph Elliott, its trustee, assign it to his children. The transaction was witnessed in Washington, D.C., by attorney José Ignacio Rodríguez and A. R. Allen. Gonzales also signed over the five-thousand-dollar bond of Mary Johnstone which he had assumed when acquiring Social Hall twenty years earlier. The bond settlement allowed Gonzales immediately to transfer ownership of Social Hall to his oldest son, who signed as Ambrose E. Gonzales. Emily Elliott was a witness to the signatures of Ambrose, Alfonso, and Gertrude, but nothing Gonzales did could appease her rancor.[38]

After almost three years Gonzales had not received his share of his aunt Juana's estate, so on June 19 he again asked his stepbrother Ignacio to forward his inheritance by way of Joseph A. Springer, the U.S. vice consul general. He indicated that he had been gravely ill and had to borrow money to prepare for a long convalescence. Brosio had been sending his father via the Adams Express Company an average of forty-five dollars monthly during the previous six months. It was part of the two thousand dollars he had been holding for years, which had by then dwindled to half the amount. Gonzales had not heard from him in a month and worried that he might have fallen sick while working as a traveling agent for the *News and Courier* in the fever-ridden lowcountry. Meanwhile, Brosio's recent contact with the African Americans of coastal South Carolina prompted him to start writing a series of tales in their Gullah dialect, which is still spoken today. The articles were later compiled in the book *The Black Border: Gullah Stories of the Carolina Coast*, containing a Gullah glossary, acclaimed by linguists. Lacking funds for travel, Gonzales borrowed money from his Cuban friends in Washington, D.C., Pedro José Guiteras Font, who was the best man at his wedding, attorney José Ignacio Rodríguez, and Aniceto García-Menocal Martín, a U.S. Navy Department civil engineer, with whom he had spent part of the summer of 1872. A week later, Gonzales stayed for a month at the Hygeia Hotel, at Old Point Comfort, Virginia, near Fort Monroe. Hotel advertisements boasted a four-story "resort for the pleasure-seeker, invalid, or resting-place for tourists . . . with accommodations for about nine hundred guests." The hotel was comfortably furnished, had two hydraulic passenger elevators, gas, water, rooms for bath, and electric bells in all the rooms with windows facing the ocean. Gonzales was greatly pleased that summer when Cincinnati artist Matthew Somerville Morgan requested his photograph, to include his likeness "in the mammoth picture of 'Sumter.'"[39]

Returning to the Mc Pherson House in late July, Gonzales again wrote to his stepbrother Ignacio for his inheritance money, "which you owe me not only as a brother but as my general proxy in whom I have deposited my confidence." He received the full amount, $1,074.94, the following month, after a three-year lapse. Gonzales replied that he had little left of his aunt Lola's ten thousand dollar bequest, nearly two thousand of which was spent on his children. The new sum allowed him to pay arrears at his boardinghouse, a physician's bill, and would not require him to receive sums from his loan to Brosio for more than a year. Although Gonzales had not received a job offer from the Cleveland administration during the previous eighteen months, he told Ignacio: "I have been waiting for employment from the present American Government. If I obtain it, it will assure in large part my future. If not, I will have to again look for some profession with which to live." He offered Brosio $150 for his three youngest children, but his son suggested that "unless you have a good position assured, and something to go upon, that it would be unwise to do so, as they are not in need, and you may be some time. I will, however, submit my proposal to them and let them decide."[40]

His youngest son Willie was the first sibling to wed. On February 2, 1887, at the age of twenty, he married seventeen-year-old Sarah Cecil Shiver, the daughter of a German merchant in Columbia. They had been engaged since the previous March. When Narciso first met Sarah, he found her to be "very sensible and well bred and was also frank, an attribute I appreciate." The lavish wedding had the largest crowd ever gathered in the First Presbyterian Church. Ambrose, Narciso, Gertrude, and Hattie were among the twenty-eight bridal party members. The guests included Gov. John P. Richardson and Willie's militia unit, the Richland Volunteer Rifle Company. The reception was held at the Shiver home on Arsenal Hill, where the newlyweds took up residence. Nuptial accounts appeared in the Columbia *Record*, the Augusta *Chronicle*, and the Charleston *News and Courier*. Cousin Leila Elliott Habersham and her family in Savannah "were quite astonished to hear of Willies engagement & speedy marriage." Gonzales, who did not receive an invitation, was informed by the newspaper clippings Narciso sent him. His presence would have marred the event for Emily and his two daughters, who remained prejudiced against him. The patriarch sent a congratulatory letter to Willie, saying that he had seen his bride at Warrenton, Virginia, in 1885, noticing "her great personal charm, and was impressed with her expression of candor and amiability."[41]

The Cuban remained at the Mc Pherson House during spring 1887 while working a temporary position. That summer, he went to the health spa at Orkney Springs, Virginia, whose seven springs were considered therapeutic for rheumatism, gout, and mental and physical exhaustion. The resort, accommodating up to eight hundred people, was owned by the Robert E. Lee family. The main four-story Orkney Springs Hotel, still standing today, had 175 bedrooms, a 50-by-100-foot ballroom, a 40-by-155-foot dining room, a reading room, and a billiard room, where Gonzales probably wielded a pool cue skillfully. A hotel guest, Flora Adams Darling, wrote to Jefferson Davis in July that the "Cuban Patriot, General

A. J. Gonzales," was staying there. The forty-seven-year-old Mrs. Darling was the widow of Louisiana Confederate officer Edward Irving Darling, mortally wounded at the battle of Franklin, Tennessee, in 1863. She was also the author of the Civil War memoirs *Mrs. Darling's Letters* and three years later founded the Society of the Daughters of the American Revolution. The former Confederate president had graced her with a testimonial "in the name of the South for standing an exponent of truth and justice."[42]

Mrs. Darling wrote to Jefferson Davis that the Cuban's "admiration for you is honest hearted and enthusiastic. He desires to send through me his most cordial regards to both you and Mrs. Davis with best wishes for your health and happiness." She described how Gonzales had sung like a patriot "the Grand Old Marseillaise that I had not heard before since the night I left the Confederacy forever." The song had been played by a band at the Confederate presidential inaugural and it symbolized their independence. Mrs. Darling informed Davis that the Orkney guests listened to Gonzales "with Varied Emotions. When he ceased there was one who could not speak. The Genl. said with tears 'I am glad you asked me to sing.' Tell President Davis God has spared his intellect to write of patriotism, and my Voice to sing, and even though we have lost our Cause we can reach mens hearts. He sings delightfully." She added that the "Old Patriot" had "lost his family and wealth, but he is rich in memorys [sic] and friends. He expects an appointment from Cleveland, but I presume he will be disappointed. President Cleveland seems the Great Disappointment." By June 1887, Cleveland had replaced 138 out of 219 consuls, but his Southern patronage was going to younger men who did not mourn the "lost cause" and had no ties with the antebellum aristocracy.[43]

When Gonzales returned to the capital in the fall, he temporarily roomed in the Hotel Windsor, and became secretary to Francisco Lainfiesta, minister of Guatemala, El Salvador, and Honduras. In late December 1887, the sixty-nine-year-old Gonzales came under the influence of thirty-two-year-old Pierre L. O. A. Keeler, a spiritual medium visiting the city. Keeler was described as having "well cut features, curly, brown hair, a small, sandy moustache, and rather worn and anxious expression; he is strongly built, about five feet eight inches high, and with rather short, quite broad, and very muscular hands and strong wrists." One of his seance participants, British scientist Dr. Alfred Russel Wallace, who formulated the theory of natural selection with Charles Darwin, saw Keeler as "a young man of the clerk or tradesman class, with only the common school education, and with no appearance of American smartness." Spiritualism, the communication with spirits through a medium by way of a seance, had spread widely in America during the 1850s, decreased during the Civil War, and was rejuvenated during the 1870s until the end of the century. The movement attracted millions from all walks of life and soon spread to England and Europe.[44]

Spiritualism proved fertile ground for many con-artists and Pierre L. O. A. Keeler exploited the perpetual grief Gonzales felt for his departed wife. The Cuban frequently attended seances, carrying rose bouquets for Hattie on their wedding anniversary, and received spurious messages from her on four-inch scraps of paper and school slates. Sometimes, a "spirit" hand that materialized over or through a curtain, wrote him a note on a pad. Dr. Wallace found Keeler's seance "peculiar," and described how "the corner of a good-sized room had a black curtain across it on a stretched cord about five feet from the ground. Inside was a small table on which was a tambourine and hand-bell. . . . Three chairs were placed close in front of this curtain on which sat the medium and two persons from the audience. Another black curtain was passed in front of them across their chests so as to enclose their bodies in a dark chamber, while their heads and the arms of the outer sitter were free."[45]

Two years earlier, Keeler had appeared before the Seybert Commission of the University of Pennsylvania, which investigated spiritualism. In their *Preliminary Report*, they described a seance held by Keeler, assisted by his wife: "Raps indicated that the Spirit, George Christy, was present. As one of those present played on the piano, the tambourine was played in the curtained space and thrown over the curtain; bells were rung; the guitar was thrummed a little." When a materialized right hand appeared, it "was put forth in this case not over the top curtain, but came from under the flap, and could easily have been the Medium's right hand were it disengaged, for it was about on a level with his shoulder and to his right." The Seybert Commission dismissed "the theory of a Spiritual origin of the hand behind Mr. Keeler's screen." Its secretary concluded that spiritualism "presents the melancholy spectacle of gross fraud, perpetrated upon an uncritical portion of the community."[46]

During sixteen months, Keeler provided Gonzales with over fifty "spirit" messages from Hattie on note papers and slates. Gonzales saw the handwriting as "identically the same" in all the messages and "very much the same" as Hattie's, "only smaller, to crowd as many words as possible into a small space." One message appeared on a slate that Keeler had "held *under his buttoned-up coat.*" Gonzales believed that "by far the greatest proof of their authenticity lies in their character and inherent quality." The first slate message from Hattie was enticing: "I will tell you all some time. No power now. I live! I live!" Other spirit communications assured Gonzales that his wife was with him "every day," that her "father sends love," and that she sent poems and asked him to return to the seances. The fee for each session was five dollars and Gonzales went to "hundreds," sometimes placing a few of his wife's mementos upon the medium's table. Hattie purportedly told Gonzales in one slate message, "I live, I breathe, I see, I hear, I know of you, and when the saddest moments of your life reigned supreme, I was near, though powerless then to cheer you." Gonzales also received from Hattie at the seances "a touch of the vanished hand and a sound of the voice that is still." On various occasions, Gonzales believed that the female hand that he clasped through the curtain had been materialized by his wife, although it probably belonged to the youthful Mrs. Keeler. One spirit note from Hattie claimed "not having had the power to kiss the previous night at a materialization seance."[47]

Seven weeks after the first message, the Cuban received a valentine from Hattie's spirit on St. Valentine's Day. The next day, in the solitude of his room, Gonzales slipped on his fingers Hattie's engagement and wedding rings, which had been enlarged to fit him. Believing that his dead wife could hear him, he said in a low voice, "With these rings I wed thee again, not as a golden or a diamond wedding, but as the wedding of the spirit, with the spirit for ever and ever and ever." The seances stopped during the summer and fall of 1888, when Keeler was absent from Washington, D.C. When they resumed in December, a spirit note from Hattie asked Gonzales to "come often, and let me clasp thy dear hand." On December 24, the slate message announced "a merry Christmas and a happy New Year." On Valentine's Day 1889, a seance message from Hattie announced, "I am your angel bride; forever are we to be together. I am so glad you hold my letters so choice, and I am so glad you come so often; it does make me *so* happy and you, too, I know." Six days later, another slate inscription begged Gonzales, "Come to-night, please, and make me still happier." A materialization seance finally occurred on March 14, 1889, with Gonzales embracing for a "few hurried moments" a "spirit" claiming to be Hattie. Gonzales continued going to seances, especially on Easter morning and on his April 17 wedding anniversary. That day, the slate message indicated, "We will combine to contribute to you, as a wedding gift, success in your livelihood efforts. You shall prosper and be well. Mr. [Abraham] Lincoln thinks you will be [employed] by the last of next week." Before Keeler departed Washington, D.C., at the end of April, the last slate writing he produced for Gonzales stated, "Darling: Have not we both been wonderfully comforted by this lovely intercommunication? Do you not feel that you have been lifted from a base condition of darkness into the light of the higher life of glory?" Two months later, an "occult telegraphy" message appeared in the New York spiritual paper *Celestial City*, addressed to "Darling Gonzie," which ended saying, "My promise was from heaven: 'Perseverance conquers every impediment.' It shall be fulfilled. Your loving bride, Harriett R. E. G."[48]

Gonzales was also victimized by spirit photography, apparently taken by the medium's brother, William M. Keeler, who charged two dollars for each sitting. When the Seybert Commission asked to investigate his phenomenon, he raised the price to three hundred dollars for three sittings, paid in advance, "whether my efforts prove satisfactory or not." He also demanded "the exclusive use of the dark room and my own instrument." The commission "regarded his terms as intentionally prohibitory" and dismissed him as a fraud. Years later, a doctor of the American Society for Psychical Research examined over four thousand Keeler spirit photographs and declared all of them fake. One of the few surviving photos of Gonzales has a superimposed image of a young woman over his left shoulder and three other blurry faces appear floating around his head.[49]

The optimistic Gonzales, who without success had enthusiastically pursued Cuban independence, political patronage, and a Confederate general's rank, became an ardent supporter of the cult which he believed had returned to him the long-departed love of his life. He seems to have drifted away from the Episcopal church, whose clergymen, along with Catholics, denounced spiritualism as lacking dignity and having a mercenary character. In September 1889, Gonzales defended spiritualism and Keeler by publishing over fifty messages from the beyond in a sixty-eight-page booklet titled *Heaven Revealed: A Series of Authentic Spirit Messages, from a Wife to her Husband, Proving the Sublime Nature of True Spiritualism.* These spirit notes to Gonzales continued for at least two more years, until he could no longer afford to attend seances. Some messages contained forgeries of the signatures of assassinated presidents Abraham Lincoln and James A. Garfield. He also received communications from the spirits of his sister Brígida, who wrote in English instead of her native Spanish, his father-in-law William Elliott, Lincoln's treasury secretary Salmon P. Chase, Dr. John Carnahan, and from coolie Wong Foo, with phony scrawl loosely resembling Chinese characters. The spirit messages to Gonzales always reflected positivity, sympathy, and encouragement.[50]

A few weeks after Gonzales published his booklet of spirit messages, he was appointed official translator of the Guatemala Legation in Washington, D.C. Minister Lainfiesta had returned to Guatemala in June due to his wife's illness, resigning his diplomatic post after she passed away. Although Lainfiesta recommended Gonzales as chargé d'affaires, Dr. Fernando Cruz, the president's cabinet chairman, was sent to fill the position. Gonzales also served as one of six translators for the Pan-American Congress held in the formal Diplomatic Room of the Department of State starting on October 2, 1889. During the next six months, he was able to revel in the diplomatic lifestyle that he had always coveted, to the chagrin of Emily Elliott. The purpose of the gathering was to promote intercommunication and commercial relations among Latin American nations and the United States. The meeting was the precursor of the Bureau of American Republics, later renamed the Pan-American Union, which in 1948 became the present Organization of American States. It was attended by twenty-seven delegates from seventeen Latin American nations, only six of whom spoke English.[51]

The Pan-American Congress, chaired by Secretary of State James Gillespie Blaine, had two secretaries, including Cuban-born New York City merchant Fidel G. Pierra, two official interpreters, one being José Ignacio Rodríguez, and six translators, including Gonzales, José R. Villalón, and J. Vicente Serrano. After the first session, the members were received in the White House by president Benjamin Harrison. Blaine tendered them a banquet that evening. Gonzalo de Quesada, a twenty-four-year-old

Cuban exile and secretary of the Argentinian delegate to the conference, was drawn to Gonzales because "his complete conversation, that night, was about his country." Quesada described him as "tall, without the years bending his rigid and elegant body nor has his beautiful head lost, crowned by gray hair, its martial posture, it was not that kind of face, and also energetic, that of a septuagenarian, the years had the clarity and the movement of youth; the nose, perfectly Roman denoted the power of command, the white mustache covered his mouth with delicate lines."[52]

Many Cubans, including Gonzales, considered Blaine "well disposed to quickly carry out any kind of adventure, no matter how risky, one of them being to seize Cuba, gradually or by force." The Cuban annexationist cause, which by 1888 had virtually disintegrated, was revived at home and abroad in early 1889. The *Philadelphia Manufacturer* favored, with reservations, Cuba joining the Union, while the *New York Evening Post* opposed it. On October 29, Cuban independence leader José Martí wrote to Quesada expressing his fears of the "high tide" of annexation enthusiasm. Martí wanted the U.S. delegation at the Pan-American Congress to issue a declaration on Cuba's right to independence, which would impede Washington from again being an ally of Spain in a future Cuban insurrection. Martí asked Quesada to influence the Cubans participating in the conference who were annexationists, but who had "the pure ideas of [José Ignacio] Rodríguez, the constant passion of the revolutionary [Ambrosio José] Gonzales and the proven and confessed interest of [Manuel R.] Moreno." Gonzales was described as "enthusiastic" by Quesada to Martí, who responded, "Of Gonzales, I know nothing, but that which is known of the López expedition, that you, remembering or asking, will know. And by a few of his lines that I read in previous days, I know that he is one of those who passionately loves this country, and would see with at least joy the annexation of ours."[53]

The Cuban independence declaration strategy failed in the Pan-American Congress. Martí believed that many of the Latin American nations were "willing to help the government of the U. States to seize Cuba." A resolution was then drafted, apparently by Rodríguez, proposing that Cuba would buy its independence from Spain with an indemnity guaranteed by the United States. Rodríguez gave the document to Gonzales, so that he could present it to his friend, Democratic Florida senator Wilkinson Call, who introduced it before Congress on December 11, 1889. Ironically, the senator's brother, George W. Call, had been the U.S. attorney for the Northern District of Florida who in 1851 prosecuted the *Pampero* steamer case for violation of the Neutrality Act. The resolution, eventually shelved by the Senate Foreign Relations Committee, was considered by Martí as "a direct mode of annexation."[54]

A few days after the Pan-American Congress ended on April 19, Gonzales received an invitation from attorney Thomas McManus to spend the summer vacation at his home in Hartford, Connecticut. McManus had been a Union major commanding a regiment at the siege of Port Hudson, Louisiana, before being honorably discharged in August 1863. After returning to Hartford, he became a judge and a state legislator. From 1887 to 1889, while he was a division chief in the U.S. Treasury Department, McManus lived in a Washington, D.C., boardinghouse, where he first met Gonzales. The former Civil War adversaries had very similar characters. McManus was devoutly religious, a Latin scholar, and "courtly and dignified, kindly and genial, he was a typical gentleman of the old school." McManus tried enticing Gonzales to be his guest in Hartford by saying: "Here all things are pleasant—the surroundings are beautiful—there are many places of historical interest, public buildings, immense libraries of reference, agreeable people, and I am sure that you will enjoy it." Gonzales apparently accepted the invitation. He departed the capital that summer and returned in the fall.[55]

In early November, Gonzales received a letter from fifty-year-old Manuel R. Moreno, a brother Freemason, Florida Republican state representative, and pediatrician in Key West. Moreno had left Havana in 1868 after studying medicine for four years. He graduated from the Medical College of South Carolina in 1878 and the following year received a medical degree from the University of New York. Moreno told Gonzales that Florida senator Wilkinson Call was returning to Washington, D.C., after recently participating in a Cuban independence rally in Key West. Call had been accompanied from his hotel to the San Carlos theater by a throng of more than two thousand Cuban exiles with torches, Bengal lights, Cuban flags, and a marching band. The reception committee included Monroe County judge Angel de Loño, a former 1851 filibuster, who was also a real estate and insurance agent. Call praised

Gonzales and Moreno, who had organized the event, and "spoke more than one hour about Cuba and its independence." Moreno asked Gonzales to go congratulate Call upon his return to the capital.[56]

During the Christmas season, Gonzales increased his visits to a spiritualist medium and was "repeatedly encouraged" by spirit messages from Hattie and her father that his condition would improve. The New Year brought glad tidings. On January 2, 1891, Gonzales received four months' salary from the Guatemalan government after a six-month delay. A week later, the State Department gave him a series of proceedings devised in Genoa, Italy, to be translated into English for the World's Fair, commemorating the four hundredth anniversary of the discovery of America. At the end of January, Gonzales received more work from the Bureau of American Republics, easing his economic anxieties. He wrote to his son Ambrose of his improvement and enclosed some spirit messages, stressing their authenticity: "It was written inside of a *sealed* envelope held in my hands *all the time,* within more blank sheets of paper and when the envelope was torn open, your mother's message was in three of them of which the enclosed in the third. I also send you one from your grandfather Elliott. Aunt Annie will recognize the signature." Gonzales never received spirit messages from Emily Elliott after she died of cancer on January 5, 1890; nor did the rest of his dead relatives refer to her joining them in heaven. Sixty-nine-year-old Annie Elliott ameliorated the family feud with Gonzales after Emily passed away. Months later, the Cuban gave his sister-in-law an affidavit needed to support her ten-thousand-dollar claim for the Elliott cotton destroyed during the Union occupation of Charleston. Annie Elliott was still fighting in federal court for the return of the confiscated family plantations Grove, Shell Point, and Ellis on Port Royal Island. Two years later, she received a final settlement of $2,623.07 from the government.[57]

Ambrose and Narciso had been "exceedingly busy" with "launching upon the troubled seas of public favor a new daily newspaper," the *State*. While Narciso acquired in Columbia equipment and staff, his brother had traveled throughout South Carolina "securing subscriptions and collecting money in advance for the unborn infant." His circular letter claimed that "it will be a Democratic newspaper, true in every line to the highest principles of the party, and its political course will be characterized by frankness and fearlessness." The publication was supported by the conservative Democratic Party faction that had opposed the recent election of populist Gov. Benjamin Ryan Tillman. Ambrose had been secretary of the Straightout Executive Committee that had endorsed the losing gubernatorial primary candidacy of conservative Alexander C. Haskell. The *State* stockholders chose nine directors who elected Haskell as their president. Ralph Elliott was working for his nephews after losing his job at the Charleston *News and Courier.* The new publication was competing against six daily newspapers circulating in South Carolina that had eight canvassing agents in the field. Gonzales, whose father had founded the first daily newspaper in Matanzas, told Ambrose that he was "glad the paper is assured and hope you will send me the first number."[58]

Gonzales received "assiduous" translation work from José Ignacio Rodríguez and William E. Curtis of the Department of State, the completion of which started affecting his eyesight in February 1891. He blamed his faulty vision on "the sedentary life and the lack of exercise and fresh air" and his "imprudence" of reading daily over half a dozen newspapers under the faint gas light of his I Street boardinghouse room. Gonzales subscribed to the *Washington Post* and the *Evening Star*, as well as three or four other American newspapers; the Havana *Diario de la Marina;* and various Cuban publications provided by expatriate José G. Delgado. The seventy-two-year-old translator was advised by a physician to have "a complete rest for a few days and complete abstinence from reading with artificial light," to avoid completely losing his sight. A few months later, Gonzales developed a progressive paralysis in his extremities, that doctors would diagnose as myelitis, a disorder of the spine marked by inflammation of the spinal cord, which made it painful to walk and difficult to write.[59]

As his health deteriorated and his finances dwindled, Gonzales sought advice from his deceased wife and relatives through spiritualist mediums, but the sessions become unaffordable. A spiritualist friend, the widower William H. Armstrong, of Milford, Pennsylvania, offered to help Gonzales communicate with Hattie through a medium named Mrs. Stratton. He proposed that the Cuban write to his "angel wife" through Mrs. Stratton since "it is evident that your dear wife *comes to her writing table* and *sits there,* and *writes,* through Mrs. Stratton's hand, *to you,* why could she not there read, or *hear Mrs. Stratton read, your letters*—and thereupon answer them." Armstrong proposed loaning Gonzales the money to resume the celestial correspondence with Hattie for the rest of the summer. He suggested paying for half of each communication if Gonzales would ask his "dear wife to tell you for me, of the conditions of progress and development in spiritual status and in happiness of my beloved wife." Armstrong was going to see Mrs. Stratton in Massachusetts and then meet mediums Pierre L. O. A. Keeler and Miss Jennie Lay in Cassadega, New York, to contact his departed wife and brother. He told Gonzales to "think, with hopefulness & cheery courage, of the many and powerful and solicitously loving friends you have in the other side—yes even around and about you wherever you are—who will do all they can to dispose the hearts and minds of friends to help you through all difficulties, to a happy and prosperous issue."[60]

The despondent Gonzales received sympathetic correspondence and frequent visits from caring friends. Cárdenas veteran Juan Manuel Macías stopped by to cheer him up and physician Manuel R. Moreno communicated that he could not forget "the old veteran, the consistent patriot, who far from his fatherland, that he never forgets, constantly yearns for her, crying like the Poles their misfortune and in a foreign land. For me, who thinks like you, who judges like you, I dream of Cuba and do not rest for one moment until I see her redeemed, you deserve my respect and consideration by building an altar in my heart to you." Moreno had just visited the island and was disappointed in what he "saw in our unfortunate fatherland. Every day I affirm more and more my belief in annexation, the only salvation that we have. How much degradation, how much hypocrisy and how much corruption. Independence can not be dreamed about there is no one who thinks of it."[61]

In July, Gonzales moved to Leonardtown, Maryland, fifty miles southeast of the national capital, to recuperate his health and escape what Delgado called "the damned Boarding Houses in Washington." The colonial hilltop village on the shore of the Potomac's Breton Bay enjoyed the cool Chesapeake breezes. Gonzales wrote his eldest son that the change made him feel better and more cheerful. Ambrose replied that Aunt Annie and the girls, Nanno, Willie, his wife Sarah and their three-year-old son Robert Elliott Gonzales, were vacationing at Glenn Springs, South Carolina. Willie, the former private secretary of Gov. John P. Richardson, had left the *News and Courier* due to cutbacks. He had gone into the hotel and real estate business in Buena Vista, North Carolina, with John R. Crapo of

Vermont, but the venture failed within a year. Thirty-one-year-old Alfonso refused to leave Oak Lawn during the fever season, made a living raising Ambrose's cattle and horses, and spent his leisure time hunting.[62]

In spite of his maladies, Gonzales tried to regain his secretariat in the Guatemala Legation by having his friends recommend him to the new minister Antonio Batres. Those who endorsed Gonzales included Cuban independence leader José Martí, his personal secretary Gonzalo de Quesada, and Aniceto García-Menocal Martín. Quesada was planing to write a biography of Narciso López after Cirilo Villaverde had recently abandoned the project. In late July, Batres informed Gonzales that a Guatemalan financial crisis prevented his immediate rehiring, but that he would try again in November "to utilize your very commendable aptitudes." In mid-October, Gonzales was in a health resort at Fortress Monroe, Virginia, when Batres notified him that "the bad financial circumstances of Guatemala will not permit that your wishes and mine be realized."[63]

In response to this latest setback, Ambrose, after consulting approval with Narciso and Willie, invited his father in late October to move to Columbia "to be in a milder climate and nearer to your sons." Ambrose offered to "arrange for you to help me with my business correspondence which you could do in your room; then too, I know that you could get some classes in languages or music, and, as you can live comfortably here more cheaply than at the North, the move will be a good one." Two weeks later, Ambrose checked his father into the luxurious Grand Central Hotel in Columbia, where he registered as "General A. J. Gonzales, of South Carolina." The Gonzales sons had been living in the same hotel, and on November 8, they were joined there by their sisters Gertrude and Hattie and their aunt Annie for the local Fair Week. Ambrose and his brothers were apparently successful in reconciling the ladies with the patriarch. The next day, Gonzales visited the offices of the *State* and greeted fifty-seven-year-old Ralph Elliott for the first time in twenty-one years. His still-resentful brother-in-law, who had been "steadily improving" from alcoholism to the point where "he hardly drinks at all," responded, "Excuse me, Sir," and refused to speak to him. Gonzales left, saying, "I did not think it," and wept at the door. Ralph later wrote to his niece Harriett that her father was "an unkempt, unshaved, very much maltreated looking, aged tramp" and that "as long as I have a memory of the sacrifices endured, and the deaths caused, by his selfish neglect of his offspring, I shall never treat him as an equal or a gentleman. I am no blood relative of his."[64]

While in Columbia, Gonzales was encouraged by Gonzalo de Quesada to write his memoirs, and provide him an account of the 1848 López meeting that designed the Cuban flag. He wanted "the narrative with all of its details, of where, who was there, why the colors were chosen, and the form, of the *Cuban flag*. Anything picturesque you remember of that incident." Gonzales was "feeling worse" by February 1892 and never started his autobiography. Two months later, Ambrose wrote to Gertrude at Oak Lawn: "I wish you girls would send some message of sympathy or something to the old gentleman. It looks cruel not to do so. He is failing rapidly and suffers intensely." Gonzales had become confined to an invalid's chair and Ambrose hired a black servant named Heyward to assist him. At his hotel residence, Gonzales felt "isolated and lacked sympathetic society," which, along with his immobility, turned his frustration into cantankerousness.[65]

The septuagenarian then decided to spend the summer at the Hot Springs resort in Virginia's Shenandoah Valley. The enterprise advertised their "naturally heated water (104°)" as "the cure" for "Rheumatism, Gout and Nervous Diseases." Ambrose accompanied him there by train. Gonzales checked into the new Virginia Hotel, which unfortunately had been built at the lowest point in the valley. It had no view from its windows and locomotive smoke from the adjacent terminal irritated guests. After ten days, Gonzales asked to return to Columbia, considering the place "a cheap country hotel." He objected to the weekly $17.50 charge, "exclusive of $1 daily for medical treatment and extra charge for meals served in my room and I became worse instead of better owing to my stubborn constipation." Ambrose decided to send his father to the mountains, but Gonzales received an invitation from physician Moreno to visit him, "because Key West is Cuba, not only its climate that you know but also more than five thousand exiled Cubans here who make up a small Republic where we can see float the glorious ensign of our fatherland. You would also have the daily visits of [Juan] Arnao, myself and innumerable Cubans who would rejuvenate you talking about the fatherland. I think that this alone would make you well." Moreno offered Gonzales beach house accommodations from where he could contemplate the sea.[66]

In July, Gonzales, accompanied by Ambrose and Heyward, departed for Key West. They rode the rails all day from Columbia to Charleston, Savannah, Jacksonville, and Tampa, where they boarded the night steamer that docked in Key West the next evening. Moreno, after receiving their telegram from Tampa, awaited at the wharf. Ambrose returned to South Carolina after helping his father settle in comfortable quarters. Gonzales was in the midst of Cuban exiles who admired and venerated him. In early September, José Martí held a meeting in Key West with the leaders of the 1868 revolution. According to Martí's newspaper, *Patria,* the patriots homaged Gonzales: "The revolutionaries of yesterday, of today, and of tomorrow, went to greet the invalid; he sat up straight in his wheelchair; his whole figure glowed, he raised his numb limbs and said with solemnity: 'I salute the redeemers of the fatherland!'"[67]

Gonzales was being treated by Dr. Manuel González Echeverría, the nephew of Havana Club secretary José Antonio Echeverría, who recommended that he move to New York, since health care facilities were nonexistent in Key West. There, he would have all the requisites of a first-class hospital and would be able to receive the hot Turkish baths recommended by the physician. Gonzales wrote to Mrs. Annie C. Bettner, an 1870s New York friend, asking for assistance in obtaining admission in "one of the great sanitary institutions . . . for chronic diseases." He then informed Ambrose on September 17 that the Confederates of Key West would pay his passage and that of his servant to New York and that "it is likely that the Masons will assist me." Gonzales closed the letter saying, "Today is the 24th Anniversary of the death of my wife." The fear of a cholera or yellow fever epidemic in Key West, which would leave him blockaded, prompted Gonzales to sail for New York City six days later on the steamer *San Marcos* of the Mallory line. He was accompanied to the dock by former filibuster Angel de Loño, and a crowd of Cubans and Americans.[68]

Gonzales lodged in the New York Hotel on Broadway and immediately notified Mrs. Bettner, Gonzalo de Quesada, and other local friends. On October 2, he was visited by Quesada and his new bride Angelina Miranda, and by Hugh Smith Thompson, comptroller of the New York Life Insurance Company, a former South Carolina governor (1882–86) and Confederate veteran. Thompson expressed sympathy and promised assistance, while Quesada said that the local Cubans would stand by him. The invalid wrote Ambrose the next day that in spite of the "anxiety to be relieved of my crippled and helpless condition and of my bodily pains and infirmities," he was not among strangers, counting on "fifty times more sympathy here that I did in Columbia." He closed on a despondent note: "I have taken great pains to set you right in regard to me not only that you may know the truth but to counteract the tendency my children have always had since the death of their mother to see only my wrong side, to which they have been trained by their aunts in order to carry out this spite of almost a lifetime. I will not live for ever and when I am gone I may hope that tardy justice will be done to my memory." Five days later, Martí's *Patria* newspaper announced Gonzales's arrival, exhorting the Cuban émigrés in New York to take care of him: "There will not be anyone who will miss his duty of accompanying the noble aged man in his solitude and in mitigating his sorrows. Gen. Ambrosio Jose Gonzalez was the first Cuban wounded in combat for the liberty of Cuba."[69]

Two weeks after arriving in New York, Gonzales was facing delays in getting into a hospital. The "Confederate Camp" was "too poor to contribute money" to place him in a paying hospital, and Mrs. Bettner, a founder of the Home for Incurables in the Bronx, proposed that he apply there. Gonzales was down to his last $50, with a pending hotel bill for $42.60, when he wrote to Ambrose, "The surgeons say they will put me on my feet and *that* is what I am enduring all this for." He received $50 from his son and asked him for another equal amount "to pay the balance of this week and what days thereafter I may be in the house, pay Heyward's return to Columbia and perhaps have a few dollars left for my contingent expenses." Narciso joined Ambrose in covering the expenses for their father. On October 17, Gonzales checked into the Roosevelt Hospital on Fifty-ninth Street and Ninth Avenue as a charity patient, "after a world of trouble, mishaps and anxiety and very nervous and feeble." Permanently bedridden by then, Gonzales sent Heyward back to South Carolina with a letter for Ambrose and the correspondence from his friends detailing his hospital arrangements. Gonzales hoped to stay at the Roosevelt temporarily before being transferred to Montefiore Hospital, which he described as "one of the foremost in the world." He expected to go to the Home for Incurables in the spring, and sent Ambrose an *Annual Report* pamphlet of the institution provided by Mrs. Bettner.[70]

His plans were once again disrupted, and by December Gonzales was interned in the Home for Incurables, today's St. Barnabas Hospital, on a twelve-acre hilltop at Third Avenue and 183rd Street in the Fordham section of the Bronx. It was America's first chronic-disease hospital, founded in 1866 by an Episcopalian minister. Admission required that the patients, "though in many instances quite poor, are, in refinement, education and social position, and, above all, religious character, so far removed from the pauper class that no Christian mind can for a moment bear the thought of dismissing them to the public charities." The home had 175 beds, usually filled with charity patients and others who could pay the minimum quota of seven dollars weekly, two months in advance. There were 84 male patients and 91 females, living in segregated wards, with an average stay of 255 days, with 43 deaths the previous year. Hemiplegia was the most common illness, affecting one-fifth of the patients. The three-story main building was surrounded by plush lawns, shady trees, and outer buildings. These included a chapel, laundry, and engine room. A volunteer ladies' board supplied the hospital linen room, "the working of the kitchens, the care of the clothing, [and] the general tidiness and cleanliness of the rooms and wards." The law firm of Stearns and Curtis, employers of Gonzalo de Quesada, legally represented the institution.[71]

Quesada went to visit Gonzales in the hospital during the Christmas holidays. As he crossed the ward, he slowly looked at the faces in the beds aligned on opposite sides of the room. Gonzales, with a gray untrimmed beard, lay sleeping in the last bed in the corner, and "the impressions of age had marked his pale face." Over the headboard hung a small religious portrait of Guido Reni's *Mater Dolorosa*. Quesada leaned over and whispered in his ear: "My General." Awakening, Gonzales responded, "I was dreaming of Cuba." A few days later, Quesada published in *Patria* a two-thousand-word biographical article on Gonzales, which began: "The man who has suffered, who has shed blood for the cause of liberty, has a holy place in our heart. The first Cuban wounded in combat by Spanish lead was Ambrosio Jose Gonzalez; that is why, even though he might have not rendered other services, his name will be remembered with love, by the grateful fatherland." Quesada reminded his readers how Gonzales had been instrumental in designing the Cuban flag, how "where ever he has lived he has aided with his word and pen the cause of Cuba," and that the noble septuagenarian invalid "does not want to die before seeing the fatherland free!"⁷²

*Patria* published on January 7, 1893, another note on "the old gentleman of liberty," along with a recent photograph. His progressive paralysis worsened the following month, his eyesight faltered, and a nurse was penning the letters to Ambrose. In mid-February, Gonzales received from his eldest son fifty-six dollars to cover his board for two months and ten dollars for personal expenses. Ambrose wrote him, "I know your existence must be very dreary and monotonous, but I know too that there must be occasional bright spots, and of these I would like to hear." In late April, Ambrose traveled from Columbia, South Carolina, to visit his father for a few days at the Home for Incurables and paid the twenty-eight dollars monthly board. He stayed at the Broadway Central Hotel and went with a Mrs. Ball to Montefiore Hospital to arrange his father's transfer there. The superintendent, a Jewish physician from Darlington, South Carolina, pledged to send his surgeon to examine Gonzales, after realizing that he and Ambrose had mutual friends back home. The doctor indicated that "they already had 150 more applications than they could take, as they did not have a single vacant bed." Ambrose then traveled to Syracuse on a business matter. The seventy-four-year-old Gonzales had less than three months to live, his health was rapidly failing, and he was lonely and despondent.⁷³

As soon as Ambrose left, his father was sending him telegrams requesting a quilt. Ambrose returned to New York on May 1, on his way back to South Carolina. He told his father that he understood that "it is hard for an invalid in your condition to exercise patience" but that he was doing everything in his power for him. Ambrose expressed that he could not neglect the responsibilities that brought him to New York, because that money helped pay the hospital expenses. Gonzales asked his son for a thirty-dollar air mattress, but Ambrose did not buy it after a nurse said that it was unnecessary. He did ask Mrs. Ball to get the mattress for his father if it was needed. When Ambrose departed, Gonzales borrowed the money for the mattress and asked his son to reimburse a Mr. Armstrong. After returning to Columbia, Ambrose wrote on May 10 to his sister Gertrude about their father: "Poor old fellow, his condition is very pitiable. The Doctor's say there's no hope of his improvement and only by the use of opiates can he be relieved of the terrible pains he suffers. He has not been able to leave his bed even to sit up for the past seven months and he cannot last much longer. I did all I could to cheer him up and arranged to give him all the attention possible." Nine days later, Ambrose reassured his father, "I am deeply interested in your welfare and would to God that I could do something to alleviate the mental and physical pains you have to endure."[74]

Gonzales continued to receive visits from Cuban exiles in New York who venerated their old patriot. Gonzalo de Quesada and his wife Angelina were frequent visitors. She informed him on May 13 of the recent failed uprising in Cuba headed by the brothers Ricardo and Félix Sartorio. Ambrose settled his father's bills, and on July 6 replied to a complaint about paying late for the air mattress: "I have sent you every dollar I could earn or borrow." He enclosed a sixty-dollar check and promised to visit soon, departing four days later and staying a week in New York. It would be the last time that Ambrose saw his father. The Cuban patriot passed away around 10:00 A.M. on July 31, 1893. His death was attributed to myelitis; asthma was cited as a contributing factor. The body was sent to local mortician J. P. Garniss, a sponsor of the Home for Incurables. Two days later, Gonzales was interred in a single hillside plot in Woodlawn Cemetery, the Bronx. The burial service was performed by an Episcopalian chaplain from the hospital. Martí and the staff of *Patria* placed a wreath at the grave, "the flowers of gratitude, the best, the most beautiful, the imperishable ones."[75]

Telegrams from New York notified his sons in Columbia, South Carolina, who published a front-page obituary on August 2 headlined "Ambrosio Jose Gonzales: Death of a Cuban and Confederate Patriot." It has the written style of Ambrose, who obtained the biographical material mostly from the recent *Patria* article. The final paragraph indicated:

> Judged by the standard of material success the life of General Gonzales does not tempt to imitation. But by other standards it is not to be held a failure. To live beyond the allotted time of man and leave behind no shame; to have striven hard and roughly with the world and gone from it with open brow and unsoiled hands; to have given some good blows for liberty's great cause; and to be conscious at the end of duty performed as seen—these are earnings greater than gold. And for an epitaph this adopted citizen would have chosen the words of Jefferson Davis applied

to him ten years ago: "A soldier under two flags but one cause; that of community independence."

Martí also penned an article praising "the old man who never gave up his first ideal: his last thoughts must have been for the [home]land. . . . Gentleman without flaw, sound Cuban, and virtuous man, in the history of his people Gonzales has written his name with characters that will never be erased."[76]

After reaching mid-life, Gonzales spent more than twenty years attempting to recuperate his financial stability and his children's affection. The sanitized family correspondence will forever hide the reason for Emily's rancor and impedes a further analysis of our subject's character flaws. As his offspring matured, they rejected their aunt's manipulation and grew closer to their father. It was not until after Emily's death that the two Gonzales daughters, who had been raised by her, made peace with their father when his life was almost over. The elderly expatriate made efforts to provide for his children, although his reduced circumstances initially made the amounts infrequent and small. After receiving his inheritance, Gonzales generously spent twenty percent on his children and loaned Brosio another twenty percent. Christian forgiveness was one of his character traits, exemplified by his buying a new home for his spiteful in-laws and his postwar defense of Jefferson Davis. Gonzales, in contrast to most Southerners, harbored no resentments about the "lost cause," befriending former Union officers Henry C. Hall, Charles Dodge, and Thomas McManus, among others.

Two passionate themes dominated Gonzales throughout his life: his love for Hattie and for his homeland. The eternal grieving for his departed wife frustrated Emily and prompted exploitation by phony spiritualists. Gonzales had pursued a diplomatic post for forty years with both Democratic and Republican administrations in the hope of furthering Cuban annexation. Eleven months before his death, his health failing rapidly, while surrounded by three generations of Cuban exiles in Key West, he conceded that independence was the only recourse available to Cuba, by embracing the future redeemers of the fatherland.

# Epilogue

During his lifetime, Ambrosio Gonzales repeatedly insisted that Cuba could achieve independence only through armed revolution because Spanish intransigence stiffened with each passing year. In January 1895, Spain imposed stricter laws on its Cuban colony. It served as a catalyst for the Cuban Revolutionary Party in exile to initiate a new independence uprising throughout the island on February 24, 1895. José Martí returned to Cuba in April and died on the battlefield on the forty-fifth anniversary of the Cárdenas disembarkment.

The independence struggle was kept alive with the same method Gonzales used nearly half a century earlier: lobbying sympathetic Americans and Cuban expatriates to raise funds, weapons, and volunteers for filibuster expeditions from the United States. In January 1896, the Spanish government appointed the ruthless Valeriano Weyler as captain general of Cuba to crush the insurrection. His *reconcentrado* policy removed entire countryside populations into concentrated military areas and declared the evacuated land a free-fire zone. Weyler's atrocities prompted the American yellow press to dub him "The Butcher" and elicited further support for the rebels with American public opinion.

After another independence insurrection exploded in the Philippines colony in 1897 and the assassination of recalcitrant Spanish conservative prime minister Antonio Cánovas del Castillo, a new Liberal government decreed autonomy for Cuba and Puerto Rico on January 1, 1898. Weyler resigned in disagreement and Spain began withdrawing its troops from the island, but some Weyler supporters refused to surrender. They rioted in Havana before departing for Spain and gathered before the American consulate to shout death threats. Consul Fitzhugh Lee, a former Confederate cavalry brigadier general and nephew of Robert E. Lee, requested that his government send a warship to protect American lives and property in the city. In response, the USS *Maine* arrived in Havana harbor on January 25. Three weeks later, a mysterious explosion destroyed the vessel, killing 264 of its 328 crew members. A U.S. Court of Inquiry reported on March 21 that an electric submarine mine had produced an external explosion; the finding was controversial, and the origins of the disaster are still being debated today. As a result of the report, on April 24, 1898, the United States Congress declared war on Spain.

South Carolina governor William H. Ellerbe quickly mustered two volunteer infantry regiments totaling some 1,840 officers and soldiers, including two independent colored companies, a volunteer battery of heavy artillery of 144 men, and a naval militia contingent of 425 enlistees. Forty-seven years earlier, Gonzales had tried raising a skeleton regiment in the state to assist the last filibuster invasion of Cuba. The naval militia did local coastal duty and provided 80 petty officers and men for the crew of the USS *Celtic*, attached to the North Atlantic squadron, which destroyed the Spanish fleet at Santiago de Cuba. The war ended on August 12, 1898, while the First Regiment was encamped near Jacksonville, Florida, and the Second Regiment was about to leave Columbia, South Carolina. The latter, as part of the 7th Army Corps commanded by Gen. Fitzhugh Lee, disembarked in Cuba on January 7, 1899. They carried the Mexican War flag of the Palmetto Regiment, which had been entrusted to them by veteran Col. James Douglass Blanding. The Second Regiment encamped at Camp Columbia, near Havana, for two months, with the army of occupation. Three of its members who succumbed to illness were interred there. The heavy artillery battalion did coastal defense duty on Sullivan's Island and Port Royal. Carolinians fighting against Spain in 1898 included Capt. Micah Jenkins of the Regiment of Rough Riders, who distinguished himself at the battle of San Juan Hill. In the Regular Army were Maj. E. A. Garlington, Captains M. C. Butler and Richard McMaster, and Lieutenants John Jenkins and George H. McMaster. In the Regular Navy were Lieutenants Bryson Patton, J. Miller Moore, and Victor Blue, who located the Spanish fleet in Santiago de Cuba harbor. Former Confederate Maj. Gen. Matthew Calbraith Butler was appointed major general of United States volunteers and served on the commission for the Spanish evacuation of Cuba. His aide was Capt. Alfred Hampton, son of Wade Hampton. Former governor John G. Evans afterward temporarily served as mayor of Havana.[1]

Gonzales would have been proud to see three of his sons, Ambrose, Narciso, and William, interrupt their work with the *State* to identify with his cause for Cuban liberty. Narciso enlisted under Cuban general Máximo Gómez and later published an account of his combat experiences in the book *In Darkest Cuba*. Ambrose was a major in the United States army of occupation, stationed at Santiago de Cuba. William was captain of Company K, Second South Carolina Volunteer Regiment, and his wife joined him at Camp Columbia. In 1913, William was appointed minister to Cuba by the Woodrow Wilson administration.

After the Gonzales brothers returned home from the war, they continued editorializing in the *State* against the Tillmanite political faction. When Lt. Gov. James H. Tillman, nephew of Benjamin Tillman, aspired to the governorship in the 1902 elections, Narciso denounced him as a "proven liar, defaulter, gambler and drunkard." A revengeful Tillman shot Gonzales at noon on January 15, 1903, in front of the state capitol, on the corner of Gervais and Main streets. A change of venue moved the trial to the Tillmanite Lexington county, where a populist jury declared the lieutenant governor not guilty of murder on the grounds of "self defense." The grateful citizens of Columbia erected an obelisk to the memory of Narciso that still stands today beside the capitol, at the corner of Senate and Sumter streets.

The youngest daughter, Ana Rosa Gonzales, would later regret that she had been denied her Cuban heritage by the Elliotts. Born in Cuba in 1869, Ana Rosa was taught to despise her father by the Elliotts, who renamed her Harriett Rutledge Elliott Gonzales. She still retained her Cuban citizenship in 1940 when she traveled to Havana with a group of journalists from South Carolina. During a luncheon given by the Havana Chamber of Commerce, one of the Carolinians told the audience that she was the daughter of Ambrosio José Gonzales, and "immediately the entire assemblage arose and cheered, showing that though nearly a century had passed, her father's patriotic services were still remembered, and still much appreciated. Miss Gonzales was naturally very touched by the demonstration; in fact so much that, retiring as was her nature, she agreed to go on the radio that night and say a few words to the Cuban people."[2] The Cuban citizens also gave Miss Gonzales a large Cuban flag, which she proudly displayed on the wall of the rear sun room at Oak Lawn. Her nephew, Dr. Ambrose Gonzales Hampton, recalled that this spirited lady rode on horseback around the plantation until the age of seventy. In 1949, Doctor Hampton took his fiancée, Ann Fripp Jones, to Oak Lawn. When Ann asked Miss Gonzales how she should address her, the octogenarian pondered and replied, "Aunt Gonzie," even though the rest of the family called her Aunt Hattie. She told the couple that she favored the name Anita Rosita and deplored that it had been changed.

The Gonzales legacy is firmly rooted in South Carolina and Cuba even though the four Gonzales sons left no male heirs. The *State* has the largest circulation of any South Carolina newspaper, and family descendants have continued the tradition of passing the patriotic name to the latest heir, the youthful Ambrose Gonzales Morris. José Martí praised Gonzales in *Patria* as a brave example for future generations of Cubans to emulate. After the island gained independence in 1902, new waves of political exiles arrived on American shores during the rest of the century. They copied the Gonzales model of fighting repressive regimes by organizing filibuster expeditions and attempting to influence American public opinion to their cause. These new endeavors from the United States, like those of 1849–51, violated the Neutrality Act and resulted in death and failure for its participants. On August 17, 1932, the Cuban Revolutionary Association in New York sent an expedition of twenty men, led by Emilio Laurent, to overthrow the Gerardo Machado government. They seized the town of Gibara, Oriente, but were defeated by overwhelming forces a day later. The Corinthia expedition, launched from Florida with twenty-seven men led by Calixto Sánchez White, landed in northern Oriente province on May 24, 1957. They were emulating Fidel Castro's eighty-two-man Granma expedition, which six months earlier had come from Mexico to depose the Fulgencio Batista regime. Sánchez White and most of his followers were murdered after surrendering to the army. Thirteen years later, former Castro guerrilla Vicente Méndez landed at Baracoa, Oriente, on April 17, 1970, with twelve exiles from Florida, vowing to restore democracy in Cuba. Méndez and three of his men were killed by Castro's troops and the rest captured. Five were later executed by firing squad and four sentenced to thirty years in prison. Indeed, the causes that had motivated Gonzales to fight a century earlier have endured in his homeland.

# Appendix A

## 167 Participants of the 1850 López Expedition

| Name | Residence | Age | Rank | Co. | Reg. | Mex. War | Result |
|---|---|---|---|---|---|---|---|
| Allen, John+ | Shelbyville | 40 | Capt. | | Ky. | 2nd Lt. | |
| Andrews, James | Cincinnati | | | | Ky. | | |
| Balser, James Henry | Cincinnati | | | H | Ky. | | |
| Balser, William | Cincinnati | | | H | Ky. | | |
| Barton, William H. | | | Lt. | | Ky. | | |
| Bates, John C. | Cincinnati | | Dr. | | Ky. | | |
| Bayne | Cincinnati | | Lt. | H | Ky. | | |
| Bell, William H. | Miss. | | Lt. Col. | | La. | Lost arm | |
| Bogges, Frances C. M. | Mobile, Ala. | | Sgt. | | La. | Pvt. | |
| Bradford | | | Lt. | E | La. | | |
| Breckenridge, Newton C. | Kentucky | | Capt. | G | La. | | |
| Bunch, M. J. | Memphis | | Lt.Col. | | Miss. | Yes | |
| Burke, V. J.* | | | Lt. | G | La. | | |
| Byrd, Theodore P. | | | Adjt. | | La. | | |
| Capers, W. C.+ | | | Lt. | A | Miss. | | |
| Colin | | | Capt. | | La. | | |
| Collins, J. | Adams Co. | | Adjt. | | Miss. | | |
| Crisler, L. | | | | | Ky. | | |
| Cruse, Henry | | | Sgt. | D | Ky. | | Killed |
| Davis, J. C. | | | Capt. | B | La. | | |

| Name | Residence | Age | Rank | Co. | Reg. | Mex. War | Result |
|---|---|---|---|---|---|---|---|
| Dear, Joseph C. | Frankfort | | Lt. | | Ky. | Lt. | |
| De la Cruz Rivero, Francisco J. | Matanzas | 46 | | | | | |
| Dennett, J. H. | | | Lt. | C | La. | | |
| Dixon | | | Maj. | | Miss. | | |
| Dumm | | | | | Ky. | | |
| Dupeau | | | Capt. | | | Yes | Deserter |
| Elliott | | | Capt. | | Staff | | |
| Ellis, Robert H.** | Washington | 21 | Lt. | | | | |
| Fayssoux, Callender Irvine++ | New Orleans | 29 | Mate | | *Creole* | | |
| Ferguson, George | | | Sgt. | | La. | | |
| Fisher, Thomas L. | | | Capt. | C | La. | | |
| Foley, James | | | Lt. | D | La. | | |
| Foster, Henry C. | | | Capt. | I | La. | | |
| Friedling, Bernard | New Orleans | | | C | Miss. | | |
| Gallup, David G. | | | | | | | |
| Garnett, James J. | Bowlg. Green | | Lt. | | Ky. | Yes | Killed |
| Gonzales, Ambrosio José | Havana | 31 | Adjt. | | Staff | | Wounded |
| Graham, Burton | Cincinnati | | | | Ky. | | |
| Greenlee | | | Lt. | | Ky. | | |
| Guy, John R. | Cincinnati | | | | Ky. | | |
| Hale | | | Capt. | D | Miss. | | |
| Hall | | | Capt. | | Staff | | |
| Hardy, Richardson | Covington | | Lt. | D | Ky. | | Wounded |
| Hardy, William | Covington | | Maj. | | Ky. | | |
| Harkins, J. | | | Lt. | | Ky. | | |
| Harnley, William | Louisville | 26 | Lt. | | Ky. | | |
| Harris | | | | | Ky. | | |
| Harris, R. A. | | | Lt. | K | Ky. | | |
| Hawkins | | | Capt. | C | Miss. | | |

Participants of the 1850 López Expedition / 371

| Name | Residence | Age | Rank | Co. | Reg. | Mex. War | Result |
|---|---|---|---|---|---|---|---|
| Hawkins, Thomas Theodore | Lexington | 30 | Maj. | | Ky. | 2nd Lt. | Wounded |
| Hayden, George B. | Arkansas | | Maj. | | La. | | |
| Henning, H. E. | Iberville | | Lt. | B | La. | | |
| Herbert, John | | | 2 Lt. | Ind. | Miss. | | |
| Hernández, José Manuel+ | Matanzas | | | | Staff | | |
| Higgins, John | Cincinnati | | | | Ky. | | |
| Hinsey, Nathaniel | Cincinnati | | | | Ky. | Yes | |
| Horton, C. O. | | | Lt. | | Ky. | | |
| Howard, J. C. | | | Capt. | | | | |
| Hoy, Thomas P. | Galveston | | Qtmstr. | | Ky. | | |
| Hunton, T. G. | | | Capt. | D | La. | | |
| Hurd | | | Lt. | I | La. | | |
| Huston | | | | | Ky. | | |
| Irish, J. | Jackson | | | | Miss. | | |
| Jackson, John | | | | Ind. | Miss. | | |
| James | | | Lt. | | Ky. | | |
| Johnson, Albert W. | Kentucky | | Capt. | | Ky. | | |
| Johnston, John Carl | Louisville | | Lt. | | Ky. | | |
| Jones, E. L. | | | Lt. | K | La. | | Wounded |
| Josephs, A. A. | | | Asst. Surg. | | La. | | |
| Keating | | | Capt. | B | Miss. | | |
| Kelly, William | Cincinnati | 19 | | | Ky. | | Deserter |
| Kennedy, Thomas J. | Kentucky | | Dr. | | Ky. | | |
| Kewen, Achilles L.+ | | | Capt. | | Miss. | | |
| Kewen, Thomas | | | Capt. | F | La. | | |
| King | | | Col. | | | | |
| Knight | Shelbyville | | Capt. | | Ky. | | |
| Knott, Clark | Frankfort | | Lt. | | Ky. | Corp. | |
| Lane, E. D. | | | Lt. | F | La. | | |
| Langan, Nicholas | Cincinnati | 21 | | | Ky. | | |

| Name | Residence | Age | Rank | Co. | Reg. | Mex. War | Result |
|---|---|---|---|---|---|---|---|
| Lawton, Thomas | | | Lt. | H | La. | | |
| Leathers, James | | 20 | Lt. | A | Miss. | | |
| Lewis | | | Capt. | | Ky. | | |
| Lewis, Armstrong Irvine | | 28 | Capt. | | *Creole* | | |
| Logan, John A. | Shelbyville | | Capt. | | Ky. | Lt. | Killed |
| López Oriola, Narciso | Venezuela | | Gen. | | | | |
| Macías, Juan Manuel | Matanzas | 23 | Aide | | Staff | | |
| Mann, Robert | | | | | Ky. | | |
| Marble, Daniel | Cincinnati | 19 | | | Ky. | | |
| March, Thomas | | | Capt. | H | La. | | |
| Marriott, James C. | | | | | Ky. | | |
| Marsh, A. W. | | | | | | | |
| Martínez, Agustín | Shelbyville | | | | Ky. | | |
| Mathews, Beverly | | | Capt. | | Staff | | |
| Mayfield | | | Pvt. | | Miss. | | |
| McCann, John McFarland | Paris, Ky. | | Chap. | | Ky. | | Killed |
| McCormeck, J. J. | | | Capt. | E | La. | | |
| McDerman, Dr. John | Boone Co. | 28 | Lt. | F | Ky. | | |
| McDonald | | | Sgt. | | Ky. | | |
| McFarland, W. A.* | | | Ord. | A | Miss. | | |
| McGreggor | Scotland | | Pvt. | | | | Deserter |
| McGuffin | | | Lt. | | Ky. | | |
| McHenry, Jonathan D. Rush** | | | Com. | | La. | | |
| McLawrin, L. | Jackson | | | | Miss. | | |
| McOunegie | | | | | La. | | |
| Mitchell | | | Lt. | E | La. | | |
| Mizell, A. | | | Capt. | Ind. | Miss. | | |
| Moore | | | Capt. | | Staff | | |
| Moore, John | | | Pvt. | H | La. | | Accidental death |
| Morgan, M. J. | | | Capt. | K | La. | | |

Participants of the 1850 López Expedition / 373

| Name | Residence | Age | Rank | Co. | Reg. | Mex. War | Result |
|---|---|---|---|---|---|---|---|
| Morris | | | Lt. | C | La. | | |
| Murry | | | Capt. | | Staff | | |
| O'Hara, Theodore | Frankfort | | Col. | | Ky. | | Wounded |
| Parish | | | Lt. | H | La. | | |
| Parker | | | Sgt. | | La. | | |
| Peabody, H. | | | Lt. | A | La. | | |
| Perkins, J. C. | | | Lt. | G | La. | | |
| Pickett, John T. | Kentucky | 28 | Lt.Col. | | Ky. | | |
| Quin, John | Columbus, Ga. | | | | | | Wounded |
| Rawlings, C. H. | | | Lt. | | Ky. | | |
| Reading, John B. | Shelbyville | | Lt. | | Ky. | Pvt. | |
| Redding, William | | | Sgt. | | Ky. | | |
| Reed, James | | | | | | | Wounded |
| Reed, John | | | | | | | |
| Robinson, Henry H.* | Cincinnati | 27 | Capt. | D | Ky. | Lt. | |
| Sánchez Iznaga, José | Trinidad | 39 | Aide | | Staff | | |
| Sartain, George F. | | | Lt. | I | La. | | |
| Saunders, Walter | Shelbyville | | | | Ky. | | |
| Sayre, Burwell B. | Frankfort | | Lt. | | Ky. | | Wounded |
| Scott, R. | | | Lt. | | | | |
| Scott, Dr. Samuel S. | Florence, Ky. | | Surg. | | Ky. | | |
| Sexias | | | Lt. | | Miss. | | Killed |
| Sherman, Hugh M. | | | Eng. | | *Creole* | | |
| Smith, Charles | Cincinnati | | | | Ky. | | |
| Smith, Edward | | | | | | | Wounded |
| Smith, G. N. L. | | | Col. | | Staff | | |
| Smith, Joseph | Louisville | | Lt. | | Ky. | | |
| Smith, Peter | Woodville | | Maj. | | Miss. | | |
| Stanley, James M. | | | Sgt. | | La. | | |
| Steede, Abner C. | | | Capt. | A | La. | | |
| Stevenson, Thomas | Scotland | | | | | | Deserter |

| Name | Residence | Age | Rank | Co. | Reg. | Mex. War | Result |
|---|---|---|---|---|---|---|---|
| Stoval | | | Sgt. | | La. | | |
| Stull | | | Surg. | | La. | | |
| Taylor, George T. | Cincinnati | | | | Ky. | | |
| Taylor, Marion C. | Shelbyville | | Pvt. | | Ky. | | |
| Thixton | | | Lt. | B | La. | | |
| Thomas, L. C. | Iberville | | Qmstr. | | La. | | |
| Thompson, James R. | Trenton, N.J. | | | | | | |
| Titus, Henry Theodore+ | Philadelphia | 27 | Adjt. | | Ky. | | |
| Triplett, R. S. | Kentucky | | Lt. | | Ky. | | |
| Velazco, Pedro | | | Servant | | | | |
| Vernon, E. | | | Lt. | A | La. | | |
| Vesey, J. | | | Capt. | | Staff | | |
| Warner, George | Evansville, Ind. | | | | | | Deserter |
| Wheat, Chatham Roberdeau+ | New Orleans | | Col. | | La. | | Wounded |
| Wheeler, Joseph | | | | | Ky. | | |
| Wheeling, Robert | | | Sgt. | D | Ky. | | Wounded |
| Williams, W. | | | | | | | Wounded |
| Williamson | | | Lt. | A | Miss. | | |
| Wilson, A. H. | Cincinnati | | | | Ky. | | |
| Wilson, Fielding C. | Cincinnati | 28 | Capt. | H | Ky. | | |
| Wilson, John | | | | | | | Wounded |
| Winston, Thomas M. | Kentucky | | Capt. | | Ky. | 2nd Lt. | |
| Woodruff, William E. | Kentucky | | Lt. | F | La. | 1st Lt. | |
| Woolfolk | | | Lt. | | Ky. | | |
| Work, Thomas | | | Cpl. | | Ky. | | |
| Wragg, Thomas | | | Sgt. | | La. | | |
| Zornes, A. J. | | | | | | | Wounded |

\* Participated in the May 1851 López expedition
\*\* Participated in the August 1851 López expedition
\+ Later joined William Walker in Nicaragua
\+\+ Callender Fayssoux was the first mate on the expedition steamer during the Round Island, Cárdenas, and last López expeditions. He later joined William Walker in Nicaragua.

# Appendix B

## 338 Participants of the 1851 López Expedition

Adeno, Bernardo
Aguedo Valdés, José
Aidelar,
Albing, James
Alfonso, Antonio Luciano
Allen, Bernard
Allen, John
Aragón, Manuel
Aragón, Ramón
Arnao Alfonso, Ramón Ignacio
Arnold, T. W.

Bachilder, John
Badneih, Emereich
Baker, J. D.
Ball, William H.
Baller, James D.
Bawder, Louis
Beach, Ransom
Bechtold, Conrad
Bell, D.
Bell, Edwin Q.
Bent, M. P.
Berry, George S.
Biro, Michael
Blumenthal,

Bontila, George
Boswell, John
Bournazal, Pierre Charles
Boyd, Franklin P.
Brady, James
Brandt, James
Breckenridge, Robert H.
Breedlove, J.
Brigham, James C.
Brown, Thomas D.
Browse, David
Bryan, Thomas
Bryce, J. O.
Bulet, James
Bulman, J.
Bush, John G.

Cabrera, Joaquin
Cajerman, James
Caldwell, Robert
Cameron, William
Canky, Robert
Carter, John
Casanovas, J.
Cay, Eugene
Chaferman, James

Charmes, William
Chassagne, Julio
Childes, John
Christi, Thomas
Ciceri, Joseph
Cichler, Conrad
Cline, John
Colchett, Alexander M.
Coleman, Patrick
Collins, E. F.
Collins, Napoleon
Collyer, B. H.
Conolly, Edward
Conrad, Peter
Consans, William
Constantine, W. S.
Cook, Cornelius
Cook, Gilman A.
Cooper, John
Costera, Andreas
Cousans, William
Craft, William H.
Cressey, Edgard
Crittenden, William Logan
Cully, Supe L.

Daily, Charles J.
Daily, Thomas
Davidson, R. C.
Davis, John N.
Denton, John
Denton, Thomas
Deville, Pedro
De Wolf, Daniel E.
Díaz, Manuel
Dillon, Patrick
Donnelly, James D.
Dorent, Pedro
Dorr, John M.

Douvren, José
Downer, Charles Augustus
Downman, Robert L.
Doyle, John
Driller, Henry
Duffy, Cornelius J.
Dunn
Dupart, Victor
Dwin, James G.

Egerton, George
Ellis, James
Ellis, Robert H.
Essex, Preston

Fagan, James Burton
Falcón, Antonio
Faust, Jacob
Fayssoux, Callender Irvine
Fiddes, James
Fischer, Thomas
Fisher, Newton. H.
Fleen, A. M.
Fleury, Manuel
Flot, Nicolas
Fonts, Jacob
Foster, George W.
Fourniquet, Henry A.
Freeborn, Isaac

Garth, Patrick Abec
Geblin, Charles
Geiger, Michael
Gibbon, Charles
Gilmore, Benjamin
Gino, David
Gonzales, Esteban
González, Andreas
González Govantes, Pedro

Gotay, Felipe N.
Graham, George A.
Grant, Reuben H.
Gray
Green, Moses
Grider, Robert M.
Guerra, Miguel
Gunst, Joseph B.

Haekel, Louis
Hagan, Lewis
Hagar, Frederick
Halpin, James
Hannd, Benjamin F.
Harbele, Jacob
Harrison, Charles
Harrison, George
Hart, Henry B.
Hartnett, Thomas
Haynes, William Scott
Hearsey, James H.
Hearsey, Thomas H.
Hefrou, Michael L.
Henry, Timothy K.
Herb, William K.
Hernández, Antonio
Hernández, Diego
Herren, Julio
Hilton, Thomas
Hodge, Charles J.
Hogan, William J.
Holdship, George
Horner, William H.
Horwell, Charles H.
Hough, Fenton B.
Howain
Howder
Hughes, Joel D.
Hundnall, Thomas

Hunter, Beverly E.
Hurd, William K.
Huston Jr., Felix

Iglesias, Francisco
Iglesias, José
Inslee, William W.
Izbert, Ibrahim

James, Thomas B.
Jasper, Henry
Jessert, Jacob
Johnson, John
Jones, L. C.
Jorge, Eleuterio
Joster, George W.

Keenan, M. J.
Kelly, James A.
Kerekes, Bela
Kerr, Victor
Koss, Alfred

Labuzan
Lacoste, Peter
La Croi, José
Lainé, Francisco Alejandro
Landa, Manuel
Landinghan,
Lee, Thomas H.
Lefrow, Michael L.
Levy, J. St.
Lewis, Armstrong Irvine
Little, Thomas
Little, William B.
Livry, James H.
López, Miguel
López, Pedro Manuel
López, Sotero

López Moreno, Víctor
Losner, William
Lowers, John A.
Ludwing, Ansell R.
Lyons, Michael

MacGath, Patrick
Mahan, Francis C.
Manville, James L.
Martínez, Manuel
McAlly, Alexander
McClelland, Thomas
McDonald, Edmund H.
McHenry, Jonathan D. Rush
McHenry, William Henry
McKensey, William H.
McKinniop, John
McLeabe, Bernard
McMullin, Peter D.
McMurray, C. A.
McNeill, Thomas L.
Melesion, Martin
Menllen, Martin
Metcalfe, George E.
Metcalfe, Henry B.
Miller, William
Mills, Samuel
Monroe, Thomas R.
Montoro, Agustín
Murphy, John
Murtigh, John
Myers, James
Myers, Joseph

Nagle, Louis
Nelson, Richard
Niesman, William
Niskos, Janos
Nolasco de Zayas, Pedro

Norriss, John
Null, Charles

Oberto Urdaneta, Ildefonso
O'Brien, John Thomas
Oglevie, James
Oleaga, Juan
Ollis, Elira J.
O'Reilly
O'Rourke, Patrick
Otis, Elias J.
Owen, James G.
Owens

Paicuriche, Antonio
Paratolt, Conrad
Parr, George
Pedro
Peteri, John
Petipiers,
Phillips, Asher J.
Phillips, Michael
Planos, Angel
Pomeroy, Augustus A.
Port, N.
Porter, J. G.
Pruitt, John T.
Purnell, Stephen Howard

Quick, George

Radnitz
Reed, Samuel
Reeves, Wilson L.
Richardson, George W.
Rives, Ramón
Robinson, Charles. A.
Robinson, John
Romero, Antonio

## Participants of the 1851 López Expedition / 379

Rosales, Juan Antonio
Ross, N.
Rousseau, David Q.
Rubio, José
Ruvira, J. B.

Salmerón, Eduardo
Salmon, James
Sanders, James
Sanford, J.
Sank, John G.
Santa Cruz, Agustín
Sayle, Henry
Scheiprt, Zyrtack
Schlesinger, Louis
Schluht, Harbo
Schmidt, George
Schmidt, Henry
Schuets, Robert
Sckneck, C.
Scott, Malbon K.
Scully, Luke
Seay, Dandridge
Sebring, Cornelius
Seilert, John
Sestar, Andrés
Sewer, Frederick S.
Shilling, William
Simpson, J. P.
Smith, Charles N.
Smith, James
Smith, John T.
Somers, Henry
Sowers, John A.
Stanmire, Henry
Stanniford, Clement
Staunton, James
Stevans, Joseph

Stevens, Theodore A.
Stewart, William H.
Stezinger
Stubbs, John

Talbot, John
Taylor, Conrad
Thomason, H. J.
Torres Hernández, Anselmo

Van Vechten, Philip S.
Vaughan, William H.
Veazy, T. B.
Velazco, Pedro
Vienne, H. T.
Villarino, Luis
Virag, Janos
Von Schlicht, H.

Weiss, Edward
West, Henry
Weymouth, J. B.
Whitcomb, T.
Wilkinson, William L.
Williams, Harney
Wilson, George
Wilson, James M.
Wilson, W.
Wilson, William
Winburn, David
Wise, Edward
Woer, Amand R.
Wragg, B. W.

Young, William

Zayas, José Antonio

# Notes

**Introduction**

1. C. Stanley Urban, "New Orleans and the Cuban Question during the Lopez Expeditions of 1849–1851: A Local Study in 'Manifest Destiny,'" *Louisiana Historical Quarterly* (Oct. 1939), 1095–167.

2. Ella Lonn, *Foreigners in the Confederacy* (Chapel Hill: University of North Carolina Press, 1940), 144.

3. Gonzalo de Quesada, "Ambrosio José Gonzalez," *Patria*, 31 Dec. 1892, 2–3; "Ambrosio Jose Gonzales," *The (Columbia, S.C.) State*. 2 Aug. 1893, 1; and U. R. Brooks, *Stories of the Confederacy* (1912; reprint, Camden, S.C.: J. J. Fox, 1991), 284–99.

4. Emeterio S. Santovenia, *Huellas de Gloria* (Havana: Editorial Trópico, 1944), 159–62; and Herminio Portell Vilá, *Vidas de la Unidad Americana: Veinte y Cinco Biografías de Americanos Ilustres* (Havana: Editorial Minerva, 1944), 369–82.

5. Lewis Pinckney Jones, "Ambrosio José Gonzales: A Cuban Patriot in Carolina," *South Carolina Historical Magazine* (Apr. 1955): 67–76; idem, "Carolinians and Cubans: The Elliotts and Gonzales, Their Work and Their Writings" (Ph.D. diss., University of North Carolina, 1952); idem, *Stormy Petrel: N. G. Gonzales and His State* (Columbia: University of South Carolina Press, 1973); "A Fighter for Freedom," *The State*, 13 Feb. 1966, 6; Fermín Peraza Sarausa, *Diccionario Biográfico Cubano* (Havana: Ediciones Anuario Bibliográfico Cubano, 1951); S. L. Latimer Jr., *The Story of The State, 1891–1969, and the Gonzales Brothers* (Columbia, S.C.: State Printing, 1970), 367–72; and Bruce S. Allardice, *More Generals in Gray* (Baton Rouge: Louisiana State University Press, 1995), 247.

6. Ambrosio José Gonzales, *Manifesto on Cuban Affairs Addressed to the People of the United States* (New Orleans: Daily Delta, 1853); idem, "On to Cuba," *New Orleans Times Democrat*, 30 Mar. 1884, 9; idem, "The Cuban Crusade: A Full History of the Georgian and Lopez Expeditions," *Times Democrat*, 6 Apr. 1884, 9; and idem, *Heaven*

*Revealed: A Series of Authentic Spirit-Messages, from a Wife to her Husband, Proving the Sublime Nature of True Spiritualism* (Washington, D.C.: McQueen and Wallace, 1889).

**Cuban Conspirator**

1. Gonzales changed the last letter of his surname from *z* to *s* after he became an American citizen. Copy of birth certificate no. 644, Ambrosio José Cándido González, in Gonzales Family Papers, on deposit at the South Caroliniana Library, University of South Carolina, hereinafter cited as GFP; and Dolores María de Ximeno y Cruz, *Aquellos Tiempos . . . Memorias de Lola María* (Havana: El Universo, 1928), 1:24.

2. "Documentos referentes a D. Ambrosio José Gonzáles," GFP; Ann Fripp Hampton, *A Family Sketchbook: Containing Genealogical Charts and Biographical Sketches* (Columbia, S.C.: n.p., 1979), 11; Dr. J. A. Treserra to Victor Barringer (husband of Gertrude Hampton), 9 Apr. and 12 Aug. 1952 and 7 Dec. 1953, in possession of Ann Fripp Hampton; and AJG to AEG, 1 Feb. 1891, GFP.

3. Ximeno, *Aquellos Tiempos*, 1:24–25, 274; Adolfo Dollero, *Cultura Cubana (La Provincia de Matanzas y su evolución)* (Havana: Imp. Seoane y Fernández, 1919), 93; and "Documentos referentes a D. Ambrosio José Gonzáles," GFP.

4. Pedro José Guiteras, *Historia de la Isla de Cuba*, 2d ed. (Havana: Cultural, S.A., 1927), ix; and Pedro Antonio Alfonso, *Memorias de un Matancero* (Matanzas: Marzal y Cía., 1854), 205–6, 230.

5. Alfonso, *Memorias*, 207–16; Francisco Calcagno, *Diccionario Biográfico Cubano* (New York: Imprenta y Librería de N. Ponce de León, 1878), 300; *Cuba en la Mano: Enciclopedia Popular Ilustrada* (Havana: Ucar, García y Cía., 1940), 893; and Dollero, *Cultura Cubana*, 69.

6. Alfonso, *Memorias*, 217–18, 228; Calcagno, *Diccionario*, 420; *Cuba en la Mano*, 958; Passenger Lists of Vessels Arriving at New York, 1820–97, list no. 166, 20 Apr. 1828, microcopy no. 237, roll 11, U.S. National Archives, Washington, D.C., hereinafter cited as NA; Richardson Hardy, *The History and Adventures of the Cuban Expedition* (Cincinnati: Lorenzo Stratton, 1850), 80; and James Grant Wilson, *The Memorial History of the City of New-York: From Its First Settlement to the Year 1892* (New York: New-York History Company, 1893), 3:361.

7. Brooks, *Stories of the Confederacy*, 284; Beauregard to Thomas Jordan, 24 Jan. 1869, Papers of P. G. T. Beauregard, reel 1, LC, Washington, D.C., hereinafter cited as PBP; and Hamilton Basso, *Beauregard: The Great Creole* (New York: Charles Scribner's Sons, 1933), 17.

8. Alfonso, *Memorias*, 218; *Cuba en la Mano*, 662; Dollero, *Cultura Cubana*, 33; Ultramar, Cuba, Gracia y Justicia, 1824–55, libro 16, 1829, no. 67, and 1830, no. 50, Archivo Histórico Nacional, Madrid, Spain, hereinafter cited as AHN; "Actuaciones del juzgado de primera instancia del norte en la ciudad de La Habana," 8 Feb. 1909, GFP; Jones, "Carolinians and Cubans: The Elliotts and Gonzales, Their Work and

Their Writings"; and Dr. J. A. Treserra to Victor Barringer, 7 Dec. 1953, letter in possession of Ann Fripp Hampton.

9. The bachelor of civil law diploma is in possession of Elliott G. McMaster, great-grandson of Ambrosio José Gonzales. Hardy, *History and Adventures*, 80; O.D.D.O. (J. C. Davis), *The History of the Late Expedition to Cuba* (New Orleans: Daily Delta, 1850), 59; *Historia Gráfica de Cuba, 1492–1925* (Miami: Trade Litho, 1976), 72; *New Orleans Picayune*, 13 May 1851, 2; Ximeno, *Aquellos Tiempos*, 1:175–76, 358; Antonio Alvarez Pedroso, *Miguel de Aldama* (Havana: Imprenta "El Siglo XX," 1948), 29; Calcagno, *Diccionario*, 386, 391; Nicholas P. Trist to Isaac E. Holmes, 6 Dec. 1848, James K. Polk Papers, LC, Washington, D.C., hereinafter cited as JPP; Nicholas Philip Trist Papers, ibid.; and Dumas Malone, ed., *Dictionary of American Biography* (New York: Charles Scribner's Sons, 1964), 9:645.

10. Alfonso, *Memorias*, 231; Hardy, *History and Adventures*, 80; O.D.D.O., *History of the Late Expedition*, 59; "Actuaciones del juzgado de primera instancia del norte en la ciudad de La Habana," 8 Feb. 1909, GFP; *Historia Gráfica de Cuba*, 72; Emilio Roig de Leuchsenring, *Los Primeros Movimientos Revolucionarios del General Narciso López, 1848–1849* (Havana: Municipio de La Habana, 1950), 71; and Levi Marrero, *Cuba: Economía y Sociedad* (Madrid: Editorial Playor, 1988), 14:109.

11. Basil Rauch, *American Interest in Cuba, 1848–1855* (New York: Columbia University Press, 1948), 45; ACB to Capitán General, Ministerio de Asuntos Exteriores, Archivo General, Madrid, Antillas Españolas, Cuba, Política Ultramar, 6 May 1847, Legajo 4628, hereinafter cited as AEA; and "The Cuban Expedition," *Brownson's Quarterly Review*, Oct. 1850, 498.

12. Biographical data for Havana Club members appears in volume 2 of Juan J. Remos Rubio, *Historia de la Literatura Cubana* (Havana: Cárdenas y Cía., 1945); Francisco Xavier de Santa Cruz y Mallen, *Historia de Familias Cubanas*, 5 vols. (Havana: Editorial Hercules, 1940–44); Peraza Sarausa, *Diccionario Biográfico*; Jorge Marbán, *José Agustín Quintero: Un enigma histórico en el exilio cubano del ochocientos* (Miami: Ediciones Universal, 2001). The economic interests of the Havana Club members appear in Hugh Thomas, *Cuba: The Pursuit of Freedom* (New York: Harper and Row, 1971), 122, 207–8. Ramiro Guerra y Sánchez et al., *Historia de la Nación Cubana* (Havana: Editorial de la Nación Cubana, S.A., 1952), 4:75; Roig de Leuchsenring, *Primeros Movimientos*, 61–62; *Cuba en la Mano*, 797; Calcagno, *Diccionario*, 413–15, 482–84; Alvarez Pedroso, *Miguel de Aldama*, 109; and Herminio Portell Vilá, *Narciso López y su época* (Havana: Cultural, S.A., 1930), 1:243–44.

13. The role of Freemasons in the Cuban independence movement is detailed in Antonio Rafael de la Cova, "Filibusters and Freemasons: The Sworn Obligation," *Journal of the Early Republic* (spring 1997): 95–120; Ambrosio José Gonzales, *Manifesto on Cuban Affairs Addressed to the People of the United States* (New Orleans: Daily Delta, 1853), 5–6; Vidal Morales y Morales *Iniciadores y primeros mártires de la revolución*

*cubana* (Havana: Cultural, S.A., 1931), 2:10, 154; "Affairs in Cuba," *New Orleans Commercial Bulletin*, 24 Aug. 1849, 2; *La Verdad*, 16 and 30 July 1848, 2, and 6 Sept. 1849, 3; Santovenia, *Huellas de Gloria*, 29; and Gonzales, *Manifesto on Cuban Affairs*.

14. Gaspar Betancourt Cisneros (28 Apr. 1803–3 Dec. 1866), a wealthy Camagüey landowner and railroad builder, educated in Philadelphia, had been exiled twice from Cuba because of his separatist ideals, which prompted the Spanish government to confiscate his property. He returned to Cuba under a general amnesty on July 7, 1861. Remos, *Historia de la Literatura Cubana*, 2:216–18, 254–55; Peraza Sarausa, *Diccionario Biográfico*, 7:13–14; and Federico Córdova, *Gaspar Betancourt Cisneros: El Lugareño* (Havana: Editorial Trópico, 1938), 118.

15. Cristóbal F. Mádan Mádan (7 July 1807–89), a Matanzas merchant, shipowner, and planter, had property assets of three hundred thousand dollars, including the La Rosa sugar mill in Cimarrones. His father, Joaquín Mádan Gutiérrez, was one of Cuba's biggest slave owners. Cristóbal Mádan had been educated in New York City and Norwich, Connecticut. He settled in Matanzas, where in November 1828 he wed María Gómez Pastrana, with whom he had four children. After Mádan became a widower, in November 1845 he married the widow Mary O'Sullivan and had two more children. She was the younger sister of John Louis O'Sullivan, editor of the *United States Magazine and Democratic Review*, who, that year, coined the phrase "Manifest Destiny." O'Sullivan's publication espoused American expansionism and Cuban annexation. Sheldon Howard Harris, "The Public Career of John Louis O'Sullivan" (Ph.D. diss., Columbia University, 1958), 275; Calcagno, *Diccionario*, 109–12, 399, 616–18; sworn statement by Cristóbal Mádan, 19 Feb. 1851, Miscellaneous Letters of the Department of State, NA, hereinafter cited as MLS; Cristóbal Mádan to Allen Owen, 6 June 1851, Dispatches from U.S. Consuls in Havana, NA, hereinafter cited as DCH; Santa Cruz y Mallen, *Historia de Familias Cubanas*, 5:166; and Manuel Moreno Fraginals, *El Ingenio* (Havana: Editorial de Ciencias Sociales, 1978), 3:234.

16. Miguel Teurbe Tolón (29 Sept. 1820–16 Oct. 1857) was a Matanzas poet, novelist, newspaper editor, and philosophy professor at the University of Havana, whose ideology forced him into exile in 1848. He left behind his wife, Emilia Teurbe Tolón, a cousin he had married four years earlier. Teurbe taught Spanish and wrote for *La Verdad* in New York City from 1848 to 1853. In 1853, he founded the newspaper *El Cubano*. A general amnesty allowed Teurbe to return to Matanzas on August 30, 1857, where he died of tuberculosis six weeks later. Dollero, *Cultura Cubana*, 167–68; Peraza Sarausa, *Diccionario Biográfico*, 7:13–14; and Calgagno, *Diccionario*, 616–18.

17. Jane McManus Storm Cazneau (6 Apr. 1807–10 Dec. 1878) was the daughter of former New York representative William Telemachus McManus. She was five feet three inches tall, with dark hair and dark complexion. Twice married, she was at the age of twenty-six rumored to be the mistress of seventy-six-year-old Aaron Burr. She wrote for the *New York Sun*, the *New York Herald*, the *New York Tribune*, and the *United*

*States Magazine and Democratic Review* and made references in her articles to her Catholic faith. Jane was a Union Democrat who advocated gradual emancipation. When the Civil War began, she returned to New York and supported the Lincoln administration. She drowned at sea when the ship she was a passenger on sank during a storm. Linda S. Hudson, *Mistress of Manifest Destiny: A Biography of Jane McManus Storm Cazneau, 1807–1878* (Austin, Tex.: Texas State Historical Association, 2001); Robert E. May, "Lobbyists for Commercial Empire: Jane Cazneau, William Cazneau, and U.S. Caribbean Policy, 1846–1878," *Pacific Historical Review* (Aug. 1979): 383–412; idem, "'Plenipotentiary in Petticoats': Jane M. Cazneau and American Foreign Policy in the Mid-Nineteenth Century," in *Women and American Foreign Policy: Lobbyists, Critics, and Insiders*, ed. Edward P. Crapol (New York: Greenwood Press, 1987), 19–44.

18. Morales, *Iniciadores*, 2:12; Roig de Leuchsenring, *Primeros Movimientos*, 67–68; Portell Vilá, *Narciso López*, 2:40; *Washington Republic*, 17 Aug. 1849, 2; and *La Verdad*, 16 July 1848, 2.

19. William Worth's biographer relies only on the Gonzales manifesto for the account of meeting with the Cubans, because the general's personal letters and papers were destroyed at the beginning of the twentieth century. Portell Vilá, *Narciso López*, 1:235, 243; Calcagno, *Diccionario*, 184; Francisco J. Ponte Domínguez, *La Masonería en la Independencia de Cuba* (Havana: Editorial "Modas Magazine," 1954), 45; Ramiro Guerra, *Manual de historia de Cuba: Desde su descubrimiento hasta 1868* (Madrid: Ediciones R, 1975), 472; Morales, *Iniciadores*, 2:13; Guerra et al., *Historia de la Nación Cubana*, 4:81; James Grant Wilson and John Fiske, eds. *Appleton's Cyclopedia of American Biography* (New York: D. Appleton, 1888), 6:615–16; Edward S. Wallace, *General William Jenkins Worth: Monterrey's Forgotten Hero* (Dallas: Southern Methodist University Press, 1953), 166, 169; *Picayune*, 27 Jan. and 2 July 1848, 2; K. Jack Bauer, *The Mexican War, 1846–1848* (New York: Macmillan, 1974), 388; *Picayune*, 22 Oct. 1851, 1; Roig de Leuchsenring, *Primeros Movimientos*, 69; Gonzales, *Manifesto on Cuban Affairs*, 6; and Wallace, *Worth*, vi, 185.

20. Portell Vilá, *Narciso López*, 1:14–16, 25; idem, *Vidas*, 295; "General Lopez: The Cuban Patriot," *United States Magazine and Democratic Review* (Feb. 1850): 97; Portell Vilá, *Narciso López*, 1:59; and Thomas W. Wilson, *An Authentic Narrative of the Piratical Descents upon Cuba Made by Hordes from the United States, Headed by Narciso Lopez, a Native of South America* (Havana: n.p., 1851), 4.

21. Roig de Leuchsenring, *Primeros Movimientos*, 15, 22; Calcagno, *Diccionario*, 642–43; and Portell Vilá, *Narciso López*, 1:298.

22. Portell Vilá, *Narciso López*, 1:110; Calcagno, *Diccionario*, 457–58; Roig de Leuchsenring, *Primeros Movimientos*, 22, 122; Portell Vilá, *Vidas*, 298–99; *La Verdad*, 7 July 1850, 1; *New Orleans Crescent*, 13 Jan. 1851, 2; *Cuba en la Mano*, 808; and Peraza Sarausa, *Diccionario Biográfico*, 1:26.

23. José María Sánchez Yznaga (24 Feb. 1811–17 Dec. 1887) was educated in Philadelphia. In 1851, he founded the Order of the Lone Star in the United States. He returned to Cuba under a general amnesty in 1856 but returned to exile in the United States in 1868. Peraza Sarausa, *Diccionario Biográfico*, 6:18–20; Roig de Leuchsenring, *Primeros Movimientos*, 37–39, 145; Ximeno, *Aquellos Tiempos*, 1:33; and Guerra et al., *Historia de la Nación Cubana*, 4:77.

24. The Cirilo Villaverde Diary of 1850–51, written in various notebooks, was sold by his son to Dr. Antonio M. Eligio de la Puente. Herminio Portell Vilá made a copy of it. I believe that Villaverde rewrote his diary after 1868 in a revisionist effort to vindicate López as an independence leader and not as an annexationist. Another famous contemporary diary, written by Mary Boykin Chesnut during the U.S. Civil War, later proved to have been rewritten. C. Vann Woodward, ed., *Mary Chesnut's Civil War* (New Haven: Yale University Press, 1981); Juan Bellido de Luna and Enrique Trujillo, *La Anexión de Cuba a los Estados Unidos* (New York: El Porvenir, 1892), 84; C. V., *General Lopez: The Cuban Patriot* (n.p., 1849), 10; Roig de Leuchsenring, *Primeros Movimientos*, 143, 147; and Morales, *Iniciadores*, 2:56.

25. U.S. Congress, Joint Committee on Printing, *Biographical Directory of the United States Congress, 1774–1989* (Washington, D.C.: Government Printing Office, 1989), 736–37. Morales, *Iniciadores*, 2:17–18; Robert Campbell to John Clayton, 7 Dec. 1849, DCH; Roig de Leuchsenring, *Primeros Movimientos*, 46, 145; and José Ahumada y Centurión, *Memoria Histórico Política de la Isla de Cuba* (Havana: A. Pego, 1874), 259.

26. Fidel Castro used a similar plot on July 26, 1953, during carnival feast days in Santiago de Cuba, to attack the Moncada garrison with 135 insurgents dressed in army uniforms. Roig de Leuchsenring, *Primeros Movimientos*, 37–38, 45, 146; Morales, *Iniciadores*, 2:19; *La Verdad*, 30 July 1848, 1; *Herald*, 29 July 1848, 1; ibid., 2 Aug. 1848, 2; Presidencia de la Comisión Militar Ejecutiva Permanente, 28 Oct. 1849, Gobierno Militar de Cuba, Expediente 1748, Archivo General Militar, Segovia, Spain; and Portell Vilá, *Narciso López*, 2:20–21. Portell Vilá claimed that López was accompanied only by "his faithful assistant" when he arrived in the U.S. Ibid., 33–35.

27. Roig de Leuchsenring, *Primeros Movimientos*, 184.

28. Ambrosio José Gonzales, "On to Cuba," *New Orleans Times Democrat*, 30 Mar. 1884, 9; Hardy, *History and Adventures*, 80; and *Picayune*, 20, 21, and 22 July and 11 Aug. 1848, 2.

29. Gonzales, "On to Cuba"; *Picayune*, 11 Aug. 1848, 2; *New Orleans Delta*, 11 Aug. 1848, 2; and William Edwards Clement, *Plantation Life on the Mississippi*, 2d ed. (New Orleans: Pelican, 1961), 129.

30. *Picayune*, 8 and 10 Aug. 1848, 2; Gonzales, "On to Cuba"; *Picayune*, 12 Aug. 1848, 4; *Charleston Courier*, 21 Aug. 1848, 2; and *Picayune*, 23 Aug. and 3 Sept. 1848, 2.

31. *Herald*, 28 Aug. 1848, 3; Gonzales, "On to Cuba"; and Wallace, *Worth*, 185. General Worth is identified as a Freemason in William R. Denslow, *10,000 Famous Freemasons* (Independence: Missouri Lodge of Research, 1957), 4:351.

32. When Gonzales wrote about these events thirty-six years later, his account did not match the sequence of events in contemporary newspaper stories. Gonzales, "On to Cuba"; Denslow, *10,000 Famous Freemasons*, 4:152; *Herald*, 3 Sept. 1848, 2; and *Tribune*, 5 Sept. 1848, 1.

33. During the Civil War, Henry Bohlen helped raise the 75th Pennsylvania Volunteer Infantry regiment in Philadelphia. He was promoted to brigadier general in April 1862 and was killed in battle four months later. Worth's biographer wrongly assumed that Lt. Col. James J. Duncan had been the general's envoy to Havana. Wallace, *Worth*, 185, 223 n. 2; Gonzales, "On to Cuba"; Ezra J. Warner, *Generals in Blue: Lives of the Union Commanders* (Baton Rouge: Louisiana State University Press, 1991), 38–39; Portell Vilá, *Narciso López*, 2:36–37; *Herald*, 10 Sept. 1848, 4; Roig de Leuchsenring, *Primeros Movimientos*, 72; and Gonzales, *Manifesto on Cuban Affairs*, 7.

34. When writing about these events thirty-six years later, Gonzales claimed that Worth also introduced him to Pres. James K. Polk. Yet, in a letter to Polk on March 4, 1849, Gonzales reminded the president they had been introduced in December by Gen. Caleb Cushing. See AJG to James K. Polk, 4 Mar. 1849, JPP. Gonzales, "On to Cuba"; Denslow, *10,000 Famous Freemasons*, 3:149; *Tribune*, 1, 7, and 13 Dec. 1848, 2; *Pittsburgh Post*, 9 Dec. 1848, 2; and *Pittsburgh Gazette*, 12 Dec. 1848, 3.

35. Nicholas P. Trist to Isaac E. Holmes, 6 Dec. 1848, JPP; U.S. Congress, *Biographical Directory*, 1205; and *Proceedings of the Most Worshipful Grand Lodge of Ancient Freemasons of South Carolina, at Their Annual Communication, November 5859* (Charleston: A. J. Burke, 1859), 58.

36. *Herald*, 19 Aug. 1849, 3; Rauch, *American Interest in Cuba*, 101; Denslow, *10,000 Famous Freemasons*, 2:47 and 4:123, 190, 208, 223, and 306; and de la Cova, "Filibusters and Freemasons."

37. Cushing did not see any action during the Mexican War and entered Mexico City with a reserve battalion months after the capital had been pacified. Cushing had on his staff Cuban-born Lt. Julius Peter Garesché. AJG to James K. Polk, 4 Mar. 1849, JPP; Denslow, *10,000 Famous Freemasons*, 1:275 and 3:353–54; Claude M. Fuess, *The Life of Caleb Cushing* (New York: Harcourt Brace, 1923), 2:48–49, 60, 79, 96, 99; and Portell Vilá, *Narciso López*, 2:36.

38. J. M. Storm to Daniel Dickinson, New York, 4 Jan. 1849, JPP; John R. Dickinson, *Speeches, Correspondence, etc. of the Late Daniel S. Dickinson, of New York* (New York: G. P. Putnam and Son, 1867), 2:437; May, "Lobbyists for Commercial Empire," idem, "'Plenipotentiary in Petticoats,'" 25; Wallace, *Destiny and Glory*, 257–58; Denslow, *10,000 Famous Freemasons*, 1:315; *Alexandria Gazette and Virginia Advertiser*, 25 Aug. 1849, 2; *Herald*, 19 Aug. 1849, 3; and *Washington Intelligencer*, 22 Aug. 1849, 3.

39. John C. Gardiner is identified as "Carlos Gardiner" in *Picayune*, 24 July 1851, 1, and as "Charles Gardiner" in *New York Times*, 6 Mar. 1854, 5. *Baltimore Argus*, 3 Feb. 1849, 2; *New Orleans Commercial Bulletin*, 18 Sept. 1849, 2; Spain, Ministry of Foreign Affairs, Consulate, South Carolina, Charleston, *Copiador General, 1848–1860*, 27 Feb. 1849, DUL, hereinafter cited as Charleston Spanish Consulate Papers; and Rafael Nieto y Cortadella, *Dignidades Nobiliarias en Cuba* (Madrid: Ediciones Cultura Hispánica, 1954), 96.

40. *Washington Daily National Whig*, 5, 7, and 26 Feb. 1849, 2; Roig de Leuchsenring, *Primeros Movimientos*, 57–58; Portell Vilá, *Narciso López*, 2:57, 171; Santa Cruz y Mallen, *Historia de Familias Cubanas*, 8:137; Dollero, *Cultura Cubana*, 55, 243, 250; Calcagno, *Diccionario*, 344–55, 398; and O.D.D.O., *History of the Late Expedition*, 61.

41. *Charleston Mercury*, 25 Aug. 1851, 2; John F. H. Claiborne, *Life and Correspondence of John A. Quitman* (New York: Harper and Brothers, 1860), 2:53; J. Franklin Jameson, ed., "Correspondence of John C. Calhoun," *Annual Report of the American Historical Association for the Year 1899* (Washington, D.C.: Government Printing Office, 1900), 2:1202–3; and Rauch, *American Interest in Cuba*, 112–13.

42. Thomas Jefferson Rusk, a South Carolinian who developed a law practice in Georgia, moved to Texas in 1835 and was a delegate to the independence convention. As secretary of war of the Texas Republic, he commanded the forces at the Battle of San Jacinto. He was subsequently appointed chief justice of the Supreme Court of Texas and major general of the Texas Militia. In 1845, Rusk presided the convention that achieved Texas annexation and thereafter was elected to the U.S. Senate. John C. Calhoun, *The Works of John C. Calhoun* (New York: D. Appleton, 1858), 4:466–67; and Gonzales, "The Cuban Crusade."

43. The meeting date is not precise. When Mrs. Davis wrote her memoirs forty years later, fraught with many time errors, she mistakenly claimed that the encounter occurred in the summer of 1848, and that within two months López was garroted, whereas this happened in 1851. Tom Chaffin, *Fatal Glory: Narciso López and the First Clandestine U.S. War against Cuba.* (Charlottesville: University Press of Virginia, 1996), 47–48, relied on this mistake to date the encounter in July 1848, when López was actually traveling to the U.S. William J. Cooper Jr., *Jefferson Davis, American* (New York: Alfred A. Knopf, 2000), 161, 179, indicated that Varina was not at the capital in 1848. She arrived with her husband on November 30, 1849, and left in late September 1850. *Republic*, 1 Dec. 1849, 2. The interview probably took place before late March 1849, when Davis left the capital, prior to the Round Island expedition. Lee had also returned to Baltimore that month. It is unlikely that Davis or Lee would have given their consideration to a project publicly denounced by a presidential proclamation in August 1849. Emory M. Thomas, *Robert E. Lee: A Biography* (New York: Norton, 1995), 148, estimated the meeting between Lee and the Cubans "in late 1849." Portell Vilá, *Narciso López*, 2:189, and William C. Davis, *Jefferson Davis: The Man and*

*His Hour* (New York: Harper Collins, 1991), 197, date the encounter in February 1850, during the winter, contradicting Mrs. Davis's recalling the summer weather. Lee's biographers and filibuster historians do not precisely track his movements in Washington, D.C., or New York City during this period. Anderson C. Quisenberry erroneously claimed that this offer was made to Davis and Lee after April 1851, for command of the second expedition. Anderson C. Quisenberry, *Lopez's Expeditions to Cuba 1850 and 1851* (Louisville, Ky.: John P. Morton, 1906), 71. Varina Davis, *Jefferson Davis: A Memoir* (New York: Belford, 1890), 1:411; Hudson Strode, *Jefferson Davis: American Patriot, 1808–1861* (New York: Hartcourt Brace, 1955), 211; "Remarks of President Davis," *Organization of the Lee Monument Association, and the Association of the Army of Northern Virginia, Richmond, Va, Nov. 3d and 4th, 1870* (Richmond: J. W. Randolph and English, 1871), 13–14; Douglas Southhall Freeman, *R. E. Lee: A Biography* (New York: Charles Scribner's Sons, 1934), 1:307; Margaret Sanborn, *Robert E. Lee: A Portrait* (Philadelphia: J. B. Lippincott, 1966), 205; U.S. Congress, *Biographical Dictionary*, 1752; and John L. O'Sullivan to Thomas J. Rusk, 13 Sept. 1849, Thomas J. Rusk Papers, Center for American History, University of Texas at Austin.

44. AJG to James K. Polk, 4 Mar. 1849, JPP; George S. Ulibarri, *Records of United States and Mexican Claims Commissions* (Washington, D.C.: National Archives, 1962), 7; *Tribune*, 4 Nov. 1848, 4; and RG 21, U.S. District Courts, naturalization file for Ambrosio José Gonzales (naturalized 26 Mar. 1849), NA.

45. *Picayune*, 4 July 1848, 3; *Intelligencer*, 1 Dec. 1849, 3; *Evening Picayune*, 24 July 1851, 1; and *Herald*, 23 Sept. 1852, 2.

46. AJG to Caleb Cushing, 29 Mar. 1849, Caleb Cushing Papers, LC, hereinafter cited as CCP; and Ulibarri, *Records of United States and Mexican Claims Commissions*, 7.

47. Narciso López to Carlos Gardiner, 10 Apr. 1849, Charles Gardiner Miscellaneous Manuscripts, Sterling Library, Yale University; *La Verdad*, May 18, 1849, 2; and Sección Ultramar, Cuba, Gobierno, Legajo 4629, no. 23, AHN.

48. *Daily National Whig*, 20 and 27 Apr. 1849, 3; O.D.D.O., *History of the Late Expedition*, 12, 46, 51; *Vicksburg Whig*, 15 Sept. 1849, 2; and U.S. House of Representatives, *Barque Georgiana*, Exec. Doc. 83, 32d Congress, 1st sess., 23 Mar. 1852 (Washington, D.C.: 1852), 144.

49. *Baltimore Argus*, 24 May 1849, 2; *Herald*, 3 July 1849, 2; *Evening Picayune*, 5 July 1849, 1; Gonzales, "On to Cuba"; Morales, *Iniciadores*, 2:193; Wallace, *Worth*, 185; *Daily National Whig*, 18 May 1849, 3; Roig de Leuchsenring, *Primeros Movimientos*, 121; Charles A. Brockaway, "Masonic Symbolism in Cuba's Flag," *Masonic Outlook*, Mar. 1931, 216; Edwin T. Tolón, "Como fue creada la bandera cubana," *Carteles*, 3 Apr. 1949; Eugene A. Atkinson, "History of the Cuban Flag," American Lodge of Research *Transactions*, 29 Jan. 1962, 449; "Homenaje a Nuestra Bandera," Tercera Conferencia Inter-Americana de la Masonería Simbólica, *Boletín Informativo* (Havana),

27 Feb. 1955, 6; and Enrique Gay-Calbo, *El Centenario de la Bandera Cubana, 1849–1949* (Havana: 1949), 8–9.

50. For a detailed account of the first López expedition, see Antonio Rafael de la Cova, "The Taylor Administration versus Mississippi Sovereignty: The Round Island Expedition of 1849," *Journal of Mississippi History* (winter 2000), 1–33. In May 1847, George William White raised the Montgomery Guard volunteers for the Mexican War. White captained what became Company A of the Louisiana Third Infantry Battalion, in the First Regiment of Gen. Persifor F. Smith's brigade of the Louisiana Volunteers, serving in Veracruz and Tampico. Three months after returning home in July 1848, White organized one thousand men in New Orleans, including Charles C. Campbell, to suppress an indigenous revolt on behalf of the governor of Yucatán. White was erroneously identified as "Joseph A. White" in Wallace, *Destiny and Glory,* 40–46, 58–59, and as "Robert M. White" in Charles H. Brown, *Agents of Manifest Destiny: The Lives and Times of the Filibusters* (Chapel Hill: University of North Carolina Press, 1980), 37. Callender Irvine Fayssoux joined William Walker in Nicaragua in April 1856 and served as aide to Walker, paymaster and captain of the schooner *Granada,* the only vessel in Walker's "navy." The identities of some of the conspiratorial pseudonyms also appear in Portell Vilá, *Narciso López,* 2:155, 191, 259. Wilson, *An Authentic Narrative,* 4–5; 1850 Louisiana Free Census, Third District in Orleans Parish, 281; *Cohen's New Orleans and Lafayette Directory, 1851* (New Orleans: Daily Delta, 1851), 198; "List of Officers of Louisiana Volunteers Distinguished by Their Service in the War with Mexico," Historical Military Data Louisiana Militia, 1847–48, Supplement, W.P.A. Project, 1941, Jackson Barracks Library, New Orleans, 209, 213; Mexican War, 1847–48, 213, ibid.; *Picayune,* 4 Aug. 1847; Evans J. Casso, *Louisiana Legacy: A History of the State National Guard* (Gretna, La.: Pelican, 1976), 69; U.S. Senate, *Message of the President of the United States, Transmitting Reports from the Several Heads of Department Relative to the Subject of the Resolution of the Senate of the 23d of May, as to Alleged Revolutionary Movements in Cuba,* Exec. Doc. 57, 31st Congress, 1st sess., 67, 69; Callender Irvine Fayssoux autobiographical manuscript, Callender I. Fayssoux Collection, Latin American Library, Tulane University; *Herald,* 8 Aug. 1849, 3; John L. O'Sullivan to Thomas J. Rusk, 13 Sept. 1849, Thomas J. Rusk Papers, Center for American History, University of Texas at Austin; "Balance de los Fondos de la Espedición de 1849," John Anthony Quitman Papers, 18 Sept. 1850, Mississippi Department of Archives and History, Jackson, hereinafter cited a JQP; Portell Vilá, *Narciso López,* 2:465–66 n. 267; U.S. Senate, Exec. Doc. 57, 88; and Roig de Leuchsenring, *Primeros Movimientos,* 67, 169.

51. *Herald,* 23, 30 July and 4, 11 Aug. 1849, 4; *Evening Picayune,* 30 Aug. 1849, 1; *New Orleans Commercial Bulletin,* 8 Sept. 1849, 2; Collins, *History of Kentucky,* 2:587; *Biographical Encyclopedia of Kentucky of the Dead and Living Men of the Nineteenth Century* (Cincinnati: J. M. Armstrong, 1878), 593; W. M. Paxton, *The Marshall*

*Family* (Cincinnati: Robert Clarke, 1885), 56; Stewart Sifakis, *Who Was Who in the Civil War* (New York: Facts on File Publications, 1988), 506; *Herald*, 16 July 1849, 3; Theodore O'Hara Papers, Filson Club, Louisville; "Col. Theodore O'Hara," *Atlanta Journal*, 29 May 1897, 15–16; Jennie C. Morton, "Theodore O'Hara," *Register of Kentucky State Historical Society*, Sept. 1903, 49; J. Stoddard Johnston, "Sketch of Theodore O'Hara," ibid., Sept. 1913, 67; Malone, *Dictionary of American Biography*, 7:4–5; Edgar Erskine Hume, "Colonel Theodore O'Hara and Cuban Independence," *Bulletin of the Pan American Union*, May 1937, 364; Richard H. Collins, *History of Kentucky* (Frankfort: Kentucky Historical Society, 1966), 1:410–11; Richard P. Weinert, "The 'Hard Fortune' of Theodore O'Hara," *Alabama Historical Quarterly* (spring–summer 1966): 33; Nathaniel Cheairs Hughes Jr. and Thomas C. Ware, *Theodore O'Hara: Poet-Soldier of the Old South* (Knoxville: University of Tennessee Press, 1998); *Intelligencer*, 16 Jan. 1850, 3; *Cincinnati Commercial*, 20 Aug. 1849; Hardy, *History and Adventures*, 20–22; and J. L. O'Sullivan to Thomas Ritchie, 25 July 1849, Ritchie-Harrison Papers, Earl Gregg Swem Library, College of William and Mary.

52. *Fauquier White Sulphur Springs* (Warrenton, Va.: Warrenton House, n.d.); *Intelligencer,* 7 June 1849, 1; *Richmond Enquirer*, 4 Sept. 1849, 4; Fauquier County Bicentennial Committee, *Fauquier County, Virginia, 1759–1959* (Warrenton, Va.: Virginia Publishing, 1959), 135–39; and Stan Cohen, *Historic Springs of the Virginias: A Pictorial History* (Charleston, W.Va.: Pictorial Histories, 1994), 47.

53. Ellwood Fisher became the editor of the *Southern Press* in Washington, D.C., in 1850–51. He died in Atlanta in October 1862 after fleeing from Nashville during the Civil War. *Mercury*, 8 Oct. 1862, 1. *Richmond Enquirer*, 24 July 1849, 2; *Alexandria Gazette and Virginia Advertiser*, 3 Aug. 1849, 2; *Washington Daily Union*, 5 Aug. 1849, 3; *Richmond Enquirer*, 7 Aug. 1849, 4; and Frederick William Franck, "The Virginia Legislature at the Fauquier Springs in 1849," *Virginia Magazine of History and Biography* (Jan. 1950): 77, 81.

54. O.D.D.O., *History of the Late Expedition*, 60; Hardy, *History and Adventures*, 80; *Macon Georgia Journal and Messenger*, 12 Sept. 1849, 1; and *Intelligencer*, 28 Aug. 1849, 3.

55. U.S. Senate, Exec. Doc. 57, 7, 67–68; Frances Calderón de la Barca, *Life in Mexico* (Berkeley and Los Angeles: University of California Press, 1982), 5–9; *Herald*, 22 Aug. 1849, 2; *Picayune*, 31 Aug. 1849, 2; *North Carolina Greensborough Patriot*, 1 Sept. 1849, 1; *Nashville True Whig*, 1 Sept. 1849, 2; and *Republic*, 23 Aug. 1849, 3.

56. The word *filibuster* derives from the Dutch *vrijbuiter*, meaning freebooter or pirate, which the Spaniards modified into *filibustero* in the sixteenth century. Filibuster now became synonymous with a member of a private military expedition organized in the United States against a Caribbean basin nation. In the 1870s, the word was used to define delaying tactics in the U.S. Senate. The *N.Y. Commercial Advertiser* quoted in *La Crónica* (New York), 17 Aug. 1849, 675–76; *La Crónica* cited in *Jackson*

Mississippian, 15 June 1849, 1; La Crónica, 12 Sept. 1849, 731; and Gaceta de La Habana quoted in "Isla de Cuba," ibid., 3 Oct. 1849, 779–80. The Daily Union indicated that it was Secretary of State John Clayton "more than it was his duty to do so" who issued the proclamation. All of Taylor's biographers overlooked that it was Clayton who wrote the presidential proclamation against the expedition. Daily Union, 16 and 28 Aug. 1849, 3; U.S. Senate, Exec. Doc. 57, 7–8; Z. Taylor to J. M. Clayton, 11 Aug. 1849, Preston Family Papers, Virginia Historical Society, Richmond; and Columbia Tri-Weekly South Carolinian, 4 Sept. 1849, 2.

57. Friend to AJG, 12 Aug. 1849, Keith Read Collection, University of Georgia.

58. Herald, 9, 20, and 25 Aug. 1849, 4; ibid., 15, 17, and 30 Aug. 1849, 2; ibid., 27 Aug. 1849, 3; Republic, 27 Aug. 1849, 2; Santovenia, Huellas de Gloria, 157; Express, 21 Aug. 1849; Republic, 23 Aug. 1849, 3; and Thomas, Robert E. Lee, 428 nn. 14, 16.

59. U.S. Senate, Exec. Doc. 57, 78–80; Rose Greenhow to John Calhoun, 29 Aug. 1849, in Jameson, "Correspondence of John C. Calhoun," 2:1203–4; Ishbel Ross, Rebel Rose: The Life of Rose O'Neal Greenhow, Confederate Spy (Marietta, Ga.: Mockingbird Books, 1992), 22–23; Mercury, 7 June and 7 Oct. 1864, 1; Courier, 5 Oct. 1864, 1; and clippings in Rose O'Neal Greenhow Papers, DUL.

60. John O'Sullivan to John Calhoun, 24 Aug. 1849, "Correspondence of John C. Calhoun," 1202–3; and Gonzales, "The Cuban Crusade," 1.

61. J. Prescott Hall to J. M. Clayton, 6 Sept. 1849, John Middleton Clayton Papers, LC, hereinafter cited as JMC; Tribune, 8 Sept. 1849, 2; U.S. Senate, Exec. Doc. 57, 15–18; Louis N. Feipel, "The Navy and Filibustering in the Fifties," United States Naval Institute Proceedings, Apr. 1918, 779; J. Prescott Hall to John M. Clayton, 15 Sept. 1849, MLS; Evening Mirror, 7 and 13 Sept. 1849, 2; Herald, 7, 8, and 24 Sept. 1849, 2; Tribune, 8 Sept. 1849, 2, 5; N.Y. Journal of Commerce, 8 Sept. 1849, 2; New Orleans Commercial Bulletin, 17 Sept. 1849, 2; Vicksburg Whig, 25 Sept. 1849, 2; La Crónica, 12 Sept. 1849, 733; Ultramar, Cuba, Gobierno, Legajo 4633, no. 26, AHN; Capitán General to Ministerio de la Gobernación, 9 Sept. 1849, Legajo 4633, AEA; Morales, Iniciadores, 194–95; Duncan Williamson, William Hughes, and Samuel Matthews v Steam Ship Fanny, Admiralty Case File no. 6617, U.S. District Courts, Eastern District of Louisiana, New Orleans, RG 21, NA; Log U.S.S. Water Witch, 30 Sept. 1849, RG 24, NA; Picayune, 5 Oct. 1849, 2; Evening Picayune, 5 and 13 Oct. 1849, 1; Vicksburg Whig, 11 Oct. 1849, 2; Ultramar, Cuba, Gobierno, Legajo 4630, no. 11, AHN; Intelligencer, 10 and 23 Oct. 1849, 3; Herald, 12 Oct. 1849, 2; La Crónica, 13 Oct. 1849, 805; V. M. Randolph to Wm. C. Nicholson, 12 Oct. 1849, Letters Received by the Secretary of the Navy from Commanders, 1804–86, RG 45, NA; William Nicholson to W. B. Preston, 12 Oct. 1849, ibid.; Foxhall Parker to W. Ballard Preston, 19 Oct. 1849, Letters Received by the Secretary of the Navy from Commanding Officers of Squadrons, 1841–86, Home Squadron, vol. 95, RG 45, NA; Portell Vilá, Narciso López, 2:176–77; Herald, 17 Oct. 1849, 4; and Picayune, 25 Oct. and 1 Nov. 1849, 2.

62. The Cuban Council document was signed by Mádan under the pseudonym "León Fragua del Calvo," Cirilo Villaverde, Pedro José Yznaga Hernández, Gaspar Agramonte, José R. Posse, Ignacio Mora, José Andrés Yznaga, Pedro M. Sánchez Yznaga, Gaspar Betancourt Cisneros, Victoriano de Arrieta, Tomás D. del Rijo, and Miguel Teurbe Tolón. Portell Vilá, *Narciso López*, 2:74–75, 86–87, 94–95, 450–52, and 457–58; *Republic*, 6 Nov. 1849, 3; Warner, *Generals in Blue*, 338; and Report of the Adjutant General of the State of Kentucky, *Mexican War Veterans* (Frankfort, Ky.: Capital Office, 1889), 116.

63. Córdova, *Gaspar Betancourt Cisneros*, 129–30; Morales, *Iniciadores*, 2:57–58; and Roig de Leuchsenring, *Primeros Movimientos*, 103.

64. Other members of the Cuban Council of Organization and Government included José Aniceto Yznaga Borrell, Ignacio Mora, Pedro M. Sánchez Yznaga, José R. Posse, Cayetano V. de Quesada, Esteban de Ayala, Francisco Castillo, Gaspar Agramonte, Carlos de Arteaga, Miguel and Pedro de Agüero, Tomás D. del Rijo, Pedro José Yznaga Hernández, Alonso Betancourt, Antonio Yznaga del Valle, Porfirio Jardines, Victoriano de Arrieta, Plutarco González, and Francisco T. Vingut. Portell Vilá, *Narciso López*, 2:96–97, 455. The Junta proclamation appeared in the *Herald*, 6 Dec. 1849, 1; *Daily Union*, 9 Dec. 1849, 3, *Republic*, 10 Dec. 1849, 2, and *La Verdad*, 13 Dec. 1849, 3. Other newspapers receiving the proclamation were the *Sun, Evening Post, Tribune*, and the Cuban annexationist *El Correo de los dos Mundos* of New York; the *Whig* and *Enquirer* of Richmond; the *Courier* and *Mercury* of Charleston; the *Chronicle* and *Journal* of Louisville; and the *Delta* and *Picayune* of New Orleans.

65. Portell Vilá claimed that Mádan had gone to Washington, D.C., on December 13 with Gaspar Betancourt Cisneros and Victoriano de Arrieta for an interview with Col. John Stuart Williams, who had been offered the leadership of the Council expedition. Hotel arrival schedules do not substantiate this. Portell Vilá, *Narciso López*, 2:97–98; *Daily Union*, 9 Dec. 1849, 3; Morales, *Iniciadores*, 2:193–95; and *Republic*, 6, 17 and 19 Dec. 1849, 3.

66. U.S. Congress, *Biographical Directory*, 1171; Denslow, *10,000 Famous Freemasons*, 2:214; Portell Vilá, *Narciso López*, 2:97, 450–51; Gonzales, "On to Cuba"; Dunbar Rowland, *Mississippi* (Atlanta: Southern Historical Publishing Association, 1907), 1:858; and Robert E. May, *John A. Quitman: Old South Crusader* (Baton Rouge: Louisiana State University Press, 1985), 65.

67. Report of the Adjutant General, *Mexican War Veterans*, 2, 28, 166; O.D.D.O., *History of the Late Expedition to Cuba*, 50; *Cincinnati Nonpareil*, 4 June 1850, 2; Hardy, *History and Adventures*, 22; Rob Morris, *The History of Freemasonry in Kentucky, in Its Relations to the Symbolic Degrees* (Louisville: Rob Morris, 1859), 562; Gonzales, "On to Cuba," 9; *Republic*, 30 Dec. 1849, 2; and ibid., 2, 3 and 4 Jan. 1850, 3.

68. *Republic*, 9 and 15 Jan. 1850, 3; *Herald*, 17 Dec. 1849, 1; and Portell Vilá, *Narciso López*, 2:100.

69. U.S. Senate, Exec. Doc. 57, 19–20; *La Verdad*, 1 Feb. 1850, 3; *Evening Picayune*, 4 Mar. 1850, 1; Portell Vilá, *Narciso López*, 2:90, 98; and D. S. Dickinson to Daniel Webster, 22 Feb. 1851, MLS.

70. The Cirilo Villaverde diary erroneously dates the Capitol meeting in March 1850. *Republic*, 1, 6, and 18 Feb. 1850, 3; "General Lopez: The Cuban Patriot," *United States Magazine and Democratic Review*, 105; Portell Vilá, *Narciso López*, 1:202; ibid., 2:119–21, 470–71; John L. O'Sullivan to Edward Curtis, 24 Feb. 1851, MLS; and *Copiador General, 1841–1860*, Charleston Spanish Consulate Papers, 1 Mar. 1850.

71. The Cirilo Villaverde diary indicates that López arrived in Louisville on February 26, but this date conflicts with hotel and steamship notices. Gonzales, "On to Cuba"; *Pittsburgh Gazette*, 13 Feb. 1850, 3; *Cincinnati Commercial*, 27 Feb. 1850, 1; *Cincinnati Enquirer*, 28 Feb. 1850, 3; Portell Vilá, *Narciso López*, 2:219, 471; *Tallahassee, Florida Sentinel*, 11 June 1850, 2; and Morales, *Iniciadores*, 2:195.

72. 1850 Mississippi Census, Free Schedules, Harrison County, 80, 86; *Cohen's New Orleans and Lafayette Directory, 1851*, 89; *Mississippian*, 17 Aug. 1849, 3; Louisiana, vol. 10, 551, R. G. Dun & Co. Collection, Baker Library, Harvard University, hereinafter cited as BLH; and Laurent J. Sigur, Death Record, 30 Sept. 1858, vol. 19, 341, Orleans Parish, City Hall, New Orleans.

73. *United States v William Hardy et al.*, Indictment, 1 Nov. 1851, U.S. Circuit Court, Southern District of Ohio, Western Division (Cincinnati), Law Records [old series], 1851–63, vol. 1 (5), U.S. District Courts, RG 21, NA; Report, *Mexican War Veterans*, 32; U.S. House, *Barque Georgiana*, 116; and Hardy, *History and Adventures*, 3, 6, 75.

74. John Anthony Quitman was grand sovereign of South West, grand inspector general of the 33d Degree of the Southern Division of the United States, for the State of Mississippi, and a founder of the Supreme Council. Quitman was also an honorary member of the Grand Lodge of South Carolina and the Grand Lodge of New York. He had been elected Grand Master of his state from 1826 to 1838, guiding the fraternity through the turbulent anti-Masonic era. As the persecution frenzy peaked in 1833, only two of the sixteen chartered lodges in Mississippi were represented at the opening session of the Grand Lodge and over half had disappeared from the roll. There was rapid Masonic expansion when Quitman was again reelected Grand Master in 1840, 1845, and 1846, the year he departed for Mexico. In his honor, Quitman U.D. Lodge was established in 1847 by American troops in Veracruz, Mexico, and there was also a Quitman Lodge No. 76 in New Orleans. By 1850, Masonic lodges had increased in Mississippi to nearly one hundred. Gonzales, "On to Cuba"; *Crescent*, 13 Jan. 1851, 2; *Delta*, 14 Jan. 1851, 2; Denslow, *10,000 Famous Freemasons*, 4:3; Robert Freke Gould et al., *The History of Freemasonry* (New York: John C. Yorston, 1889), 4:489–90; Albert Pike, "John Anthony Quitman," *Proceedings of the Grand Lodge*

of Mississippi, Ancient, Free, and Accepted Masons (Jackson: Clarion Steam Printing Establishment, 1882), x–xvi, 626; *Delta* 3 Jan. 1851, 3; Vicksburg Council no. 2 to John A. Quitman, 27 Feb. 1850, JQP; and Francis Gildart et. al. to John A. Quitman, 7 Sept. 1850, ibid.

75. In 1861, Cotesworth Pinckney Smith was appointed to the staff of Gen. Reuben Davis, commanding Corinth, and died on Nov. 11, 1862. Gonzales, "On to Cuba"; John Ray Skates Jr., *A History of the Mississippi Supreme Court, 1817–1948* (Jackson: 1973), 88–91; Denslow, *10,000 Famous Freemasons*, 4:125; Jeanne Hand Henry, *1819–1849 Abstradex of Annual Returns Mississippi Free and Accepted Masons* (New Market, Ala.: Southern Genealogical Services, 1969), 428, 430; *Proceedings of the Grand Lodge of the Most Ancient and Honorable Fraternity of Free and Accepted Masons, of the State of Mississippi, 1826* (Natchez: 1826), 22–23; Dunbar Rowland, *Courts, Judges, and Lawyers of Mississippi, 1798–1935* (Jackson: Press of Hederman Bros., 1935), 87–89, 92–94; idem, *Mississippi*, 1:285; and *Biographical and Historical Memoirs of Mississippi* (Chicago: Goodspeed, 1891), 1:126, 416–17.

76. Claiborne, *Life and Correspondence of John A. Quitman*, 2:57–58, 383–85; May, *Quitman*, 238; *Georgia Journal and Messenger*, 6 Mar. 1850, 3; and AJG to John A. Quitman, 20 Mar. 1850, JQP.

77. AJG to John A. Quitman, 23 Mar. 1850, JQP.

**Freedom Fighter**

1. AJG to John A. Quitman, 23 Mar. 1850, JQP.

2. Portell Vilá, *Narciso López*, 2:125, 265; *Picayune*, 4 and 5 Apr. 1850, 2; *Mississippian*, 6 Sept. 1850, 3; AJG to John A. Quitman, 5 Apr. 1850, JQP; Gonzales, "On to Cuba"; U.S. House, *Barque Georgiana*, 143; and 1850 Mississippi Free Census, Hinds County, 122.

3. AJG to John A. Quitman, 5 Apr. 1850, JQP.

4. Ibid; Biographical sketch of Chatham Roberdeau Wheat by his father John Thomas Wheat, John Thomas Wheat Papers, Southern Historical Collection, University of North Carolina, Chapel Hill, hereinafter cited as SHC; Thomas R. Wolfe to John Thomas Wheat, 9 May 1850, ibid.; *National Cyclopedia of American Biography* (New York: J. T. White, 1893), 9:168; *Picayune*, 8 July 1848, 2 (Wheat's biographer erroneously places his date of arrival in New Orleans in "the fall of 1848"); Charles L. Dufour, *Gentle Tiger: The Gallant Life of Roberdeau Wheat* (Baton Rouge: Louisiana State University Press, 1957), 33–34; This work, while citing some of Wheat's motivations for joining the Cuban expedition, mistakenly attributes his father's account to his brother Leo and omits the Masonic inducement. Dufour, *Gentle Tiger*, 37.

5. Henry Titus became a proslavery leader in Bleeding Kansas against abolitionist John Brown in 1856. The following year he fought with William Walker's filibusters

in Nicaragua and was assistant quartermaster of the Florida Militia at the start of the Civil War but did not enter Confederate service. In 1867, he settled in Sand Point, Florida, became its postmaster, and changed the name of the town to Titusville, presently the gateway to the John F. Kennedy Space Center. Titus, confined to a wheelchair because of rheumatoid arthritis, became a wealthy developer, citrus grower, and administrator of a general store and the Titus House, "one of the finest combinations of saloons and hotels on the east coast of Florida." "Early Recollections of Minnie Titus Ensey, Youngest Daughter of Col. Henry Theodore Titus, as Told to Her Daughter Fedora Ensey Grey," in Henry Theodore Titus Collection, North Brevard Public Library, Titusville, Florida; "Henry Theodore Titus: Famous or Infamous," ibid.; U.S. Department of State, *Register of all Officers and Agents, Civil, Military, and Naval, in the Service of the United States* (Washington, D.C.: J. and G. S. Gideon, 1845), 432; and Elaine Murray Stone, *Brevard County: From Cape of the Canes to Space Coast* (Northridge, Calif.: Windsor Publications, 1988), 21. F. C. M. Boggess, *A Veteran of Four Wars* (Arcadia, Fla.: Champion Job Rooms, 1900), 8–9; 1850 Louisiana Free Census, First Municipality of New Orleans, 7; *Picayune*, 11 and 12 Apr. 1850, 2; Hardy, *History and Adventures*, 9, 46, 75; O.D.D.O., *History of the Late Expedition*, 50; and *Evening Picayune*, 11 Apr. 1850, 1.

    6. *Picayune*, 18, 20, and 24 Apr. 1850, 2; Portell Vilá, *Narciso López*, 2:215, 472, n. 348; and Dollero, *Cultura Cubana*, 148.

    7. Survey of Federal Archives in Louisiana, Work Projects Administration, *Ship Registers and Enrollments of New Orleans, Louisiana, 1841–1850* (Baton Rouge: Louisiana State University, 1942), 4:65; 1850 Louisiana Free Census, Third District in Orleans Parish, 362, 368; *Picayune*, 26 July 1849, 2; *U.S. v Narciso Lopez et al.*, U.S. District Courts, Eastern District of Louisiana, New Orleans Circuit Court, General Case Files (E-121), Case 1965, "Enrollment No. 183," RG 21, NA, hereinafter cited as *U.S. v Lopez*; and *Crescent*, 13 Jan. 1851, 2.

    8. Augustin's name is misspelled as "Donahen Augusten" in Rauch, *American Interest in Cuba*, 148, and the mistake was repeated in Brown, *Agents of Manifest Destiny*, 69. Morales, *Iniciadores*, 2:196; *Cohen's New Orleans and Lafayette Directory, 1851*, 6, 22; 1850 Louisiana Free Census, First Municipality of New Orleans, 393, and Third District in Orleans Parish, 179; Stanley Clisby Arthur, *Old Families of Louisiana* (New Orleans: Harmanson, 1931), 49–51; *Evening Picayune*, 17 June 1850, 1; *Courier*, 17 June 1850; *New Orleans Bee*, 18 June 1850; *New Orleans True Delta*, 18 June 1850; *Picayune*, 26 Apr. 1850, 3; *U.S. v Lopez*, RG 21, NA; Enrollment no. 5, 10 May 1851, MLS; *Cincinnati Nonpareil*, 17 June 1850, 2; *Crescent*, 13 Jan. 1851, 2; "Col. M. C. Taylor's Diary in Lopez Cardenas Expedition, 1850," *Register of the Kentucky State Historical Society*, Sept. 1921, 81; Louisiana, vol. 9, 90, R. G. Dun & Co. Collection, BLH; and *Barque Georgiana*, 96, 116, 149, 157–58.

9. *Evening Picayune*, 28 June 1850, 1; Hardy, *History and Adventures*, 16, 18–19; "Col. M. C. Taylor's Diary," 80–81; *Cincinnati Nonpareil*, 17 June 1850, 2; *Barque Georgiana*, 96, 116; *Delta*, 14 Jan. 1851, 2; and *Crescent*, 11 Jan. 1851, 2.

10. *Barque Georgiana*, 100; *Picayune*, 27 Apr. 1850, 2; *Delta*, 14 Jan. 1851, 2; Thomas R. Wolfe to John Thomas Wheat, 9 May 1850, John Thomas Wheat Papers, SHC; May, *Quitman*, 238; and *Evening Picayune*, 23 May 1850, 1.

11. U.S. Senate, Exec. Doc. 57, 23–24; and *Picayune*, 26 Apr. 1850, 2.

12. *Picayune*, 1 May 1850, 2; Portell Vilá, *Narciso López*, 2:221–22 (which includes a copy of the twenty-thousand-dollar Cuban bond signed by Judge Cotesworth Pinckney Smith); and John Henderson to John A. Quitman, 9 May 1850, JQP.

13. O.D.D.O. *History of the Late Expedition to Cuba*, 4–7; Boggess, *A Veteran of Four Wars*, 9; Hardy, *History and Adventures*, 31–32; *Evening Picayune*, 28 June 1850, 1; Theodore O'Hara to Quitman, 17 Dec. 1854, JQP; *U.S. v Lopez*, RG 21, NA; *Picayune*, 3 May 1850, 3; and Thomas R. Wolfe to John Thomas Wheat, 9 May 1850, John Thomas Wheat Papers, SHC.

14. *Evening Picayune*, 23 May 1850, 1; and Gonzales, "On to Cuba."

15. James Robb, a Pennsylvania native, had moved to New Orleans in 1837 and established the Bank of James Robb. He became president of the New Orleans Gaslight and Banking Company and in 1844 organized the Spanish Gaslight Company, which provided public lighting for Havana, in partnership with Spain's Queen María Cristina. Robb was one of the securities for the consular bond of Robert Campbell in Havana, who described him as "perhaps the most wealthy Banker of New Orleans." "Robb Papers Discovered," *Historic New Orleans Collection Newsletter* (winter 1986): 1–4; Robert Campbell to John Clayton, 17 Apr. 1850, DCH; and U.S. Senate, Exec. Doc. 57, 49–51.

16. *Barque Georgiana*, 100, 142; 1850 Louisiana Census, Third District in Orleans Parish, 192, NA; *Delta*, 8 May 1851, 1; ibid., 7 Jan. 1851, 2; Portell Vilá, *Narciso López*, 2:274, 307–8; and *La Verdad*, 7 July 1850, 1.

17. John Henderson to John A. Quitman, 9 May 1850, JQP; *Picayune*, 16 June 1850, 2; *Crescent*, 13 Jan. 1851, 2; *Delta*, 14 Jan. 1851, 2; *Barque Georgiana*, 100; "Soldiers of the Liberating Expedition of Cuba!," broadside in *U.S. v Lopez*, RG 21, NA; and Hardy, *History and Adventures*, 31–32.

18. Hardy, *History and Adventures*, 31; O.D.D.O., *History of the Late Expedition*, 24–25; *Barque Georgiana*, 100, 142, 153, 169; and *Crescent*, 13 Jan. 1851, 2.

19. O.D.D.O., *History of the Late Expedition*, 27; and Hardy, *History and Adventures*, 31, 80–81.

20. *Evening Picayune*, 28 June 1850, 1; "Col. M. C. Taylor's Diary," 82–85; Hardy, *History and Adventures*, 22–28, 34; O.D.D.O., *History of the Late Expedition*, 30–34, 39, 42–44, 47; *Barque Georgiana*, 97–98, 100, 117, 141; 1850 Kentucky Free Census, Meade

County, 173; Robert Granville Caldwell, *The Lopez Expeditions to Cuba, 1848–1851* (Princeton: Princeton University Press, 1915), 62; *Proceedings of the Grand Lodge of Kentucky, at a Grand Annual Communication in the City of Lexington, Commencing August Twenty-Sixth, 5850* (Frankfort: A. G. Hodges, 1850), 62; *Delta*, 12 Jan. 1851, 1; *La Verdad*, 7 July 1850, 1; and Henry, *1819–1849 Abstradex*, 431.

21. *La Verdad* claimed that 629 men landed in Cárdenas, Captain Simeon Pendleton estimated the number at 630, José Sánchez Yznaga put the figure at 604 filibusters, and U.S. attorney William R. Hackley determined there were 610 expeditionaries. *Steamer Creole*, 19 June 1850, U.S. District Courts, Southern District of Florida, Key West Admiralty Records, 4:443–50, RG 21, NA; Hardy, *History and Adventures*, 35–36; Boggess, *A Veteran of Four Wars*, 11; *Evening Picayune*, 28 June 1850, 1; "Col. M. C. Taylor's Diary," 84; O.D.D.O., *History of the Late Expedition*, 44–45; *Crescent*, 11 Jan. 1851, 2; *Delta*, 14 Jan. 1851, 2; *Barque Georgiana*, 97, 101, 117, 142, 167, 170; *Cincinnati Nonpareil*, 17 June 1850, 2; Justo Zaragoza, *Las Insurreciones en Cuba* (Madrid: Imprenta de Manuel G. Hernandez, 1872), 1:595; and *La Verdad*, 7 July 1850, 1.

22. O.D.D.O., *History of the Late Expedition*, 54, 64; Hardy, *History and Adventures*, 34; Carlos Hellberg, *Historia Estadística de Cárdenas* (1893; reprint, Cárdenas: Comité Pro-Calles de Cárdenas, 1957), 62; Gonzales, "On to Cuba"; "Col. M. C. Taylor's Diary," 84; and Boggess, *A Veteran of Four Wars*, 10–11.

23. O.D.D.O., *History of the Late Expedition*, 61.

24. Ibid., 58; "Col. M. C. Taylor's Diary," 84; Hardy, *History and Adventures*, 35–36; *Intelligencer*, 25 May 1850, 4; *Florida Sentinel*, 11 June 1850, 2; and Gonzales, "On to Cuba."

25. Robert Campbell to John Clayton, 19 May and 4 June 1850, DCH; "Recortes de periódicos de Cuba sobre los sucesos de Cárdenas," *Boletín del Archivo Nacional*, July–Dec. 1920, 193–94; *Picayune*, 24 May 1850, 2; *Diario de la Marina* (Havana), 1 July 1850, 2; and *Cincinnati Nonpareil*, 26 Aug. 1850, 2.

26. Portell Vilá, *Narciso López*, 2:302; *Barque Georgiana*, 58–59, 85, 111–15, 167–68; "Al Ministro de Estado," 27 May 1850, Legajo 43, no. 53, Archivo Nacional, Havana, Cuba, hereinafter cited as AN; Zaragoza, *Las Insurreciones en Cuba*, 1:596; *Picayune*, 26 May 1850, 1; *Delta*, 11 Jan. 1851, 2; and *Crescent*, 11 Jan. 1851, 2.

27. O.D.D.O., *History of the Late Expedition*, 63; Hardy, *History and Adventures*, 37–39; "Col. M. C. Taylor's Diary," 85; Portell Vilá, *Narciso López*, 2:301, 306; and *Intelligencer*, 29 May 1850, 3 (the "officer of rank" who provided this account was probably Lt. Col. M. J. Bunch).

28. *Cuba en la Mano*, 43; Hellberg, *Historia Estadística de Cárdenas*, 28, 59, 68; Herminio Portell Vilá, *Historia de Cárdenas* (Havana: Talleres Gráficos "Cuba Intelectual," 1928), 55, 74–75, 95; idem, *Narciso López*, 2:304–5; *Picayune*, 2 Sept. 1849, 1; "Reseña oficial de lo ocurrido en Cárdenas cuando la invasión del General Narciso

López," *Boletín del Archivo Nacional,* Jan.–Dec. 1926, 152; and Robert Campbell to John Clayton, 3 Dec. 1849, DCH.

29. Sumario, 22 May 1850, Legajo 43, no. 52, AN; Gonzales, "On to Cuba"; Hardy, *History and Adventures,* 39; O.D.D.O., *History of the Late Expedition,* 65; "Col. M. C. Taylor's Diary," 85; Portell Vilá, *Narciso López,* 2:306–7; "Callender Irvine Fayssoux" autobiographical manuscript, Callender I. Fayssoux Collection, Tulane University; *Picayune,* 26 May 1850, 1; *Intelligencer,* 29 May 1850, 3; and "Recortes de periódicos de Cuba sobre los sucesos de Cárdenas," *Boletín del Archivo Nacional,* July–Dec. 1920, 193–94. This contemporary account in sequence identifies the port captain as Patricio Montojo, while Portell Vilá, citing Hellberg, calls him Miguel Baldasano y Ros, and erroneously refers to "True Fayssoux." Portell Vilá, *Narciso López,* 2:306–7.

30. *Picayune,* 26 May 1850, 1; *La Verdad,* 7 July 1850, 1; *Evening Picayune,* 8 June 1850, 1; Hellberg, *Historia Estadística de Cárdenas,* 57, 62, 64; and Robert Campbell to John Clayton, 22 May 1850, DCH.

31. *Evening Picayune,* 28 June 1850, 1; *Intelligencer,* 29 May 1850, 3; "Primera página de la historia de la revolución de Cuba," *La Verdad,* 7 July 1850, 1; John H. Goddard to Alexander H. H. Stuart, 12 Dec. 1850, Records Concerning the Cuban Expedition, 1850–51, RG 48, NA; Hellberg, *Historia Estadística de Cárdenas,* 63; *Evening Picayune,* 28 June 1850, 1; O.D.D.O., *History of the Late Expedition,* 65–66; Hardy, *History and Adventures,* 7, 39; "Col. M. C. Taylor's Diary," 85; and Portell Vilá, *Narciso López,* 2:308–10.

32. *Evening Picayune,* 28 June 1850, 1; O.D.D.O., *History of the Late Expedition,* 65–66, 68; Hardy, *History and Adventures,* 39; *Intelligencer,* 29 May 1850, 3; and Gonzales, "On to Cuba."

33. The description of the Capitular House and the Alejandro Rodríguez-Capote y de la Cruz genealogy was provided by his great-grandson, Dr. Leopoldo Cancio, a Washington, D.C., resident. O.D.D.O., *History of the Late Expedition,* 66; Hardy, *History and Adventures,* 40; Portell Vilá, *Narciso López,* 2:300; and Hellberg, *Historia Estadística de Cardenas,* 51–53, 63.

34. Hardy, *History and Adventures,* 39–40, 75–78; O.D.D.O., *History of the Late Expedition,* 66–67; *Evening Picayune,* 28 June 1850, 1; Portell Vilá, *Narciso López,* 2:308–10; idem, *Historia de Cárdenas,* 90; Boggess, *A Veteran of Four Wars,* 14; *Evening Picayune,* 8 June 1850, 1; *Intelligencer,* 29 May 1850, 3; *Cincinnati Nonpareil,* 17 June 1850, 2; Ahumada y Centurión, *Memoria Histórico Política,* 272; and Gonzales, "On to Cuba."

35. The three jailed soldiers who joined the filibusters were Miguel Ancejo, Carlos Arlandes, and Francisco Grau. *Intelligencer,* 29 May 1850, 3; Hardy, *History and Adventures,* 40–41, 76; "Primera página de la historia de la revolución de Cuba," *La Verdad,* 7 July 1850, 1; Gonzales, "On to Cuba"; and Portell Vilá, *Narciso López,* 2:311–12, 340–42.

36. *Evening Picayune*, 28 June 1850, 1; Boggess, *A Veteran of Four Wars*, 13–14; O.D.D.O., *History of the Late Expedition*, 67; Hardy, *History and Adventures*, 40; Dufour, *Gentle Tiger*, 48–49; Portell Vilá, *Narciso López*, 2:313–15, 483 n. 491; and U.S. Congress, *Biographical Directory*, 2084.

37. Boggess, *A Veteran of Four Wars*, 15; O.D.D.O., *History of the Late Expedition*, 68; *Evening Picayune*, 28 June 1850, 1; *La Verdad*, 7 July 1850, 1; and Hardy, *History and Adventures*, 77.

38. O.D.D.O., *History of the Late Expedition*, 68–69; *Intelligencer*, 29 May 1850, 3; Hardy, *History and Adventures*, 76–77; Gonzales, "On to Cuba"; *Evening Picayune*, 28 June 1850, 1; and War with Mexico, 1846–48, Muster Rolls, Wars Vertical File, Jackson Barracks, New Orleans.

39. *Evening Picayune*, 28 June 1850, 1; *Intelligencer*, 29 May 1850, 3; Boggess, *A Veteran of Four Wars*, 15; *La Verdad*, 7 July 1850, 1; O.D.D.O., *History of the Late Expedition*, 69; Gonzales, "On to Cuba"; and *Cincinnati Nonpareil*, 17 June 1850, 2.

40. Gonzales, "On to Cuba"; O.D.D.O., *History of the Late Expedition*, 69, 72; Hardy, *History and Adventures*, 79; and Portell Vilá, *Narciso López*, 2:314.

41. *Intelligencer*, 29 May 1850, 3; *La Verdad*, 7 July 1850, 1; "Recortes de periódicos de Cuba sobre los sucesos de Cárdenas," *Boletín del Archivo Nacional*, July–Dec. 1920, 194; Zaragoza, *Las Insurreciones en Cuba*, 1:596; *Evening Picayune*, 8 June 1850, 1; *Tribune*, 30 May 1850, 6; Boggess, *A Veteran of Four Wars*, 15–16; Hardy, *History and Adventures*, 41; and Portell Vilá, *Narciso López*, 2:314.

42. Hardy, *History and Adventures*, 45; "Col. M. C. Taylor's Diary," 85; Boggess, *A Veteran of Four Wars*, 17; *Intelligencer*, 29 May 1850, 3; Portell Vilá, *Narciso López*, 2:314–15; *Evening Picayune*, 28 June 1850, 1; and Robert Campbell to John Clayton, 22 May 1850, DCH.

43. O.D.D.O., *History of the Late Expedition*, 71; Hardy, *History and Adventures*, 41; *Evening Picayune*, 8 and 28 June 1850, 1; Allen Owen to Daniel Webster, 2 Dec. 1851, DCH; Boggess, *A Veteran of Four Wars*, 18; and Portell Vilá, *Narciso López*, 2:323, 364.

44. Consul Robert Campbell reported that the money expropriated was $2,400 from the Custom House and about $1,600 from City Hall. Robert Campbell to John Clayton, 22 May 1850, DCH; Zaragoza, *Las Insurreciones en Cuba*, 1:596; "Recortes de periódicos de Cuba sobre los sucesos de Cárdenas," *Boletín del Archivo Nacional*, July–Dec. 1920, 194; *Evening Picayune*, 8 and 28 June 1850, 1; *Intelligencer*, 29 May 1850, 3; *Tribune*, 30 May 1850, 6; *Picayune*, 26 May 1850, 1; *La Verdad*, 25 Sept. 1850, 2; Portell Vilá, *Narciso López*, 2:317, 371, 374; Hellberg, *Historia Estadística de Cárdenas*, 64–65; W. Grayson to John Clayton, 25 May 1850, MLS; "Reseña oficial de lo ocurrido en Cárdenas," 155; Sumario, 22 May 1850, Legajo 43, no. 52, AN; O.D.D.O., *History of the Late Expedition*, 70–71; and Hardy, *History and Adventures*, 44.

45. Boggess, *A Veteran of Four Wars*, 17; Hardy, *History and Adventures*, 41; "Recortes de periódicos de Cuba sobre los sucesos de Cárdenas," *Boletín del Archivo*

*Nacional*, July–Dec. 1920, 195; "Reseña oficial de lo ocurrido en Cárdenas," 153; and Portell Vilá, *Narciso López*, 2:324.

46. *La Verdad*, 25 Sept. 1850, 2; Zaragoza, *Las Insurreciones en Cuba*, 596; "Personal Narrative of Louis Schlesinger, of Adventures in Cuba and Ceuta," *Democratic Review*, Sept. 1852, 22; Boggess, *A Veteran of Four Wars*, 16; Portell Vilá, *Narciso López*, 2:312, 324, 326–28, 342–47; Hellberg, *Historia Estadística de Cárdenas*, 48–49, 65; *Evening Picayune*, 8 June 1850, 1; and Hardy, *History and Adventures*, 41–42.

47. O.D.D.O., *History of the Late Expedition*, 72; Hardy, *History and Adventures*, 42; *Picayune*, 19 Oct. 1849, 2; *Florida Sentinel*, 11 June 1850, 2; and Sarah L. A. Doyle to John Clayton, 26 June and 6 July 1850, MLS.

48. The Cárdenas Company of the León Infantry Regiment members who joined López were Pedro Almerillo, Miguel Ancejo, Carlos Arlandes, Manuel Barrera, Ambrosio Castaño, Manuel Coya, José Esteves, Jacinto Gaite, Francisco Grau, Francisco Iglesias, Juan López, Miguel López, Felipe Merino, Juan Rodríguez, Felipe Román, José Ronquillo, Felipe Sainz, Francisco Sainz, Juan Sanderra, Andrés Sestar, Manuel Sila, Antonio Valdespino, Luis Villarino, Luis Viñas, and Tomás Yañez. *La Verdad*, 27 Aug. 1850, 2–3; Portell Vilá, *Narciso López*, 2:333–34, 338, 370; O.D.D.O., *History of the Late Expedition*, 70; and "Col. M. C. Taylor's Diary," 85.

49. *La Verdad*, 7 July 1850, 1; Robert Campbell to John Clayton, 22 May 1850, DCH; *Evening Picayune*, 8 and 28 June 1850, 1; Hellberg, *Historia Estadística de Cárdenas*, 67; *Tribune*, 30 May 1850, 6; and O.D.D.O., *History of the Late Expedition*, 71–72.

50. Portell Vilá, *Narciso López*, 2:328–29; and Hellberg, *Historia Estadística de Cárdenas*, 65.

51. Robert Campbell to John Clayton, 31 May 1850, DCH; Hellberg, *Historia Estadística de Cárdenas*, 67; *Evening Picayune*, 8 June 1850, 1; "Reseña oficial de los ocurrido en Cárdenas," 154; Portell Vilá, *Narciso López*, 2:317, 324, 362; John Thrasher to William Marcy, 7 Nov. 1853, MLS; and Sumario, 22 May 1850, Legajo 43, no. 52, AN.

52. *Mississippian*, 6 Sept. 1850, 3; *Intelligencer*, 29 May 1850, 3; Hardy, *History and Adventures*, 42; O.D.D.O., *History of the Late Expedition*, 72; and "Recortes de periódicos de Cuba sobre los sucesos de Cárdenas," *Boletín del Archivo Nacional*, July–Dec. 1920, 195.

53. *Mississippian*, 6 Sept. 1850, 3; Boggess, *A Veteran of Four Wars*, 17–18; *Cincinnati Nonpareil*, 17 June 1850, 2; *Evening Picayune*, 28 June 1850, 1; Portell Vilá, *Historia de Cárdenas*, 111, 113; and idem, *Narciso López*, 2:349.

54. Portell Vilá, *Narciso López*, 2:349; O.D.D.O., *History of the Late Expedition*, 72–73; *Evening Picayune*, 28 June 1850, 1; and Robert Campbell to John Clayton, 22 May 1850, DCH.

55. *Evening Picayune*, 28 June 1850, 1; "Col. M. C. Taylor's Diary," 85–86; Portell Vilá, *Narciso López*, 2:356; Boggess, *A Veteran of Four Wars*, 20; and Hardy, *History and Adventures*, 42.

56. *Evening Picayune*, 28 June 1850, 1.

57. Ibid.; *La Verdad*, 25 Sept. 1850, 2; Hardy, *History and Adventures*, 44; *Intelligencer*, 29 May 1850, 3; *Mississippian*, 6 Sept. 1850, 3; and Portell Vilá, *Narciso López*, 2:351.

58. *Evening Picayune*, 28 June 1850, 1; Hardy, *History and Adventures*, 42–43; Zaragoza, *Las Insurreciones en Cuba*, 1:597; and Portell Vilá, *Narciso López*, 2:353–56.

59. Hardy, *History and Adventures*, 43; Hellberg, *Historia Estadística de Cárdenas*, 66–67; Calcagno, *Diccionario Biográfico Cubano*, 398; and Portell Vilá, *Narciso López*, 2:357, 486 n. 568.

60. Portell Vilá, *Narciso López*, 2:351–52; *Evening Picayune*, 28 June 1850, 1; and Hardy, *History and Adventures*, 22, 43.

61. *Evening Picayune*, 28 June 1850, 1; Hardy, *History and Adventures*, 43; and Portell Vilá, *Narciso López*, 2:357.

62. *Evening Picayune*, 31 May and 28 June 1850, 1; Hardy, *History and Adventures*, 78; and Boggess, *A Veteran of Four Wars*, 19.

63. Hardy, *History and Adventures*, 44; *Evening Picayune*, 28 June 1850, 1; and *Cincinnati Nonpareil*, 17 June 1850, 2.

64. The obelisk honoring the Spaniards who died at Cárdenas on May 19, 1850, reads: "Here lie the ashes of the soldiers Vicente Pérez, Antonio Martínez, Francisco López, Ramón Caballero and Galo Tejedor of the León Infantry Regiment and of First Corp. Ginés Ibáñez, and the soldiers of the King's Lancers Cavalry Regiment Feliciano Carrasco, Roque Blanco, José Crespo and Francisco Valenzuela." Portell Vilá, *Narciso López*, 2:361–62; Hellberg, *Historia Estadística de Cárdenas*, 66–67; Robert Campbell to John Clayton, 22 May 1850, DCH; *Evening Picayune*, 28 June 1850, 1; Hardy, *History and Adventures*, 43, 45–46; "Col. M. C. Taylor's Diary," 86; Caldwell, *The Lopez Expeditions to Cuba*, 72; and W. R. Hackley to Solicitor of the Treasury, 22 May 1850, MLS.

65. Sumario, 22 May 1850, Legajo 43, no. 52, AN; *Cincinnati Nonpareil*, 17 June 1850, 2; *Intelligencer*, 29 May 1850, 3; Gonzales, "On to Cuba"; Hardy, *History and Adventures*, 44–45; O.D.D.O., *History of the Late Expedition*, 76–77; "Col. M. C. Taylor's Diary," 86; Boggess, *A Veteran of Four Wars*, 20; and Portell Vilá, *Narciso López*, 2:375–76.

66. Hardy, *History and Adventures*, 46–47, 57, 59; O.D.D.O., *History of the Late Expedition*, 77; *Evening Picayune*, 28 June 1850, 1; *Intelligencer*, 29 May 1850, 3; *Florida Sentinel*, 11 June 1850, 2; and "Col. M. C. Taylor's Diary," 86.

67. *Evening Picayune*, 28 June 1850, 1; "Col. M. C. Taylor's Diary," 86; O.D.D.O., *History of the Late Expedition*, 77; *Barque Georgiana*, 47; Portell Vilá, *Narciso López*, 2:382; Gonzales, "On to Cuba"; Boggess, *A Veteran of Four Wars*, 21; Zaragoza, *Las Insurreciones en Cuba*, 1:597; *Intelligencer*, 29 May 1850, 3; *Florida Sentinel*, 11 June 1850, 2; and Robert Campbell to John Clayton, 22 May 1850, DCH.

68. *Barque Georgiana*, 123, 127; *Evening Picayune*, 8 June 1850, 1; Robert Campbell to John Clayton, 31 May 1850, DCH; *Tribune*, 11 and 12 June 1850, 4; "Expediente

promovido por el contratista de la conducción de cadáveres del Hospital Militar de Matanzas al Cementerio General de aquella ciudad . . . ," *Boletín del Archivo Nacional*, Jan.–Dec. 1951, 401–2; and Portell Vilá, *Narciso López*, 2:326, 362.

69. Gonzales, "On to Cuba"; *Picayune*, 26 May 1850, 1; "Col. M. C. Taylor's Diary," 86; O.D.D.O., *History of the Late Expedition*, 77–79; Boggess, *A Veteran of Four Wars*, 20; Hardy, *History and Adventures*, 46–47; and Robert Campbell to John Clayton, 22 May 1850, DCH.

70. "Col. M. C. Taylor's Diary," 86; Boggess, *A Veteran of Four Wars*, 22–24; Hardy, *History and Adventures*, 47; O.D.D.O., *History of the Late Expedition*, 79–80; *Intelligencer*, 29 May 1850, 3; *Picayune*, 31 May 1850, 1; *Barque Georgiana*, 44; and Gonzales, "On to Cuba."

71. Hardy, *History and Adventures*, 47–48; O.D.D.O., *History of the Late Expedition*, 80–81; *Picayune*, 31 May 1850, 1; *Barque Georgiana*, 44; and Boggess, *A Veteran of Four Wars*, 24.

72. *Evening Picayune*, 28 June 1850, 1; Gonzales, "On to Cuba"; Hardy, *History and Adventures*, 48–49; and O.D.D.O., *History of the Late Expedition*, 81.

73. O.D.D.O., *History of the Late Expedition*, 81; Gonzales, "On to Cuba"; Joseph T. Durkin, *Confederate Navy Chief: Stephen R. Mallory* (Columbia: University of South Carolina Press, 1987), 32–33; and *Tallahassee Floridian and Journal*, 18 Aug. 1849, 2.

74. *Steamer Creole*, RG 21, NA; *Barque Georgiana*, 44–46; and Hardy, *History and Adventures*, 49–50.

75. "Col. M. C. Taylor's Diary," 86; Boggess, *A Veteran of Four Wars*, 26, 28; Hardy, *History and Adventures*, 50–51; and W. R. Hackley to Solicitor of the Treasury, 22 May 1850, MLS.

76. Theodore O'Hara to William Nelson, 18 Mar. 1854, JQP; Theodore O'Hara to Quitman, 17 Dec. 1854, ibid.; and Hardy, *History and Adventures*, 57, 72.

**Filibuster Fiascos**

1. Attorney Robert M. Charlton was elected to the U.S. Senate in 1852 and died two years later. U.S. Senate, Exec. Doc. 57, 31st Congress, 1st sess., 46; David H. Galloway, *Directory of the City of Savannah for the Year, 1850* (Savannah: Edward C. Councell, 1849), 14; Henry Williams to John Clayton, 26 and 27 May 1850, MLS; *Proceedings of the Most Worshipful Grand Lodge of Georgia, at Its Annual Communication for the Year 5850* (Macon, Ga.: S. Rose, 1851), 51; *N.Y. Tribune*, 31 May 1850, 8; *Evening Picayune*, 1 June 1850, 1; *Savannah Morning News*, 10 Mar. 1935; William J. Northen, ed., *Men of Mark in Georgia* (Spartanburg, S.C.: Reprint Company, 1974), 2:295–97; *Memoirs of Georgia* (Atlanta: Southern Historical Association, 1895), 1:247; and Allen D. Candler and Clement A. Evans, eds., *Georgia* (Atlanta: State Historical Association, 1906), 1:348–49.

2. Henry Williams to John Clayton, 26 and 27 May 1850, MLS; *Evening Picayune*, 1 June 1850, 1; *Tribune*, 31 May 1850, 8; John H. Goddard to Thomas Ewing, 17 June

1850, Records Concerning the Cuban Expedition, 1850–51, RG 48, NA; U.S. Senate, Exec. Doc. 57, 31st Congress, 1st sess., 46–47; *Tribune*, 10 June 1850, 6; and *Picayune*, 2 and 5 June 1850, 2.

    3. Prosecutor Logan Hunton, a forty-four-year-old Virginia native, had served in the Kentucky legislature, in 1838 established a law office in St. Louis, Missouri, and two years later had been a Whig convention delegate. He moved to New Orleans in 1844 and got his judiciary appointment from the Taylor administration. Logan Hunton to John Clayton, 1 and 6 June 1850, MLS; *Picayune*, 8 June 1850, 1; *National Cyclopedia of American Biography*, vol. 3 generally and 7:477–79; Malone, *Dictionary of American Biography*, 8:191–92; Rowland, *Mississippi*, 2:466–77; George L. Prentiss, ed., *A Memoir of S.S.* Prentiss (New York: Charles Scribner, 1855), 2:554–57; Joseph D. Shields, *The Life and Times of Seargent Smith Prentiss* (Philadelphia: J. B. Lippincott, 1884), 421–29; Dallas C. Dickey, *Seargent S. Prentiss: Whig Orator of the Old South* (Baton Rouge: Louisiana State University Press, 1945), 400–401; and *Intelligencer*, 16 July 1850, 4.

    4. Dufour, *Gentle Tiger*, 53; *Evening Picayune*, 28 June 1850, 1; Gonzales, "On to Cuba"; Durkin, *Confederate Navy Chief*, 34; William Marvin, "Autobiography of William Marvin," *Florida Historical Quarterly* (Jan. 1958), 201; *Proceedings of the Grand Lodge of the Most Ancient and Honorable Fraternity of Free and Accepted Masons of the State of Florida, 1846* (Tallahassee: Office of the Floridian, 1846), 54; *Proceedings of the M. W. Grand Lodge of the State of Florida, 1849* (Tallahassee: Office of the Floridian and Journal, 1849), vi; *Evening Picayune*, 28 June 1850, 1; and *Tribune*, 25 June 1850, 4.

    5. The New Orleans federal grand jury that indicted Gonzales, and the filibuster leadership was composed of: Cornelius Fellows, J. M. Lapeyre, William Laughlin, James L. McLean, George M. Pinckard, William Tufts (foreman), L. D. C. Wood, all commission merchants; Horace Bean, banker; Charles F. Caruthers, cotton broker; A. F. Dunbar, boot and shoe store owner; Louis E. Forstall, commission agent; Levi H. Gale, ship broker; L. J. Harris, grocer; Henry Hopkins and Joseph Lallande, no occupation listed. *Cohen's New Orleans and Lafayette Directory, 1851; Proceedings of the M. W. Grand Lodge of Free and Accepted Masons of the State of Louisiana, from 22d June 5850 to February 25th, 5851;* Logan Hunton to John Clayton, 17 and 22 June 1850, MLS; "U.S. District Court," *Evening Picayune*, 17–21 June 1850, 1; *Intelligencer*, 21 June 1850, 3; *U.S. v Lopez*, RG 21, NA; *La Verdad*, 7 July 1850, 4; Claiborne, *Life and Correspondence of John A. Quitman*, 2:59–60; and Morales, *Iniciadores y primeros mártires*, 2:121.

    6. John Henderson to John A. Quitman, 2 July 1850, JQP.

    7. *Barque Georgiana*, 66–67; Ida Young, Julius Gholson, and Clara Nell Hargrove, *History of Macon, Georgia* (Macon, Ga.: Lyon, Marshall, and Brooks, 1950), 146; *Savannah Georgian*, 22 July 1850, 2; *Savannah Morning News*, ibid.; Minutes Book, 1849–59,

Solomon's Lodge No. 1, Free and Accepted Masons, roll X-0940-05, Georgia Historical Society, Savannah, hereinafter cited as GHS; *Mercury,* 24 July 1850, 3; and *Herald,* 8 and 17 Aug. 1850, 2.

8. *Savannah Morning News,* 22 July 1850, 2; *Tribune,* 27 July 1850, 4; *Picayune,* 30 July 1850, 2; *Cincinnati Nonpareil,* 31 July 1850, 2; Gonzales, *Manifesto on Cuban Affairs,* 9; *Gaceta de La Habana,* 7 Sept. 1850; and *Barque Georgiana,* 37–38, 104, 137–38, 148, 151–52, 173.

9. John Lama has been erroneously identified as a Cuban émigré in Portell Vilá, *Narciso López,* 2:59, 410. Minutes Book, 1849–59, Solomon's Lodge No. 1, Free and Accepted Masons, roll X-0940-05, GHS; *Proceedings of the Most Worshipful Grand Lodge of Georgia, 5850,* 41–42; John Lama affidavit, William Coolidge Papers, GHS; John Lama, Hartridge Collection, ibid.; John Lama, General Index to Keepers' Record Books, 1853–1938, Catholic Cemetery, Savannah; Robert N. Rosen, *The Jewish Confederates* (Columbia: University of South Carolina Press, 2000), 13; "List of Letters Remaining in the Post Office Savannah," *Savannah Georgian,* 2 Sept. 1842, 3; *Proceedings of the Most Worshipful Grand Lodge of Georgia, at Its Annual Convention, for the Year 5849* (Macon, Ga.: S. Rose, 1849), 63; *Savannah Morning News,* 14 July 1879, 3; Galloway, *Savannah Directory, 1850,* 6, 25, 31, 34; *Savannah Morning News,* 2 Oct. 1851, 1; Tax Digests, 1850, 69, City Hall, Savannah; Georgia, vol. 28, 97, R. G. Dun & Co. Collection, BLH; "Soneto" *La Verdad,* 12 Aug. 1850, 4; *Marriages of Chatham County, Georgia,* vol. 1, *1748–1852* (Savannah: Georgia Historical Society, n.d.), 156, 180; and Annual Report of the Republican Blues, 25 Aug. 1848, Georgia Military Records, Chatham County, Georgia Department of Archives and History, Atlanta.

10. Minutes Book, 1849–59, Solomon's Lodge No. 1; Portell Vilá, *Narciso López,* 1:77; *Savannah Morning News,* 13 Nov. 1850, 2; and Minutes Book, 1849–59, Solomon's Lodge No. 1, Free and Accepted Masons, roll X-0940-05, GHS.

11. The Cuban *tinajones* are massive oval clay tubs that collect rain water. Annual Report of the Georgia Hussars, 6 Oct. 1848, Georgia Military Records, Chatham County, Georgia Department of Archives and History, Atlanta; 1850 Georgia Free Census, Chatham County; ibid., McIntosh County, 127, 195; U.S. Congress, *Biographical Directory,* 1849; Galloway, *Directory of Savannah, 1850,* 28; E. Merton Coulter, *Thomas Spalding of Sapelo* (Baton Rouge: Louisiana State University Press, 1940); and Huxford, *Pioneers of Wiregrass Georgia,* 5:407.

12. *Evening Picayune,* 17 Dec. 1850, 2; John A. Quitman to H. J. Harris, 2 Oct. 1850, MLS; *Picayune,* Dec.18, 19, 24, and 25 1850, 2; 1850 Louisiana Census, Parish of Orleans, 363; and Logan Hunton to Daniel Webster, 26 Dec. 1850, MLS.

13. *Evening Picayune,* 2, 6, 9, and 11 Jan. 1851, 1; *Picayune,* 3, 7, and 11 Jan. 1851, 2; *Delta,* 3, 7, 8, 10, and 11 Jan. 1851, 2; *Bee,* 3, 7, 8, and 10 Jan. 1851, 1; *Crescent,* 13 Jan. 1851, 2; Pierce Butler, *Judah P. Benjamin* (Philadelphia: George W. Jacobs, 1906),

179–82; Eli N. Evans, *Judah P. Benjamin: The Jewish Confederate* (New York: Free Press, 1988), 41–43; *Barque Georgiana*, 111, 168; and Quisenberry, *Lopez's Expeditions*, 122.

14. *Picayune*, 12 Jan. 1851, 2; *Crescent*, 13 Jan. 1851, 2; and *Delta*, 14 Jan. 1851, 2.

15. *Crescent*, 13 Jan. 1851, 2.

16. Ibid.; and *Delta*, 14 Jan. 1851, 2.

17. *Picayune*, 12 Jan. 1851, 2.

18. *Crescent*, 13 Jan. 1851, 2; *Picayune*, 12 Jan. 1851, 2; and *Delta*, 14 Jan. 1851, 2.

19. *Delta*, 12 Jan. 1851, 3; *Picayune*, 12 Jan. 1851, 2; and *Evening Picayune*, 14 Jan. 1851, 2 .

20. *Evening Picayune*, 13, 15, and 16 Jan. 1851, 1; *Crescent*, 14 Jan. 1851, 2; *Delta*, 16 Jan. 1851, 2; *Picayune*, 21 and 22 Jan. 1851, 2; *Delta*, 22 Jan. 1851, 1; *Bee*, 13 and 22 Jan. 1851; and Logan Hunton to Daniel Webster, 22 Jan. 1851, MLS.

21. *Bee*, 23 Jan. and 4 Feb. 1851, 1; *Picayune*, 1 Feb. 1851, 1; Horatio J. Harris to Millard Fillmore, 6, 17, and 25 Jan. and 4 Feb. 1851, MLS; *Evening Picayune*, 5 and 7 Feb. 1851, 1; *Delta*, 8 Feb. 1851, 2; and *Bee*, 8 Feb. 1851, 1.

22. Fifty-year-old Georgia governor George Washington Towns was the son of a Revolutionary War veteran and a Macon attorney who had served as a colonel in the state militia. Elected to both branches of the state legislature as a Democrat, he opposed nullification in 1832 and was later elected to Congress in Washington, D.C., on three different occasions. When Towns became governor 1847, he was a staunch fire-eater objecting to the Wilmot Proviso and the spread of abolitionism. Towns retired from politics in November 1851 to operate a cotton plantation with many slaves and died in Macon on July 15, 1854. Two years later, a new Georgia county was named after him. *Delta*, 9, 11, and 12 Feb. 1851, 2; *Picayune*, 11 and 12 Feb. 1851, 2; *Evening Picayune*, 11 Feb. 1851, 2; *Bee*, 12 Feb. 1851, 1; Logan Hunton to Daniel Webster, 12 Feb. 1851, MLS; Roig de Leuchsenring, *Los Primeros Movimientos*, 148; Portell Vilá, *Narciso López*, 3:107, 114–15, 217; Lucian Lamar Knight, *Georgia's Landmarks, Memorials, and Legends* (Atlanta: Byrd, 1913), 1:969–70; idem, *A Standard History of Georgia and Georgians* (Chicago: Lewis, 1917), 5:2716–17; Kenneth Coleman and Charles Stephen Gurr, *Dictionary of Georgia Biography* (Athens: University of Georgia Press, 1983), 2:996–97; and *Daily Union*, 21 July 1854, 3.

23. *Picayune*, 2, 4, and 5 Mar. 1851, 2; *Evening Picayune*, 4 Mar. 1851, 1; *Bee*, 5 Mar. 1851, 1; and May, *Quitman*, 251; *Picayune*, 7 Mar. 1851, 2; *Delta*, 8 Mar. 1851, 2; *Evening Picayune*, 7 Mar. 1851, 1; *Bee*, 7 and 8 Mar. 1851, 1; and "The Late Cuba State Trials," *United States Magazine and Democratic Review* (Apr. 1852), 307.

24. *United States v Henry Robinson and William A. McEwen*, Indictment, 1 Nov. 1851, U.S. Circuit Court, Southern District of Ohio, Western Division (Cincinnati), Law Records [old series], 1851–63, vol. 1 (5), RG 21, U.S. District Courts, NA; and AJG to M. B. Lamar, 14 Mar. 1851, Mirabeau Buonaparte Lamar Papers, Texas State Archives, Austin, hereinafter cited as TSA.

25. AJG to M. B. Lamar, 14 Mar. 1851, Mirabeau Buonaparte Lamar Papers, TSA; *Mississippian*, 24 Aug. 1849, 2; and Robert J. Travers to Mrs. R. L. Downman, 24 Sept. 1851, Robert L. Downman Papers, Alabama Department of Archives and History, Montgomery.

26. John Forsyth later became editor of the *Mobile Register*. AJG to M. B. Lamar, 14 Mar. 1851, Mirabeau Buonaparte Lamar Papers, TSA; 1850 Georgia Free Census, Muscogee County, 304, 309; ibid., City of Macon, Bibb County, 144; ibid., McIntosh County, 220; "Autobiography of Isaac Scott," Scott Genealogical Folder, Georgia Department of Archives and History, Atlanta; *Mississippian*, 24 Aug. 1849, 2; Gonzales, "The Cuban Crusade"; *Intelligencer*, 25 Apr. 1851, 3; and Huxford, *Pioneers of Wiregrass Georgia*, 5:407.

27. *Intelligencer*, 21 Apr. 1851, 3; *Tribune*, 2 May 1851, 5; L. M. Perez, "Lopez's Expeditions to Cuba, 1850–51," *Publications of the Southern History Association*, Nov. 1906, 360; and Portell Vilá, *Narciso López*, 3:191–92.

28. Deposition of William W. Boyle, *United States v John O'Sullivan et al.*, Case File 1–267, U.S. District Courts, Southern District of New York, RG 21, NA.; and Perez, "Lopez's Expeditions to Cuba, 1850–51," 348–53, 360.

29. Daniel Henry Burtnett eventually received twelve thousand dollars for his services, two-thirds of it drawn from the Cuban treasury. A personal tour of the Woodbine plantation was provided to this writer by its owners, Mack and Maryann McKenzie, in February 1994. David Bailey eventually had five children with Isabella Lang, who died in 1860. Four years later, he married her thirty-eight-year-old sister Catherine. Bailey was buried in 1889 in the Cambray plantation graveyard, which contains the remains of slaves and the latest Lang family generation. 1850 Georgia Census, Camden County, 171; 1860 Georgia Census, Camden County, 16; Cemetery Records of Camden County and Camden County Marriage Records, in the Bryan-Lang Historical Library, Woodbine, Georgia. The main house of Cambray plantation, built in the 1820s, is still inhabited by descendant Kevin Lang, who provided this writer a tour of the plantation and its cemetery in February 1994. Woodbine Plantation Papers, Bryan-Lang Historical Society, Woodbine, Georgia; Marguerite Reddick, *Camden's Challenge: A History of Camden County, Georgia* (Jacksonville: Paramount Press, 1976), 57, 121–22; Camden County Historical Commission, *Camden County, Georgia* (Camden, Ga.: n.p., 1992), 10; Francisco Stoughton to Capitán General, 26 Apr. 1851, AN; ACB to Capitán General, 14 Apr. 1851, ibid.; and Hamilton Fish to Millard Fillmore, 26 Apr. 1851, MLS.

30. Warren Akin to Sir, 5 Apr. 1851, MLS; A. F. Owen to Daniel Webster, 10 Apr. 1851, ibid.; U.S. Congress, *Biographical Directory*, 1600; *Rome (Ga.) Courier*, 10 Apr. 1851, 2; and J. Renean to the president of the United States, 10 Apr. 1851, MLS.

31. William R. Manning, *Diplomatic Correspondence of the United States: Inter-American Affairs, 1831–1860* (Washington, D.C.: Carnegie Endowment for International

Peace, 1939), 7:433; M. B. Lamar to General Gonzales, 12 Apr. 1851, Mirabeau Buonaparte Lamar Papers, TSA; and Isaac Scott to Daniel Webster, 13 Apr. 1851, MLS.

32. ACB to Daniel Webster, 14 Apr. 1851, in Manning, *Diplomatic Correspondence*, xi, 587–88; ACB to Capitán General, 15 Apr. 1851, Legajo 217, no. 17, AN; Expense sheet of Cuba Emigrants, Robert L. Downman Papers, ADAH; and *Picayune*, 22 Apr. 1851, 2.

33. J. Prescott Hall to Daniel Webster, 22 and 24 Apr. 1851, MLS; *United States v John O'Sullivan et al.*; C. McKnight Smith to Hugh Maxwell, 23 Apr. 1851, MLS; and C. McKnight Smith to Daniel Webster, 25 Apr. 1851, ibid.

34. Recently discovered evidence shows that the filibusters encamped at Burnt Fort in 1851, on the present-day Wildwood plantation, Camden County, Georgia. Allen Drury, a mechanic and metal-detecting hobbyist from Folkston, Georgia, uncovered in that spot more than two hundred unfired musket balls, a few brass trigger guards, an eighteen-pound cannon ball, a Georgia Militia belt buckle, shirt buttons with the Georgia state seal, and fifteen Spanish coins. Forty-six-year-old Hiram Roberts had been appointed collector of customs in Savannah by President Taylor in August 1849. He was an officer in the Georgia Hussars militia, a dry goods merchant, and owned sixty-four slaves on his rice plantation. Roberts was a bank president from 1854 to 1865 and was later a cotton factor and commission merchant. *Daily Union*, 5 Aug. 1849, 3; 1850 Georgia Slave Schedules, Chatham County, 84; *Courier*, 4 and 19 Mar. 1864, 1; *Savannah Morning News*, 8 Sept. 1880, 3; Gonzales, "The Cuban Crusade"; *Savannah Georgian*, 16 May 1850, 1; Portell Vilá, *Narciso López*, 3:210–11; Hiram Roberts to W. L. Hodge, 3 May 1851, MLS; and Fae Oemler Smith, "A Sea-Island Plantation: 'China Grove,' Wilmington Island, Ga.," Wilmington Island Collection, Savannah Public Library, Savannah.

35. *Jacksonville, Florida Republican*, 1 May 1851, 3; *Tribune*, 2 May 1851, 5; *Floridian and Journal*, 6 Oct. 1849, 2; and Perez, "Lopez's Expeditions to Cuba, 1850–51," 359–60.

36. Gonzales, "The Cuban Crusade"; *Savannah Morning News*, 21 May 1883, 4; 1850 Georgia Free Census, Chatham County, 252; *Proceedings of the Most Worshipful Grand Lodge of Georgia, 1850,* 79; William W. Evarts to Daniel Webster, 26, 28, and 29 Apr. 1851, MLS; *United States v John O'Sullivan et al.*; H. F. Tallmadge to Daniel Webster, 30 May 1851, MLS; and J. Prescott Hall to H. F. Tallmadge, 30 May 1851, ibid.

37. The Barstow plantation was sold in 1907 to Thomas Bourke Floyd. The remains in the Barnard burial vault were transferred to Laurel Grove Cemetery in Savannah and the structure collapsed by 1936. Today the former plantation is the Sheraton Savannah Resort and Country Club. Deposition of Marmaduke H. Floyd Sr., July 31, 1936, folder 91, M. H. and D. B. Floyd Papers, GHS. Gonzales, "The Cuban Crusade"; *Savannah Morning News*, 10 Mar. 1935; and Smith, "A Sea-Island Plantation: 'China Grove.'"

38. Virginia Louise Glenn, "James Hamilton, Jr., of South Carolina: A Biography" (Ph.D. diss. University of North Carolina, 1964); N. Louise Bailey, Mary L. Morgan, and Carolyn R. Taylor, *Biographical Dictionary of the South Carolina Senate, 1776–1985* (Columbia: University of South Carolina Press, 1986), 1:641–45; *Cyclopedia of Eminent and Representative Men of the Carolinas of the Nineteenth Century* (Madison, Wis.: Brant and Fuller, 1892), 1:560–61; Frampton Erroll Ellis, *Some Historic Families of South Carolina*, 2d ed. (Atlanta: n.p., 1962), 66–67; 1850 Georgia Free Census, Chatham County; Galloway, *Directory of Savannah, 1850*, 28; and Gonzales, "The Cuban Crusade."

39. Portell Vilá, *Narciso López*, 3:211; William L. Hodge to the president, 3 May 1851, MLS; "Warrant to Arrest," 2 May 1851, U.S. District Courts, Southern District of Georgia, Savannah, Case File D-15, RG 21, NA; *Savannah Morning News*, 13 May 1851, 2; *Savannah Republican*, 13 and 22 May 1851, 2; and Henry Williams to Daniel Webster, 23 Oct. 1851, MLS.

40. Hiram Roberts to William L. Hodge, 5 May 1851, MLS; Portell Vilá, *Narciso López*, 3:211; Galloway, *Directory of Savannah, 1850*, 37; *Proceedings of the Most Worshipful Grand Lodge of Georgia, 5849*, 96; and *Picayune*, 10 May 1851, 2.

41. Camden County Historical Commission, *Camden County, Georgia*, 7; and Gonzales, "The Cuban Crusade."

42. Collector John Hardee Dilworth died on June 16, 1853, and an obelisk monument, etched with the Masonic square-and-compass emblem, marks his grave in Oak Grove Cemetery, St. Marys, Georgia. Dilworth Street, in St. Marys, was named after him. Gonzales, "The Cuban Crusade"; *Washington Union*, 9 Aug. 1849, 2; Huxford, *Pioneers of Wiregrass Georgia*, 5:118; and *Proceedings of the Most Worshipful Grand Lodge of Georgia, at Its Annual Communication for the Year 5852* (Macon, Ga.: S. Rose, 1852), 140.

43. Portell Vilá, *Narciso López*, 3:228–29; and Gonzales, "The Cuban Crusade."

44. N. L. to D. German [Ambrosio Gonzales], 23 June and 2 July 1851, Keith Read Collection, University of Georgia; *Florida Republican*, 3 July 1851, 3; and *Evening Picayune*, 19 July 1851, 1.

45. Portell Vilá, *Narciso López*, 3:458–61; Jorge Juárez Cano, *Hombres del 51* (Havana: Imprenta "El Siglo XX," 1930), 7, 30, 58; *Gaceta de La Habana*, 16 July 1851, 1; *Evening Picayune*, 18 and 21 Aug. 1851, 1; Peraza Sarausa, *Diccionario Biográfico Cubano*, 1:26.

46. *Evening Picayune*, 14 July 1851, 1; and Portell Vilá, *Narciso López*, 3:465, 482.

47. Alejandro Angulo Guridi was born in Santo Domingo in 1826 and later moved to Havana, where in 1841 he requested permission to publish the novel *La Joven Carmela*. He went into exile in the United States in 1848 and died in Nicaragua in 1906. Carlos M. Trelles y Govin, *Bibliografía Cubana del Siglo 19* (Matanzas: Imprenta de Quirós y Estrada, 1912), 2; Schlesinger, "Personal Narrative," 212; *Florida Republican*, 7 Aug.

1851, 3; Portell Vilá, *Narciso López* 1:76–77 and 3:483, 656; and Louisville *Courier*, 25 and 28 July 1851, 3.

48. *Evening Picayune*, 25 and 31 July and 7 Aug. 1851, 1; William Freret, *Correspondence between the Treasury Department, &c., in Relation to the Cuban Expedition, and William Freret, Late Collector* (New Orleans: Alex, Levy, 1851), 8, 45; and Schlesinger, "Personal Narrative," 212–15. The Schlesinger account gave the erroneous date of Saturday, August 1, when the first day of the month was actually a Friday.

49. Gonzales, "On to Cuba"; and Schlesinger, "Personal Narrative," 568–69.

50. *Picayune*, 9 Aug. 1851, 2; "Expedición del General Narciso López," *Boletín de los Archivos de la República de Cuba*, Jan.–Feb. 1904, 13–14; Schlesinger, "Personal Narrative," 214–15; *Picayune*, 9 Aug. 1851, 2; and Foxhall Parker to Daniel Webster, 12 Sept. 1851, MLS.

51. William R. Hackley to Daniel Webster, 23 Aug. 1851, MLS; Schlesinger, "Personal Narrative," 217–18; "Expedición del General Narciso López," Jan.–Feb. 1906, 16; *New York Times*, 23 Sept. 1851, 3; *Picayune*, 22 Aug. 1851, 2; "The Cuban Expedition," 3; Allen Owen to Daniel Webster, 16 Aug. 1851, DCH; and Schlesinger, "Personal Narrative," 219–23, 354.

52. *Picayune*, 22 Aug. 1851, 2; Allen Owen to Daniel Webster, 16 Aug. 1851, DCH; Expedición del General Narciso López," Jan.–Feb. 1904, 17; and Schlesinger, "Personal Narrative," 224, 353–56. Both of these accounts vary slightly as to the hours and the number of carts and expeditionaries sent back to Crittenden.

53. Ibid., 357–58, 362; *Savannah Morning News*, 30 Aug. 1851, 2; and James H. Brigham to JD, 7 Apr. 1853, DCH.

54. Kentuckians Robert H. Breckenridge and Ransom Beach, who fled with Crittenden, were "captured in an open boat, some thirty miles or more from the island" a month later by a Spanish warship and sent to prison in Havana. Allen Owen to Daniel Webster, 16 and 31 Aug., 18 Sept., and 1 Oct., 1851, DCH; Leslie Combs to Daniel Webster, 7 Nov. 1851, MLS; "Expedición del General Narciso López," 13–14; Schlesinger, "Personal Narrative," 363, 366–67, 558–59; J. Wilson to Daniel Webster, 27 Oct. 1851, DCH; Quisenberry, *Lopez's Expeditions*, 93; Enclosure A, 25 Sept. 1851, frames 312–14, MLS; Correspondence on the Lopez Expedition to Cuba, 1849–51, State Department Miscellaneous Correspondence, 1784–1906, box 1, entry 121, RG 59, NA; *Evening Picayune*, 21 Aug. 1851, 1; Boggess, *A Veteran of Four Wars*, 29; Portell Vilá, *Historia de Cárdenas*, 75; Foxhall Parker to Daniel Webster, 6 Sept. 1851, MLS; John A. Watkins to Daniel Webster, 6 Nov. 1851, ibid.; James T. Donaldson to L. W. Powell, 19 Sept. 1851, MLS; and Portell Vilá, *Narciso López*, 3:617.

55. Schlesinger, "Personal Narrative," 561, 566; and "Expedición del General Narciso López," 18–19.

56. Portell Vilá, *Narciso López*, 3:658; *Louisville Courier*, 22 Aug. 1851, 3; *Daily Union*, 22 Aug. 1851, 3; and *Courier*, 25 Aug. 1851, 2.

57. *Mercury*, 25 Aug. 1851, 2. The Gonzales letter was reprinted in the *Savannah Morning News*, 26 Aug. 1851, 3; *Evening Picayune*, 30 Aug. 1851, 1, and the *Jacksonville News*, 6 Sept. 1851, 1.

58. *Courier*, 25 and 26 Aug. 1851, 2; *Evening Picayune*, 29 and 30 Aug., 1851, 1; Gonzales, *Manifesto on Cuban Affairs*, 10; *Herald*, 22 Sept. 1852, 4; Portell Vilá, *Narciso López*, 3:659–60; and *Florida Republican*, 28 Aug. 1851, 2.

59. *Evening Picayune*, 20 Aug. 1851, 1; Portell Vilá, *Narciso López*, 3:658; *Savannah Morning News*, 25 Aug. 1851, 2; *United States v The Steamer* Pampero, U.S. District Courts, North District of Florida, Opinion and Decision on Libel and Information for Violation of the Revenue Laws, 12 Dec. 1851, Solicitor of the Treasury, Letters Received, U.S. Attorneys, Clerks of Courts, and Marshals, Florida 1846–Apr. 1863, box 19, RG 206, NA; *Florida Republican*, Nov. 20, 1851, 1; and *Jacksonville News*, 6 Sept. 1851, 2.

60. Antonio de la Cova, "Cuban Filibustering in Jacksonville in 1851," *Northeast Florida History*, vol. 3, 1996; *News*, 13 Sept. 1851, 2; *Florida Republican*, 20 Nov. 1851, 1; *Courier*, 4 Sept. 1851, 2; and Gonzales, *Manifesto on Cuban Affairs*, 10–11.

61. The four filibusters released by the captain general were Col. William Scott Haynes, Captains James A. Kelly and Henry Somers, and Lt. Philip Van Vechten. Comdr. Foxhall Parker indicated there were 176 prisoners: 135 sent to Spain, 25 in the hospital, and 16 in jail waiting transport. Foxhall Parker to Daniel Webster, 12 Sept. 1851, MLS. Filibuster tally in Correspondence on the Lopez Expedition to Cuba, 1849–51, State Department Miscellaneous Correspondence, 1784–1906, box 1, entry 121, RG 59, NA. Allen Owen to Daniel Webster, 5 Sept. 1851, DCH; *Florida Republican*, 27 Nov. 1851, 1; Julia W. Howe, *A Trip to Cuba* (Boston: Ticknor and Fields, 1860). Howe is best known for writing the *Battle Hymn of the Republic* two years later. Portell Vilá, *Narciso López*, 3:708; "List of Prisoners," 11 Sept. 1851, MLS; D. F. Miller to A. C. Dodge, 3 May 1853, DCH; and William H. Robertson to William Marcy, 6 July 1853, ibid.

62. *Courier*, 5 Sept. 1851, 2; *Savannah Morning News*, 8, 10, and 17 Sept. 1851, 2; *Savannah Republican*, 13 Sept. 1851; Charles A. Labuzan to Millard Fillmore, 11 Sept. 1851, MLS; Foxhall Parker to Daniel Webster, 12 Sept. 1851, ibid.; Isaiah D. Hart to Thomas Corwin, 13 Sept. 1851, ibid.; *United States v The Steamer* Pampero; *Florida Republican*, 18 Dec. 1851, 3; *Jacksonville News*, 20 Dec. 1851, 3; *Florida Republican*, 22 Jan. 1852, 2; and "Payments to Titus, 3 April 1852," *Florida Senate Journal*, 6th sess., 22 Nov, 1852, 116.

63. U.S. District Courts, Southern District of Georgia, Savannah, Case File D-15, RG 21, NA; Henry Williams to Daniel Webster, 23 Oct. 1851, MLS; *Savannah Morning News*, 15 Nov. 1851, 2; *Cincinnati Nonpareil*, 17 Nov. 1851, 1; *Delta*, 18 Nov. 1851, 1; *Florida Republican*, 27 Nov. 1851, 1; *United States v William Hardy et al.*; and Deposition of Ambrosio J. Gonzales, 15 Dec. 1851, U.S. District Courts, Southern District of New York, Case File Crim. 1–267, RG 21, NA.

## Patronage Pursuer

1. Portell Vilá, *Narciso López*, 3:148; Pew Owners, Dec. 1850, Christ Church Records, Vestry Minutes, vol. 1, 1839–59, Savannah, GHS; *Proceedings of the Most Worshipful Grand Lodge of Georgia, 5850*, 41–42; and C. M. Manigault to Miss Elliott, Apr. 20 [1867], Elliott-Gonzales Family Papers, SHC, hereinafter cited as EGP.

2. James Henry Rice Jr., *Cheeha-Combahee*, typed manuscript, 1932, 4, claimed that Gonzales was at Oak Lawn *after* his arrest order was issued in April 1851. The Rice data is garbled, wrongly stating that Gonzales was wounded in the López insurrection of 1853 and that he married Harriette [sic], Elliott's daughter by a *second* marriage. Rice, a transplanted Northerner described as a "literary gadfly," was on the staff of the Columbia newspaper *The State* from the 1890s to 1935, becoming a close friend of the Gonzales sons and their neighbor on his Chehaw Creek plantation. Jones, *Stormy Petrel*, 8, 271–72; Beverly Scafidel, "The Letters of William Elliott" (Ph.D. diss., University of South Carolina, 1978); Pew Owners, Dec. 1850, Christ Church Records, GHS; *Courier*, 4 Feb. 1852, 2; *Savannah Morning News*, 3 Mar. 1852, 2; and Mrs. Mary W. Wayne to Mrs. Elliott, 18 Aug. 1852, EGP.

3. Lewis Pinckney Jones, "William Elliott, South Carolina Nonconformist," *Journal of Southern History*, Aug. 1951, 361–81; Suzanne C. Linder, *Historical Atlas of the Rice Plantations of the Ace River Basin, 1860* (Columbia: South Carolina Department of Archives and History, 1995), 112–14; Bailey, Morgan, and Taylor, *Biographical Directory of the South Carolina Senate*, 469–72; State Board of Agriculture of South Carolina, *South Carolina* (Charleston, S.C.: Walker, Evans, and Cogwell, 1883), 27; Chalmers Gaston Davidson, *The Last Foray; the South Carolina Planters of 1860: A Sociological Study* (Columbia: University of South Carolina Press, 1971), 196; Hutchinson to EEL, 14 July 1873, EGP; MAE to HRG, 23 May [1869], EGP; Federal Census, 1860, Slave Schedule, South Carolina; and Willie Lee Rose, *Rehearsal for Reconstruction: The Port Royal Experiment* (New York: Oxford University Press, 1964), 119.

4. *150 Years of Orange Lodge, No. 14* (Charleston, S.C.: n.p., 1939), 64; and *Proceedings of Freemasons of South Carolina, November 5859*, 68.

5. Mary W. Wayne to Mrs. Elliott, 18 Aug. 1852, EGP; *Picayune*, 13 May 1851, 2; and Jones, *Stormy Petrel*, 6.

6. In January 1996, the author was unable to locate the grave of Capt. Armstrong Irvine Lewis in Mobile, Alabama. The Mobile City Archive is missing all death records from 1852 to 1854. Lewis is not listed in the Magnolia Cemetery files, and all Catholic Cemetery records prior to 1892 have vanished. *Courier*, 27 Mar. 1852, 2; *Savannah Morning News*, 26 Apr. 1852, 2; U.S. Congress, *Biographical Directory*, 562; Kenneth E. Shewmaker and Kenneth R. Stevens, eds. *The Papers of Daniel Webster: Diplomatic Papers*, vol. 2, *1850–1852* (Hanover, N.H.: University Press of New England, 1987), 426–27; *Mobile Daily Advertiser*, 10 Aug. 1852, 2; *Savannah Georgian*, 26 June 1852, 2; *Daily Union*, 29 July 1852, 2; and AJG to Caleb Cushing, 27 May 1853, CCP.

7. Gonzales, *Manifesto on Cuban Affairs*. When the manifesto was published in pamphlet form in 1853, Gonzales sent Caleb Cushing a copy, now in the LC, with the inscription: "To Hon. Caleb Cushing—With the compliments of the author."

8. Ibid., 4, 5.

9. Ibid., 6.

10. *Florida Republican*, 7 Oct. 1852, 2; *Savannah Morning News*, 11 Oct. 1852, 1; and *Union*, 16 Oct. 1852, 3.

11. *Union*, 24 Sept. 1852, 2; and Manning, *Diplomatic Correspondence*, xi, 664.

12. *Savannah Morning News*, 11 Oct. 1852, 1; Ivor Debenham Spencer, *The Victor and the Spoils: A Life of William L. Marcy* (Providence: Brown University Press, 1959), 223; AJG to John A. Quitman, 25 Jan. 1853, JQP; *Herald*, 27 Oct. 1852, 1; AJG to Caleb Cushing, 27 May 1853, CCP; and *Savannah Morning News*, 3 Nov. 1852, 2.

13. *Savannah Morning News*, 17 Feb. 1853, 2; *Herald*, 21 Nov. 1852, 4; *Savannah Morning News*, 24 Nov. 1852, 1; *Picayune*, 3, 18, and 19 Dec. 1852, 2; and Charleston Spanish Consulate Papers, Copiador General, 1841–60, 22 Dec. 1852, DKU.

14. AJG to John A. Quitman, 25 Jan. 1853, JQP; *Picayune*, 15 Aug. 1849, 3; ibid., 3 and 23 Dec. 1852, 2; and *Cohen's New Orleans Directory for 1851*, 143.

15. *Cohen's New Orleans Directory for 1851*, 143.

16. The reasons enumerated by Gonzales for his political appointment were "1° Sympathy for a sufferer in the cause of American annexation; 2° The qualifications, as knowledge of languages, of the world, address at; 3° The policy of propitiating the Cuban people and encouraging the annexationist party; 4° That of demonstrating to the world the equality of all citizens irrespective of birth; 5° If sent to Spanish America, knowledge of language character & manners—identity of political pursuit—policy of showing them, now that Europe seeks to prejudice them against the US, the course pursued towards adopted citizens in opening to them; 6° The avenues to distinction on the same footing as the rest; 6° [sic] The higher the post the greater the prestige given to me in the eyes of the Cubans; the greater my usefulness when the time for action comes; 7° It will propitiate the poor Cuban element throughout the country; 8° It will enable me to render greater services at the South in future political contests; 9° On the acquisition of the island something will by this means be done towards bringing her into the democratic ranks; 10° In case of war I would be available at once for any governmental measures; 11° Significance to Spain & the European powers; 12° Relief to them from their dread of expeditions and proof that the government does not seek to encourage them." AJG to Caleb Cushing, 11 Feb. 1853, CCP; *Picayune*, 8 and 9 Feb. 1853, 2; ibid., 9 Aug. 1851, 2; William Scote Haynes to John A. Quitman, 27 Feb. 1853, JQP; Morales, *Iniciadores y primeros mártires*, 2:208; and *Republic*, 19 Feb. 1853, 3.

17. "Remarks submitted to Genl. Cushing by Genl. A. J. Gonzales," 3 Mar. 1853, CCP; and *Republic*, 23 and 26 Feb. 1853, 3.

18. Inaugural Speech, Franklin Pierce Papers, series 3, reel 6, LC.

19. Nichols, *Franklin Pierce*, 249; Spencer, *The Victor and the Spoils*, 222–24; Cooper, *Jefferson Davis*, 247; and *Florida Republican*, 21 Apr. 1853, 1.

20. Nichols, *Franklin Pierce*, 251; AJG to Caleb Cushing, 22 Mar. 1853, CCP; and J. A. Reinecke Jr., "The Diplomatic Career of Pierre Soulé," *Louisiana Historical Quarterly* (Apr. 1932): 286.

21. *Republic*, 21 Feb. 1853, 3; Spencer, *The Victor and the Spoils*, 226; A. G. Penn to Marcy, 24 Mar. 1853, William L. Marcy Papers, LC; and AJG to Caleb Cushing, 15 May 1853, CCP.

22. The government owned over a quarter million acres with live oaks in Florida, Alabama, Mississippi, and Louisiana that ensured an adequate high-quality wood supply for U.S. Navy ships. The position of live-oak agent had been created to stop poachers who illegally seized the trees and the submerged timber deposits and sold them to the British. AJG to Caleb Cushing, 27 May and 30 June 1853, CCP; and Laurence C. Walker, *The Southern Forest: A Chronicle* (Austin: University of Texas Press, 1991), 161–62.

23. AJG to Caleb Cushing, 6 July 1853, CCP.

24. May, *Quitman*, 272; *Picayune*, 26 Apr. 1853, 2; and AJG to Laurent Sigur, 14 Dec. 1854, KRC.

25. *Florida Republican*, 14 July 1853, 1; AJG to JD, 18 Dec. 1853, Lynda Lasswell Crist, ed., *The Papers of Jefferson Davis*, vol. 5, *1853–1855* (Baton Rouge: Louisiana State University Press, 1985), 54; and *Union*, 15 and 18 Feb. 1854, 3.

26. Capt. Duncan N. Ingraham, a South Carolinian who commanded the sloop *St. Louis* in Turkish waters, had demanded that an Austrian warship surrender Martin Koszta, a seized Hungarian refugee facing execution. Koszta had lived in America for two years and had declared his intention of becoming a U.S. citizen. Ingraham, claiming that Koszta was entitled to protection, prepared for a naval battle. The dispute was settled when it was agreed that the French consul at Constantinople should take care of Koszta until the issue was resolved. Secretary of State Marcy denounced the Austrian kidnapping as "an atrocious outrage" and shielded Koszta. The public was pleased by this action, but it aroused the jealousy of some of Marcy's partisans. *Intelligencer*, Oct. 1 1853; Allan Nevins, *Ordeal of the Union: A House Dividing, 1852–1857* (New York: Charles Scribner's Sons, 1947), 61; AJG to President Ulysses S. Grant, 8 Jan. 1876, Letters of Application and Recommendation during the Administration of Ulysses S. Grant, 1869–77, RG 59, NA; *New York Times*, 6 Aug. 1853; *Mobile Daily Advertiser*, 18 Aug. 1853, 2; *Harper's*, Oct. 1853, 692; Reinecke, "The Diplomatic Career of Pierre Soulé," 286–87; Amos A. Ettinger, *The Mission to Spain of Pierre Soulé, 1853–1855; A Study in the Cuban Diplomacy of the United States* (New Haven: Yale University Press, 1932), 167–77; Morgan Dix, *Memoirs of John Adams Dix* (New York: Harper and Brothers, 1883), 1:277; Franklin Pierce to John A. Dix, 3 Sept. 1853, John

A. Dix Collection, Columbia University; John A. Dix to Franklin Pierce, 7 Sept. 1853, ibid.; AJG to Caleb Cushing, 20 Oct. 1853, CCP; *United States v William Hardy et al.*; William H. Robertson to William Marcy, 23 Nov. 1853, DCH; T. W. Downs to William Marcy, 23 Dec. 1853, MLS; and E. Warren Moise to William Marcy, 24 Dec. 1853, ibid.

27. AJG to Caleb Cushing, 18 Dec. 1853, CCP; AJG to JD, 18 Dec. 1853, Crist, *The Papers of Jefferson Davis*, 5:54; and *Washington Daily Evening Star*, 5 Jan. 1854, 4.

28. "By the President of the United States: A Proclamation," 18 Jan. 1854, Attorney General's Papers, Letters Received, State Department, 1850–58, NA; and Stephen Mallory to Caleb Cushing, 19 Jan. 1854, CCP.

29. Santiago Bombalier was banished from Cuba by decree of Capt. Gen. José de la Concha in April 1851. In Madrid, he sued in vain for justice and then moved to University Place, in New York City, in April 1853. He started publishing *El Mulato* in 1854. The Cuban filibuster veterans who joined William Walker in Nicaragua included John Allen, W. C. Capers, Domingo de Goicouría, Callender Irvine Fayssoux, José Manuel Hernández Canalejos, Achilles L. Kewen, Francisco Alejandro Lainé, Louis Schlesinger, Henry Theodore Titus, and Chatham Roberdeau Wheat. William L. Marcy to Pierre Soulé, 3 Apr. 1854, Manning, *Diplomatic Correspondence*, xi, 175–78; and Santiago Bombalier to William Marcy, 2 May 1853, MLS.

30. C. Stanley Urban, "The Africanization of Cuba Scare, 1853–1855," *Hispanic American Historical Review* (Feb. 1957): 29–45; Thomas, *Cuba*, 223; and *Union*, 25 Apr. 1854, 3.

31. *Union*, 3 May 1854, 3.

32. Ibid., 7 May 1854, 3.

33. Ibid., 1 June 1854, 2; AJG to Caleb Cushing, 20 June 1854, CCP; *Intelligencer*, 22 June 1854; *Union*, 22 June 1854, 3; and AJG to Laurent Sigur, 14 Dec. 1854, KRC.

34. John T. Pickett was later appointed general of the Hungarian Army. In 1861, he was Confederate commissioner to Mexico and later became assistant adjutant general to Confederate general John Breckenridge. In 1863, he was secretary of the First Confederate Peace Mission and afterward special envoy extraordinary to Mexico (1864–65). After the war, he practiced law in Washington, D.C., where he died on October 18, 1884, of cerebral hemorrhage, after a six-year illness, and was interred in the Congressional Cemetery. *New York Times*, 22 and 26 June 1854, 4; and *Picayune*, 4 July 1854, 2.

35. John Sidney Thrasher, after the Quitman expedition disbanded, traveled throughout Latin America as a newspaper correspondent and returned to New York to edit *El Noticioso*, a Spanish-American journal. In November 1860, he moved to Texas and married a wealthy woman. Thrasher was appointed in March 1863 as superintendent for the Associated Press in Atlanta. He returned to New York City after the war and edited Frank Leslie's *Ilustración Americana*. Thrasher then moved to Galveston, Texas, where he died on Nov. 10, 1879. *Picayune*, 2 and 4 July 1854, 2;

*Courier*, 11 Mar. 1863, 2; and Wilson and Fiske, *Appleton's Cyclopedia of American Biography*, vol. 6.

36. Appointment Records, Applications and Recommendations for Office, Registers of Application for Appointment to Miscellaneous Federal Offices, 1813–23 and 1834–89, vol. 3, entry 763, RG 59, NA; AJG Memorandum [June 1889], GFP; WE to MAE, 2 Mar. 1855, EGP; Mary W. Wayne to MAE, 7 Aug. 1854, EGP; and JD to AJG, 28 Oct. and 20 Nov. 1854, Jefferson Davis Papers, DUL, hereinafter cited as JDD.

37. *Union*, 19 Nov. 1854, 3; and JD to AJG, 27 Nov. 1854, KRC.

38. AJG to Laurent Sigur, 14 Dec. 1854, KRC.

39. Mary E. Pinckney wrote Harriett Elliott in January 1856 that her brother remembered General Gonzales "as an agreeable gentleman he met at Oak Lawn, a year or two ago." Mary E. Pinckney to Harriett Gonzales, 31 Jan. 1856, EGP. Diego González, *Historia Documentada de los Movimientos Revolucionarios por la Independencia de Cuba de 1852 a 1867* (Havana: Imprenta "El Siglo XX," 1939), 2:232–33; May, *Quitman*, 289–90; and *Picayune*, 19 and 23 Jan., 14 and 17 Feb. 1855, 2.

40. The letters between Gonzales and Emily Elliott have not been found and were probably destroyed years later by her or her relatives. Correspondence between them existed, and Gonzales made reference to Hattie of the way Emmie dated her letters. AJG to Harriett Elliott, 13 Jan. 1856, EGP; Caroline Elliott to EEL, 2 July [1853], ibid.; WE to MAE, 2 Mar. 1855, ibid.; EEL to MAE, 31 July 1855, ibid.; and MAE to WE, 24 Oct. 1859, ibid.

41. EEL to Harriett Elliott, [13 May 1855], EGP; and WE to Mrs. Phoebe Elliott, 12 May 1855, ibid.

42. William Elliott's mother, eighty-four-year-old Phoebe Elliott, passed away two days after his departure for Europe. WE to Mrs. Phoebe Elliott, 12 May 1855, EGP; Harriett Elliott to EEL, 13 June [1855], ibid.; and MAE to Quatitle [William Elliott Jr.], 2 July 1855, ibid.

43. EEL to MAE, 31 July 1855, EGP; WEJ to REE, 30 Aug. 1855. EGP; and Cohen, *Historic Springs of the Virginias*, 170, 177.

44. WE to MAE, 1 Jan. 1855, EGP; Gonzales, *Heaven Revealed*, 7; AJG to Harriett Elliott, 13 and 27 Jan. 1856, EGP; MBJ to MAE [Dec. 1855], ibid.; and S. T. Shugert to Robert McClelland, 29 Aug. 1855, Department of Interior, Appointments Division, Miscellaneous Letters Received, 1849–1907, box 1, entry 23., RG 48, NA.

45. Emily Elliott's perpetual love-hate attitude toward Gonzales has been difficult to decipher. All of their correspondence, as well as most of her letters deriding him to his children, were excised from the family correspondence before the collection was donated to the University of North Carolina at Chapel Hill. William E. Gonzales informed Prof. R. L. Meriwether that he had burned the family correspondence.

Jones, *Stormy Petrel*, 318. Cooper, *Jefferson Davis*, 260; and AJG to Harriett Rutledge Elliott, 20 and 27 Jan., 24 Feb., and [ca. Feb.] 1856, EGP.

46. Harriett Elliott to AJG, 10 Feb. [1856], GFP; *Courier*, 14 and 15 Apr., 1856, 4; Pedro J. Guiteras madrigals, 19 Apr. 1856, EGP; James Petigru Carson, *Life, Letters, and Speeches of James Louis Petigru: The Union Man of South Carolina* (Washington, D.C.: W. H. Lowdermilk, 1920), 317; John B. O'Neall to WE, 15 Apr. 1856, EGP; "Marriage Settlement of Ambrosio Jose Gonzales and Harriet Rutledge Elliott, 17 Apr. 1856, ibid.; and wedding notice in the *Courier*, 24 Apr. 1856, 2, and *Savannah Morning News*, 25 Apr. 1856, 2. The seven hundred dollars' annual interest on Hattie's dowry was paid in 1857, 1858, 1859, 1860, 1863, and 1864.

47. EEL to HRG, [31 Aug. 1856], EGP; HRG to ANE, 11 July [1856] ibid.; Mears and Turnbull, *The Charleston Directory* (Charleston, S.C.: Walker, Evans), 166; and *Daily Evening Star*, 2 July 1856, 2.

48. HRG to ANE, 11 July [1856], EGP; WE to MAE, July 1856, ibid.; MBJ to HRG 27 July [1856], ibid.; and Cohen, *Historic Springs of the Virginias*, 99.

49. Nevins, *Ordeal of the Union*, 510; HRG to MAE, 23 Oct. and 2 Dec. [1856], EGP; and HRG to ANE, 6 and 23 Dec. [1856] and 9 Jan. [1857], ibid.

50. The "ailing" Pedro José Guiteras Font outlived Emmie Elliott by one month and was probably at her deathbed and funeral. He passed away on Feb. 3, 1890, at the age of seventy-five, in the Charleston Hotel, where he had resided for the previous two months. His remains were sent to his native Matanzas. See City of Charleston Death Certificates, reel 27, Pedro José Guiteras, Feb. 3, 1890, Charleston County Library, Charleston, S.C. HRG to ANE, 23 Dec. [1856] EGP; EEL to HRG, 21 and 30 Dec. [1856] ibid.; and ANE to HRG, 16 Jan. [1857], ibid.

51. EEL to HRG, 29 Jan. and 9 Feb. [1857] EGP; MAE to HRG, 17 Feb. [1857], ibid.; WE to MAE, [Mar. 1857], ibid.; WE to Sigr. Alfonso, 21 Mar. 1857, ibid.; Caroline Elliott to EEL 28 Mar. 1857, ibid.; and "A Trip to Cuba," *Russell's Magazine,* Oct. 1857, 59–63; Nov. 1857, 116–23; Dec. 1857, 235–39; Jan. 1858, 322–27; Feb. 1858, 439–45; Apr. 1858, 60–69. The March 1858 issue of the magazine is missing from the microfilm, but Gonzales later reproduced it in "Natural Wealth of Cuba," *Detroit Daily Free Press*, 9 Dec. 1858.

52. EEL to "Dear Crusoe" [William Elliott Jr.], 3 June [1857], EGP; and WE to MAE, 28 July 1857, ibid.

53. AJG to HRG, 7 Sept. 1857, EGP.

54. *The State*, 2 Aug. 1893, 1; and Brooks, *Stories of the Confederacy*, 290–91.

55. *Daily Evening Star*, 25 and 28 Sept. 1857, 4; William Carl Klunder, *Lewis Cass and the Politics of Moderation* (Kent: Kent State University Press, 1996), 188, 272; and WE to MAE, 6 Sept. and 4 and 27 Oct. 1857, EGP.

56. WE to MAE, 27 Oct. 1857 and [Oct. 1857], EGP.

57. "Register of St. John-in-the-Wilderness, Flat Rock," *South Carolina Historical Magazine* (Apr. 1962), 109; *Daily Evening Star*, 18 Dec. 1857, 4; Durkin, *Confederate Navy Chief*, 96; and AJG to WE, 31 Mar. 1858, EGP.

58. AJG to James Henry Hammond, 9 May 1858, James Henry Hammond Papers, LC; "Trip to Cuba," *Russell's Magazine*, Apr. 1858, 60–69; "Poverty and Prostration of Business in Jamaica," *Evening Picayune*, 29 Mar. 1850, 1; and Robert Debs Heinl Jr., and Nancy Gordon Heinl, *Written in Blood: The Story of the Haitian People, 1492–1971* (Boston: Houghton Mifflin, 1978), 95.

59. An 1874 hurricane destroyed all the Eddingsville buildings except three. Another storm eleven years later swept the remains into the sea. Mary Wayne to WE, 20 Apr. 1858, EGP; WE to MAE, 16 Aug. 1857 and 28 July 1858, ibid.; EEL to REE, 20 Sept. and 29 Oct. [1858], ibid.; *News and Courier*, 2 Nov. 1930; Nell S. Graydon, *Tales of Edisto* (Columbia, S.C.: R. L. Bryan, 1955), 140; and Clara Childs Puckette, *Edisto: A Sea Island Principality* (Cleveland: Seaforth Publications, 1978), 19.

60. *Herald*, 8 Aug. 1858.

61. AJG to REE, 5 Aug. 1858, EGP; A. E. Gonzales, *In Darkest Cuba* (Columbia, S.C.: The State, 1922), 7; and Caroline Elliott to WE, 11 Aug. [1858], EGP.

62. WE to EEL, 26 Aug. 1858, EGP; EEL to WE, 1 Sept. [1858], ibid.; and WE to MAE, 3 Sept. 1858, ibid.

63. AJG Memorandum [June 1889], GFP. The ten Gonzales articles in the *Detroit Free Press* appeared on 25 and 30 Nov., 1858, 1, 7, 9, 18, 21, 25, and 31 Dec. 1858, and 4 Jan. 1859.

64. Elbert B. Smith, *The Presidency of James Buchanan* (Lawrence: University Press of Kansas, 1988), 77; and Allan Nevins, *The Emergence of Lincoln: Douglas, Buchanan, and Party Chaos, 1857–1859* (New York: Charles Scribner's Sons, 1950), 445–50.

65. EEL to Caroline Elliott, 28 Aug. [1859] EGP; EEL to WE, 21 Sept. [1859], ibid.; MAE to WE, 24 Oct. 1859, ibid.; EEL to REE, 26 Oct. [1859], ibid.; Francis Butler Simkins and Robert Hilliard Woody, *South Carolina during Reconstruction* (Chapel Hill: University of North Carolina Press, 1932), 240–41; AJG to W. G. Freeman, 10 Oct. 1859, Maynard Arms Company Collection, Museum of American History, Smithsonian Institution, Washington, D.C., hereinafter cited as MAC; William P. McFarland to W. C. Bestor, 5 Dec. 1859, ibid.; *The Maynard Rifle* (Washington, D.C.: G. S. Gideon, 1860), 10–11; WE to MAE, 18 Dec. 1859, ibid.; REE to WE, 25 Nov. 1860, ibid.; William C. Bee to WE, 25 Feb. 1860, ibid.; H. M. Stewart to REE, 15 Mar. 1860, ibid.; and Walter B. Edgar, ed., *Biographical Directory of the South Carolina House of Representatives*, vol. 1, *Session Lists, 1692–1973* (Columbia: University of South Carolina Press, 1974), 384. In 1860, William Elliott also bought for seven thousand dollars the Bee Hive plantation, formerly the property of a Dr. Hamilton. Bank of Charleston to WE, 30 June 1860, EGP.

66. AJG to Bestor, 5 Mar. 1860, MAC.

67. Ibid.; *Savannah Morning News*, 7 May 1860, 1; and AJG to McFarland, 7 May 1860, MAC.

68. AJG to Bestor, 10 May 1860, MAC.

69. AJG to Bestor, 16 July and 20 Aug. 1860, MAC; and McFarland to Bestor, 15 Oct. 1860, MAC.

70. *New York Times*, 7 Jan. 1853, 6; and A. E. Gonzales, *In Darkest Cuba*, 7.

71. *Mercury*, 19 Nov. 1860, 3; and U.S. War Department, *The War of the Rebellion: A Compilation of the Official Records of the Union and Confederate Armies*, vol. 1 (28), pt. 1 (Washington, D.C.: Government Printing Office, 1890), 4, hereinafter cited as OR.

72. AJG to William Gist, 19 Nov. 1860, Compiled Service Records of Confederate General and Staff Officers and Nonregimental Enlisted Men, reel 108, Ambrosio Jose Gonzales, RG 109, NA, hereinafter cited as CMR; Lonn, *Foreigners in the Confederacy*, 144; AJG to Gen. G. T. Beauregard, 17 Apr. 1861, EGP; William McFarland to Bestor, 12 Dec. 1860, MAC; Mears and Turnbull, *The Charleston Directory, 1859,* 228; and William P. McFarland to W. C. Bestor, 29 Dec. 1860, MAC.

73. OR, 1 (1), 110; *Courier*, 20 Dec. 1860, 2; AJG to G. T. Beauregard, 17 Apr. 1861, EGP; and E. Milby Burton, *The Siege of Charleston, 1861–1865* (Columbia: University of South Carolina Press, 1990), 2, 4.

74. OR, 1 (1), 108–9, 124; and ibid., (28), pt. 1, 4.

**Davis Dispute**

1. OR, 1 (1), 260; *Courier*, 4 Mar. 1861, 2; T. Harry Williams, *P. G. T. Beauregard: Napoleon in Gray* (Baton Rouge: Louisiana State University Press, 1954), 50–51; John S. Wise, *The End of an Era* (Boston: Houghton Mifflin, 1902), 330; Department of State, Civil War Amnesty, and Pardon Records, 1863–67, Amnesty Oaths, 1864–66, South Carolina, box 117, entry 1001, RG 59, NA; Woodward, ed., *Mary Chesnut's Civil War*, 52, 143; Denslow, *10,000 Famous Freemasons*, 1:73; and AJG to PGTB, 17 Apr. 1861, EGP.

2. OR, 1 (1), 11, 25, 190, 201, 251, 266, 272, 289, 291, 301, 304; PGTB to Dr. W. N. Mercer, 13 Apr. 1861, PBP; and Burton, *Siege of Charleston*, 35.

3. Sixty years later, son Ambrose E. Gonzales wrote an erroneous version of these events, claiming that his father was "too impatient to wait for the train, rode on horseback thirty miles to Charleston to offer his sword to the State of his adoption." A. E. Gonzales, *In Darkest Cuba*, 8. *Mercury*, 13 Apr. 1861, 2; AJG to PGTB, 17 Apr. 1861, EGP; *Courier*, 22 July 1861, 4; Rolls of South Carolina Staff and Confederate Officers, vol. 1, South Carolina Department of Archives and History, Columbia, hereinafter cited as SCA; *Courier*, 13 Apr. 1861, 2; *Mercury*, 16 Apr. 1861, 2; Alfred Roman, *The Military Operations of General Beauregard in the War between the States, 1861 to 1865*

(New York: Harper and Brothers, 1884), 1:42; Robert S. Holzman, *Adapt or Perish: The Life of General Roger A. Pryor, C.S.A.* (1976; reprint, Biloxi, Miss.: Beauvoir Press, 1992), 57; *Courier,* 12 Apr. 1861, 2; MBJ to MAE, 15 [Apr. 1861], EGP; and *Courier,* 4 May 1861, 1.

4. The location of the headquarters on Morris Island appears in OR, 1 (1), 229, 244; ibid., 293, 295, 296, 302, 308, 312; ibid., (53), 140; AJG to PGTB, 17 Apr. 1861, EGP; AJG to PGTB, 18 Apr. 1861, RG 109, vol. 2 (256), 183; Roman, *Military Operations,* 1:41, 44; U.S. Congress, *Biographical Directory,* 641; Edward McCrady, *Cyclopedia of Eminent and Representative Men of the Carolinas of the Nineteenth Century* (Madison, Wis.: Brant and Fuller, 1892), 1:88–90; *Courier,* 22–24 Apr. 1861, 2; Johnson Hagood, *Memoirs of the War of Secession* (Columbia, S.C.: The State, 1910), 31; and Patricia L. Faust, ed., *Historical Times Illustrated Encyclopedia of the Civil War* (New York: Harper Perennial, 1991), 70.

5. OR, 1 (1), 29, 309, 313–14.

6. PGTB to Samuel Cooper, 27 Apr. 1861, reel 8, PBP; OR, 1 (1), 317; *Mercury,* 16 Apr. 1861, 2; *Courier,* 16 and 18 Apr. 1861, 1; and ibid., 2 May 1862, 4.

7. AJG to PGTB, 17 Apr. 1861, EGP; RG 109, 9 (90), petition no. 1888; and Jefferson Davis to AJG, 7 May 1861, in possession of Elliott G. McMaster.

8. *Annual Report of the Adjutant General to the Governor of the State of Louisiana, November 1861.* (Baton Rouge: J. M. Taylor, State Printer, 1861), 44; William Frayne Amann, *Personnel of the Civil War,* vol. 1, *The Confederate Armies* (New York: Thomas Yoseloff, 1968), 40; Compiled Service Records of Confederate Soldiers Who Served in Organizations from the State of Louisiana, M320, roll 389, RG 109, NA; and Cuban Rifles Company, Confederate Units Vertical File, Jackson Barracks, New Orleans; Vault Book 14 and 25, ibid.

9. José Agustín Quintero, son of a wealthy Cuban tobacco planter and an English mother, was born in Havana in 1829. He was educated at Harvard College, where he became friends with Prof. Henry Wadsworth Longfellow, the popular American poet, and essayist Ralph Waldo Emerson; he translated much of their work. Returning to Cuba in 1848, Quintero was imprisoned for his role in an independence conspiracy in 1852. He escaped to New Orleans and was active in filibuster activities. Quintero studied law in Richmond, Texas, under Mirabeau Buonaparte Lamar, and later took up the legal practice. Residing in New York City as the editor of *El Noticioso* when the Civil War began, he returned to Texas and enlisted as a private in the Quitman Rifles in Austin. He went to Virginia with that unit until Jefferson Davis assigned him a diplomatic mission to Mexico from 1861 to 1864. See Marbán, *José Agustín Quintero*; J. A. Quintero to Henry W. Longfellow, 21 Nov. 1860, Henry Wadsworth Longfellow Papers, Houghton Library, Harvard University; Confederate States of America, containers 8, 11, 17, 58, and 60, LC; Clement A. Evans, ed., *Confederate Military History,* vol. 13, *Louisiana* (1899; reprint, Wilmington, N.C.: Broadfoot, 1988), 556–57; Glenn

R. Conrad, *A Dictionary of Louisiana Biography* (New Orleans: Louisiana Historical Association, 1988) 2:669–70; *Picayune*, 8 Sept. 1885, 4; James Morton Callahan, *The Diplomatic History of the Southern Confederacy* (Baltimore: Johns Hopkins Press, 1901), 76; Lonn, *Foreigners in the Confederacy*, 81–83; Frank Lawrence Owsley, *King Cotton Diplomacy: Foreign Relations of the Confederate States of America* (Chicago: University of Chicago Press, 1959), 113–33; James W. Daddysman, *The Matamoros Trade: Confederate Commerce, Diplomacy, and Intrigue* (Newark: University of Delaware Press, 1984), 43–64, 86–92, 138–41, 182–83; Remos y Rubio, *Historia de la literatura cubana*, 2:270–76; and Donald E. Herdeck, *Caribbean Writers: A Bio-Biographical-Critical Encyclopedia* (Washington, D.C.: Three Continents Press, 1979), 852.

10. Paul Francis de Gournay was born in Cuba on March 15, 1828, of French parents, Joseph B. de Gournay and Marie La Fevre. When he was twenty-three years old, he became a secret courier for Gen. Narciso López, carrying messages for the independence conspirators between Havana and New Orleans. After López was captured and executed in 1851, de Gournay settled in New Orleans, married a local woman, and worked as a reporter for the *Picayune*. When Louisiana seceded from the Union in January 1861, de Gournay enlisted in the state militia and was appointed quartermaster of the Orleans Artillery in Fort Jackson. Within two months, he was commissioned a captain in the Orleans Independent Artillery and transferred to Pensacola, Florida. During the Virginia Peninsula Campaign, de Gournay's artillery unit helped repulse the Union advance led by Gen. George Brinton McClellan. As a result, he was promoted to lieutenant colonel and given command of the Louisiana 12th Heavy Artillery Battalion. In 1862, his unit was transferred to Port Hudson, Louisiana, to impede the Union navy from gaining control of the Mississippi River. De Gournay's artillery battalion fought against Union ironclads for nine months during the siege of Port Hudson, until the Confederate surrender at Vicksburg. De Gournay spent the next two years as a prisoner of war in various camps. After being paroled in 1865, he lived in France for two years and then moved to Baltimore. He was editor of the *Catholic Mirror* newspaper, wrote articles, translated books, taught French and Spanish, and served as vice consul to France for seven years. He died on July 26, 1904, and is buried in the Confederate Section H-8 of Loudon Park Cemetery in Baltimore. Mauriel Joslyn, "Well-Born Lt. Col. Paul Francois de Gournay was the South's adopted 'marquis in gray,'" *America's Civil War*, Sept. 1995, 8; 1860 Louisiana Free Census, New Orleans, Fifth Ward, 252, NA; Maryland Biographical File, Enoch Pratt Free Library, Baltimore; *Woods's Baltimore City Directory, 1870* (Baltimore: John W. Woods, 1870), 149; and Certificate of Death, B71001, 26 July 1904, Baltimore City Deaths, reel 48126, Maryland State Archives, Annapolis.

11. Confederate first lieutenant Enrique B. D'Hamel was in Company G, 33rd Regiment, Texas Cavalry. His brief memoirs, *Adventures of a Tenderfoot*, were published in 1914.

12. Loreta Janeta Velázquez, *The Woman in Battle: A Narrative of the Exploits, Adventures, and Travels of Madame Loreta Janeta Velazquez, Otherwise Known as Lieutenant Harry T. Buford, Confederate States Army* (Richmond: Dustin, Gilman, 1876); and Richard Hall, *Patriots in Disguise: Women Warriors of the Civil War* (New York: Marlowe, 1994), 107–17.

13. Julius Peter Garesché du Rocher was born on a plantation near Havana in 1821. His family moved to New York six years later, and he entered the military academy at West Point in 1837. Garesché graduated in 1841 and was promoted to second lieutenant of artillery. While stationed in Fort Monroe, Virginia, in 1842, he earned an A.M. degree from Georgetown College. Garesché fought in the Mexican War (1846–48) as a first lieutenant of artillery. He was later garrisoned in various posts in Texas and New Mexico, and was promoted in 1855 to captain and assistant adjutant general. In 1861, Garesché was a major in the adjutant general's office in Washington, D.C. He was promoted in July 1862 to lieutenant colonel and chief of staff of Maj. Gen. William S. Rosecrans, commanding the Army of the Cumberland. During the Battle of Stones River, Tennessee, on December 31, 1862, when Rosecrans and his staff galloped to the front lines during a critical moment, Garesché was hit in the face by an artillery shell and decapitated above the jaw. His engraved officer's sword is at the Smithsonian Museum of American History. See Louis Garesché, *Biography of Lieut. Col. Julius P. Garesché, Assistant Adjutant-General, U.S. Army* (Philadelphia: J. B. Lippincott, 1887); George W. Cullum, *Biographical Register of the Officers and Graduates of the U.S. Military Academy at West Point, N.Y. from Its Establishment, in 1802, to 1890* (Boston: Houghton Mifflin, 1891), 2:81–82; Robert Underwood Johnson and Clarence Clough Buel, eds., *Battles and Leaders of the Civil War*, vol. 3 (New York: Century Company, 1887), 623; William M. Lamers, *The Edge of Glory: A Biography of General William S. Rosecrans, U.S.A.* (New York: Harcourt, Brace, and World, 1961), 232–34; and James Lee McDonough, *Stones River: Bloody Winter in Tennessee* (Knoxville: University of Tennessee Press, 1980), 115–16.

14. Federico Fernández Cavada was born in Cienfuegos, Cuba, in 1831. Years later, after his father's death, the family, which now included brothers Adolfo and Emilio, moved to their mother's native city of Philadelphia. When the Civil War broke out, Adolfo joined as a private in Montgomery's Company, Commonwealth Heavy Artillery, Pennsylvania, and three months later was named captain, Company C, of the 23rd Pennsylvania Infantry, where Federico had volunteered as a captain of Company K. Federico was promoted in September 1862 to lieutenant colonel of the 114th Pennsylvania Infantry, which he led during the Battle of Gettysburg, and was taken prisoner when the Confederates overran the Peach Orchard. Adolfo was also at Gettysburg, on the staff of Gen. Andrew A. Humphreys. He ended the war with the brevet rank of lieutenant colonel. Federico spent six months in Richmond's Libby prison, until he was exchanged, and then wrote a memoir titled *Libby Life*. He resigned from the army on

June 19, 1864, and was appointed U.S. vice consul in Trinidad, Cuba. Federico gave up this post on Feb. 9, 1869, and with Adolfo joined the Cuban rebels fighting for independence. He held the rank of general when he was captured by the Spaniards and executed on July 1, 1871. Adolfo, also a general, died six months later of yellow fever. Adolfo Fernández Cavada Diary, 1861–63, in possession of Fernando Fernández Cavada, Santo Domingo, Dominican Republic; Index to Compiled Service Records of Union Soldiers Who Served in Organizations from Pennsylvania, M-554, reel 19, NA; Samuel P. Bates, *History of the Pennsylvania Volunteers, 1861–5* (Harrisburg: B. Singerly, 1870), 1183–88; Harry W. Pfanz, *Gettysburg: The Second Day* (Chapel Hill: University of North Carolina Press, 1987); and Portell Vilá, *Vidas de la Unidad Americana*, 117–29. Ella Lonn, *Foreigners in the Union Army and Navy* (Baton Rouge: Louisiana State University, 1951), 314, mentions Garesché as being Catholic, but not as a Cuban, and makes omission of the Fernández Cavada brothers.

15. RG 109, vol. 2 (256), 183, NA; ibid., (263), 181–82; OR, 1 (28), 1:69; ibid., (53), 148, 152, 157–58; Lynda Lasswell Crist, *The Papers of Jefferson Davis,* vol. 7, *1861* (Baton Rouge: Louisiana State University Press, 1992), 109; *Courier,* 20 Apr. 1861, 1; ibid., 22–24 Apr. 1861, 2; ibid., 27 Apr. 1861, 4; *Mercury,* 29 Apr. 1861, 2; and WE to MAE, 7 [May 1861], EGP.

16. "Extracts from Inspector General A. J. Gonzales' Report to Gen. Beauregard Relative to Troops in Morris Island, S.C., May 8th, 1861," Ambrosio Jose Gonzales Papers, South Caroliniana Library, University of South Carolina, hereinafter cited as AGP; OR, 1 (53), 153, 174–75; Hagood, *Memoirs of the War of Secession,* 30–31, 36; and *Courier,* 30 Apr. and 31 May 1861, 1.

17. *Courier,* 11 and 15 May 1861, 2; *Mercury,* 13 and 15 May 1861, 2; OR, 1 (53), 165, 167–68; Virginia C. Holmgren, *Hilton Head: A Sea Island Chronicle* (Hilton Head Island, S.C.: Hilton Head Island Publishing, 1959), 80; and Burton, *Siege of Charleston,* 84.

18. OR, 1 (53), 171; PGTB to JD, 17 May 1861, P. G. T. Beauregard Papers, Emory University; and AJG to JD, 18 May 1861, Jefferson Davis Papers, Tulane University, hereinafter cited as JDP.

19. RG 109, vol. 2 (263), 267; *Mercury,* 28 May 1861, 2; *Courier,* 28 and 31 May 1861, 2; Gilbert Sumter Guinn, "Coastal Defense of the Confederate Atlantic Seaboard States, 1861–1862: A Study in Political and Military Mobilization" (Ph.D. diss., University of South Carolina, 1973), 75; and Burton, *Siege of Charleston,* 64.

20. *Courier,* 7 and 10 June 1861, 2; *Union,* 20 Apr. 1854, 2; Gonzales Requisition, 10 June 1861, Ordnance Department Vouchers, 1860–65, box 2, Confederate Records, SCA; *Courier,* 13 June 1861, 1; and ibid., 18 June 1861, 2.

21. *Courier,* 24 June 1861, 2; *Mercury,* 24 June 1861, 2; *Courier,* 25 June 1861, 2; Evans, *Confederate Military History,* 5:849; OR, 1 (6), 278; U.S. War Department, *Official Records of the Union and Confederate Navies in the War of the Rebellion,* series 1, vol.

12 (Washington, D.C.: Government Printing Office, 1901), 405, hereinafter cited as ORN; and AJG to F. W. Pickens, 24 June 1861, E. M. Law Papers, SHC.

22. AJG to F. W. Pickens, 24 June 1861, E. M. Law Papers, SHC; *Courier*, 24 June 1861, 2; *Mercury*, 24 June 1861, 2; *Mercury*, 24 June 1861, 1; and *Courier*, 26 June 1861, 1.

23. REE to Callie Elliott, 12 Aug. 1861, EGP; WEJ to Mrs. Anne Elliott, 7 Oct. 1861, EGP; *Herald*, reprinted in *Courier*, 19 July 1861, 4; *Mercury*, 16 July 1861, 1; and ORN, 1 (12), 202.

24. *Courier*, 24 July 1861, 2; and *Mercury*, 20 July 1861, 1.

25. *Courier*, 27 July 1861, 1; *Mercury*, 27 July 1861, 1; ibid., 29 July 1861, 2; ibid., 5 and 12 Aug. 1861, 2; and OR, 1 (6), 267.

26. *Mercury*, 8 Oct. 1861, 1; *Courier*, 25 Apr. 1863, 1; OR, 1 (53), 179; and Woodward, ed., *Mary Chesnut's Civil War*, 143, 149, 157.

27. Ibid., 177; *Courier*, 2 Sept. 1861, 4; ibid., 5 Sept. 1861, 1, 2; ibid., 12 Sept. 1861, 2; OR, 1 (6), 278–83; and ORN, 1 (12), 271.

28. Confederate Ordnance Bureau, *The Field Manual for the Use of the Officers on Ordnance Duty* (Richmond: Ritchie and Dunnavant, 1862), 3; Warren Ripley, *Artillery and Ammunition of the Civil War* (Charleston, S.C.: Battery Press, 1984), 17; Jack Coggins, *Arms and Equipment of the Civil War* (Wilmington, N.C.: Broadfoot, 1990), 61–65; Dean S. Thomas, *Cannons: An Introduction to Civil War Artillery* (Gettysburg, Pa.: Thomas Publications, 1985), 2; and Albert Mauncy, *Artillery through the Ages: A Short Illustrated History of Cannon, Emphasizing Types Used in America* (Washington, D.C.: Government Printing Office, 1990), 17–20.

29. OR, 1 (6), 280–83.

30. Brooks, *Stories of the Confederacy*, 298.

31. Davis, *Jefferson Davis*, 435–36; Steven E. Woodworth, *Jefferson Davis and His Generals: The Failure of Confederate Command in the West* (Lawrence: University Press of Kansas, 1990), 26–30; Ezra J. Warner, *Generals in Gray: Lives of the Confederate Commanders* (Baton Rouge: Louisiana State University Press, 1959), 242–43; Faust, *Historical Times*, 590; and Richard E. Beringer, Herman Hattaway, Archer Jones and William N. Still Jr., *Why the South Lost the Civil War* (Athens: University of Georgia Press, 1986), 43.

32. AJG to JD, 16 Sept. 1861, Charles Coffin Papers, New England Historic Genealogical Society, Boston.

33. AJG to REE, 23 Sept. 1861, EGP; *Courier*, 23 Sept. 1861, 2; ibid., 25 Apr. 1863, 1; and OR, 1 (51), pt. 2, 322.

34. *Mercury*, 8 Oct. 1861, 1; *Courier*, 8 Oct. 1861, 2; Dunbar Rowland, ed., *Jefferson Davis Constitutionalist: His Letters, Papers, and Speeches* (Jackson: Mississippi Department of Archives and History, 1923), 5:143–44; and RG 109, vol. 1 (47), 9 Oct. 1861.

35. TSE to WE, 15 [Sept. 1861], EGP; WEJ to MAE, 7 Oct. 1861, ibid.; and WE to WEJ, 24 Sept. 1861, ibid.

36. Rowland, *Jefferson Davis Constitutionalist*, 143–44.

37. Ibid.

38. In November 1861, Thomas Drayton commanded the unsuccessful defense of Port Royal and was later harshly criticized by his superior officers for inefficiency. Ibid., 144; Warner, *Generals in Gray*, 75–76; *Courier*, 17 Aug. 1861, 1; OR, 1 (53), 181; and Francis W. Pickens to JD, 24 Nov. 1861, Gov. Francis W. Pickens Papers, box 1, SCA.

39. Crist, *The Papers of Jefferson Davis*, 7:317; and Davis, *Jefferson Davis*, 366.

40. David Potter, "Jefferson Davis and the Political Factors in Confederate Defeat," in *Why the North Won the Civil War*, ed. David Donald (Baton Rouge: University of Louisiana Press, 1960), 91–114; Nevins, *The Emergence of Lincoln*, 141, 264; Jefferson Davis, *The Rise and Fall of the Confederate Government* (New York: Da Capo Press, 1990), 1:ii; Clement Eaton, *Jefferson Davis* (New York: Free Press, 1977), 272; Davis, *Jefferson Davis*, 435, 442; Woodworth, *Jefferson Davis and His Generals*, 315; and John H. Reagan, *Memoirs, With Special Reference to Secession and the Civil War* (New York: Neale, 1906), 120.

41. *Courier*, 31 Oct. 1861, 2; Burton, *Siege of Charleston*, 69; Robert Carse, *Department of the South: Hilton Head Island in the Civil War* (Columbia, S.C.: State Printing, 1961), 9; *Courier*, 5 Nov. 1861, 2; *Mercury*, 6 Nov. 1861, 2; R. S. Ripley to AJG, 16 May 1862, RG 109, Letters Received by the Confederate Adjutant and Inspector General, 1862, G551, NA, hereinafter cited as AIG; *Courier*, 14 Feb. 1862, 2; Woodward, ed., *Mary Chesnut's Civil War*, 227; OR, 1 (6), 11, 185, 285; ORN, 1 (12), 229–31; and TSE to MAE [11 Nov. 1861], EGP.

42. South Carolina, vol. 6, 212, R. G. Dun & Co. Collection, BLH; J. C. Hemphill, *Men of Mark in South Carolina*, vol. 1 (Washington, D.C.: Men of Mark Publishing, 1907), 186; Mears and Turnbull, *Charleston Directory, 1859*, 106; Ordnance Department Vouchers, 1860–65, box 3, Confederate Records, SCA; OR, 1 (6), 324, 326; AJG to Judah P. Benjamin, 22 Nov. 1861, RG 109, Letters Received by the Confederate Secretary of War, 1861–65, roll 16, NA, hereinafter cited as CSW; R. S. Ripley to AJG, 16 May 1862, G551, AIG; Yates Snowden, *History of South Carolina* (Chicago: Lewis, 1920), 2:703; *Courier*, 16 Oct. 1861, 2; ibid., 15 Jan. 1862, 2; ibid., 25 Apr. 1863, 1; and *Cyclopedia of Eminent and Representative Men of the Carolinas*, 1:148–49.

43. Filibuster activist Randolph Spalding died of pneumonia in Savannah on March 17, 1862. *Mercury*, 19 Mar. 1862, 4. OR, 1 (6), 3–29, 187, 317; ibid., (28), 1:37; ibid., (53), 185; ORN, 1 (12), 261–319, 351–52; *Mercury*, 8 Nov. 1861, 2; *Courier*, 8 Nov. 1861, 1; ibid., 18 Nov. 1861, 2; ibid., 22 and 28 Nov. 1861, 1; Rose, *Rehearsal for Reconstruction*, 104; and Simkins and Woody, *South Carolina during Reconstruction*, 14.

44. TSE to MAE [11 Nov. 1861], EGP; "Schedule of Losses Sustained by Thomas R. S. Elliott of the Town of Beaufort in the State of South Carolina, by the Invasion of the Enemy, in the Years 1861 and 1862," Miscellaneous Claims for Damages, 1861–64, Confederate Records, SCA; WE to William Whaley, Chairman of the Commission to Ascertain the Losses on the Seaboard Islands, 1 Jan. 1862, ibid.; MBJ to MAE [15 June 1862], EGP; WE to REE, 15 Dec. 1861, ibid.; Rose, *Rehearsal for Reconstruction*, 16, 106; Simkins and Woody, *South Carolina during Reconstruction*, 3; and Julie Saville, *The Work of Reconstruction: From Slave to Wage Laborer in South Carolina, 1860–1870* (Cambridge: Cambridge University Press, 1994), 34.

45. *Mercury*, 8 Nov. 1861, 2; Clement A. Evans, ed., *Confederate Military History*, vol. 6, *South Carolina* (1899; reprint, Wilmington, N.C.: Broadfoot, 1988), 37–38; Robert E. Lee, *Recollections and Letters of General Robert E. Lee* (Garden City, N.Y.: Doubleday, Page, 1924), 54; Freeman, *R. E. Lee*, vol. 1, 608–10, 613; OR, 1, (6), 199, 203, 208, 211, 214, 309, 312–13, 327–29, 344–45, 386; ibid. (53), 195; ORN, 1 (12), 321–23, 392–93, 399, 405–6; Guinn, "Coastal Defense of the Confederate Atlantic Seaboard States, 1861–1862," 66; *Mercury*, 3 Jan. 1862, 2; and Burton, *Siege of Charleston*, 79–80.

46. *Courier*, 16 Oct. 1861, 2; Royal," ibid., 11 and 14 Nov. 1861, 1; ibid., 15 Nov. 1861, 2; and *Mercury*, 15 Nov. 1861, 2.

47. The author visited the Huguenin plantation earthworks in the summer of 1994. The steamer *Lady Davis*, named in honor of Confederate First Lady Varina Davis, carried a twenty-four-pounder howitzer and a rifled cannon. *Courier*, 14 Mar. 1861, 2. *Courier*, 11 Nov. 1861, 1; OR, 1 (6), 324, 326; ibid., (53), 196; Evans, *Confederate Military History*, vol. 6, *South Carolina*, 36; Thomas Abram Huguenin Autobiography, DUL; REE to MAE, 17 Nov. 1861, EGP; REE to EEL, 27 Nov. 1861, ibid.; MBJ to MAE, 15 [Dec. 1861], ibid.; and *Mercury*, 29 Nov. 1861, 2.

48. AJG to Judah P. Benjamin, 22 Nov. 1861, roll 16, CSW; RG 109, 2 (42), 181–82; and ORN, 1 (12), 379–80.

49. WE to REE, 15 Dec. 1861, EGP; MBJ to MAE, 15 [Dec. 1861], ibid.; AJG to JD, 20 Mar. 1862, JDD; Warner, *Generals in Gray*, 133; Faust, *Historical Times*, 358; and OR, 1 (53), 190–91, 761–63.

50. RG 109, 2 (185), 184, 186; *Mercury*, 18 Dec. 1861, 2; and *Courier*, 27 Dec. 1861, 2.

51. OR, 1 (6), 44–75; *Mercury*, 3, 4, 6, 7, 8, 10, and 21 Jan. 1862, 2; *Courier*, 3, 4, 6, and 10 Jan. 1862, 1; ibid., 7 Jan. 1862, 4; ibid., 10 Jan. 1862, 1; John H. Kinsler to Capt. A. C. Haskell, 16 Jan. 1862, Alexander Cheves Haskell Collection, SHC; and TSE to WE, 4 Feb. 1862, EGP.

52. OR, 1 (6), 77–82, 89–90, 210–13, 225, 353; and "A Sketch of the Life of Thomas Abram Huguenin, Written at the Request of My Family," Manuscript Collection, Fort Sumter National Monument, Charleston, S.C.

53. Special Orders no. 16, 28 Jan. 1862, Department of South Carolina and Georgia, RG 109, vol. 2 (42), NA; OR, 1 (14), 181; Special Orders no. 43, 18 Mar. 1862, RG 109, vol. 2 (42), 219–20; *Courier*, 7 Jan. 1862, 4; ibid., 5 and 10 Feb. 1862, 2; ibid., 24 Feb. 1862, 3; *Mercury*, 4 Mar. 1862, 2; *Courier*, 11 Nov. 1862, 4; TSE to WE, 7 Feb. 1862, EGP; WEJ to EEL, 11 Feb. and 6 Mar. 1862, ibid.; OR, 1 (6), 373–74; and RG 109, General Orders no. 16, 12 May 1862, Department of South Carolina, Georgia, Florida, General Orders 1861–64, box 65, NA.

54. Woodward, ed., *Mary Chesnut's Civil War*, 290; and Charles E. Cauthen, *Journals of the South Carolina Executive Councils of 1861 and 1862* (Columbia: South Carolina Archives Department, 1956), 94–95, 174.

55. W. W. Harlee to James Chesnut Jr., 25 Feb. 1862, Dearborn Collection, Confederate Officers, Houghton Library, Harvard University, hereinafter cited as DHU; *Mercury*, 13 Feb. 1862, 2; *Courier*, 25 Apr. 1863, 1; and MBJ to MAE, 10 Mar. [1862], EGP.

56. *Mercury*, 5 Mar. 1862, 2; Michael B. Ballard, *Pemberton: A Biography* (Jackson: University Press of Mississippi, 1991), 90; John C. Pemberton, *Pemberton: Defender of Vicksburg* (Chapel Hill: University of North Carolina Press, 1942), 28; Burton, *Siege of Charleston*, 91–93; General Orders no. 9, 25 Apr. 1862, RG 109, vol. 2 (42), 75; OR, 1 (6), 394, 402, 414, 420–21, 425, 427; ibid., (14), 86; ibid., (28), pt. 1, 5; "Sketch of Lucas' Battalion of Heavy Artillery," Confederate Historian, Miscellaneous Historical Sketches, 1866–98, SCA; and Hagood, *Memoirs*, 60–62.

57. OR, 1 (6), 248, 255–58, 263; Faust, *Historical Times*, 376; Warner, *Generals in Blue*, 244; Rose, *Rehearsal for Reconstruction*, 144; *Courier*, 14 Mar. 1862, 2; *Mercury*, 19 Mar. 1862, 2; and 22 Mar. and 4 Apr. 1862, 1.

58. AJG to JD, 20 Mar. 1862, JDD.

59. Warner, *Generals in Gray*, 75–76, 225, 309–10; Woodward, ed., *Mary Chesnut's Civil War*, 315–16; WE to James Louis Petigru, 27 Apr. 1862, EGP; and Carson, *Life, Letters, and Speeches of James Louis Petigru*, 444.

60. A. L. Long, *Memoirs of Robert E. Lee* (New York: J. M. Stoddart, 1886), 143; Warner, *Generals in Gray*, 191; John C. Pemberton to Samuel Cooper, 27 May 1862, CMR; AJG to Samuel Cooper, 8 May 1862, ibid.; AJG to Samuel Cooper, 28 May 1862, G551, AIG; OR, 1 (14), 556; and MBJ to MAE, [15 June 1862], EGP.

## Confederate Colonel

1. OR, 1 (14), 16, 20–28, 556; Capers, *South Carolina*, 41; *Mercury*, 21, 23, 24, 27, and 28 May 1862, 2; and Burton, *Siege of Charleston*, 99–101.

2. The Rev. Stiles Mellichamp was rector of St. James Episcopal Church. OR, 1 (14), 575, 595–96, 610.

3. OR, 1 (14), 561; and AJG to R. G. M. Dunovant, 29 May 1863, AGP.

4. AJG to John Pemberton, 10 June 1862, Ambrosio José Gonzales Manuscript, Chicago Historical Society, Chicago, vol. 3; RG 109, vol. 2 (42), 256, 288–89, 301,

310; Aide-de-Camp J. C. Taylor to AJG, 14 June 1862, OR, 1 (14), 565; Special Orders no. 79, 15 June 1862, RG 109, vol. 2 (42), 291; and OR, 1 (14), 522.

5. OR, 1 (14), 42–46, 48–50, 52, 59. 79, 95; and *Courier*, 25 Apr. 1863, 1.

6. Col. Thomas Gresham Lamar died in Charleston, of malaria fever, on Oct. 18, 1862. OR, 1 (14), 42–43 47, 51–52, 62, 65–75, 83, 88, 90, 96, 107; and Burton, *Siege of Charleston*, 104–11.

7. *Courier*, 16 and 27 June 1862, 2.

8. General Orders no. 36, 18 July 1862, RG 109, vol. 2 (42), 178, NA; *Mercury*, 16 and 19 July 1862, 2; and *Courier*, 17, 18, and 19 July 1862, 1.

9. Form no. 21 and Officers, Pay Account, 19 Aug. 1862, CMR; WE to WEJ, 26 Aug. 1862, EGP; WEJ to EEL, 13 Sept. [1862], ibid.; Stewart Sifakis, *Compendium of the Confederate Armies: South Carolina and Georgia* (New York: Facts on File, 1995), 58; and *Mercury*, 19 July 1862, 1.

10. *Courier*, 19 Nov. 1861, 2; *Mercury*, 28 Feb. 1862, 2; *Courier*, 21 Feb. and 15, 18, and 27 Mar. 1862, 2; *Mercury*, 27 Mar. and 3 Apr. 1862, 4; *Courier*, 2 and 7 Apr. 1862, 1; Simkins and Woody, *South Carolina during Reconstruction*, 9; John C. Pemberton to George W. Randolph, 11 Aug. 1862, CMR; and General Orders no. 45, 16 Aug. 1862, RG 109, vol. 2 (43), 6, NA.

11. Register of Appointments, Confederate States Army, Col. A. J. Gonzales, 29 Aug. 1862, CMR; Officer's Pay Account, 12 May 1863, ibid.; and Harriett Gonzales to AJG, 2 Sept. [1862], GFP.

12. *Courier*, 1 Aug. 1862, 2; Ballard, *Pemberton*, 107–10; Basso, *Beauregard*, 202; OR, 1 (14), 601; Williams, *P. G. T. Beauregard*, 164; *Courier*, 25 Aug. 1862, 2; ibid., 2 and 4 Sept. 1862, 1; AJG to PGTB, 6 Sept. 1862, reel 5, frame 497, PBP; RG 109, vol. 2 (35), 458, NA; and AJG to PGTB, 6 Sept. 1862, AGP.

13. The court martial had been correct in its appraisal that the incident would not occur again. Two months after John Dunovant's dismissal, Governor Pickens appointed him colonel of the South Carolina 5th Cavalry Regiment. When the unit was sent to Virginia in March 1864, Dunovant fought in numerous engagements, distinguishing himself at Cold Harbor and the Petersburg Campaign, prompting Jefferson Davis to appoint him brigadier general on August 22, 1864. Dunovant was killed five weeks later while leading a cavalry charge in Virginia. Warner, *Generals in Gray*, 78–79; Faust, *Historical Times*, 230; RG 109, vol. 2 (42), 6, 356–57, 363; ibid., (30), 138; and Johnson Hagood, AJG and James M. Cullough to President Davis, 23 Nov. 1862, RG 109, Compiled Service Records of Confederate Generals and Staff Officers and Non-regimental Enlisted Men, roll 81, John Dunovant, NA.

14. The South Carolina units stationed at Adams Run during this time were Capt. Stephen Elliott Jr.'s Beaufort Volunteer Artillery; four companies of Col. John L. Black's South Carolina 1st Cavalry Regiment; Maj. Robert J. Jefford's South Carolina 17th Cavalry Battalion; Capt. William L. Trenholm's Rutledge Mounted Riflemen Cavalry

Battalion; Col. William C. Heyward's South Carolina 11th Infantry Regiment; and Maj. Benjamin B. Smith Jr.'s South Carolina 2d Infantry Battalion Sharpshooters. Sifakis, *Compendium of the Confederate Armies: South Carolina and Georgia*, 12, 38–39, 48, 58. WE to WEJ, 26 Aug. 1862, EGP; Harriet Gonzales to AJG, 2 Sept. [1862], GFP; A. E. Gonzales, *In Darkest Cuba*, 9; and Inmates of the Convalescent Hospital to Miss Elliott, 1 July 1864, EGP.

15. The gun carriage department of the Noble Brothers and Company cannon foundry in Rome, Georgia, had been swept by fire in August 1862, and due to faulty casting, the Confederacy and the State of Georgia had canceled that summer their orders for heavy Columbiads. Larry J. Daniel and Riley W. Gunter, *Confederate Cannon Foundries* (Union City, Tenn.: Pioneer Press, 1977), 42–43. *Mercury*, 15 Sept. 1862, 2; Williams, *P. G. T. Beauregard*, 166; O.R., vol. 1 (14), 604; ibid., (28), 1:96; John Johnson, *The Defense of Charleston Harbor* (Charleston: Walker, Evans, and Cogswell, 1890), 22; *Savannah Republican*, 22 Sept. 1862, 1; Burton, *Siege of Charleston*, 116; and *Mercury*, 22 Sept. 1862, 2.

16. Warner, *Generals in Gray*, 232–33; RG 109, vol. 2 (22), 290–92; Requisition, May 19, 1863, CMR; Beauregard Order, 28 Sept. 1862, Pierre Gustave Toutant Beauregard Papers, DUL; and RG 109, vol. 2 (20), 176–77.

17. AJG to Thomas Jordan, 18 Oct. 1862, AGP; RG 109, vol. 2 (29), 269; RG 109, vol. 2 (29), 370; Thomas Jordan to R. S. Ripley, 6 Nov. 1862, Thomas Jordan Papers, DUL; James H. Rion to AJG, 4 Oct. 1862, James Henry Rion Papers, SCL; AJG to PGTB, 8 Oct. 1862, AGP; RG 109, vol. 2 (20), 172–73; and WEJ to EEL, 13 Oct. 1862, EGP.

18. RG 109, vol. 2 (29), 308; George W. Randolph to PGTB, 10 Nov. 1862, PBP; and OR, 1 (14), 673.

19. OR, 1 (14), 144–52, 155, 178, 180–88.

20. Chief Engineer David B. Harris graduated from West Point in 1833 and taught engineering there for two years. He was a planter from 1845 until joining the Confederacy in February 1862. He was sent to Charleston, with the rank of major, on October 7, 1862, to serve on Beauregard's staff and promoted to lieutenant colonel in May 1863 and to colonel five months later. He died from yellow fever a year later in Summerville, S.C. General Beauregard regarded Harris as "one of my best and most valued friends." *Mercury*, 11 and 12 Oct. 1864, 2; *Courier*, 12 and 24 Oct. 1864, 1; RG 109, vol. 2 (20), 178, 425; ibid., (30), 11, 19; OR, 1, (28), pt. 1, 4; WEJ to Father, 31 Oct. 1862, EGP; RG 109, vol. 2 (22), 243–44; and Thomas Jordan to H. W. Mercer, 13 Nov. 1862, Thomas Jordan Papers, DUL.

21. Thomas Jordan to H. W. Mercer, 4 Nov. 1862, DHU; Faust, *Historical Times*, 487; RG 109, vol. 2 (20), 30, 301; and TSE to EEL, 13 Nov. 1863, EGP.

22. Warren Ripley, ed., *Siege Train: The Journal of a Confederate Artilleryman in the Defense of Charleston* (Columbia: University of South Carolina Press, 1986), 311–16; Sifakis, *Compendium of the Confederate Armies: South Carolina and Georgia*, 10; PGTB

to AJG, 13 Nov. 1862, RG 109, vol. 2 (30), 22; PGTB to R. S. Ripley, 13 Nov. 1862, ibid., 19–20; PGTB to Captains P. C. Warrick and J. L. Fraser, 27 Sept. 1864, DHU; PGTB to JD, 13 Dec. 1864, reel 7, PBP; Capt. T. A. Huguenin to Lt. Col. A. Roman, 14 Dec. 1864, Pierre Gustave Toutant Beauregard Papers, SCL; Warner, *Generals in Gray*, 257; OR, 1 (6), 366; RG 109, vol. 2 (20), 185, 191, 198; ibid., vol. 2 (30), 353; and *New Orleans Directory, 1851*, 144.

23. RG 109, vol. 2 (30), 40, 43; ibid., (20), 180, 182, 196; and ibid., (22), 269.

24. AJG to General Ripley, 4 Dec. 1862, RG 109, vol. 2 (30), 165; ibid., 141, 186–87, 226; AJG to Adjutant and Inspector General's Office, 9 Dec. 1862, RG 109, vol. 2 (30), 140; RG 109, vol. 2 (20), 184; OR, 1 (14), 726–27; and AJG to A. N. Toutant Beauregard, 13 Dec. 1862, CMR.

25. Capt. Franklin B. Du Barry died of consumption on his way to Europe on May 27, 1864, and was buried at sea. *Mercury*, 10 June 1864, 2; Col. Olin M. Dantzler was killed while leading a charge at Petersburg, Virginia, in June 1864. *Courier*, 8 June 1864, 1. OR, 1 (35), 2:110; *Courier*, 9 Dec. 1862, 2; RG 109, vol. 2 (43), 178; ibid., (30), 188–89, 197, 242; and ibid., vol. 2, (20), 185.

26. AJG to L. Jaquelin Smith, 19 Dec. 1862, roll 85, C-362, CSW; and RG 109, vol. 2, (30), 196.

27. F. L. Childs to J. Gorgas, 19 Dec. 1862, roll 85, C-362, CSW.

28. Register of Burials, bk. 1, 60, 61, and Lot Book, lot 1172 and 1173, Magnolia Cemetery, Charleston, S.C.; "Sacred to the Memory of Phoebe Caroline Elliott," 20 Dec. 1862, EGP; REE to MAE, 31 Jan. [1863] and [1 Feb. 1863], ibid.; and REE to "My Dear Sisters," [2 Feb. 1863], ibid.

29. The *New Ironsides* was described by Rear Admiral John A. Dahlgren as "a powerful but most impracticable vessel," whose great draft impeded its approach to enemy targets. Its ports allowed limited gun elevation, and its ends were not armored. OR, 1 (28), 2:55; ORN, 1 (15), 529; *Mercury*, 30 Nov. 1861, 1; ibid., 6 Mar. 1862, 1; Burton, *Siege of Charleston*, 135; and Stephen R. Wise, *Gate of Hell: Campaign for Charleston Harbor, 1863* (Columbia: University of South Carolina Press, 1994), 23–25.

30. *Courier*, 29 May 1861, 1; OR, 1 (28), 2:306; and Daniel and Gunter, *Confederate Cannon Foundries*, 66.

31. OR, 1 (14), 745–47.

32. RG 109, vol. 2 (22), 334, 389–90; and OR, 1 (14), 745–47.

33. The gunboats *Chicora* and *Palmetto State* each carried two nine-inch Dahlgren guns, two seven-inch Brooke rifles, and a crew of about sixty men and officers. Callender Irvine Fayssoux had joined filibuster Gen. William Walker in Nicaragua in 1855. He commanded the armed vessel *Granada* during a naval battle on July 23, 1856, with the Costa Rican brig *Once de Abril*, which he destroyed. Fayssoux returned to New Orleans in 1857, where he remained until October 1862. A year later, he enlisted as a private in Company E, South Carolina 17th Infantry Regiment. In

December 1863, he was honorably discharged on Sullivan's Island due to physical disability. He returned to New Orleans after the war, where he died on August 30, 1897, and was buried in Lafayette Cemetery no. 1. Evans, *Confederate Military History*, vol. 13, *Louisiana*, 409–10. OR, 1 (14), 199–210; ibid., (35), 2:287–88; and Burton, *Siege of Charleston*, 120–23, 127–29.

34. RG 109, vol. 2 (20), 194–97; ibid., (25), 23–24, 51–52, 68, 132; ibid. (30), 419, 426, 430, 434; PGTB to Adjutant and Inspector General's Office, 27 Jan. 1863, RG 109, vol. 2 (30), 391; and PGTB to AJG, 12 Feb. 1863, RG 109, vol. 2 (25), 162.

35. RG 109, vol. 2 (20), 200; ibid., (25), 120, 130; *Courier*, 20, 24, and 27 Feb. and 2 Mar. 1863, 4; MBJ to EEL, 2 Mar. [1863], EGP; and WEJ to MAE, 19 Mar. 1863, ibid.

36. OR, 1 (14), 810, 828; RG 109, vol. 2 (20), 418–19; ibid., (31), 1–2; *Courier*, 3, 7, 10, 11, 13, and 31 Mar. 1863, 4; TSE to EEL, 27 Mar. 1863, EGP; and Ordnance Voucher, 24 Mar. 1863, CMR.

37. R. S. Ripley to Department Headquarters, 22 Mar. 1863, ibid., (25), 273–74; Thomas Jordan to AJG, 1 Apr. 1863, RG 109, vol. 2 (31), 31; and Stephen Elliott Jr., to EEL, 31 Dec. 1863, EGP.

38. RG 109, vol. 2 (27), 63–64.

39. Ibid., (25), 316.

40. RG 109, vol. 2 (25), 300; G. T. Beauregard, "Defense of Charleston, South Carolina," *North American Review* (May 1886), 432; and *Courier*, 25 Apr. 1863, 1.

41. Samuel Jones, *The Siege of Charleston* (New York: Neale, 1911), 168–69; Daniel Ammen, *The Navy in the Civil War: The Atlantic Coast* (New York: Scribner, 1883), 91–92, 101; and Burton, *Siege of Charleston*, 136–37.

42. OR, 1 (14), 241–54; Jones, *Siege of Charleston*, 170–78; and Ammen, *The Navy in the Civil War*, 93–99.

43. OR, 1 (14), 241–49; ORN, (14), 230; Burton, *Siege of Charleston*, 138–48; and PGTB to Samuel Cooper, 19 May 1863, reel 8, frame 726, PBP.

44. AJG to Thomas Jordan, 13 Apr. 1863, G-666, AIG; and ORN, (14), 230.

45. The artillery officers in the department were Lt. Col. William J. Saunders, assistant inspector of light artillery; Maj. Andrew Burnet Rhett, chief of artillery, General Gist's command; Lt. Col. Delaware Kemper, chief of artillery, Gen. Hagood's command; Capt. Stephen Elliott Jr., chief of artillery, Gen. William Stephen Walker's command; Lt. Col. Charles C. Jones, chief of artillery, Gen. Hugh W. Mercer's command; Maj. M. Stanley, chief of artillery, Gen. Howell Cobb's command. The ordnance officers in the department were Maj. John G. Barnwell, C.S.A., assistant chief of ordnance; Capt. L. Jaquelin Smith, assistant chief of ordnance; Captain H. Laurens Ingraham, in charge of the Department of Chief of Ordnance; Capt. Franklin B. Du Barry, ordnance officer, First Military District; Lt. Isaac W. Hayne, ordnance officer, Second Military District; Lt. William Waight Elliott, ordnance officer, Third Military

District; Lt. John L. Boatwright, ordnance officer, Gen. James Heyward Trapier's command; Lt. A. F. Cunningham, ordnance officer, Gen. Mercer's command; Lt. W. D. Harden, ordnance officer, General Mercer's command; Lt. Thomas E. Buckman, ordnance officer, East Florida; Capt. A. F. Pope, ordnance officer, Middle Florida; Capt. Charles Pinckney, Lieutenants Charles M. Creswell, Joseph J. Legare, J. R. Russell, and Capt. Franklin B. Du Barry, General Ripley's ordnance officers; Maj. Andrew Burnet Rhett, General Gist's ordnance officer; Lieutenant Ashe, Gen. Thomas L. Clingman's ordnance officer; and Maj. J. J. Pope, inspector of ordnance, Georgia. *Courier*, 13 and 22 Apr. 1863, 2; Thomas Jordan to AJG, 16 Apr. 1863, RG 109, vol. 2 (31), 79; John Otey to AJG, 23 Apr. 1863, RG 109, vol. 2 (31), 96; and ibid., (20), 509.

46. RG 109, vol. 2 (20), 507; and ibid., (25), 316.

47. *Courier*, 25 Apr. 1863, 1.

48. Thomas Jordan to R. S. Ripley, 28 Apr. 1863, RG 109, vol. 2, (31), 111, 165; ibid., 121–22, 140; RG 109, AJG to Thomas Jordan, 2 May 1863, CMR; RG 109, vol. 2 (20), 546; *Courier*, 28 May 1863, 1; and RG 109, Department of South Carolina, Georgia, and Florida, Special Orders 1863–64, Special Order no. 97, box 66, NA.

49. *Mercury*, 6 Apr. 1861, 2; ibid., 20 June 1861, 2; ibid., 10 Oct. 1861, 2; "Gen. Beauregard and Staff," ibid., 25 Dec. 1861, 2; "General Robert E. Lee," ibid., 3 June 1863, 2; "Grinevald's Pictures of the Battles of Charleston," ibid., 7 Aug. 1863, 2; and Rose, *Rehearsal For Reconstruction*, 121.

50. OR, 1 (14), 945–46; ibid., vol. 2 (19), 145; RG 109, vol. 2 (25), 434, NA; ibid., (26), 13, 18, 36, 49, 70; ibid., (34), 126, 128; RG 109, Adjutant and Inspector General's Office, Register of Letters Received, Apr.–July 1863, vol. 30, chap. 1, ibid.; and General Orders no. 72, 27 May 1863, RG 109, Department of South Carolina, Georgia, and Florida, General Orders 1861–64, box 65, ibid.

51. The officers commanded by Gonzales at headquarters were Majors John G. Barnwell and William J. Saunders; Captains Franklin B. Du Barry, H. Laurens Ingraham, L. Jaquelin Smith, and Willis Wilkinson; Lieutenants Issac W. Hayne, William Waight Elliott, John L. Boatwright, W. W. Legare, J. R. Russell, and Charles M. Creswell. The Siege Train officers were Maj. Charles Alston Jr.; First Lt. David W. Edwards; Second Lieutenants John H. Gardner, F. M. Godbold, and Charles Elwegg Jr.; Asst. Surgeon Joseph Winthrop; Company A: Benjamin C. Webb, William H. Chapman, and James A. Brux; Company B: 1st Lt. F. W. Wilson and 2d Lt. S. P. Smith; Company C: Capt. M. B. Stanly and 1st Lt. Thomas E. Gregg. "Roster of Officers under Command of Col. A. J. Gonzales," May 1863, CMR; Vouchers, C. A. Coste and servant Elias, 3 June 1863, CMR; Ordnance Stores receipt, 3 June 1863, ibid.; Requisition for Forage, 3 June 1863, ibid.; Special Requisition no. 40, 6 and 14 July 1863, ibid.; and Chief of Staff to AJG, 19 June 1863, RG 109, vol. 2 (26), 91, NA.

52. RG 109, vol. 2 (26), 35, 40, 53, 77, 99, NA; ibid., (34), 127, 130; RG 109, Adjutant and Inspector General's Office, Register of Letters Received, Apr.–July 1863, 2 June 1863, chap. 1, vol. 55, ibid.; and OR, 1 (28), 1:7, 56.

53. Thomas Jordan to R. S. Ripley, 22 June 1863, RG 109, vol. 2 (31), 221, NA; and Ripley, *Siege Train*, vii–viii, 253.

54. RG 109, vol. 2 (26), 128, 147–48, 154, NA; and ibid., (34), 130.

55. RG 109, vol. 2 (31), 281, 293, 299, NA; MBJ to EEL, 28 June [1863], EGP; OR, 1 (28), 2:184–85; and *Courier*, 10 July 1863, 1.

56. Burton, *Siege of Charleston*, 153–55; Rose, *Rehearsal For Reconstruction*, 255; RG 109, vol. 2 (26), 161, NA; OR, 1 (28), 1:349, 529–31, and ibid., 2:3, 10, 12–13, 185–89, 537–41.

57. Ripley, *Siege Train*, 1; Maj. Edward Manigault Diary, 10 July 1863, in possession of Peter Manigault, Charleston, S.C.; and RG 109, vol. 2 (26), 161, NA.

58. Milton F. Perry, *Infernal Machines: The Story of Confederate Submarine and Mine Warfare* (Baton Rouge: Louisiana State University Press, 1965), 58–60, relied on the OR for its version of the origin of the land mines at Battery Wagner, which omits the Gonzales recommendation. Burton, *Siege of Charleston*, 157–59, 176, also overlooked the role of Gonzales. PGTB to R. S. Ripley, 27 Jan. 1863, RG 109, vol. 2 (30), 387, NA; ibid., vol. 2 (20), 193; OR vol. 1 (28), 1:72–73, 310–12, 352–56, 370–71, 470; ibid., 2:186, 191, 195, 213.

59. RG 109, vol. 2 (26), 168, NA; ibid., (31), 293, 299; ibid., (34), 131; OR, 1 (28), 1:74, 371; and ibid., 2:192, 197, 200, 207–8.

60. Batteries Ryan, Tatom, and Haskell were named after Captains William H. Ryan, William T. Tatom, and Charles T. Haskell Jr., killed at Battery Wagner. OR, 1 (28), 1:79–80, 273; ibid., 2:203–5, 224, 233, 236, 244, 299; and M. L. Bonham to President Davis, 17 July 1863, DHU.

61. Burton, *Siege of Charleston*, 161–71; Perry, *Infernal Machines*, 58; Rose, *Rehearsal for Reconstruction*, 256; and OR, 1 (28), 1:15–16, 75–77, 210, 346–48, 360–63, 373, 416–20, 454, 525, and 2:23, 29–30, 41, 201, 206–10, 406, 550.

62. Burton, *Siege of Charleston*, 172–73, fails to give Gonzales credit for sending the shotguns to Battery Wagner. Manigault Diary, 26 July 1863; RG 109, vol. 2 (26), 200, 225, NA; OR, 1 (28), pt. 1, 94–96, 479, 552–53; and ibid., 2:232, 240, 242, 256.

63. Batteries Simkins and Cheves were named after Lt. Col. John C. Simkins and Capt. Langdon Cheves, killed while defending Battery Wagner. OR, 1 (28), 1:553; ibid., 2:224, 226, 237–39, 260, 373; RG 109, vol. 2 (21), 218, NA; and ibid. (31), 383.

64. In August 1863, the First Military District of the Department of South Carolina, Georgia, and Florida, commanded by Brig. Gen. Roswell S. Ripley, was divided into five areas: First Sub-Division, James Island and St. Andrews Parish, Brig. Gen.

William B. Taliaferro; Second Sub-Division, Sullivan's Island, Brig. Gen. Nathan G. Evans; Third Sub-Division, Morris Island, Brig. Gen. Alfred H. Colquitt; Fourth Sub-Division, Fort Sumter, Col. Alfred Rhett; Fifth Sub-Division, Charleston, Brig. Gen. Wilmot Gibbes DeSaussure. The Second Military District, headquartered at Adams Run, was led by Col. Hugh K. Aiken. The Third Military District was commanded by Brig. Gen. William Stephen Walker from McPhersonville, and the Fourth Military District was led by Brig. Gen. James Heyward Trapier from Georgetown. Today, only two of the five batteries on the New Lines survive, Batteries no. 1 and no. 5, in the Seaside Plantation development. RG 109, Department Special Orders 1863–64, no. 152, box 66, NA; ibid., (189), 25; OR, 1 (28), 1:82, 378, 421–25; and ibid., 2:230, 259, 264, 268, 325–27.

65. RG 109, vol. 2 (26), 270, NA; ibid., (34), 135–36; and OR, 1 (28), 1:549–50 and 2:266–67, 284–85.

66. RG 109, vol. 2 (26), 229, 259, NA; ibid., (34), 133; and OR, 1 (28), 2:269–70.

67. Capt. William M. Ramsay was tried by court martial in March 1864 on undisclosed charges and expelled from the army the following month. RG 109, vol. 2 (26), 254; ibid., (31), 415; Richard Jacques to Tute, 12 Aug. 1863, Richard Jacques Papers, Center for American History, University of Texas at Austin, hereinafter cited as RJP; Manigault Diary, 14 Aug. 1863; and OR, 1 (28), 1:378, and 2:284–85.

68. Burton, *Siege of Charleston*, 184, neglects to credit Gonzales for strengthening the Fort Sumter casemates with cotton bales and sand. OR, 1 (28), 1:3, 75, 84, 216, 574–77, 598–600, 611–13, 621, 654; ibid., 2:200; and AJG to JD, Feb. 7, 1864, roll 113, G-666, AIG.

69. OR, 1 (28), 1:535; ibid., 2:290–91, 294, 388; and AJG to James A. Seddon, Aug. 19, 1863, CMR.

70. AJG to JD, 25 Aug. 1863, CMR.

71. Tute to Richard Jacques, 21 Aug. 1863, RJP; Richard Jacques to Tute, 21 Aug. 1863, ibid.; Clyde Bresee, *How Grand a Flame: A Chronicle of a Plantation Family, 1813–1947* (Chapel Hill: Algonquin Books, 1992), 16; and idem, *Sea Island Yankee* (Columbia: University of South Carolina Press, 1995), 48.

72. Manigault Diary, 20 Aug. 1863; AJG to George W. Rains, 21 Aug. 1863, AIG, 1863, roll 65, frame 768, NA; OR, 1 (28), 1:30–32, 85, 219, 425, 676, and 2:57–59, 593; RG 109, vol. 2 (26), 284; and Burton, *Siege of Charleston*, 177, 186, 251–52.

73. Manigault Diary, 25 Aug. 1863; OR, 1 (28), 1:555–59; ibid., 2:289–90, 314, 323, 388; RG 109, vol. 2 (26), 369–70, NA; ibid., (34), 139; Warner, *Generals in Gray*, 105; Faust, *Historical Times*, 311; and Sarah Woolfolk Wiggins, ed., *The Journals of Josiah Gorgas, 1857–1878* (Tuscaloosa: University of Alabama Press, 1995), 61.

### Honey Hill

1. William Wallace McLeod had moved his family to Greenwood, South Carolina, in 1861 and left a slave driver in charge of his plantation and its slave force.

He joined the Charleston Light Dragoons and died in 1864 while in service. Fillmore G. Wilson, "Historic Summary for McLeod Plantation, James Island, S.C.," Meadors Construction Corporation, 16 Apr. 1993, Historic Charleston Foundation, Charleston, S.C.; Richard Jacques to Tute, 10 Sept. 1863, RJP; Special Requisition no. 40, 4 and 11 Sept. 1863, CMR; and Voucher to Abstract 1, no. 38, 7 Sept. 1863, ibid.

2. Col. Lawrence Massillon Keitt, a thirty-nine-year-old former Democrat U.S. representative from South Carolina, was killed nine months later while leading a charge on Union breastworks at the Battle of Cold Harbor, near Richmond, Virginia. *Courier*, 6 June 1864, 1; *Mercury*, 6 June 1864, 2; ibid., 20 June 1864, 1; U.S. Congress, *Biographical Directory*, 1293; OR, 1 (28), 1:3, 26–27, 36, 89, 204, 206–7, 296–97, 310–12, 327, 479, 483–88, 512; ibid., 2:84–86, 96, 343–44; and Wise, *Gate of Hell*, 201–2.

3. AJG to G. T. Beauregard, 7 Sept. [1863], RG 109, Reports, James Island, June 1863 to Oct. 1864, box 40, entry 74, NA; Special Orders 1863–64, no. 1–291, box 66, ibid.; and OR, 1 (28), 2:351, 359–63.

4. AJG to John Pemberton, 17 Sept. 1863, JCP.

5. RG 109, vol. 2 (26), 413, 439, NA; ibid., (32), 63–64; ibid., (34), 140, 147; ibid., (187), 110, 113; ibid., vol. 3, (9), 338–39; OR, 1 (28), 2:350, 377; and Manigault Diary, 26 Sept. 1863.

6. RG 109, vol. 2 (34), 143, 144, NA; ibid., (187), 128, 190; and OR, 1 (28), 1:120, 386, 400.

7. William C. Bee to Cousin, 24 Sept. 1863, EGP; EEL to Mother, [Sept. 1863], ibid.; OR, 1 (28), 2:421; and AJG to EEL, 3 Oct. 1863, ibid.

8. RG 109, vol. 2 (32), 107, NA; ibid., (34), 144; ibid., (187), 143; ibid., vol. 3 (9), 338–39; and OR, 1 (28), pt. 1, 751, and 2:593.

9. OR, 1 (28), 1:369; Manigault Diary, 12 Oct. 1863; RG 109, vol. 2 (32), 165, NA; and ibid., (187), 169.

10. Battery Wampler was named after Eng. Capt. J. Morris Wampler, killed at Battery Wagner. The Martello Tower Battery was renamed two months later as Battery Harleston. OR, 1 (28), 1:146–48; and ibid., 2:288, 403–10, 414–16, 430–31, 526, 577.

11. OR, 1 (28), 2:111, 436–38.

12. Richard Jacques to Tute, 20 Oct. 1863, RJP.

13. OR, 1 (28), pt. 1, 630–37, and 2:441, 486, 584.

14. OR, 1 (28), 2:385, 444–45; RG 109, vol. 2 (32), 235, NA; and ibid., (187), 233, 237–38, 242, 257, 305.

15. Manigault Diary, 2 and 4 Nov. 1863; John M. Laren to MAE, 26 Oct. 1863, EGP; RG 109, vol. 2 (32), 235, NA; ibid., (187), 256; OR, 1 (28), 1:156; ibid., 2:484–86; AJG to JD, 7 Feb. 1864, roll 113, G-666-1864, AIG; and Burton, *Siege of Charleston*, 202.

16. Manigault Diary, 5, 7, and 8 Nov. 1863; RG 109, vol. 2 (32), 258, NA; ibid., (34), 152–53; ibid., (187), 305, 312; OR, 1 (28), 1:143, 145, 149, 155; and ibid., 2:505–6, 513–14.

17. The number of guns established on James Island before Gonzales departed were: Fort Pemberton, eight; Battery Tynes, five; Battery Pringle, seven; Secessionville, sixteen; Battery Glover, three; Battery Wampler, two; Martello Tower Battery, five; Fort Johnson, four; Battery Cheves, four; Battery Simkins, three; Redoubt no. 1, four; Mellichamp's plantation, one; Battery Ryan, three; Battery Tatom, two; Legare's House, two; Battery Haskell, twelve, for a total of eighty-one artillery pieces. OR, 1 (28), 2:478–82, 499, 515–19, 530, 538–39; "Official," 14 Nov. 1863, Confederate States of America, Army Records, Headquarters, First Military District, Department of South Carolina, Georgia, and Florida, 1861–65, South Carolina Historical Society, Charleston; G. T. Beauregard to AJG, 23 Nov. 1863, RG 109, vol. 2 (34), 497, NA; ibid., (45), 99; AJG to Thomas Jordan, 24 Nov. 1863, CMR; and REE to MAE, [13 Dec. 1863], EGP.

18. John M. Laren to MAE, 26 Oct. 1863, EGP; WEJ to MAE, 13 Nov. [1863], ibid.; EEL to WEJ, 19 Nov. [1863] ibid.; WEJ to EEL, 21 Nov. 1863, ibid.; REE to MAE, 21 Nov. and 8 Dec. 1863, ibid; *Mercury*, 2 Dec. 1863, 2; and OR, 1 (28), 2:470.

19. RG 109, Military Departments, Special Orders 1863–64, no. 258, box 66, NA; Commutation Form no. 21, 27 Feb. 1864, CMR; OR, 1 (28), 1:682–85; ibid., 2:539; RG 109, vol. 2 (32), 445, NA; ibid., (34), 498–99; ibid., (45), 133; ibid., (187), 427, 478; and Richard Jaques to Tute, 8 Dec. 1863, RJP.

20. AJG to G. T. Beauregard, 28 Dec. 1863, CMR; WEJ to EEL, 4 Jan. 1864, EGP; and Stephen Elliott Jr., to EEL, 31 Dec. 1863, ibid.

21. AJG to G. T. Beauregard, 30 Dec. 1863, CMR; Circular, 31 Dec. 1863, RG 109, vol. 2 (32), 478, NA; and Thomas Jordan to AJG, 1 Jan. 1864.

22. *Courier*, 4 Jan. and 10 Mar. 1864, 1; AJG to Henry A. Wise, 1 Jan. [1864], Ambrosio José Gonzalez Manuscript, Henry E. Huntington Library, San Marino, Calif.; RG 109, vol. 2 (33), 172, 176, NA; ibid., (187), 555; Henry A. Wise to MAE, 19 Jan. [1864], EGP; and REE to MAE, [13 Dec. 1863], ibid.; and *Mercury*, 24 Feb. 1864, 2.

23. *Courier*, 6 Jan. 1864, 2; *Mercury*, 8, 9, 14, and 18 Jan. 1864, 2; *Courier*, 11 Jan. and 17 Feb. 1864, 1; RG 109, vol. 2 (33), 173, 175, NA; and ibid., (187), 556.

24. *Courier*, 8 Feb. 1864, 1; and AJG to JD, 7 Feb. 1864, roll 113, G-666-1864, AIG.

25. Ibid.; and OR, 1 (28), pt. 2, 129.

26. William C. Bee was also on the board of directors of the Charleston Gas Light Company and the Charleston and Savannah Railroad. Special Orders no. 40, 10 Feb. 1864, RG 109, NA; Phoebe Elliott to EEL, 14 Feb. 1864, EGP; T. R. S. Elliott to EEL, 29 Feb. 1864, ibid.; *Courier*, 22 and 23 Feb. 1864, 1; and *Mercury*, 22 Feb. 1864, 2.

27. Maj. Andrew Burnet Rhett was the son of Robert Barnwell Rhett Sr. REE to EEL, 14 Mar. 1864, ibid.; RG 109, vol. 2 (33), 184–85, 191, 195, NA; and OR, 1 (35), 2:329, 366, 382–88, 413–22.

28. OR, 1 (35), 2:382–88.

29. The Gonzales report to the secretary of war listed the following field officers of artillery, regiments, and battalions serving in the Department of South Carolina, Georgia, and Florida: South Carolina 1st Artillery Regiment: Col. Alfred Moore Rhett, Lt. Col. Joseph A. Yates, Maj. Ormsby Blanding; South Carolina 2d Artillery Regiment: Col. A. D. Frederick, Lt. Col. J. Wesley Brown, Maj. Frederick F. Warley; South Carolina Palmetto Light Artillery Battalion: Col. Edward B. White (not appointed under act of January 22, [18]62); Maj. William H. Campbell; South Carolina 1st Infantry Regiment (acting heavy artillery): Col. William Butler; Lt. Col. Robert De Treville, Maj. Thomas Abram Huguenin; Lucas' Heavy Artillery Battalion: Maj. James Jonathan Lucas; Georgia 18th Infantry Battalion (acting heavy artillery): Maj. William S. Basinger; Georgia 22d Siege Artillery Battalion: Lt. Col. William R. Pritchard, Maj. Mark J. McMullen; Georgia 12th Heavy Artillery Battalion: Lt. Col. Henry D. Capers, Maj. George M. Hanvey; and Georgia 1st Infantry Regiment (acting heavy artillery): Col. Charles H. Olmstead, Lt. Col. William S. Rockwell, Maj. Martin J. Ford. The list of field officers of artillery not attached to any regiment or battalion included Col. A. J. Gonzales, chief of artillery of the department; Lt. Col. John R. Waddy, chief ordnance officer of the department; Lt. Col. Delaware Kemper, chief of artillery of Sixth Military District; Lt. Col. Charles C. Jones Jr., chief of artillery of the Department of Florida; Lt. Col. Stephen Elliott, commanding Fort Sumter; Maj. John G. Barnwell, assistant to chief ordnance officer; Maj. George Upshur Mayo, inspector of artillery; Maj. Andrew Burnet Rhett, chief of artillery, Second Military District; Major L. F. Terrell, chief of artillery, Seventh Military District; Maj. J. J. Pope, inspector of artillery of the District of Georgia; Maj. Cleland Kinloch Huger, assistant chief ordnance officer; Maj. Edward Manigault, South Carolina Siege Train; Maj. George Lamb Buist, Georgia Siege Train. RG 109, vol. 2 (33), 190–91, 197, NA; OR, 1 (35), 2:463–68; *Mercury*, 22 Mar. 1864, 1; ibid., 6 Apr. 1864, 2; Williams, *Beauregard*, 203–4; and Fred A. Palfrey to G. T. Beauregard, 16 Apr. 1864, roll 95, 1092-B-1864, AIG.

30. OR, 1 (35), 2:364–65, 378, 398, 423, 434, 443–45, 490, 513; Williams, *Beauregard*, 207; Warner, *Generals in Gray*, 166; *Mercury*, 13 Apr. 1864, 2; *Courier*, 20 Apr. 1864, 1; REE to MAE, 21 and 24 Apr. 1864, EGP; WEJ to MAE, 3 Apr. 1864, ibid.; WEJ to EEL, 28 Apr. 1864, ibid.; *Mercury*, 3 May 1864, 2; and *Courier*, 3 June 1864, 1.

31. RG 109, vol. 2 (33), 193, 195, 197, 429, NA; ibid., (50), 133; *Mercury*, 26 May 1864, 2; OR, 1 (35), 2:489, 493–94, 497–99; and AJG to H. W. Feilden, 26 May 1864, CMR.

32. OR, 1 (35), 2:67, 79, 98, 101, 104, 106, 156, 514; Warner, *Generals in Blue*, 157; *Mercury*, 2 May, 11 and 14 June 1864, 2; *Courier*, 10 June 1864, 1; and RG 109, vol. 2 (33), 197, 428, NA.

33. *Courier*, 29 July 1864, 2; REE to MAE, 15 June 1864, EGP; MBJ to MAE, [June 1864], ibid.; *Courier*, 16, 17, and 20 June 1864, 1; *Mercury*, 17 June 1864, 2; *Courier*, 18 June 1864, 2; ibid., 20 and 25 June 1864, 1; Murder of Mr. Andrew Johnstone, AJP; and OR, 1 (35), 2:521.

34. OR, 1 (35), 2:184, 197, 541–42, 546, 552–59, 569, 575, 577–78; Manigault Diary, 2, 3, and 8 July 1864; RG 109, vol. 2 (33), 427, NA; *Mercury*, 4, 6, and 9 July 1864, 2; and *Courier*, 6 and 7 July 1864, 1.

35. After the death of Capt. John C. Mitchel, command of Fort Sumter passed to Capt. Thomas Abram Huguenin. RG 109, vol. 2 (33), 428, 431; OR, 1 (35), 2:581, 588; *Mercury*, 21 and 22 July 1864, 2; *Courier*, 21 and 22 July 1864, 1; RG 109, Special Orders no. 184, 21 July 1864, box 66, NA; AJG to Lt. Col. Fd. A. Palfrey, 30 July 1864, roll 110, E-433, frame 44, AIG; and OR, 1 (28), 2:424.

36. AJG to Samuel Cooper, 13 Aug. 1864, CMR; Beauregard to Cooper, 6 Aug. 1864, ibid.; Pemberton to Seddon, 16 Aug. 1864, ibid.; *Courier*, 24 May 1864, 2; Freeman, *Pemberton*, 262; and Wiggins, *The Journals of Josiah Gorgas*, 132.

37. Wiggins, *The Journals of Josiah Gorgas*, 115, 131; Rose, *Rehearsal for Reconstruction*, 153; Mary Barnwell Elliott to Annie Elliott, 5 [Sept. 1864], EGP; RG 109, vol. 2 (33), 203, NA.; and OR, 1 (35), 2:276, 284, 295.

38. Beauregard's new command, the Military Division of the West, comprised the Department of Tennessee and Georgia, under Gen. John Bell Hood, and the Department of Alabama, Mississippi, and East Louisiana, under Lt. Gen. Richard Taylor. AJG to T. B. Roy, 31 Oct. 1864, P. G. T. Beauregard Papers (1861–65), NA; *Mercury*, 21 Sept., 1 and 10 Oct. 1864, 1; *Courier*, 26 Sept. and 6 Oct. 1864, 1; AJG to Hattie, 29 Sept. 1864, EGP; Williams, *Beauregard*, 240–41; Nathaniel Cheairs Hughes Jr., *General William J. Hardee: Old Reliable* (1965; reprint, Wilmington, N.C.: Broadfoot, 1987), 250–51; RG 109, vol. 2 (258'), General Orders no. 76, 11 Oct. 1864, NA; and Burke and Boinest, *Charleston Directory, 1866* (New York: M. B. Brown, 1866), 38.

39. General Orders no. 18, on October 17, 1864, changed the Fourth Military District into Sub-District 1, the First Military District into Sub-District 2, the Seventh Military District into Sub-District 3, the Second and Sixth Military Districts into Sub-District 4, and the Third Military District into Sub-District 5. *Courier*, 19 Oct. 1864, 2; OR, 1 (35), 2:318; RG 109, vol. 2 (33), 252–53, NA; AJG to Lt. Col. T. B. Roy, 31 Oct. 1864, AGP; and MBJ to MAE, 27 Oct. [1864], EGP.

40. RG 109, vol. 2 (33), 253, NA; and AJG to Lt. Col. T. B. Roy, 11 Nov. 1864, AGP.

41. OR, 1, (44), 506, 517–18, 525, 535, 666, 902; RG 109, Papers of Various Confederate Notables, Maj. Gen. Sam Jones, Telegrams and Letters Received, 1864–65, box 9-A, NA; and AJG to Lt. Col. T. B. Roy, 28 Nov. 1864, AGP.

42. OR, 1 (35), 2:517; ibid., (44), 420–21, 547, 564; *Courier*, 1 Dec. 1864, 1; *New York Times*, 9 Dec. 1864, 1; Colcock Jr., "The Battle of Honey Hill"; Charles J. Colcock to Charles C. Jones Jr., 5 Nov. 1867, Charles Colcock Jones Jr., Papers, DUL; and William R. Scaife, "Battle of Honey Hill," *Blue and Gray* (Dec. 1989): 30.

43. ORN, 79; OR, 1 (44), 422, 436–38; Colcock Jr., "The Battle of Honey Hill"; John Jenkins's Narrative of Honey Hill, n.d., John Jenkins Papers, South Carolina Department of Archives and History, Columbia, S.C.; and Scaife, "Battle of Honey Hill," 31.

44. *New York Times*, 9 Dec. 1864, 1; OR, 1 (44), 422, 586–87; and Scaife, "Battle of Honey Hill," 31.

45. Gustavus Woodson Smith was a Kentucky native who had graduated from the U.S. Military Academy in 1842 and fought in the Mexican War. In 1848, he was assistant professor in engineering and the art of war at West Point, when General Worth introduced him to Ambrosio Gonzales. Smith resigned his position in December 1854 and moved to New Orleans, where he joined Gen. John A. Quitman in a Cuba filibuster expedition that was neutralized by the Pierce administration. In September 1861, Smith was appointed major general in the Confederate Army but quit seventeen months later, claiming that President Davis interfered with his command and promoted six officers over his head. Smith then served briefly as a volunteer aide to Beauregard in Charleston. Since June 1864 he had been commanding the First Division of Georgia Militia, with the rank of major general. Leonne M. Hudson, *The Odyssey of a Southerner: The Life and Times of Gustavus Woodson Smith* (Macon, Ga.: Mercer University Press); Gustavus W. Smith, "The Georgia Militia during Sherman's March to the Sea," *Battles and Leaders of the Civil War* (New York: Thomas Yoseloff, 1956), 4:667; OR, 1 (44), 906, 908–9; "John Jenkins's Narrative of Honey Hill"; Colcock Jr., "The Battle of Honey Hill"; and Scaife, "Battle of Honey Hill," 30.

46. Col. Charles Jones Colcock married Agnes Bostick on December 3, 1864. *Mercury*, 6 Dec. 1864, 2; Charles J. Colcock to Charles C. Jones Jr., 5 Nov. 1867, Charles Colcock Jones Jr., Papers, DUL; Colcock Jr., "The Battle of Honey Hill"; OR (44), 415, 911, 913; and Scaife, "Battle of Honey Hill," 31.

47. The ruins of the Honey Hill fortifications were inspected by the author in December 1991, guided by Gregory Lane, a Civil War buff from Yemassee, South Carolina. Scaife, "Battle of Honey Hill," 31; *Courier*, 3 Dec. 1864, 1; *Mercury*, 5 Dec. 1864, 1; and OR, 1 (44), 415.

48. OR, 1 (44), 416; *Courier*, 5 Dec. 1864, 1; and Scaife, "Battle of Honey Hill," 31.

49. Charles J. Colcock to Charles C. Jones Jr., 5 Nov. 1867, Charles Colcock Jones Jr., Papers, DUL.

50. ibid.; OR, 1 (44), 420, 422; George W. Williams, *A History of Negro Troops in the War of the Rebellion, 1861–1865* (New York: Harper and Brothers, 1888), 210; Colcock Jr., "The Battle of Honey Hill"; and Scaife, "Battle of Honey Hill," 31.

51. Confederate forces participating in the Battle of Honey Hill consisted of the following: Infantry: First Brigade Georgia State Militia, Colonel Willis; Georgia 2nd Infantry Regiment, State Line, Col. James Wilson; Georgia 17th Infantry Battalion, Lt. Col. Edwards; Georgia 32d Infantry Regiment, Lt. Col. Edwin H. Bacon Jr.; Georgia Militia Athens Battalion, Major Cook; Augusta Battalion, Major Jackson. Cavalry: Companies B and E, and detachments from Company C and the Rebel Troop, all belonging to the South Carolina 3d Cavalry Regiment, commanded by Maj. John Jenkins. Artillery: a section of the Beaufort Volunteer Artillery, Capt. Henry M. Stuart; a section of William Lambert De Pass's South Carolina Palmetto Light Artillery Battalion, Company G; and a section of the Lafayette Artillery. "The News," *Mercury*, 2 Dec. 1864, 1; Colcock Jr., "The Battle of Honey Hill"; C. C. Jones Jr., "The Battle of Honey Hill," *Southern Historical Society Papers* (Jan.–Dec. 1885), 364; OR, 1 (44), 422–23; Luis F. Emilio, *History of the Fifty-Fourth Regiment of Massachusetts Volunteer Infantry, 1863–1865* (Boston: Boston Book Company, 1894), 241–242; and Carse, *Department of the South*, 140.

52. *Mercury*, 5 Dec. 1864, 1; Colcock Jr., "The Battle of Honey Hill"; and Scaife, "Battle of Honey Hill," 28.

53. Colcock Jr., "The Battle of Honey Hill"; and *Savannah Republican*, 3 Dec. 1864, 1. This article was reprinted in the *Courier* two days later.

54. *Mercury*, 2 and 5 Dec. 1864, 1; *Courier*, 2 and 3 Dec. 1864, 1; Colcock Jr., "The Battle of Honey Hill"; OR, 1 (44), 416, 420, 914, 920; *Savannah Republican*, 2 Dec. 1864, 1; Carse, *Department of the South*, 141–42; and Scaife, "Battle of Honey Hill," 28.

55. OR, 1 (44), 16, 420–21, 438–48, 655–56, 665–66, 699, 708, 718, 723–24, 934–37, 941–47, 956.

56. OR, 1 (44), 702, 707, 713, 728, 741, 743, 911, 925, 940, 959; PGTB to AJG, 15 Dec. 1864, PBP; *Mercury*, 8 Dec. 1864, 2; ibid., 16 Dec. 1864, 1; and Williams, *P. G. T. Beauregard*, 247.

57. In 1897, Capt. Henry M. Stuart wrote a six-page account of the Battle of Honey Hill, challenging General Smith's official report, published in the OR in 1893, from which he is omitted. The captain alleged receiving "no order from Gen. Gonzales," nor seeing him at the front. Stuart, a native of Beaufort, was Gen. Stephen Elliott Jr.'s brother-in-law. His antagonism toward Gonzales could be a result of the Elliott family animosity, his resentment at not appearing in the OR accounts of Honey Hill, or the disdain of a junior officer toward his superior staff officer. OR, 1 (44), 57, 689, 772, 781, 783, 786, 792, 967, 970, 974; "Confidential Circular," Head Quarters, Savannah, Lieutenant General Hardee, 19 Dec. 1864, JDP; S. M. Bowman and R. B. Irwin, *Sherman and His Campaigns: A Military Biography* (New York: Charles B. Richardson, 1865), 297; and Hughes, *Hardee*, 271.

58. OR, 1 (44), 875, 970–71, 975; "Memorandum of Orders to Be Issued by Lieut. Genl. Hardee Immediately after the Evacuation of Savannah," Pocotaligo, 20 Dec. 1864, G. T. Beauregard, General, JDP; and Hardee telegram to Beauregard, 21 Dec. 1864, P. G. T. Beauregard Papers, DUL.

59. OR, 1 (44), 972, 975, 985; "Memorandum of Orders to Be Issued by Lieut. Genl. Hardee Immediately after the Evacuation of Savannah," Pocotaligo, 20 Dec. 1864, G. T. Beauregard, General, JDP; Beauregard to Hardee, 21 Dec. 1864, PBP; and Hughes, *Hardee*, 274.

60. OR, 1 (44), 7, 797–800, 994; and J. M. Otey to Hardee, 27 Dec. 1864, JDP.

61. Davis to PGTB, 30 Dec. 1864, ibid.; PGTB to Hardee, 31 Dec. 1864, ibid.; OR, 1 (44), 1009; John F. Marszalek, *Sherman: A Soldier's Passion for Order* (New York: Free Press, 1993), 320; and Hughes, *Hardee*, 275–76.

62. AJG to MAE, 18 Jan. 1865, EGP.

63. William P. McFarland to W. C. Bestor, 12 Dec. 1860, MAC; Mears and Turnbull, *The Charleston Directory, 1859*, 228; *Courier*, 1 Feb. 1864, 2; *Nassau Guardian*, 7 Mar. 1863; Stephen R. Wise, *Lifeline of the Confederacy: Blockade Running during the Civil War* (Columbia: University of South Carolina Press, 1988), 209–10; AJG to Mrs. A. H. Elliott, 18 Jan. 1865, EGP; Hattie Gonzales to Annie Elliott, [4 Aug. 1867], ibid.; and Walter L. Monteith to Ann Elliott, 6 Jan. 1893, ibid.

64. Allan Macfarlan died of consumption in Charleston on March 21, 1869, and was buried in St. Davids Cemetery in Cheraw, S.C. EEL to MAE, 29 Jan. [1865], EGP; AJG to Col. Allan Mcfarlan, 10 Feb. 1865, Allan Macfarlan Papers, SCL; Col. Allan Macfarlan Collection, Darlington County Historical Commission, Darlington, S.C.; *Courier*, 18 Nov. 1861, 2; Davidson, *The Last Foray*, 224; and Edgar, *Biographical Directory of the South Carolina House of Representatives*, 378, 382.

65. Freeman, *Pemberton*, 264; and Hardee to Sam Cooper, 10 Feb. 1865, RG 109, CMR.

66. J. Gorgas to Secretary of War, 20 Feb. 1865, CMR.

67. Gov. Andrew Gordon Magrath was arrested by Union troops in Columbia and spent seven months imprisoned in Fort Pulaski. OR, 1 (44), 1012; Thomas Benton Roy Diary, 17 Feb. 1865, copy in possession of Nathaniel Cheairs Hughes, Chattanooga, Tenn.; Burton, *Siege of Charleston*, 320–22; and Hughes, *Hardee*, 278.

68. Roy Diary, 20 Feb. 1865.

69. A. R. Barlow, *Company G: A Record of the Services of One Company of the 157th N.Y. Vols. in the War of the Rebellion* (Syracuse, N.Y.: A. W. Hall, 1899), 215; EEL to MAE, 15 Apr. [1866], EGP; and Simkins and Woody, *South Carolina during Reconstruction*, 318.

70. United Daughters of the Confederacy, South Carolina Division, *Recollection and Reminiscences, 1861–1865* (South Carolina: n.p., 1990), 546; Copy of Gen. Wilmot G.

DeSaussure Diary, Henry William and Wilmot Gibbes DeSaussure Papers, DUL; Gilbert Govan and James Livingwood, *General Joseph E. Johnston, C.S.A.: A Different Valor* (1856; reprint, New York: Konecky and Konecky, 1993), 347–48; and Mark L. Bradley, *Last Stand in the Carolinas: The Battle of Bentonville* (Campbell, Calif.: Savas Woodbury, 1996), 31–33.

71. Eliza Cowan Ervin and Horace Fraser Rudisill, *Darlingtoniana: A History of People, Places, and Events in Darlington County, South Carolina* (Columbia, S.C.: R. L. Bryan, 1964), 92, 391. A photograph of the Gibson house appears on page 388.

72. Albertus Chambers Spain to Davia Donella Hardeman, 11 Mar. 1865, Maj. Albertus Chambers Spain Collection, Darlington County Historical Commission, Darlington, S.C.; and EEL to Rev. Dr. Bachman, 2 May 1865, EGP.

73. Govan and Livingwood, *General Joseph E. Johnston*, 350; Craig L. Symonds, *Joseph E. Johnston: A Civil War Biography* (New York: W. W. Norton, 1992), 344; AJG Memorandum [June 1889], GFP; and 20 Mar. 1865, reel 7, frame 244, PBP.

74. William R. Trotter, *Silk Flags and Cold Steel, The Civil War in North Carolina: The Piedmont* (Winston-Salem, N.C.: John F. Blair, 1988), 297, 299.

75. OR, 1 (47), pt. 3, 725–26; and A.A.G. to G. L. Dudley, 31 Mar. 1865, PBP.

76. Parole List, Head Quarters, Chief of Artillery, Hardee's Corps, near Greensboro, N.C., 30 Apr. 1865, CMR; Muster Rolls and Lists of Confederate Troops Paroled in Greensboro, N.C., microfilm 1781, roll 2, RG109, NA; Williams, *Beauregard*, 256; Simkins and Woody, *South Carolina during Reconstruction*, 10; Ernest McPherson Lander Jr., *A History of South Carolina, 1865–1960* (Columbia: University of South Carolina Press, 1970), 5; and Randolph W. Kirkland Jr., *Broken Fortunes* (Charleston, S.C.: South Carolina Historical Society, 1995), xiv, 412.

**Reconstruction Retailer**

1. EEL to Rev. John Bachman, 2 May 1865, EGP; and United Daughters of the Confederacy, *Recollections and Reminiscences, 1861–1865*, 546.

2. MBJ to REE, 9 July 1865, EGP.

3. REE to MAE, 11 July 1865, EGP.

4. Stephen Elliott to EEL, 2 Aug. 1865, EGP; and Simkins and Woody, *South Carolina during Reconstruction*, 239–40.

5. Rose, *Rehearsal for Reconstruction*, 215, 288, 327–28, 330, 349–51; Eric Foner, *Reconstruction: America's Unfinished Revolution, 1863–1877* (New York: Harper and Row, 1988), 158–61; Joel Williamson, *After Slavery: The Negro in South Carolina during Reconstruction, 1861–1877* (Chapel Hill: University of North Carolina Press, 1965), 59–60, 79–80; Thomas Holt, *Black over White: Negro Political Leadership in South Carolina during Reconstruction* (Urbana: University of Illinois Press, 1979), 25; Simkins and Woody, *South Carolina during Reconstruction*, 31, 33–34, 227; Saville, *The Work of Reconstruction*, 18–19, 44, 72, 81; and Richard De Treville to Mrs. Ann H. Elliott, 30 May 1866, EGP.

6. Department of State, Civil War Amnesty, and Pardon Records, 1863–67, Amnesty Oaths, 1864–66, South Carolina, box 117, entry 1001, nos. 2, 508, and 682, RG 59, NA; REE to EEL, 15 Oct. 1865, EGP; Lander, *A History of South Carolina, 1865–1960*, 6–7; and Robert H. Jones, *Disrupted Decades: The Civil War and Reconstruction Years* (New York: Charles Scribner's Sons, 1973), 417–18.

7. The Reverend John B. Bachman (1790–1874), born in New York, moved to South Carolina after being ordained in 1815, where he "began his notable ministry to the Negro population in Charleston." He was a member of the Association for the Advancement of Science. During the Civil War, Bachman organized a Confederate Soldiers' Relief Association. His son, Capt. William K. Bachman, belonged to the German Artillery, Hampton's Legion. South Carolina Synod of the Lutheran Church in America, *A History of the Lutheran Church in South Carolina* (Columbia: R. L. Bryan, 1971), 163–69, 833; *Intelligencer*, 15 May 1850, 3; *Courier*, 23 Aug. 1861, 17 Mar. 1862, 7 Oct. 1863, 1, and 4 Feb. 1864, 1. John Bachman to EEL, 3 Aug. 1865, EGP; Simkins and Woody, *South Carolina during Reconstruction*, 6; Walter J. Fraser Jr., *Charleston! Charleston!: The History of a Southern City* (Columbia: University of South Carolina Press, 1989), 275; Robert N. Rosen, *Confederate Charleston: An Illustrated History of the City and the People during the Civil War* (Columbia: University of South Carolina Press, 1994), 142; *Courier*, 12 Aug. 1865, 2; and Elliott Johnstone to Mrs. Anne Elliott, 5 Sept. 1865, EGP.

8. The plantations in Colleton District were plundered worse than any other area. Saville, *The Work of Reconstruction*, 76; EEL to MBJ, 21 Sept. [1865], EGP; REE to MAE, 25 Sept. 1865, ibid.; *Courier*, 5 Sept. 1865, 2; EEL to Tom Elliott, 8 Oct. 1865, Thomas Rhett Smith Elliott Papers, DUL, hereinafter cited as TRE; Richard De Treville to MAE, [30 May 1866], EGP; and Rose, *Rehearsal for Reconstruction*, 397.

9. Gen. Rufus Saxton was replaced on 20 Jan. 1866 by Gen. Robert K. Scott, who in November 1867 was elected governor of South Carolina. Richard De Treville to MAE, 9 and 16 Nov. and 9 Dec. 1865, EGP; Restoration Order, Headquarters, Assistant Commissioner, Bureau of Refugees, Freedmen and Abandoned Lands, R. Saxton, 9 Dec. 1865, ibid.; and Richard De Treville to MAE, [30 May 1866], ibid.

10. *Courier*, 12 Dec. 1865, and 1 Jan. 1866, 2; OR, 1 (1), 57–58; ibid., (14), 265; Jeanne A. Calhoun and Martha A. Zierden, *Charleston's Commercial Landscape, 1803–1860* (Charleston: Charleston Museum Archaeological Contributions 7, Sept. 1984); and *Charleston News and Courier*, 13 Apr. 1948, 13, and 16 Apr. 1979, 1-B.

11. *Courier*, 22 Jan. 1866, 2; 1860 Free Population Census, South Carolina, Second Ward of Charleston, 241, NA; Mears and Turnbull, *The Charleston Directory, 1859*, 64, 86; South Carolina, vol. 7, 482, R. G. Dun & Co. Collection, BLH; and *Courier*, 5 Jan. 1866, 4.

12. REE to MAE, 13 Jan. 1866, EGP; M. W. Clement to Mrs Ann Elliott, 16 Jan. 1866, ibid.; REE to EEL, 28 Jan. 1866, ibid.; WEJ to MAE, 25 Mar. 1866, ibid.; Rose, *Rehearsal for Reconstruction,* 357; and Williamson, *After Slavery*, 39–42, 84.

13. AJG to REE, 7 Feb. 1866, EGP; and *Courier*, 1 Feb. 1866, 4.

14. Cuba Commission Report, *A Hidden History of the Chinese in Cuba* (Baltimore: Johns Hopkins University Press, 1993), 21–22; Duvon Clough Corbitt, *The Chinese in Cuba, 1847–1947* (Wilmore, Ky.: Asbury College, 1971), 5; *Courier*, 10 Sept. 1866, 2; Alrutheus Ambush Taylor, *The Negro in South Carolina during the Reconstruction* (1924; reprint, New York: AMS Press, 1971), 20, 58; MAE to HRG, 29 June [1869], EGP; MBJ to EEL, 16 Mar. [1866], ibid.; Richard De Treville to MAE, 6 Aug. 1866, ibid.; HRG to EEL, [late Apr. 1866], ibid.; Williamson, *After Slavery*, 319; *Courier*, 17 and 29 May 1866, 2; Richard De Treville to MAE, 16 Nov. 1865, EGP; and MBJ to MAE, 6 Apr. [1866], ibid.

15. WEJ to MAE, 13 and 25 Mar. 1866, EGP; and WEJ to REE, 2 Apr. 1866, ibid.

16. Williamson, *After Slavery*, 88–89, 92; George Brown Tindall, *South Carolina Negroes, 1877–1900* (Columbia: University of South Carolina Press, 1952), 97; EEL to MAE, [7 and 14 Apr. 1866], EGP; MBJ to MAE, 24 Apr. [1866], ibid.; Saville, *The Work of Reconstruction*, 73, 83, 98; Simkins and Woody, *South Carolina during Reconstruction*, 231; and Martin Abbott, *The Freedmen's Bureau in South Carolina, 1865–1872* (Chapel Hill: University of North Carolina Press, 1967), 59, 68, 70, 73.

17. Simkins and Woody, *South Carolina during Reconstruction*, 60, 333; EEL to MAE, [14 Apr. 1866], EGP; and EEL to MAE, 15 Apr. [1866], ibid.

18. Charleston-Savannah Railroad conductor William Crovatt was born in Charleston in 1832 and owned Otranto Plantation near Goose Creek Church. He moved to Brunswick, Georgia, in 1869, where he died in December 1898 and was interred in Magnolia Cemetery, Charleston. EEL to MAE, [7 Apr. 1866], EGP; EEL to MAE, [14 Apr. 1866], ibid.; EEL to MAE, 15 Apr. [1866], ibid.; and Williamson, *After Slavery*, 45–46, 87, 107, 111.

19. HRG to EEL, [late Apr. 1866], EGP; Williamson, *After Slavery*, 35–36, 112; Simkins and Woody, *South Carolina during Reconstruction*, 319–20; MBJ to MAE, 1 May 1866, EGP; and MAE to Sons, 15 Aug. [1866], ibid.

20. *Courier*, 7 June 1866, 2; ibid., 9 and 11 June 1866, 3; EEL to MAE, 29 Apr. and 1 May [1866], EGP; and MAE to Sons, 15 Aug. [1866], ibid.

21. Spain, Ministry of Foreign Affairs, Consulate, Charleston, S.C., *Registro de pasaportes, 1861–1867*, May 1866, DUL; Charleston City Taxes, Monthly Returns, Nov. 1865–Feb. 1866, Charleston Library Society, Charleston, S.C.; *Courier*, 29 Jan., 1 and 24 Feb., 1, 5, 6, 9, 13, and 28 Mar., 2, 7, 11, 13, 18, 19, 23, and 28 Apr., 3, 8, 11, 18, 23, 30, and 31 May 1866, 4; Gonzales, Woodward, May 1866, Internal Revenue Assessment, Lists for South Carolina, 1864–66, M789, roll 1, frame 568, RG 58, NA; and *New York Times*, 13 June 1866, 8.

22. *Courier*, 25 June 1866, 2; W. E. B. Du Bois, *Black Reconstruction in America* (1935; reprint, New York: Atheneum, 1992), 168; Taylor, *The Negro in South Carolina during the Reconstruction*, 45; Tindall, *South Carolina Negroes, 1877–1900*, 7; Simkins

and Woody, *South Carolina during Reconstruction*, 48–49; *Journal of the House of Representatives of the State of South Carolina* (special sess., July 6, 1868), 169–73; Williamson, *After Slavery*, 72–74, 258–59; Lander, *A History of South Carolina, 1865–1960*, 8; Bernard E. Powers Jr., *Black Charlestonians: A Social History, 1822–1885* (Fayetteville: University of Arkansas Press, 1994), 76–77; Head Quarters, Second Military District, Register of Letters Received, 12 July 1866, vol. 521, pt. 1, no. 4108, RG 393, NA; and *Courier*, 27 July 1866, 4.

23. Chapman's Bluff Deed, 9 June 1866, in Ambrosio J. Gonzales, in Bankruptcy File, 15 Dec. 1868, U.S. District Courts, Eastern District of South Carolina, box 55, RG 21, NA, hereinafter cited as AJG Bankruptcy File; Warren Ripley, *The Battle of Chapman's Fort* (Green Pond, S.C.: Ashepoo Plantation, 1978), 51; *George Page and Co. v A. J. Gonzales*, 5 Nov. 1867, U.S. District Courts, South Carolina District, roll 368, docket 189, box 8, RG 21, NA; *Courier*, 30 July 1866, 4; WEJ to MAE, 13 July 1866, EGP; WEJ to MAE, 26 July [1866], ibid.; and WEJ to EEL, 5 Jan. 1867, ibid.

24. MAE to Sons, 15 Aug. [1866], EGP; *Courier*, 20 Aug. 1866, 4; ibid., 4 Sept. 1866, 3; *Charleston City Directory for 1867–68* (Charleston: Jno. Orrin Lea, 1867), 92; *Courier*, 21 June 1866, 2; ibid., 5 Feb. 1867, 3; Washington Lodge No. 5, Return for 1869, Return Files, Grand Lodge of South Carolina, Ancient Free Masons, Columbia, S.C.; and WEJ to EEL, 16 Sept. 1866, EGP.

25. REE to TSE, 25 Oct. [1866], EGP; Ambrosio J. Gonzales Bond, 31 Oct. 1866, ibid.; and HRG to MAE, 19 May [1867], ibid.

26. Matanzas planter Benigno Gener Junco is identified as the owner of the Sabanilla Rail Road in Vicente Baez, ed., *La Enciclopedia de Cuba* (Madrid: Playor, 1974), 7:687. *Courier*, 10 Nov. 1866, 3; ibid., 31 July, 8 and 9 Nov. 1866, 2; "Two Hundred and Fifty Barrels of Molasses v. United States," Circuit Court, District South Carolina, June Term 1869, case no. 14293, United States Series, *The Federal Cases Comprising Cases Argued and Determined in the Circuit and District Courts of the United States*, bk. 24 (St. Paul: West Publishing, 1896), 437–43; and AJG to HRG, 30 Nov. 1866 and 6 Dec. 1866, GFP.

27. AJG to HRG, 21 Dec. 1866, GFP.

28. A Cuban wooden box of sugar had an average sixteen *arroba* capacity. The *arroba* is a Cuban weight of 25.36 pounds, which would have made this shipment weigh around 405.76 pounds. Manuel Moreno Fraginals, *The Sugarmill: The Socioeconomic Complex of Sugar in Cuba* (New York: Monthly Review Press, 1976), 35. AJG to HRG, 21 Dec. 1866, 3 Jan. 1866 [1867], and 6 Jan. 1867, GFP; and MAE to WEJ, 20 Jan. [1867], TRE.

29. The Gracchi were two Roman brothers, reformers, and statesmen during 134–121 B.C., killed during political riots. When their mother heard that her son Tiberius had been killed by a mob of senators and their followers, she asked her

younger son Gaius not to be obsessed with revenge. WEJ to EEL, 19 Dec. 1866, EGP; Ann and EEL to MAE, 10 Jan. 1867, ibid.; Tom Elliott to REE, 22 Jan. 1867, ibid.; and REE to Sisters, [22 Jan. 1867], ibid.

30. The cemetery monument to William Elliott was erected 130 years later by descendant Dr. Ambrose Gonzales Hampton with an epitaph written by Ann Fripp Hampton. MAE to WEJ, 20 Jan. [1867], TRE; [William Amory] to Anne Elliott, 6 Feb. 1867, EGP; Ann Lears Amory to Miss Elliott, 25 Mar. [1867], ibid.; and E. Henderson Otis to REE, 14 May 1867, ibid.; W. Amory to Anne Elliott, 6 May 1867, ibid.; REE to Mrs. E. Henderson Otis, 11 July 1867, ibid.; W. Amory to Miss Elliott, 4 June 1867, ibid.; and MAE to HRG, 14 June [1867], ibid.

31. *Courier*, 13 Feb. 1867, 4; 1870 Federal Census, South Carolina, Second Ward of Charleston, 63; *Charleston City Directory for 1867–68*, 106; "Mortgage of Real Estate," 25 Feb. 1867, EGP; "Memorandum of an Agreement between Ralph E. Elliott and George H. Hoppock," 1 Mar. 1867, ibid.; REE to Willie Elliott, 11 Mar. 1867, ibid.; REE to TSE, 13 Mar. 1867, TRE; C. K. Prioleau to Miss Elliott, 1 Nov. 1867, EGP; HRG to ANE, [29 Nov. 1867], ibid.; and HRG to MAE, 10 Dec. [1868], ibid.

32. Simkins and Woody, *South Carolina during Reconstruction*, 64; Mary P. Manigault to EEL, 26 Feb. [1867], EGP; Social Hall Deed, 18 Aug. 1874, GFP; AJG to Asger Hamerik, 15 June 1872, EGP; Colleton County Auditor's Tax Return, 1868, A. J. Gonzales, box 5, file 19, South Carolina Department of Archives and History, Columbia, S.C.; AJG Bankruptcy File; and Williamson, *After Slavery*, 154.

33. HRG to MAE, Easter [22 Apr. 1867], EGP; HRG to EEL, 9 Mar. [1867], ibid.; HRG to EEL, 27 Mar. [1867], ibid.; HRG to MAE, 11 [Nov. 1867], ibid.; HRG to Alfonso Beauregard Gonzales, [3 May 1867], ibid.; and HRG to MAE, 3 May [1867], ibid.; and Simkins and Woody, *South Carolina during Reconstruction*, 278.

34. HRG to EEL, 9 and 27 Mar. [1867], and 26 [July 1867], EGP; HRG to MAE, 1 and 3 May [1867], [28 July 1867], and [3 Aug. 1867], ibid.; and HRG to ANE, 6 May [1867], ibid.

35. HRG to MAE, Easter [22 Apr. 1867], and 18 Nov. [1867], EGP; HRG to ANE, [4 Aug. 1867], ibid.; HRG to EEL [2 May 1867], ibid.; and NGG to AEG, 8 Nov. 1873, ibid.

36. HRG to EEL, 9 Mar. [1867], EGP; HRG to Alfonso Beauregard Gonzales, [3 May 1867], ibid.; HRG to Sister, [17 Mar. 1867], ibid.; HRG to MAE, 19 May [1867], ibid.; HRG to EEL, [17 May 1867], ibid.; and 1870 Federal Census, South Carolina, Fourth Ward of Charleston, 258.

37. HRG to MAE, 19 May [1867], EGP; HRG to Sister, [17 Mar. 1867], ibid.; and HRG to MAE, Easter [22 Apr. 1867], ibid.

38. HRG to Sister, [17 Mar. 1867], EGP; HRG to EEL, [2 May 1867], ibid.; HRG to MAE, 3 May [1867], ibid.; Saville, *The Work of Reconstruction*, 113, 119; Simkins and

Woody, *South Carolina during Reconstruction*, 237; and Taylor, *The Negro in South Carolina during the Reconstruction*, 57, 60.

39. Spartanburg planter David Golightly Harris at this time also regarded his contracted black laborers more reliable than his hired white men. Saville, *The Work of Reconstruction*, 127. HRG to EEL, [2 May 1867], EGP; HRG to Alfonso Beauregard Gonzales, [3 May 1867], ibid.; HRG to MAE, 3 May [1867], ibid.; HRG to ANE, 6 May [1867], ibid.; HRG to MAE, 19 May [1867], ibid.; and Foner, *Reconstruction*, 275–76.

40. HRG to MAE, 3 May [1867], EGP; HRG to MAE, [28 May 1867], ibid.; HRG to EEL, 9 June [1867], ibid.; HRG to ANE, 7 Oct. [1867], ibid.; HRG to EEL, 15 July [1867], ibid.; and HRG to MAE, 19 May [1867], ibid.

41. Northern abolitionists who became local planters also complained about the freedmen's poor work habits, lying, and pilfering. Rose, *Rehearsal for Reconstruction*, 136, 364–69, 372; HRG to MAE, [28 May 1867], EGP; HRG to EEL, 9 June [1867], ibid.; HRG to MAE, 12 July [1867], ibid.; HRG to EEL, 15 July [1867], ibid.; HRG to ANE, 24 June [1867], ibid.; Simkins and Woody, *South Carolina during Reconstruction*, 225, 237, 332–33; Williamson, *After Slavery*, 322–24; Saville, *The Work of Reconstruction*, 68, 129; Tindall, *South Carolina Negroes, 1877–1900*, 261–62; and Taylor, *The Negro in South Carolina during the Reconstruction*, 122.

42. HRG to Sister, [17 Mar. 1867], EGP; HRG to EEL, 27 Mar. [1867], ibid.; HRG to EEL, [2 May 1867], ibid.; HRG to MAE, [13 May 1867], ibid.; HRG to EEL, 9 June [1867], ibid.; HRG to EEL, 9 Mar. [1867], ibid.; and HRG to Alfonso Beauregard Gonzales, [3 May 1867], ibid.

43. John Fraser and Company was owned by George A. Trenholm, Theodore Wagner, James Welsman, Charles K. Prioleau, and John B. Lafitte. The federal government auctioned off the company's real estate to pay for the 1861–65 import duties. Eighteen months before going bankrupt, the company was worth three million dollars and owned a very large amount of real estate. South Carolina, vol. 7, 409, R. G. Dun & Co. Collection, BLH; Ethel S. Nepveux, *George Alfred Trenholm and the Company that Went to War, 1861–1865* (Anderson, S.C.: Electric City, 1994), 97; HRG to EEL, [1 June 1867], EGP; AJG Bankruptcy File; Williamson, *After Slavery*, 133–34; HRG to EEL, 9 June [1867], EGP; and MAE to HRG, 14 June [1867], ibid.

44. HRG to EEL, 9 June [1867], EGP; and HRG to MAE, [28 July 1867], ibid.

45. MAE to HRG, 14 June [1867], EGP; Abbott, *The Freedmen's Bureau*, 42, 76; and REE to Tom Elliott, 7 July 1867, TRE.

46. HRG to EEL, 15 July [1867], EGP; HRG to ANE, 6 May [1867], ibid.; HRG to MAE, 12 July [1867], ibid.; HRG to EEL, [7 Sept. 1867], ibid.; HRG to ANE, 24 June [1867], ibid.; HRG to EEL, 30 June [1867], ibid.; and HRG to MAE, [28 July 1867], ibid.

47. HRG to ANE, [4 Aug. 1867], EGP; Williamson, *After Slavery*, 246; HRG to EEL, [15 Aug. 1867] and [7 Sept. 1867], ibid.; HRG to MAE, 2 Oct. [1867], ibid.; HRG to MAE, 18 Nov. [1867], ibid.; MAE to HRG, [28 Oct. 1867], ibid.; and MAE to HRG, 14 Aug. [1869], ibid.

48. HRG to MAE, 2 Oct. [1867], EGP; HRG to MAE, [27 Oct. 1867], ibid.; AJG Bankruptcy File; and *New York Times*, 1 Oct. 1867, 1.

49. MAE to HRG, [28 Oct. 1867], EGP; Williamson, *After Slavery*, 65, 165, 168; Saville, *The Work of Reconstruction*, 114, 136; Simkins and Woody, *South Carolina during Reconstruction*, 45, 225, 290–91; Tom Elliott to MAE, 20 Oct. 1867, TRE; HRG to EEL, [18 Sept. 1867], EGP; and HRG to ANE, 7 Oct. [1867], ibid.

50. HRG to MAE, 16 Oct. [1867], EGP; MAE to HRG, [28 Oct. 1867], ibid.; and EEL to HRG, 11 Nov. [1867], ibid.

51. HRG to MAE, 16 Oct. [1867], EGP; AJG to HRG, 18 Oct. 1867, GFP; HRG to MAE, [23 Oct. 1867], EGP; HRG to MAE, 25 Oct. [1867], ibid.; HRG to MAE, [27 Oct. 1867], ibid.; HRG to ANE, [mid-Nov. 1867], ibid.; and HRG to MAE, 18 Nov. [1867], ibid.

52. AJG to HRG, 1 Nov. 1867, GFP; and HRG to MAE, 18 Nov. [1867], EGP.

53. EEL to HRG, 11 Nov. [1867], EGP; HRG to ANE, 9 Nov. [1867], ibid.; HRG to MAE, [23 Oct. 1867], ibid.; HRG to MAE, [27 Oct. 1867], ibid.; HRG to MAE, 25 Oct. [1867], ibid.; and HRG to MAE, 18 Nov. [1867], ibid.

54. HRG to MAE, 2 Oct. [1867], EGP; HRG to MAE, 18 Nov. [1867], ibid.; MAE to HRG, [16 Oct. 1867], ibid.; TSE to MAE, 20 Oct. 1867, TRE; HRG to ANE, 7 Oct. [1867], EGP; and HRG to ANE, 9 Nov. [1867], ibid.

55. HRG to ANE, 9 Nov. [1867], EGP; *George Page and Co. v A. J. Gonzales*; 1870 Federal Census, South Carolina, Second Ward of Charleston, 54; *Mercury*, 5 Oct. 1861 and 11 Mar. 1862, 2; ibid., 31 Dec. 1864, 1; U.S. Congress, *Biographical Directory*, 1583; Mary Doline O'Connor, *The Life and Letters of M. P. O'Connor* (New York: Dempsey and Carroll, 1893), 31–32; and HRG to MAE, 18 Nov. [1867], EGP.

56. HRG to MAE, 18 Nov. [1867]; EGP; HRG to EEL, 18 Sept. [1867], ibid.; and HRG to EEL, 26 Nov. [1867], ibid.

57. HRG to ANE, [mid-Nov. 1867], EGP; HRG to ANE, [29 Nov. 1867], ibid.; HRG to MAE, 6 Dec. [1867], ibid.; and HRG to EEL, 26 Nov. [1867], ibid.

58. HRG to MAE, 6 Dec. [1867], EGP; *Charleston City Directory, 1867–68*, 86; and HRG to MAE, 16 Dec. [1867], ibid.

59. HRG to MAE, 16 Dec. [1867], EGP; and AJG Bankruptcy File.

60. HRG to EEL, 26 Nov. [1867], EGP; HRG to MAE, 6 Dec. [1867], ibid.; and HRG to MAE, 16 Dec. [1867], ibid.

61. HRG to EEL, 8 Jan. [1867], EGP; Simkins and Woody, *South Carolina during Reconstruction*, 89; Fraser, *Charleston! Charleston!*, 286; MBJ to MAE, 29 Feb. [1868], EGP; AJG Bankruptcy File; and 1870 Federal Census, Schedule 1, South Carolina, Charleston, Ward 2, 58.

Notes to Pages 281–287 / 449

62. MAE to HRG, [28 Oct. 1867], EGP; HRG to ANE, 7 Oct. [1867], ibid.; EEL to HRG, 11 Nov. [1867], ibid.; HRG to EEL, 26 [July 1867], ibid.; HRG to MAE, 16 Oct. [1867], ibid.; HRG to MAE, 11 [Nov. 1867], ibid.; and MBJ to MAE, 29 Feb. [1868], ibid.

63. Birkett Davenport Fry was a veteran of the Mexican War and had been a filibuster in Nicaragua with William Walker. He was severely wounded four times during major engagements of the Civil War, the last being Pickett's charge at the Battle of Gettysburg, where he was taken prisoner. A special exchange returned him to duty as a district commander in South Carolina and Georgia. Warner, *Generals in Gray*, 95–96; *The Southern Enterprise* (Greenville, S.C.), 22 Apr. 1868, 2; and Cooper, *Jefferson Davis, American*, 575.

64. Spain, Ministry of Foreign Affairs, Consulate, Charleston, S.C., *Registro de pasaportes espedidos por el consulado, 1868–1871*, DUL; Simkins and Woody, *South Carolina during Reconstruction*, 109; AJG Bankruptcy File; HRG to AEG, 13 Aug. [1868], EGP; NGG to AEG, 13 Aug. 1868, ibid.; HRG to MAE, 26 Dec. [1868], ibid.; 1870 Federal Census, South Carolina, Sixth Ward of Charleston, 490; *Charleston City Directory for 1867–68*, 143; *Charleston Post and Courier*, 13 May 1987, 1; Simkins and Woody, *South Carolina during Reconstruction*, 50; and Annie Johnstone to MAE, 11 Oct. 1868, EGP.

65. Annie Johnstone to MAE, 11 Oct. 1868, EGP; HRG to MAE, 4 Nov. [1868], ibid.; and Simkins and Woody, *South Carolina during Reconstruction*, 344.

66. PGTB to AJG, 10 Dec. 1868, GFP; Guerra, *Manual de historia de Cuba*, 683–90; and Thomas, *Cuba*, 245.

67. PGTB to Thomas Jordan, 24 Jan. 1869, reel 1, frame 594, PBP.

68. AJG Bankruptcy File; 1870 Federal Census, Second Ward of Charleston, 63; Simkins and Woody, *South Carolina during Reconstruction*, 100; and *George Page and Co. v A. J. Gonzales*.

69. HRG to MAE, 26 Dec. [1868], EGP; and Ebet Burnet to EEL, 5 Jan. 1869, ibid.

70. *Registro de pasaportes espedidos por el consulado, 1868–1871*, 29 Dec. 1868, Spanish Consulate Papers, Charleston; *Courier*, 31 Dec. 1868, 2; and Ebet Burnet to EEL, 5 Jan. 1869, EGP.

**Family Feud**

1. Leila Elliott Habersham (1830–1901) was the widow of Lt. F. A. Habersham, killed at the Battle of Chancellorsville on May 3, 1863. HRG to MAE, 14 Jan. [1869], EGP; 1870 Federal Population Census, Savannah, 167; *A History of Cedar Key* (Cedar Key, Fla.: Cedar Key Beacon, 1994), 20; and *Historic Old Cedar Key* (Cedar Key, Fla.: Cedar Key Historic Society, 1994), 2.

2. HRG to MAE, 23 Jan. and 27 Feb. [1869], EGP; Mary P. Manigault to EEL, 14 Feb. 1869, ibid.; MBJ to MAE, 20 Feb. [1869], ibid.; Mary P. Manigault to MAE, 28 Feb. [1869], ibid.; and Thomas, *Cuba*, 252.

3. HRG to MAE, 23 Jan. [1863], EGP; HRG to MAE, 29 Apr. [1869], ibid.; HRG to MAE, Mar. [1869], ibid.; and HRG to MAE, 27 Feb. [1869], ibid.

4. HRG to MAE, 27 Feb. [1869], EGP; Alvarez Pedroso, *Miguel de Aldama*, 89; Dollero, *Cultura Cubana*, 244; A. E. Gonzales, *In Darkest Cuba*, 12–13; HRG to MAE, 9 July [1869], EGP; MAE to HRG, 14 Aug. [1869], ibid.; HRG to MAE, 3 Sept. [1869], ibid.; and HRG to MAE, Mar. [1869], ibid.

5. HRG to MBJ, 14 Apr. [1869], EGP; HRG to MAE, 29 Apr. [1869], ibid.; HRG to MAE, [18 May 1869], ibid.; HRG to MAE, 8 June [1869], ibid.; AEG to MAE, 12 Aug. [1869], ibid.; and Ximeno, *Aquellos Tiempos*, 357–58.

6. *New York Times*, 5 Apr. 1868, 10; and Ximeno, *Aquellos Tiempos*, 2:192.

7. HRG to MAE, 29 Apr. [1869], EGP; HRG to MAE, [18 May 1869], ibid.; Baez, ed., *La Enciclopedia de Cuba*, 7:679; Carlos M. Trelles y Govin, *Matanzas en la Independencia de Cuba* (Havana: Imprenta "Avisador Comercial," 1928), 29; Francisco J. Ponte y Domínguez, *Matanzas: Biografía de una provincia* (Havana: Imprenta "El Siglo XX," 1959), 196; HRG to MAE, 8 June [1869], EGP; HRG to MAE, 1 Aug. [1869], ibid.; AEG to MAE, 12 Aug. [1869], ibid.; HRG to MAE, 14 Aug. [1869], ibid.; HRG to MAE, 9 July [1869], ibid.; and Dollero, *Cultura Cubana*, 71.

8. HRG to MAE, [May 1869], EGP; HRG to MAE, 8 June [1869], ibid.; MAE to HRG, 29 June [1869], ibid.; HRG to MAE, 1 Aug. [1869], ibid.; HRG to MAE, 9 July [1869], ibid.; Ebet Burnet to EEL, [21 Nov. 1869], ibid.; and Dollero, *Cultura Cubana*, 71.

9. AEG to MAE, 12 Aug. [1869], EGP; HRG to MAE, 9 July [1869], ibid.; MAE to HRG, 29 June [1869], ibid.; HRG to MAE, 1 Aug. [1869], ibid.; and MAE to HRG, 14 Aug. [1869], ibid.

10. HRG to MAE, 9 July [1869], EGP; HRG to MAE, 1 Aug. [1869], ibid.; HRG to MAE, 3 Sept. [1869], ibid.; AEG to MAE, 12 Aug. [1869], ibid.; and HRG to MAE, 14 Aug. [1869], ibid.

11. MAE to HRG, 14 Aug. [1869], EGP; MAE to AEG, [mid-May 1870], ibid.; Elliott Johnstone to MAE, 31 Aug. 1869, ibid.; Simkins and Woody, *South Carolina during Reconstruction*, 175–78; Lander, *A History of South Carolina, 1865–1960*, 12; Williamson, *After Slavery*, 149–51, 158, 226, 382–84; Rose, *Rehearsal for Reconstruction*, 383–84; and Taylor, *The Negro in South Carolina during the Reconstruction*, 148, 176–79, 192.

12. MAE to HRG, 13 Sept. [1869], EGP; HRG to MAE, 3 Sept. [1869], ibid.; Mrs. Poujaud to [?], 25 Sept. 1869, ibid.; Francis Peyre Porcher, M.D., *Yellow Fever in Charleston, 1871, with Remarks upon its Treatment*, President's Address: Before South Carolina Medical Association (Charleston, S.C.: Walker, Evans, and Cogswell, 1872), 13; *Patria*, 31 Dec. 1892, 2–3; NGG to AEG, 26 Aug. 1870, EGP; AJG to Ulysses S. Grant, 8 Jan. 1876, Letters of Application and Recommendation during the Administration of Ulysses S. Grant, 1869–77, M968, roll 23, frames 283–286, RG 59, NA; and Cecilia M. de Poujaud to MAE, 28 Jan. 1871, EGP.

13. Trelles y Govin, *Matanzas en la Independencia de Cuba*, 29; Ximeno, *Aquellos Tiempos*, 2:76, 111; Ponte y Domínguez, *Matanzas*, 197; Ambrosio Jose Gonzales y Rufin Passport, 11 Nov. 1869, GFP; AEG to N. G. Gonzales, 3 Dec. [1869], EGP; Leila Habersham to EEL, 19 [Nov. 1869], ibid.; Mary P. Manigault to EEL, 19 Nov. 1869, ibid.; Alonzo B. Luce, 22 Nov. 1869; Georgia, vol. 28, 57, R. G. Dun & Co. Collection, BLH; and *Courier*, 22 Nov. 1869, 4.

14. *Patria*, 31 Dec. 1892, 3; EEL to "My Dear Boys," [24 May 1872], EGP; EEL to AEG, 4 June 1870, 25 June [1872], 8 Oct. [1872], 6 [June 1873], ibid.; and *Charleston Daily News*, 15 Mar. 1873, 2.

15. A. Toomer Porter, *The History of a Work of Faith and Love in Charleston, South Carolina* (New York: D. Appleton, 1882), 11; Colyer Meriwether, *History of Higher Education in South Carolina with a Sketch of the Free School System* (Washington, D.C.: Government Printing Office, 1889), 50; Karen Greene, *Porter-Gaud School: The Next Step* (Easley, S.C.: Southern Historical Press, 1982), 12; Simkins and Woody, *South Carolina during Reconstruction*, 419; MBJ to EEL, [Feb. 1870], EGP; MBJ to MAE, 16 Feb. [1870], EGP; and AEG to AJG, 1 Mar. [1870], ibid.

16. EEL to AEG, 19 Feb. [1870], EGP; Mary P. Manigault to EEL, 28 Feb. 1870, ibid.; MAE to AEG, [mid-May 1870], ibid.; and GdMama to Grandsons, 9 Mar. 1870, ibid.

17. NGG to AJG, 10 Feb. 1870, EGP; NGG to Grandmama, 28 Mar. 1870, ibid.; and NGG to AJG, 29 Mar. 1870, GFP.

18. AJG to REE, 20 Apr. 1870, EGP; and MAE to AJG, [late Apr. 1870], ibid.

19. AJG to MAE, 16 May 1870, EGP; Long, *Memoirs of Robert E. Lee*, 471; Simkins and Woody, *South Carolina during Reconstruction*, 344; ANE to AEG, [18 May 1870], EGP; MAE to AEG, [mid-May 1870], ibid.; and EEL to AEG, 18 May [1870], ibid.

20. EEL to AEG, 20 May [1870], and AEG, to EEL, 24 May [1870], ibid.; EEL to AEG, 4 June 1870, ibid.; AEG to EEL, 29 May [1870], ibid.; and EEL to AEG, 31 May [1870], ibid.

21. EEL to AEG, 7 June [1870], EGP; Tom Elliott to AJG, 12 July 1870, GFP; and AEG to EEL, 11 June [1870], EGP.

22. EEL to AEG, 14 June [1870], EGP; and AEG to EEL, 11 June [1870], ibid.

23. 1870 Population Census, Savannah, 104, NA; AEG to AJG, 14 June [1870], GFP; and AJG to AEG, 17 June 1870, EGP.

24. NGG to AJG, 2 July 1870, GFP; AJG to NGG, 8 July 1870, EGP; and EEL to AEG, 12 July [1870], ibid.

25. Tom Elliott to AJG, 12 July 1870, GFP.

26. Alfonso B. Gonzales to AJG, 25 July 1870, GFP; and NGG to EEL, 29 Aug. 1870, EGP.

27. NGG to EEL, 29 July 1870, EGP; EEL to "My Dear Boys," 16 Aug. 1870, ibid.; AEG to AJG, 20 Aug. [1870], GFP; NGG to AJG, 24 Aug. 1870, ibid.; Mrs. Y. B. Dalcour to MAE, 15 Sept. 1870, EGP.

28. 1870 Federal Census, Colleton County, S.C., 12; C. M. Manigault to Miss Elliott, [Sept. 1870], EGP; AEG to AJG, 18 Sept. 1870, GFP; Lena L. Cary to MAE, [1872], EGP; E. F. Belt to EEL, 7 Oct. [1870], ibid.; E. F. Belt to EEL, 10 Nov. [1870], ibid.; Mrs. Belt to EEL, 27 Feb. 1871, ibid.; Mary H. Schoolcraft to EEL, 10 Apr. 1871, ibid.; William Elliott to EEL, 22 Nov. 1872, ibid.; Gertrude Gonzales [written by Emily Elliott] to NGG, 8 Oct. [1870], ibid.; NGG to Gertrude Gonzales, 28 Sept. 1870, ibid.; and AEG to AJG, 20 Aug. [1870], GFP.

29. NGG to AJG, 27 Sept. 1870, GFP; AJG to AEG, 2 Oct. 1870, EGP; AJG to AEG, 15 July 1872, GFP; AEG ro AJG, 28 July [1872], ibid.; and *Courier*, 14 Oct. 1870, 4.

30. Edward Lafitte to AEG, 18 Jan. 1871, EGP; and AJG to AEG, 15 July 1872, GFP.

31. *Tribune*, 7 Nov. 1870, 6; and *New York Times*, 11 July 1871, 5.

32. 1870 Record of Assessments, Fifteenth Ward, Municipal Archives, City of New York, roll 116, 68; and Matthew Hale Smith, *Sunshine and Shadow in New York* (Hartford: J. B. Burr, 1868), 214–18.

33. Rev. Bruce Forbes, Interview with the Author, 10 May 1992, St. Bartholomew's Episcopal Church, New York City, who confirmed that choir members received a salary. E. Clowes Chorley, *The Centennial History of Saint Bartholomew's Church in the City of New York, 1835–1935* (New York: private publisher, 1935), 3; and Leonard Young, *A Short History of St. Bartholomew's Church in the City of New York, 1835–1960* (New York: private publisher, 1960), 4–5.

34. NGG to MAE, 19 Nov. 1870, EGP; Ambrosio Gonzales Jr., to AJG, 28 July [1872], GFP; MBJ to MAE, 10 Dec. [1870] EGP; and NGG to AJG, 24 Aug. 1870, GFP.

35. AJG to Ambrosio José Gonzales Jr., 15 July 1872, GFP; George H. Hoppock, City of Charleston Death Certificates, Charleston County Library, Charleston, S.C.; MBJ to MAE, 27 Nov. [1870], EGP; Edward Lafitte to AEG, 18 Jan. 1871, ibid.; Cecilia M. de Poujaud to MAE, 28 Jan. 1871, ibid.; C. C. Pinckney to MAE, 9 Feb. 1871, ibid.; and MBJ to MAE, [15 Feb. 1871], ibid.

36. AJG to AEG, 27 Mar. 1871, EGP; MBJ to MAE, 14 Feb. [1872], ibid.; and Simkins and Woody, *South Carolina during Reconstruction*, 417.

37. AJG to John D. Warren, 7 Apr. 1871, John D. Warren Papers, SCL; 1870 Federal Census, Colleton County, S.C., 302; AJG to Ambrosio José Gonzales Jr., 15 July 1872, GFP; AJG to William C. Bee, 25 Feb. 1872, EGP; William C. Bee to AJG, 6 Mar. 1872, GFP; and Unendorsed Check no. 43 "to the Bearer or A. J. Gonzales," from Rafael Carrasco, 3 June 1871, National City Bank, ibid.

38. AJG to Ambrosio José Gonzales Jr., 15 July 1872, GFP; 1870 Federal Census, South Carolina, Fourth Ward of Charleston, 266; *New York Times*, 23 Aug. 1871, 5; MBJ to MAE, 14 Feb. [1872], EGP; and Elliott Johnstone to MBJ, 20 July 1871, ibid.

39. AJG to AEG, 15 July 1872, GFP; G. Decourt to AJG, 8 Nov. 1871, ibid.; Simkins and Woody, *South Carolina during Reconstruction*, 145, 242–47; and Williamson, *After Slavery*, 114–15, 118–19, 332–35.

40. AJG to MAE, 29 May 1872, GFP; Oliver Perry Chitwood, *John Tyler: Champion of the Old South* (New York: Russell and Russell, 1964), 478; *Mercury*, 1 Aug. 1861, 1; AJG to William C. Bee, 25 Feb. 1872, EGP; MBJ to MAE, 25 Nov. [1868], ibid.; Emmie Johnstone to MAE, 27 Jan. [1869], ibid.; Emma E. Johnstone, Membership Application, 19 Oct. 1910, Archives of the Maryland Historical Society, Baltimore; *Wood's Baltimore City Directory, 1871* (Baltimore: John W. Woods, 1871), 541; and John Dorsey, *Mount Vernon Place* (Baltimore: Maclay and Associates, 1983), 59, 65.

41. Francis F. Beirne, *St. Paul's Parish Baltimore: A Chronicle of the Mother Church* (Baltimore: Horn-Shafer, 1967), 103–5, 125, 126; Porter, *The History of a Work of Faith and Love*, 20; and *The Baltimore Saturday Night*, 9 Mar. 1872, 2.

42. In 1887, the rooming house became the Woman's Industrial Exchange. In 1979, it became part of the National Register of Historic Places. Today, it operates a WIE storefront outlet for handmade wares and a rear dining room serves breakfast and lunch. Rooms are still rented on the upper floors to "ladies in reduced circumstances." The author had lunch there and viewed the second floor quarters. Land Records, G.R. 591, folio 117, Record Office, Baltimore City Courthouse; *Baltimore Sun*, 1 Apr. 1979; Vertical Clipping File, "Woman's Industrial Exchange," Enoch Pratt Library, Baltimore; Beirne, *St. Paul's Parish Baltimore*, 128; *Wood's Baltimore City Directory, 1872*, 251; *The Baltimore Saturday Night*, 27 Jan. 1872, 2; AEG to MAE, 14 May 1872, GFP; *Proceedings of the Grand Lodge of Ancient Free and Accepted Masons of Maryland, 1872* (Baltimore: Henry E. Huber and Son, 1873), 104–5; and *Wood's Baltimore City Directory, 1871*, 134.

43. Dorsey, *Mount Vernon Place*, 42–43, 51; *Wood's Baltimore City Directory, 1872*, 175, 350, 361; *Baltimore Sun*, 30 Apr. 1875, 4; *Baltimore Evening Sun*, 8 May 1911; ibid., 7 Sept. 1949 and 10 Sept. 1950; MBJ to MAE, 25 Nov. [1868], EGP; Name Index File and Vertical Clipping File, "Schools: Mme. Lefebvre's," Enoch Pratt Library, Baltimore; and Sarah A. Kummer Papers, Maryland Historical Society Library, Baltimore.

44. MBJ to MAE, 3 Jan. 1872 and Sunday Night [Jan. 1872], EGP; May Johnstone to MAE, 14 Feb. [1872], ibid.; John D. McAulay, *Civil War Breech Loading Rifles* (Lincoln, R.I.: Andrew Mobray, 1987), 110; and EEL to AEG, 8 Feb. [1872], EGP.

45. 1870 Federal Census, Beaufort, S.C., 14; C. A. Hamilton to MAE, 1 May 1872, EGP; MBJ to AEG, 6 Feb. [1872], ibid.; EEL to AEG, 8 Feb. [1872], ibid.; and *New York Times*, 12 Mar. 1871, 8.

46. AJG to William C. Bee, 25 Feb. 1872, EGP; William C. Bee to MAE, 29 Feb. 1872, ibid.; William C. Bee to AJG, 6 Mar. 1872, ibid.; and EEL to AEG, 27 Feb. [1872] and 5 Mar. [1872], ibid.; AEG to NGG, 2 Mar. 1872, ibid.; C. A. Hamilton to MAE, 1 May 1872, ibid.; NGG to EEL, 28 May 1872, ibid.; AEG to EEL, 28 May 1872, 6 June [1872], and 15 June [1872], ibid.; and NGG to Alfonso Gonzales, 6 June 1872, ibid.

47. EEL to "My Dear Boys," [24 May 1872], EGP; Social Hall Contract [1872], GFP; AJG to MAE, 29 May 1872, ibid.; and AEG to EEL, 15 June [1872], EGP.

48. AJG to Asger Hamerik, 15 June 1872, EGP; and AEG to AJG, 28 July [1872], GFP.

49. MBJ to MAE, 16 June [1872], EGP; Thomas, *Robert E. Lee*, 106, 149, 161; Avery Craven, ed., *"To Markie": The Letters of Robert E. Lee to Martha Custis Williams* (Cambridge: Harvard University Press, 1933); MBJ to EEL, 20 Feb. [1877], EGP; EEL to "My Dear Boys," 21 June [1872], ibid.; and EEL to AEG, 25 June [1872], ibid.

50. AJG to AEG, 15 July 1872, GFP; William C. Bee to MAE, 24 July 1872, EGP; and AEG to AJG, 28 July [1872], GFP.

51. MBJ to MAE, 28 July [1872], EGP; AJG to AEG, 28 Aug. 1872, ibid.; A. G. Menocal to AJG, 8 Aug. 1886, GFP; "Aniceto Garcia Menocal," *The National Cyclopedia of American Biography*, 14:354; Santa Cruz y Mallen, *Historia de Familias Cubanas*, 1:167; and Portell Vilá, *Vidas de la Unidad Americana*, 179–91.

52. MBJ to MAE, 28 July [1872], EGP; AJG to AEG, 28 Aug. 1872, ibid.; AJG to NGG, 31 Aug. 1872, ibid.; MAE to W. W. Corcoran, [mid-Aug. 1873], ibid.; "William Wilson Corcoran," *The National Cyclopedia of American Biography*, 3:153; AEG to MAE, 22 Sept. [1872] and 9 Mar. [1873], EGP; EEL to AEG, 8 Oct. [1872], ibid.; and H. H. Elliott to EEL, [mid-Oct. 1872] and 26 Nov. 1872, ibid.

53. AJG to MAE, 30 Jan. 1873, EGP; NGG to AEG, [mid-Nov. 1872], ibid.; and EEL to AEG, 24 Jan. [1873] and 7 Feb. [1873], ibid.

54. A. B. Stockwell was president of the Pacific Mail Steamship Company. MBJ to MAE, 12 Feb. [1873], EGP; AEG to MAE, 9 Mar. [1873], ibid.; NGG to AEG, 10 and 23 Feb. 1873 and 1 Aug. 1873, ibid.; EEL to AEG, 15 Mar. [1873], ibid.; and NGG to AEG, 1 Aug. 1873, ibid.

55. MBJ to MAE, 28 June [1873], EGP; EEL to AEG, 18 June [1873], ibid.; MAE to William Wilson Corcoran, [mid-Aug. 1873], ibid.; and AEG to NGG, 9 Aug. 1873, GFP.

56. Ralph K. Andrist, ed., *The American Heritage History of the Confident Years, 1865–1916* (New York: Crown, 1987), 84–85; and William C. Bee to REE, 30 Sept. 1873, EGP.

57. Mr. Johnston to MAE, 18 Sept. 1873, EGP; NGG to MAE, 4 Nov. and 23 Dec. 1873, ibid.; NGG to AEG, 8 Nov. 1873, ibid.; Anne Ward Crocker, "St. Timothy's Protestant Episcopal Church," Herndon Historical Society, Herndon, Va.; and Virginia Carter Castleman, "Reminiscences of an Oldest Inhabitant," Herndon Historical Society, Herndon, Va.

58. A full account of the Virginius incident appears in Richard H. Bradford, *The Virginius Affair* (Boulder: Colorado Associated University Press, 1980). AJG to Ulysses S. Grant, 8 Jan. 1876, Letters of Application and Recommendation during the Administration of Ulysses S. Grant, 1869–77, RG 59, NA; *New York Times*, 10 Nov. 1873, 8; ibid., 16 and 27 Nov. 1873, 1; William F. McFeely, *Grant: A Biography* (New York:

W. W. Norton, 1981), 350; Herminio Portell Vilá, *Historia de Cuba en sus Relaciones con los Estados Unidos y España* (1938; reprint, Miami: Mnemosyne Publishing, 1969), 2:428–47; and Thomas, *Cuba*, 262.

59. AJG to Ulysses S. Grant, 8 Jan. 1876, Letters of Application and Recommendation during the Administration of Ulysses S. Grant, 1869–77, RG 59, NA.

60. AEG to REE, 18 Jan. 1874 and 8 Mar. 1874, EGP; NGG to EEL, 17 Feb. 1874, [12 Mar. 1874] and 5 July 1874, ibid.; NGG to AEG, 21 Feb. 1874, ibid.; REE to MAE, 25 [Jan. 1874] and 26 Apr. [1874]. ibid.; Simkins and Woody, *South Carolina during Reconstruction*, 137–40, 155, 180–81; Williamson, *After Slavery*, 153; Lander, *A History of South Carolina, 1865–1960*, 25; Taylor, *The Negro in South Carolina during the Reconstruction*, 208; Title under Order of Court, 29 Oct. 1879, EGP; Carlos Tracy to William Elliott, Esq., 12 Apr. 1878, ibid.; Carlos Tracy to J. B. Campbell, 17 June 1878, ibid.; RG 109, General Orders no. 16, 12 May 1862, Department South Carolina, Georgia, and Florida, General Orders 1861–64, box 65, NA; and Mortgage Bond for Bluff plantation, 26 Apr. 1886, GFP.

61. MAE to REE, Ash Wednesday [late Feb. 1874], EGP; Social Hall Delinquent Land Certificate, 18 May 1874, GFP; John D. Warren to AJG, 6 June 1874, John D. Warren Papers, SCL; ANE to NGG, 11 June [1874], EGP; William C. Bee to MAE, 22 June 1874, ibid.; Elliott Johnstone to MAE, 16 Aug. 1874, ibid.; William Elliott, Esq., to "My Dear Cousin," 20 Aug. 1874, ibid.; Certificate of Release, 17 July 1875, ibid.; William Elliott, Esq., to EEL, 3 June 1881, ibid.; and Williamson, *After Slavery*, 86.

62. A. E. Gonzales, *In Darkest Cuba*, 18; AEG to EEL, 9 Oct. 1874, EGP; AEG to NGG, 25 Oct. 1874; ibid.; NGG to MAE, 26 Sept. 1875, ibid.; John F. Trow, *Trow's New York City Directory, for the Year ending May 1, 1875* (New York: Trow City Directory Company, 1875), 488; Charles C. Dodge to AJG, 26 Jan. 1875, GFP; "New York," *New York Times*, 28 Jan. 1875, 8.

63. Portell Vilá, *Historia de Cuba*, 2:500–503; AJG to Ulysses S. Grant, 8 Jan. 1876, Letters of Application and Recommendation during the Administration of Ulysses S. Grant, 1869–77, RG 59, NA; NGG to "Dear Ladies," 14 Feb. 1876, EGP; NGG to MAE, 25 Apr. 1876, ibid.; AEG to MAE, 11 Oct. 1876, EGP; Foner, *Reconstruction*, 570–75; A. E. Gonzales, *In Darkest Cuba*, 19, 22; Richard Mark Zuczek, "State of Rebellion: People's War in Reconstruction South Carolina, 1865–1877" (Ph.D. diss., Ohio State University, 1993), 451–57; Warner, *Generals in Gray*, 59–60; William Arthur Sheppard, *Red Shirts Remembered: Southern Brigadiers of the Reconstruction Period* (Atlanta: Ruralist Press, 1940), 47–50; Williamson, *After Slavery*, 410–11; Simkins and Woody, *South Carolina during Reconstruction*, 497–504; Edmund L. Drago, *Hurrah for Hampton! Black Red Shirts in South Carolina during Reconstruction* (Fayetteville: University of Arkansas Press, 1998), 4, 8–10; Tindall, *South Carolina Negroes, 1877–1900*, 34; and William J. Cooper Jr., *The Conservative Regime: South Carolina, 1877–1890* (Baltimore: Johns Hopkins Press, 1968), 22.

64. Tom Elliott to "My Dear Nephew," 30 May 1876, EGP; and S. G. Elliott to MAE, 23 Aug. 1876, ibid.

65. Ambrosio José Gonzales Address Book in GFP; Holzman, *Adapt or Perish*, 107–8; and *New York Times,* 17 Nov. 1876, 8.

66. William C. Bee to AJG, 13 Dec. 1876, GFP; William C. Bee to AJG, 8 Jan. 1877, ibid.; AEG to AJG, 21 Feb. 1877, ibid.; MBJ to Miss Williams, [12 Feb. 1877], EGP; MBJ to EEL, 20 Feb. [1877], ibid.; and Gertrude Gonzales to EEL, 28 Jan. 1877 and 11 Mar. [1877], ibid.

67. AEG to AJG, 21 Feb. 1877, GFP; REE to AEG, 21 [Feb.] 1877; EGP; Charles C. Pinckney to AEG, 23 [Feb.] 1877, ibid.; C. S. Gadsden to Conductor, 23 [Feb.] 1877, ibid.; Lena L. Cary to MBJ, 12 Mar. [1877], EGP; AEG to REE, 14 Mar. 1877, ibid.; and AEG to AJG, 2 and 24 Apr. 1877, GFP.

68. Taylor, *The Negro in South Carolina during the Reconstruction*, 260; Simkins and Woody, *South Carolina during Reconstruction*, 514; and Cooper, *The Conservative Regime,* 21, 39–40, 209–10.

69. AEG to AJG, 23 June 1877, GFP; Gertrude Gonzales to EEL, 9 July 1877, EGP; and NGG to AJG 1 July 1877, GFP.

70. George M. Barbour, *Florida for Tourists, Invalids, and Settlers* (1882; reprint, Gainesville: University of Florida Press, 1964), 68–69; *Savannah Morning News,* 13 Sept. 1877, 1; Elliott Johnstone to "My Dear Aunts," 13 Sept. 1877, EGP; Gertrude Gonzales to EEL, 21 Oct. 1877, ibid.; C. to EEL, 7 Nov. 1877, ibid.; and MBJ to EEL, 28 Apr. [1878], ibid.

**Paralytic Patriot**

1. NGG to EEL, 6 Dec. [1877], 3 Feb. [1878] and 7 May [1878], EGP; A. E. Sholes, *Shole's Directory of the City of Savannah, January 1879* (Charleston: Walker, Evans, and Cogwell, 1879), 212; NGG to REE, 9 Oct. 1881, EGP; and REE to ANE, 26 June [1878], ibid.

2. Although correspondence indicates that Narciso and Brosio wore their Hampton red shirts in Beaufort in 1878, Brosio erred forty-four years later when writing that it had occurred in 1876. A. E. Gonzales, *In Darkest Cuba*, 19. William Elliott, Esq., to EEL, 17 June 1878, EGP; 1880 Federal Census, Lowndes County, Ga., 190; NGG to AJG, 13 Jan. 1879, GFP; NGG to Trudie and Plug, 9 Aug. 1878, EGP; REE to EEL, 3 Sept. 1878, ibid.; NGG to AEG, 7 Sept. 1878, ibid.; and NGG to EEL, 3 Nov. 1878, ibid.

3. AJG to AEG, 27 Dec. 1878 and 6 Jan. 1879, EGP.

4. MBJ to EEL, Thursday [Jan. 1879], EGP.

5. John D. Warren to AEG, 22 Jan. 1879, EGP; and NGG to AJG, 8 and 13 Jan., 23 Mar., and 24 Aug. 1879, GFP.

6. NGG to AJG, 23 Mar. 1879, GFP; MBJ to EEL, 23 [Apr. 1879], EGP; and Gertrude Gonzales to EEL, 24 May 1879, ibid.

7. NGG to AJG, 23 Mar. 1879, GFP.

8. NGG to EEL, 9 Jan. 1881, EGP; Prospero Gonzales Certificate, 15 Dec. 1869, GFP; AJG to Ignacio Gonzalez Gauffreau, 10 Sept. 1886, ibid.; AJG to Ignacio Gonzales, 18 Sept. 1883 and 17 Nov. 1884, ibid.; and NGG to AJG, 23 Mar., 5 May, 9 June, and 24 Aug. 1879, ibid.

9. NGG to EEL, 26 July 1879, EGP; NGG to AJG, 24 Aug. and 12 Oct. 1879, GFP; and NGG to EEL, 12 Oct. 1879, ibid.

10. Charles C. Pinckney to AJG, 5 Mar. 1880, GFP; NGG to AJG, 7 Mar. 1880, ibid.; and Greene, *Porter-Gaud School*, 12.

11. NGG to AJG, 14 Mar. 1880, GFP; and NGG to EEL, 28 Mar. 1880, EGP.

12. AEG to EEL, 8 June 1880, EGP; 1880 Federal Census, Colleton County, S.C., 45; 1880 Federal Census, Lowndes County, Ga., 190; REE to Harriet R. Gonzales, 14 Aug. 1880, EGP; NGG to EEL, 6 Aug. 1880, ibid.; AEG to EEL, 15 Oct. 1880, ibid.; EEL to WEG, [Apr. 1881], ibid.; NGG to WEG [Apr. 1881], ibid.; and AEG to WEG, 26 Apr. 1881, ibid.

13. NGG to EEL, 9 Jan. 1881, EGP; Gertrude Gonzales to EEL, 6 Feb. 1881, ibid.; NGG to WEG, 29 Mar. 1881, ibid.; Gertrude Gonzales to WEG, 20 Feb. 1881, ibid.; and NGG to AEG, 15 Mar. 1881, ibid.

14. Matthew Calbraith Butler (1836–1909), a Greenville antebellum lawyer and son-in-law of Gov. Francis Wilkinson Pickens, resigned his seat in the state legislature in 1861 for a captain's commission in the Hampton Legion. As a result of the battle of Brandy Station in June 1863, his right foot was amputated and he was promoted to general. In 1871, he created a lottery to raise funds to acquire lands for immigrants and pay their ship passages. He was elected as a Conservative Democrat to the U.S. Senate in 1876. Warner, *Generals in Gray*, 40–41; NGG to AEG, 15 and 19 Mar. 1881, EGP, and 22 Mar. 1881, GFP; Simkins and Woody, *South Carolina during Reconstruction*, 245–47; Tindall, *South Carolina Negroes, 1877–1900*, 178, 180, 182–84; Taylor, *The Negro in South Carolina during the Reconstruction*, 71; NGG to WEG, 29 Mar. EGP; and REE to EEL, 10 Sept. 1881, ibid.

15. AEG to "My Dear Folks," 23 July and 5 Sept. 1881, EGP; *New York Times*, 1 Sept. 1881, 5; REE to EEL, 10 Sept. 1881, EGP; AEG to REE, 5 Sept. 1881, ibid.; and AEG to WEG, 15 Oct. 1881, ibid.

16. *New York Times*, 23 June 1881, 5; NGG to AJG, 13 Sept. 1881, GFP; NGG to EEL, 15 Sept. 1881, EGP; and NGG to REE, 18 Sept. 1881, ibid.

17. AEG to REE, 16 and 27 Sept. and 9 Oct. 1881, EGP; AEG to WEG, 15 and 26 Oct. 1881, ibid.; NGG to REE, 9 Oct. 1881, ibid.; and AEG to his aunts, [9 Oct. 1881], ibid.

18. NGG to REE, 9 Oct. 1881, EGP; NGG to WEG, 16 Oct. 1881, ibid.; and NGG to REE, 21 Oct. [1881], ibid.

19. AEG to his aunts, [9 Oct. 1881], EGP; NGG to EEL, 30 Oct. 1881, ibid.; and NGG to AJG, 6 Nov. 1881, GFP.

20. Robert Smalls later avoided Narciso when he published another article calling him a "Convicted Bribe-Taker and Stalwart Pot." *News and Courier*, 31 Jan. 1883, 1. Smalls was found guilty in court of bribing the House and Senate clerks to falsify his $2,250 claim. He returned to Congress in 1884 but was defeated for reelection two years later by attorney William Elliott of Beaufort. NGG to WEG, 18 Dec. 1881, GFP; NGG to EEL, 25 Dec. 1881 and 8 Jan. 1882, EGP; Edward A. Miller Jr., *Gullah Statesman: Robert Smalls from Slavery to Congress, 1839–1915* (Columbia: University of South Carolina Press, 1995), 137; Rose, *Rehearsal for Reconstruction*, 149–50, 391; AEG to EEL, 12 Jan. 1882, EGP; and NGG to EEL, 25 Jan. 1882, ibid.

21. Irene Espinosa to AJG, 12 Jan. 1882, GFP; Gertrude Gonzales to EEL, 19 Mar. 1882, EGP; NGG to EEL, 27 Mar. and 9 Apr. 1882, ibid.; AJG to Irene Espinosa, 13 Apr. 1882, GFP; and *New York Times*, 18 Apr. 1882, 5.

22. REE to AEG, 3 Oct. [1881], EGP; AEG to REE, 26 Sept. 1882, ibid.; AEG to EEL, 9 Oct. 1882, 19 Oct. 1883 and 2 and 26 Nov. 1883, ibid.; Social Hall Title, 19 Jan. 1883, GFP; AJG to Ignacio Gonzales Gauffreau, 5 and 19 Aug. 1883, not in the family correspondence, but mentioned in Jones, *Stormy Petrel*, 56; and *New York Times*, 17 May 1882, 8.

23. NGG to REE, 17 Aug. 1882, EGP; and NGG to EEL, 22 Aug. 1882, ibid.

24. NGG to Gertrude Gonzales, 1 Dec. 1882, EGP; NGG to EEL, 3 Jan. 1883, ibid.; Ambrosio José Gonzales Address Book in GFP; and *New York Times*, 1 Aug. 1883, 8.

25. AEG to "Dear Folks," 7 July 1883, EGP; AEG to "Dear Folks," 16 Aug. 1883, GFP; WEG to ANE, 21 Aug. 1883, EGP; AEG to EEL, 29 Aug. and 24 Sept. 1883, ibid.; and WEG to EEL, 12 Sept. 1883, GFP.

26. WEG to EEL, 27 Sept. 1883, EGP; AJG to Ignacio Gonzales Gauffreau, 18 Sept. 1883, GFP; Ignacio Gonzales Gauffreau to AJG, 5 July 1884, ibid.; AEG to EEL, 2 Nov. 1883, EGP; NGG to Gertrude Gonzales, 20 Sept. 1883, EGP; WEG to EEL, 8 Oct. 1883, ibid.; and WEG to AJG, 3 Nov. 1883, GFP.

27. NGG to Gertrude Gonzales, 20 Sept. 1883, EGP; WEG to EEL, 8 Oct. 1883, ibid.; and WEG to AJG, 3 Nov. 1883, GFP.

28. *News and Courier*, 20 and 21 Nov. 1883, 4; WEG to EEL, 27 Nov. 1883, 7 Dec. 1883 and 24 Jan. 1884, EGP; AEG to EEL, 29 Nov. 1883, ibid.; NGG to Gertrude Gonzales, 8 Apr. 1884, ibid.; and Jones, *Stormy Petrel*, 89.

29. U.S. Navy physician Daniel Guiteras Gener and his wife had three sons, all of whom became engineers and fought with the U.S. Armed Forces in France during the First World War. One son, Lt. Julián Guiteras, died there on Oct. 12, 1918. Santa Cruz y Mallen, *Historia de Familias Cubanas*, 8:175–76; Dollero, *Cultura Cubana*, 94; and AJG, Address Book, GFP.

30. AJG, Address Book, GFP.

31. AEG to [?], 13 Feb. 1884, EGP; "On to Cuba," *New Orleans Times-Democrat*, 30 Mar. 1884, 9; "The Cuban Crusade," ibid., 6 Apr. 1884, 9; Petition to the Trustees of the University of Louisiana, 28 May 1884, GFP; *Announcement of the University of Louisiana, 1884-5: Catalogue of the Academical Department, Sixth Session, 1883-4* (New Orleans: G. T. Lathrop, 1884), 6; and *Tulane University of Louisiana: Catalogue of Academical Department, 1884-85* (New Orleans: L. Graham and Son, 1885), 15.

32. AJG Address Book, GFP; AJG to Ignacio Gonzales Gauffreau, 17 Nov. 1884, GFP; Richard E. Welch Jr., *The Presidencies of Grover Cleveland* (Lawrence: University Press of Kansas, 1988), 43; Tindall, *South Carolina Negroes, 1877-1900*, 43; *News and Courier*, 21 Jan. 1885, and a note from Gonzales to Davis in Jefferson Davis Collection, Museum of the Confederacy, Richmond; John F. Marszalek, *Sherman: A Soldier's Passion for Order* (New York: Free Press, 1993), 473-74; Davis, *Jefferson Davis*, 679-80; Warner, *Generals in Blue*, 219-20; and Wise, *Gate of Hell*, 238.

33. *News and Courier*, 21 Jan. 1885; Allan Nevins, *Grover Cleveland: A Study in Courage* (New York: Dodd, Mead, 1966), 51-52; and Welch, *The Presidencies of Grover Cleveland*, 24, 62.

34. AJG to William Porcher Miles, 23 May 1885, William Porcher Miles Papers, SHC; WEG to EEL, 15 Feb. 1885, EGP; Nevins, *Grover Cleveland*, 1-3; and NGG to [?], [Mar. 1885], EGP.

35. NGG to EEL, 5 Apr. 1885, EGP; NGG to REE, 16 Apr. 1885, ibid.; AJG to William Porcher Miles, 23 May 1885, William Porcher Miles Papers, SHC; "William Porcher Miles," *The National Cyclopedia of American Biography*, 11:35; AEG to Gertrude Gonzales, 27 Apr. 1885, GFP; and AEG to EEL, 13 Oct. 1885, ibid.

36. AJG to Thomas F. Bayard, 24 May 1885, Letters of Application and Recommendation during the Administrations of Franklin Pierce and James Buchanan, 1853-61, RG 59, NA; Welch, *The Presidencies of Grover Cleveland*, 177-79; Nevins, *Grover Cleveland*, 207, 404; and *New York Times*, 19 June 1885, 4.

37. Fauquier County Bicentennial Committee, *Fauquier County, Virginia, 1759-1959*, 173; AJG to José Ignacio Rodríguez, 31 Aug. 1885, Jose Ignacio Rodriguez Papers, LC; Oscar Fay Adams, *A Dictionary of American Authors* (Boston: Houghton Mifflin, 1904), 552; and EEL, Last Will and Testament, 10 Nov. 1885, EGP.

38. AJG to D. S. Lamont, 24 Jan. 1886, Grover Cleveland Presidential Papers, series 2, reel 29, LC; D. S. Lamont to AJG, 15 May 1886, GFP; Marriage Settlement, 29 May 1886, ibid.; Title to Real Estate, Social Hall plantation, 29 May 1886, ibid.; and Bond Discharge and Release, May 1886, ibid.

39. The first Gullah tale of Ambrose E. Elliott appeared in the *Charleston Sunday News*, 2 Sept. 1886. The tales were later compiled in AEG, *The Black Border: Gullah Stories of the Carolina Coast*. Columbia, S.C.: The State, 1922. Matthew Somerville Morgan (b. 1839 London, d. 1890 New York) a cartoonist, imported to New York in 1870 by Frank Leslie for his *Illustrated Newspaper*, to counter Thomas Nast's pro-Republican

cartoons. In 1872 and 1873, Morgan published various sketches depicting life in the Manhattan slums where Gonzales had lived. AJG to Ignacio Gonzales, 19 June 1886, GFP; AJG to José Ignacio Rodríguez, 26 June and 15 July 1886, José Ignacio Rodríguez Papers, LC; Aniceto García Menocal to AJG, 8 Aug. 1886, GFP; and W. Miller Owen to AJG, 27 Aug. 1886, ibid.

40. AJG to Ignacio Gonzales Gauffreau, 28 July and 10 Sept. 1886, GFP; Ignacio Gonzales Gauffreau to AJG, 20 Aug. 1886, ibid.; and AEG to AJG, 1 Oct. 1886, ibid.

41. WEG to Gertrude Gonzales, 19 Mar. 1886, EGP; NGG to Aunt, 12 Apr. 1886, ibid.; and *News and Courier*, 6 Feb. 1887, 3; Leila Habersham to Cousin, 19 Jan. 1887, EGP; and AJG to WEG, 7 Feb. 1887, GFP.

42. Cohen, *Historic Springs of the Virginias*, 84–89; Flora Adams Darling to JD, 19 July 1887, JDP; and *The National Cyclopedia of American Biography*, 19:138–39.

43. Flora Adams Darling to JD, 19 July 1887, JDP; Cooper, *Jefferson Davis*, 373; and Nevins, *Grover Cleveland*, 251, 323.

44. Advocates of spiritualism included Queen Victoria; Mary Todd Lincoln, who held seances in the White House; New York newspaper editors Horace Greeley and William Cullen Bryant; Cornelius Vanderbilt, who sought stock manipulation advice from the ghost of the murdered financier James Fisk; U.S. senators Nathaniel Tallmadge and Benjamin Wade; U.S. representative Joshua Gidding; New York Supreme Court judge John Worth Edmonds; New York City school superintendent Henry Kiddle; Thomas Wentworth Higginson; William Lloyd Garrison; James Fenimore Cooper; Henry Wadsworth Longfellow; Washington Irving; Robert Dale Owen; James Greenleaf Whittier; Victoria Woodhull; Lydia Maria Child; and Harriet Beecher Stowe, who became interested in communicating with spirits after the death of her first son in 1857. Mrs. William H. Boyd, *Boyd's Directory of the District of Columbia, 1888* (Washington, D.C.: Wm. Ballantyne and Son, 1887), 427; *Preliminary Report of the Commission Appointed by the University of Pennsylvania to Investigate Modern Spiritualism in Accordance with the Request of the Late Henry Seybert* (Philadelphia: J. B. Lippincott, 1887), 82–83; Alfred Russel Wallace, *My Life: A Record of Events and Opinions* (London: Chapman and Hall, 1905), 2:341–42; A. J. Gonzales, *Heaven Revealed*, 11; Ronald Pearsall, *The Table-Rappers* (New York: St. Martin's, 1972), 59; Russell M. and Clare R. Goldfarb, *Spiritualism and Nineteenth-Century Letters* (Cranbury, N.J.: Associated University Presses, 1978), 26, 33, 41–45; Ruth Brandon, *The Spiritualists: The Passion for the Occult in the Nineteenth and Twentieth Centuries* (New York: Alfred A. Knopf, 1983), 40–41; R. Laurence Moore, *In Search of White Crows: Spiritualism, Parapsychology, and American Culture* (New York: Oxford University Press, 1977), 3; and Howard Kerr, *Mediums, and Spirit-Rappers, and Roaring Radicals* (Urbana: University of Illinois Press, 1972), 162.

45. Gonzales, *Heaven Revealed*, 5–10; and Wallace, *My Life*, 342.
46. *Preliminary Report*, 23, 26, 85–86.
47. Gonzales, *Heaven Revealed*, 6–8, 11–14, 17, 30.
48. Ibid., 18, 21, 24, 35, 37, 52, 63–64, 67–68.
49. Walter Franklin Prince, "Supplementary Report on the Keeler-Lee Photographs," *Proceedings of the American Society for Psychical Research*, Mar. 1920, 529–87; *Preliminary Report*, 91–92; *Encyclopedia of Occultism and Parapsychology* (Detroit: Gale Research, 1978–82), 2:716 and 3:1261–62; Brandon, *The Spiritualists*, 103–4, 222–23; Pearsall, *The Table-Rappers*, 118–25; and Kerr, *Mediums, and Spirit-Rappers, and Roaring Radicals*, 119.
50. Goldfarb, *Spiritualism and Nineteenth-Century Letters*, 35; and Gonzales, *Heaven Revealed*. The spirit messages in slips of paper are in GFP.
51. E. Martínez Sobral to AJG, 21 Sept. 1889, GFP; Francisco Lainfiesta to AJG, 3 July 1889, ibid.; AJG Memorandum [June 1889], ibid.; U.S. Senate, *International American Conference: Reports of Committees and Discussions Thereon*, vol. 1, Exec. Doc. 232, pt. 1, 51st Congress, 1st sess. (Washington, D.C.: Government Printing Office, 1890), 53; F. G. Pierra to William E. Curtis, 27 and 29 Aug. 1889, International Conferences, Commissions, and Expositions, First International Conference of American States, 1889–90, Letters Received by the Executive Officer, 24 July 1889–25 Apr. 1890, box 1, RG 43, NA; Samuel Guy Inman, *Inter-American Conferences, 1826–1954: History and Problems* (Washington, D.C.: University Press of Washington, D.C., 1965), 34–36, 45; Ernesto Quesada, *Primera Conferencia Panamericana* (Buenos Aires: Imprenta Schenone, 1919), 5; and David Saville Muzzey, *James G. Blaine: A Political Idol of Other Days* (Port Washington, N.Y.: Kennikat Press, 1963), 432.
52. Inman, *Inter-American Conferences, 1826–1954*, 38–40; José R. Villalón to William E. Curtis, International Conferences, Commissions, and Expositions, First International Conference of American States, 1889–90, Letters Received by the Executive Officer, 24 July 1889–25 Apr. 1890, box 1, RG 43, NA; Muzzey, *James G. Blaine*, 432–33; and Gonzalo de Quesada, "Ambrosio Jose Gonzalez," *Patria*, 31 Dec. 1892, 3.
53. José Ignacio Rodríguez, *Estudio Histórico Sobre el Origen, Desenvolvimiento y Manifestaciones Prácticas de la Idea de la Anexión de la Isla de Cuba á los Estados Unidos de América* (Havana: Imprenta La Propaganda Literaria, 1900), 249–53; Portell Vilá, *Historia de Cuba*, 3:57; and Ramon Cernuda, ed., *La Gran Enciclopedia Martiana* (Miami: Editorial Martiana, 1978), 2:185–90.
54. José Martí to Serafín Bello, 16 Nov. 1889, Cernuda, *La Gran Enciclopedia Martiana*, 2:195; Rodríguez, *Estudio Histórico*, 261–63; Portell Vilá, *Historia de Cuba*, 3:64–65; *Herald*, 12 Dec. 1889, 6; and José Martí, *Política de Nuestra América* (Mexico: Siglo 21 Editores, 1977), 162.

55. Thomas McManus to AJG, 18 Apr. 1890, GFP; and "Thomas McManus," in *The National Cyclopedia of American Biography*, 27:71–72.

56. Angel de Loño had been residing in Key West since February 1876. Angel de Loño to George F. Drew, 6 May 1877, Governor Marcellus L. Stearns: Appointments, Resignations, and Removals, 1873–77, RG 101, series 577, box 4, Florida State Archives, Tallahassee; Secretary's Office, Commissions, 3 Jan. 1877 to 17 Feb. 1877, Florida Secretary of State, ibid.; Jefferson B. Browne, *Key West: The Old and the New* (1912; reprint, Gainesville: University of Florida Press, 1973), 69, 118–21; Manuel R. Moreno to AJG, 3 Nov. 1890, GFP; Rowland H. Rerick, *Memoirs of Florida* (Atlanta: Southern Historical Association, 1902), 2:629–30; J. Roy Crowther, *The Grand Lodge of Florida Free and Accepted Masons History, 1830–1898*, vol. 1 (Jacksonville: Drummond Press, 1988), 267, 303; Office of the Clerk, Florida House of Representatives, *The People of Lawmaking in Florida, 1822–1995* (Tallahassee: Florida House of Representatives, 1995), 82; U.S. Congress, *Biographical Directory*, 730; *Jacksonville Florida Times-Union*, 1 Nov. 1890, 1; and ibid., 2 Nov. 1890, 3.

57. AJG to AEG, 1 Feb. 1891, GFP; Anne Elliott to W. J. Monteith, 2 Dec. 1892, ibid.; James Lowndes to AEG, 14 July 1892, EGP; *Anne H. Elliott v The United States*, in the Court of Claims, Direct Tax Case no. 17388, ibid.; AEG to Gertrude Gonzales, 10 May 1893, ibid.; and James Lowndes to Ann H. Elliott, 11 May 1893, ibid.

58. AEG to Gertrude Gonzales, 8 Feb. 1891, EGP; Latimer, *The Story of* The State, 13; Lander, *A History of South Carolina, 1865–1960*, 34, 36; Simkins and Woody, *South Carolina during Reconstruction*, 549, 552; Tindall, *South Carolina Negroes, 1877–1900*, 51–52; Pierce, *Palmettos and Oaks*, 16; and AJG to AEG, 1 Feb. 1891, GFP.

59. AJG to José Ignacio Rodríguez, 10 Feb. 1891, JIR; and William H. Armstrong to AJG, 3 July 1891, GFP.

60. Ibid.

61. Manuel R. Moreno to AJG, 5 June 1891, GFP.

62. Jose G. Delgado to AJG, 9 July 1891, GFP; Ambrose E. Gonzales to AJG, 11 July 1891, ibid.; AEG to AJG, 8 Aug. 1891, ibid.; E. Culpepper Clark, *Francis Warrington Dawson and the Politics of Restoration: South Carolina, 1874–1889* (Tuscaloosa: University of Alabama Press, 1980), 211; and *The State*, 24 and 28 Nov. 1891, 2.

63. William H. Armstrong to AJG, 3 July 1891, GFP; Gonzalo de Quesada to AJG, 19 Aug. and 29 Sept. 1891, ibid.; Antonio Batres to AJG, 27 July 1891, ibid.; and Antonio Batres to AJG, 15 Oct. 1891, ibid.

64. AEG to AJG, 28 Oct. 1891, GFP; NGG to Aunt, 2 Nov. 1888, EGP; *The State*, 9 Nov. 1891, 24; and REE to Hattie Gonzales, 21 Nov. 1891, ibid.

65. Gonzalo de Quesada to AJG, 28 Nov. 1891, GFP; J. S. Duvall to AJG, 11 Feb. 1892, ibid.; AEG to AJG, 7 Apr. 1892, EGP; and AEG to AJG, 25 Sept. 1892, GFP.

66. Juan Arnao Alfonso (1812–1901) was a writer and poet born in Matanzas, who participated in the Narciso López conspiracy of 1848. He landed in Cuba on Oct. 4,

1869, with the expedition led by Domingo de Goicouría and became an insurrectional leader in Matanzas. Arnao returned to the United States after the 1878 Cuban rebel peace treaty. He was a founder of the Cuban Revolutionary Party headed by José Martí. His brother Ramón Ignacio Arnao Alfonso landed in Cuba with the 1851 López expedition, was captured and imprisoned in Spain, and later joined William Walker in Nicaragua. AEG to AJG, 25 Sept. 1892, GFP; Cohen, *Historic Springs of the Virginias*, 60–61; AJG to AEG, 3 Oct. 1892, GFP; and Manuel R. Moreno to AJG, 14 May 1892, ibid.

67. Manuel R. Moreno to AJG, 2 July 1892, GFP; AEG to AJG, 25 Sept. 1892, ibid.; and *Patria*, 31 Dec. 1892, 3.

68. AEG to AJG, 5 Sept. 1892, GFP; AJG to AEG, 17 and 23 Sept. 1892, ibid.; Angel de Loño to AJG, 28 Sept. 1892, ibid.; and *Patria*, 8 Oct. 1892, 3.

69. Gonzalo de Quesada to AJG, 1 Oct. 1892, GFP; AJG to AEG, 3 Oct. 1892, ibid.; "Hugh Smith Thompson," *The National Cyclopedia of American Biography*, 24:78; and *Patria*, 8 Oct. 1892, 3.

70. AJG to AEG, 10, 13, 16, and 17 Oct. 1892, GFP; and AEG to AJG, 19 Oct. 1892, ibid.

71. St. Barnabas Hospital, *1991 Annual Report* (New York: Wolk Press, 1992), 4; and Home for Incurables, *Twenty-Sixth Annual Report* (New York: Eugene D. Croker, Stationer and Printer, 1892), 15, 26, 30, 76.

72. *Patria*, 31 Dec. 1892, 3.

73. *Patria*, 7 Jan. 1893, 3; AEG to AJG, 14 Feb. 1893, GFP; and AEG to AJG, 17 and 27 Apr. 1893, ibid.

74. AEG to AJG, 1 and 19 May 1893, GFP; and AEG to Gertrude Gonzales, 10 May 1893, EGP.

75. AEG to AJG, 1 May 1993, EGP; AEG to AJG, 6 July 1893, ibid.; *The State*, 11 July 1893, 8; ibid., 18 July 1893, 5; Certificate and Record of Death no. 28502, Ambrosio J. Gonzales, 31 July 1893, Municipal Archives, City of New York; Register of Interment no. 43587, folio 655, 2 Aug. 1893, lot no. A, range 131, grave 20, Woodlawn Cemetery, Bronx, New York; *Twenty-Sixth Annual Report of the Home for Incurables*, 36, 44; and *Patria*, 5 Aug. 1893, 2.

76. *The State*, 2 Aug. 1893, 1; and *Patria*, 5 Aug. 1893, 2.

## Epilogue

1. J. W. Floyd, *Historical Roster and Itinerary of South Carolina Volunteer Troops Who Served in the Late War between the United States and Spain, 1898*. (Columbia, S.C.: R. L. Bryan, 1901), 14–15, 23–24, 28, 148, 180–81, 247, 253, 263–64; Lander, *A History of South Carolina, 1865–1960*, 44; and Warner, *Generals in Gray*, 41.

2. S. A. Latimer Jr., "From across the Editor's Desk," *The State*, 25 Jan. 1959.

# BIBLIOGRAPHY

## Primary Sources

*Cuba*
Archivo Nacional, Havana

*Spain*
Archivo Histórico Nacional, Madrid
Ultramar, Cuba, Gracia y Justicia, 1824–55
Archivo General Militar, Segovia
Gobierno Militar de Cuba, Presidencia de la Comisión Militar Ejecutiva Permanente
Ministerio de Asuntos Exteriores, Archivo General, Madrid
Antillas Españolas, Cuba, Política Ultramar

*U.S. National Archives and Records Administration*

*Record Group 21, District Courts of the United States*
Ambrosio J. Gonzales Bankruptcy File, 15 Dec. 1868. U.S. District Courts, Eastern District of South Carolina, Box 55.
Duncan Williamson, William Hughes, and Samuel Matthews v Steam Ship Fanny, U.S. District Courts, Eastern District of Louisiana, New Orleans, Admiralty Case File No. 6617.
George Page and Co. v A. J. Gonzales, 5 Nov. 1867. U.S. District Courts, South Carolina District, Roll 368, Docket 189, Box 8.
Robert A. Parrish, Jr. v The Steamship New Orleans, 14 Sept. 1849, U.S. District Courts, Southern District of New York, Admiralty Case File No. 7–372.
Steamer Creole, 19 June 1850, U.S. District Courts, Southern District of Florida, Key West, Admiralty Records, 4:443–50.
United States v John O'Sullivan, Louis Schlesinger, and A. Irvine Lewis, 24 May 1851, U.S. District Courts, Southern District of New York Case File 1–267.
United States v Miguel T. Tolón, Case File 1–227, U.S. District Courts, Southern District of New York.

*United States v Narciso López, Ambrosio Gonzales, Capt. Robert Young, Capt. Samuel J. Coockygy, Capt. William B. McLean, Lieut. Richard Ralston, Lieutenant Rogers, Sergeant Brown, Sergeant Howard, Lieut. Chapman, and Others,* U.S. District Courts, Southern District of Georgia, Savannah, Case File D-15.

*United States v Narciso López et al.* U.S. District Courts, Eastern District of Louisiana, New Orleans Circuit Court, General Case Files (E-121), 1965–70.

*United States v William Hardy, William A. McEwen, and Henry Robinson,* Indictment, 1 Nov. 1851, U.S. Circuit Courts, Southern District of Ohio, Western Division (Cincinnati), Law Records [old series], 1851–63, Vol. 1 (5).

U.S. District Courts of the United States, Naturalization File for Ambrosio José Gonzales (Naturalized 26 Mar. 1849).

"Warrant to Arrest," 2 May 1851, U.S. District Courts, Southern District of Georgia, Savannah, Case File D-15.

*Record Group 24, Bureau of Naval Personnel*
List of Logbooks of U.S. Navy Ships, Stations, and Miscellaneous Units, 1801–1947.
Log Book, Log USS *Water Witch,* 24 May 1849–14 May 1850.

*Record Group 29, Bureau of the Census*
Free Population Schedules. Florida, 1850.
———. Georgia, 1850, 1860, 1870, 1880.
———. Kentucky, 1850.
———. Louisiana, 1850, 1860.
———. Mississippi, 1850.
———. South Carolina, 1860, 1870, 1880.

Slave Schedules. Florida, 1850.
———. Georgia, 1850.
———. South Carolina, 1860.

*Record Group 36, U.S. Customs Service*
Passenger Lists of Vessels Arriving at New York, 1820–97. Microfilm 237.

*Record Group 43, International Conferences, Commissions, and Expositions*
First International Conference of American States, 1889–90.

*Record Group 45, Naval Records Collection*
Letters Received by the Secretary of the Navy from Commanders, 1804–86.
Letters Received by the Secretary of the Navy from Commanding Officers of Squadrons 1841–86, Home Squadron.

*Record Group 48, Office of the Secretary of the Interior*
Records of the Appointments Division, Miscellaneous Letters Received, 1849–1907.
Records Concerning the Cuban Expedition, 1850–51, Box 145, Entry 142.

*Record Group 58, Internal Revenue Service*
Internal Revenue Assessment, Lists for South Carolina, 1864–66, Microfilm 789.

*Record Group 59, Department of State*
Appointment Records, Applications, and Recommendations for Office, Registers of Application for Appointment to Miscellaneous Federal Offices, 1813–23 and 1834–89, Vol. 3, Entry 763, 1854.
Attorney General's Papers, Letters Received, State Department, 1850–58.
Civil War Amnesty and Pardon Records, 1863–67. Amnesty Oaths, 1864–66, South Carolina, Box 117, Entry 1001.
Consular Dispatches, Havana.
Correspondence on the López Expedition to Cuba, 1849–51. State Department Miscellaneous Correspondence, 1784–1906, Box 1, Entry 121.
Letters of Application and Recommendation during the Administration of Ulysses S. Grant, 1869–77.
Letters of Application and Recommendation during the Administrations of Franklin Pierce and James Buchanan, 1853–61.
Miscellaneous Letters.

*Record Group 109, War Department Collection of Confederate Records*
War Department Collection of Confederate Records. Military Departments, Department of South Carolina, Georgia, and Florida.
Endorsements, May–November 1862, Chap. 2, Vol. 29.
Endorsements, 1863, Chap. 2, Vol. 25.
Compiled Service Records of Confederate General and Staff Officers, and Nonregimental Enlisted Men.
Compiled Service Records of Confederate Soldiers Who Served in Organizations from the State of Louisiana.
General Orders, 1861–64, Department South Carolina, Georgia, and Florida.
Index to Compiled Service Records of Union Soldiers Who Served in Organizations from Pennsylvania.
Letters Received by the Confederate Adjutant and Inspector General, 1861–65.
Letters Received by the Confederate Secretary of War, 1861–65.
Letters Sent, May–September 1863, Chap. 2, Vol. 31.
Letters and Telegrams Sent, General Beauregard, Mar. 3, 1861–May 27, 1861, Chap. 2, Vol. 263.
Muster Rolls and Lists of Confederate Troops Paroled in Greensboro, N.C., Microfilm 1781.
Papers of Various Confederate Notables, Maj. Gen. Sam Jones, Telegrams, and Letters Received, 1864–65, Box 9-A.
Special Orders, 1863–64, Box 66.
Telegrams Sent, July 1863–February 1864, Chap. 2, Vol. 45.

*Record Group 206, Solicitor of the Treasury*

United States v The Steamer Pampero, District Court of the United States for the North District of Florida, Opinion and Decision on Libel and Information for Violation of the Revenue Laws, 12 December 1851, Solicitor of the Treasury, Letters Received, U.S. Attorneys, Clerks of Courts, and Marshals, Florida. 1846–April 1863, Box 19.

*Record Group 393, U.S. Army Continental Commands, 1821–1920*

Head Quarters, Second Military District, Register of Letters Received, Vol. 521, Part 1.

*Government Documents*

Annual Report of the Adjutant General to the Governor of the State of Louisiana, November 1861. Baton Rouge: J. M. Taylor, 1861.

Confederate Ordnance Bureau. *The Field Manual for the Use of the Officers on Ordnance Duty.* Richmond: Ritchie and Dunnavant, 1862.

*Journal of the House of Representatives of the State of South Carolina.* Special Sess., 6 July 1868.

Office of the Clerk, Florida House of Representatives. *The People of Lawmaking in Florida, 1822–1995.* Tallahassee: Florida House of Representatives, 1995.

Report of the Adjutant General of the State of Kentucky. *Mexican War Veterans.* Frankfort, Ky.: Capital Office, 1889.

Survey of Federal Archives in Louisiana. Work Projects Administration. *Ship Registers and Enrollments of New Orleans, Louisiana.* Baton Rouge: Louisiana State University, 1942.

Ulibarri, George S. *Records of United States and Mexican Claims Commissions.* Washington, D.C.: National Archives, 1962.

U.S. Congress. House. *Barque Georgiana and Brig Susan Loud.* Exec. Doc. 83, 32d Cong., 1st sess., 23 March 1852.

U.S. Congress. House. *Special Committee in the Gardiner Investigation.* Report No. 1. Serial Set: #656, vol. 1. 32d Cong., 1st sess., 7 December 1852.

U.S. Congress. Joint Committee on Printing. *Biographical Directory of the United States Congress, 1774–1989.* Washington, D.C.: Government Printing Office, 1989.

U.S. Department of State. *Register of all Officers and Agents, Civil, Military, and Naval, in the Service of the United States.* Washington, D.C.: J. and G. S. Gideon, 1845.

U.S. Senate. *International American Conference: Reports of Committees and Discussions Thereon,* Vol. 1, Exec. Doc. 232, Part 1, 51st Cong., 1st sess. Washington, D.C.: Government Printing Office, 1890.

U.S. Senate. *Message of the President of the United States, Transmitting Reports from the Several Heads of Department Relative to the Subject of the Resolution of the Senate of the 23d of May, as to Alleged Revolutionary Movements in Cuba.* Exec. Doc. 57, 31st Cong., 1st sess., 19 June 1850.

U.S. Series. *The Federal Cases Comprising Cases Argued and Determined in the Circuit and District Courts of the United States,* Book 24. St. Paul: West Publishing, 1896.

U.S. War Department. *Official Records of the Union and Confederate Navies in the War of the Rebellion.* 31 vols. Washington, D.C.: Government Printing Office, 1894–1927.

U.S. War Department. *The War of the Rebellion: A Compilation of the Official Records of the Union and Confederate Armies.* 128 vols. Washington, D.C.: Government Printing Office, 1880–1901.

*Manuscript and Record Collections*

Alabama Department of Archives and History, Montgomery, Ala.
    Robert L. Downman Papers

Baltimore City Courthouse, Record Office, Baltimore, Md.
    Land Records

Bryan-Lang Historical Society, Woodbine, Ga.
    Camden County Marriage Records
    Cemetery Records of Camden County
    Woodbine Plantation

Charleston County Library, Charleston, S.C.
    City of Charleston Death Certificates

Charleston Library Society, Charleston, S.C.
    Charleston City Taxes, Monthly Returns, November 1865–February 1866

Chicago Historical Society, Chicago, Ill.
    Ambrosio Jose Gonzales Manuscript

College of William and Mary, Williamsburg, Va.
    Ritchie-Harrison Papers

Columbia University, New York, N.Y.
    John A. Dix

Darlington County Historical Commission, Darlington, S.C.
    Allan Macfarlan Collection
    Albertus Chambers Spain Collection

Duke University, Durham, N.C.
    Pierre Gustave Toutant Beauregard
    Jefferson Davis
    Henry William and Wilmot Gibbes DeSaussure
    Thomas Rhett Smith Elliott
    Rose O'Neill Greenhow
    Thomas Abram Huguenin Autobiography
    Thomas Jordan
    Spanish Consulate, Charleston, S.C.

Emory University, Atlanta, Ga.
    P. G. T. Beauregard Papers
Enoch Pratt Free Library, Baltimore, Md.
    Maryland Biographical File
    Name Index File
    Schools—Mme. Lefebvre's File
    Woman's Industrial Exchange File
Filson Club, Louisville, Ky.
    Theodore O'Hara
Florida Department of Military Affairs, State Arsenal, St. Augustine, Fla.
    Florida Militia Muster Rolls
Florida State Archives, Tallahassee, Fla.
    Gov. Marcellus L. Stearns—Appointments, Resignations, and Removals, 1873–77
Fort Sumter National Monument, Charleston, S.C.
    A Sketch of the Life of Thomas Abram Huguenin, Manuscript Collection
Georgia Department of Archives and History, Atlanta, Ga.
    Georgia Militia Records
    Scott Genealogical Folder
Georgia Historical Society, Savannah, Ga.
    Christ Church Records, Vestry Minutes, Vol. 1, 1839–59
    William Coolidge
    M. H. and D. B. Floyd
    Hartridge Collection
    Solomon's Lodge No. 1, Free and Accepted Masons, Minutes Book, 1849–59
Grand Lodge of South Carolina, Ancient Free Masons, Columbia, S.C.
    Return Files, Washington Lodge No. 5, Return for 1869
Harvard University, Cambridge, Mass.
    Baker Library
    R. G. Dun & Co. Collection
    Houghton Library
    Dearborn Collection, Confederate Officers
    Henry Wadsworth Longfellow Papers
Henry E. Huntington Library, San Marino, Calif.
    Ambrosio José Gonzales
Historic Charleston Foundation, Charleston, S.C.
    Historic Summary for McLeod Plantation, James Island, S.C.
Historic New Orleans Collection, New Orleans, La.
    James Robb Collection
Jackson Barracks Library, New Orleans, La.
    Historical Military Data Louisiana Militia, 1847–48

Mexican War, 1847–48
Vault Books
Library of Congress, Manuscript Division, Washington, D.C.
P. G. T. Beauregard
John Middleton Clayton
Grover Cleveland
Caleb Cushing
James Henry Hammond
James Monroe
Franklin Pierce
James K. Polk
José Ignacio Rodríguez
Martin Van Buren
Maryland Historical Society Library, Baltimore, Md.
Archives of the Maryland Historical Society
Sarah A. Kummer
Maryland State Archives, Annapolis, Md.
Baltimore City Deaths
Mississippi Department of Archives and History, Jackson, Miss.
John A. Quitman
Museum of the Confederacy, Richmond, Va.
Jefferson Davis Collection
New England Historic Genealogical Society, Boston, Mass.
Charles Coffin Papers
New York City Municipal Archives, New York, N.Y.
Certificate and Record of Death
Record of Assessments, 15th Ward, 1870
North Brevard Public Library, Titusville, Fla.
Henry Theodore Titus Collection
Savannah City Hall, Savannah, Ga.
Tax Digests, 1850
Savannah Public Library, Savannah, Ga.
Wilmington Island Collection
Smithsonian Institution, Museum of American History, Washington, D.C.
Maynard Arms Company Collection
South Carolina Department of Archives and History, Columbia, S.C.
Colleton County Auditor's Tax Return, 1868
Confederate Historian, Miscellaneous Historical Sketches, 1866–98
John Jenkins
Miscellaneous Claims for Damages, 1861–64, Confederate Records

　　　　Ordnance Department Vouchers, 1860–65, Confederate Records
　　　　Governor Francis W. Pickens Papers
　　　　Quartermaster Department Vouchers, 1860–65, Confederate Records
　　　　Rolls of South Carolina Staff and Confederate Officers
　　South Carolina Historical Society, Charleston, S.C.
　　　　Confederate States of America. Army Records. Headquarters, First Military District, Department of South Carolina, Georgia, and Florida, 1861–65.
　　South Caroliniana Library, University of South Carolina, Columbia, S.C.
　　　　Pierre Gustave Toutant Beauregard
　　　　Ambrosio Jose Gonzales
　　　　Gonzales Family (on deposit)
　　　　Alexander Cheves Haskell
　　　　Allan Macfarlan
　　　　James Henry Rion
　　　　John D. Warren
　　Texas State Archives, Austin, Tex.
　　　　Mirabeau Buonaparte Lamar
　　Tulane University, New Orleans, La.
　　　　Jefferson Davis
　　　　Callender Faissoux Collection
　　University of Georgia, Athens, Ga.
　　　　Keith Read Collection
　　University of North Carolina, Chapel Hill, N.C.
　　　　Elliott-Gonzales Family
　　　　E. M. Law
　　　　William Porcher Miles
　　　　John Thomas Wheat
　　University of Texas, Austin, Tex.
　　　　Richard Jacques
　　　　Thomas J. Rusk
　　Virginia Historical Society, Richmond, Va.
　　　　Preston Family
　　Yale University, Sterling Library, New Haven, Conn.
　　　　Charles Gardiner Miscellaneous Manuscripts

*Cemetery Records*
Catholic Cemetery, Savannah, Ga.
　　General Index to Keepers' Record Books, 1853–1938
Laurel Grove Cemetery, Savannah, Ga.
　　General Index to Keepers' Record Books, 1852–1938

Magnolia Cemetery, Charleston, S.C.
  Lot Book
  Register of Burials
Woodlawn Cemetery, The Bronx, N.Y.
  Register of Interment

*Diaries*

Adolfo Fernández Cavada Diary, 1861–63, in possession of Fernando Fernández Cavada, Santo Domingo, Dominican Republic.
Maj. Edward Manigault Diary, in possession of Peter Manigault, Charleston, S.C.
Thomas Benton Roy Diary, 17 Feb. 1865, copy in possession of Nathaniel Cheairs Hughes, Chattanooga, Tenn.

*Periodicals*

Alexandria Gazette and Virginia Advertiser
Atlanta Journal
Baltimore Argus
Baltimore Evening Sun
Baltimore Saturday Night
Baltimore Sun
Boletín del Archivo Nacional de Cuba
Carteles (Havana)
Charleston (S.C.) Courier
Charleston (S.C.) Evening Post
Charleston (S.C.) Mercury
Charleston (S.C.) News and Courier
Charleston (S.C.) Sunday News
Cincinnati Commercial
Cincinnati Enquirer
Cincinnati Nonpareil
Columbia (S.C.) State
Columbia Tri-Weekly South Carolinian
Columbus (Ga.) Times
La Crónica (New York)
Detroit Daily Free Press
Diario de la Marina (Havana)
Gaceta de La Habana
Greensborough (N.C.) Patriot
Greenville (S.C.) Southern Enterprise
Harper's Magazine
Jackson Mississippian
Jacksonville, Florida Republican
Jacksonville, Florida Times-Union
Jacksonville News
Louisville Courier
Louisville Journal
Macon, Georgia Journal and Messenger
Mobile Daily Advertiser
Morning Courier and New York Enquirer
Nashville True Whig
Nassau Guardian (Bahamas)
New Orleans Bee
New Orleans Commercial Bulletin
New Orleans Crescent
New Orleans Delta
New Orleans Picayune
New Orleans Times Democrat
New Orleans True Delta
New York Evening Mirror
New York Express
New York Herald
New York Journal of Commerce
New York Sun
New York Times
New York Tribune
Patria (New York)
Pittsburgh Gazette
Pittsburgh Post

Richmond Enquirer
Richmond Whig
Rome (Ga.) Courier
Russell's Magazine
Savannah Georgian
Savannah Morning News
Savannah Republican
Tallahassee Florida Sentinel
Tallahassee Floridian and Journal

United States Magazine and
    Democratic Review
La Verdad (New York)
Vicksburg (Miss.) Whig
Washington, D.C. Daily Evening Star
Washington, D.C. Daily National Whig
Washington, D.C. Daily Union
Washington, D.C. National Intelligencer
Washington, D.C. Republic

*Biographical Directories*

Adams, Oscar Fay. *A Dictionary of American Authors.* Boston: Houghton Mifflin, 1904.

Bailey, N. Louise, Mary L. Morgan, and Carolyn R. Taylor. *Biographical Directory of the South Carolina Senate, 1776–1985.* Vol. 1. Columbia: University of South Carolina Press, 1986.

*Biographical and Historical Memoirs of Mississippi.* Chicago: Goodspeed, 1891.

*Biographical Encyclopedia of Kentucky of the Dead and Living Men of the Nineteenth Century.* Cincinnati: J. M. Armstrong, 1878.

Calcagno, Francisco. *Diccionario Biográfico Cubano.* New York: Imprenta y Librería de N. Ponce de León, 1878.

Conrad, Glenn R., ed. *A Dictionary of Louisiana Biography.* New Orleans: Louisiana Historical Association, 1988.

*Cuba en la Mano: Enciclopedia Popular Ilustrada.* Havana: Ucar, García y Cía., 1940.

Cullum, George W. *Biographical Register of the Officers and Graduates of the U.S. Military Academy at West Point, N.Y., from Its Establishment, in 1802, to 1890.* Boston: Houghton Mifflin, 1891.

*Cyclopedia of Eminent and Representative Men of the Carolinas of the Nineteenth Century.* Madison: Brant and Fuller, 1892.

Edgar, Walter B., ed., *Biographical Directory of the South Carolina House of Representatives.* Vol. 1, *Session Lists, 1692–1973.* Columbia: University of South Carolina Press, 1974.

Ellis, Frampton Erroll. *Some Historic Families of South Carolina.* 2d ed. Atlanta: n.p., 1962.

Faust, Patricia L., ed. *Historical Times Illustrated Encyclopedia of the Civil War.* New York: Harper Perennial, 1991.

Hemphill, J. C. *Men of Mark in South Carolina.* Vol. 1. Washington, D.C.: Men of Mark Publishing, 1907.

Herdeck, Donald E., ed. *Caribbean Writers: A Bio-Biographical-Critical Encyclopedia.* Washington, D.C.: Three Continents Press, 1979.

Malone, Dumas, ed. *Dictionary of American Biography*. 23 vols. New York: Charles Scribner's Sons, 1964.

McCrady, Edward. *Cyclopedia of Eminent and Representative Men of the Carolinas of the Nineteenth Century*. Madison: Brant and Fuller, 1892.

*National Cyclopedia of American Biography*. New York: J. T. White, 1893.

Nieto y Cortadella, Rafael. *Dignidades Nobiliarias en Cuba*. Madrid: Ediciones Cultura Hispánica, 1954.

Northen, William J., ed. *Men of Mark in Georgia*. 1910; reprint, Spartanburg, S.C.: Reprint Company, 1974.

Peraza Sarausa, Fermín. *Diccionario Biográfico Cubano*. Havana: Ediciones Anuario Bibliográfico Cubano, 1951.

Rowland, Dunbar. *Courts, Judges, and Lawyers of Mississippi, 1798–1935*. Jackson: Press of Hederman Brothers, 1935.

Santa Cruz y Mallen, Francisco Xavier de. *Historia de Familias Cubanas*. 5 vols. Havana: Editorial Hercules, 1944.

Sifakis, Stewart. *Who Was Who in the Civil War*. New York: Facts on File Publications, 1988.

Wilson, James Grant, and John Fiske, eds. *Appleton's Cyclopedia of American Biography*. 8 vols. New York: D. Appleton, 1888.

*City Directories*

Boyd, Mrs. William H. *Boyd's City Directory of the District of Columbia, 1888*. Washington, D.C.: Wm. Ballantine and Son, 1887.

Burke and Boinest. *Charleston Directory, 1866*. New York: M. B. Brown and Company, 1866.

*Charleston City Directory for 1867–68* Charleston, S.C.: Jno. Orrin Lea and Company, 1867.

*Cohen's New Orleans and Lafayette Directory, 1851*. New Orleans: Daily Delta, 1851.

Galloway, David H. *Directory of the City of Savannah for the Year 1850*. Savannah: Edward C. Councell, 1849.

Mears, Leonard, and James Turnbull. *The Charleston Directory, 1859*. Charleston: Walker, Evans, and Company Printers, 1859.

Sholes, A. E. *Shole's Directory of the City of Savannah, January 1879* Charleston, S.C.: Walker, Evans, and Cogwell Printers, 1879.

Trow, John F. *Trow's New York City Directory, 1875*. New York: Trow City Directory Company, 1875.

*Woods's Baltimore City Directory, 1870*. Baltimore: John W. Woods, 1870.

*Woods's Baltimore City Directory, 1871*. Baltimore: John W. Woods, 1871.

*Woods's Baltimore City Directory, 1872*. Baltimore: John W. Woods, 1872.

## Freemasonry

Atkinson, Eugene A. "History of the Cuban Flag." *Transactions,* 29 January 1962.

Brockaway, Charles A. "Masonic Symbolism in Cuba's Flag." *Masonic Outlook,* March 1931, 216.

Crowther, J. *Roy Crowther: The Grand Lodge of Florida Free and Accepted Masons History, 1830–1898.* Vol. 1. Jacksonville: Drummond Press, 1988.

de la Cova, Antonio Rafael. "Filibusters and Freemasons: The Sworn Obligation. "*Journal of the Early Republic* (spring 1997): 95–120.

Denslow, William R. *10,000 Famous Freemasons.* 4 vols. Independence, Mo.: Missouri Lodge of Research, 1957.

Gould, Robert Freke, et al. *The History of Freemasonry.* 4 vols. New York: John C. Yorston and Company, 1889.

Henry, Jeanne Hand. *1819–1849 Abstradex of Annual Returns, Mississippi Free and Accepted Masons.* New Market, Ala.: Southern Genealogical Services, 1969.

"Homenaje a Nuestra Bandera." *Tercera Conferencia Inter-Americana de la Masonería Simbólica. Boletín Informativo* (Havana), 27 February 1955.

Morris, Rob. *The History of Freemasonry in Kentucky, In Its Relations to the Symbolic Degrees.* Louisville: Rob Morris, 1859.

*150 Years of Orange Lodge, No. 14.* Charleston, S.C.: n.p., 1939.

Pike, Albert. "John Anthony Quitman." In *Proceedings of the Grand Lodge of Mississippi: Ancient, Free, and Accepted Masons.* Committee of the Grand Lodge, Jackson: Clarion Steam Printing Establishment, 1882.

Ponte Domínguez, Francisco J. *La Masonería en la Independencia de Cuba.* Havana: Editorial "Modas Magazine," 1954.

*Proceedings of the Grand Lodge of Ancient Free and Accepted Masons of Maryland, 1872.* Baltimore: Henry E. Huber and Son, 1873.

*Proceedings of the Grand Lodge of Kentucky at a Grand Annual Communication in the City of Lexington, Commencing August Twenty-Sixth, 5850.* Frankfort: A. G. Hodges and Company, 1850.

*Proceedings of the Grand Lodge of the Most Ancient and Honorable Fraternity of Free and Accepted Masons of the State of Florida, 1846.* Tallahassee: Office of the Floridian, 1846.

*Proceedings of the Grand Lodge of the Most Ancient and Honorable Fraternity of Free and Accepted Masons of the State of Mississippi, 1826.* Natchez: n.p., 1826.

*Proceedings of the M. W. Grand Lodge of Free and Accepted Masons of the State of Louisiana from 22d June 5850 to February 25th, 5851.* New Orleans: J. L. Sollée, 1851.

*Proceedings of the M. W. Grand Lodge of the State of Florida, 1849.* Tallahassee: Office of the Floridian and Journal, 1849.

*Proceedings of the Most Worshipful Grand Lodge of Ancient Freemasons of South Carolina at the Annual Communication, November 5859.* Charleston, S.C.: A. J. Burke, 1859.

*Proceedings of the Most Worshipful Grand Lodge of Georgia at Its Annual Communication for the Year 5849.* Macon, Ga.: S. Rose and Company, 1849.
*Proceedings of the Most Worshipful Grand Lodge of Georgia at Its Annual Communication for the Year 5850.* Macon, Ga.: S. Rose and Company, 1851.
*Proceedings of the Most Worshipful Grand Lodge of Georgia at Its Annual Communication for the Year 5852.* Macon, Ga.: S. Rose and Company, 1852.

Genealogies

Hampton, Ann Fripp. *A Family Sketchbook: Containing Genealogical Charts and Biographical Sketches.* Columbia, S.C.: n.p., 1979.
*Marriages of Chatham County, Georgia.* Vol. 1, *1748–1852.* Savannah: Georgia Historical Society, n.d.
Paxton, W. M. *The Marshall Family.* Cincinnati: Robert Clarke and Company, 1885.

Books, Pamphlets, and Articles

Alfonso, Pedro Antonio. *Memorias de un Matancero.* Matanzas: Marzal y Cía, 1854.
Ammen, Daniel. *The Navy in the Civil War: The Atlantic Coast.* New York: Scribner, 1883.
*Announcement of the University of Louisiana, 1884–5: Catalogue of the Academical Department, Sixth Session, 1883–4.* New Orleans: G. T. Lathrop, 1884.
Barlow, A. R. *Company G: A Record of the Services of One Company of the 157th N.Y. Vols. in the War of the Rebellion.* Syracuse: A. W. Hall, 1899.
Bates, Samuel P. *History of the Pennsylvania Volunteers, 1861–5.* Harrisburg, Pa.: B. Singerly, 1870.
Beauregard, G. T. "Defense of Charleston, South Carolina." *North American Review* (May 1886): 419–36.
Beirne, Francis F. *St. Paul's Parish, Baltimore: A Chronicle of the Mother Church.* Baltimore: Horn-Shafer, 1967.
Bellido de Luna, Juan, and Enrique Trujillo. *La Anexión de Cuba a los Estados Unidos.* New York: El Porvenir, 1892.
Boggess, F. C. M. *A Veteran of Four Wars.* Arcadia, Fla.: Champion Job Rooms, 1900.
Bowman, S. M., and R. B. Irwin. *Sherman and His Campaigns: A Military Biography.* New York: Charles B. Richardson, 1865.
Brooks, U. R. *Stories of the Confederacy.* 1912. Reprint, Camden, S.C.: J. J. Fox, 1991.
Calderón de la Barca, Frances. *Life in Mexico.* Berkeley and Los Angeles: University of California Press, 1982.
Calhoun, John C. *The Works of John C. Calhoun.* New York: D. Appleton, 1858.
Carson, James Petigru. *Life, Letters, and Speeches of James Louis Petigru: The Union Man of South Carolina.* Washington, D.C.: W. H. Lowdermilk and Company, 1920.
Castleman, Virginia. *Reminiscences of an Oldest Inhabitant.* Herndon Historical Society, Herndon, Va.: Herndon Historical Society, 1976.

Cauthen, Charles E. *Journals of the South Carolina Executive Councils of 1861 and 1862.* Columbia: South Carolina Archives Department, 1956.

Claiborne, John F. H. *Life and Correspondence of John A. Quitman, Major-General, U.S.A., and Governor of the State of Mississippi.* 2 vols. New York: Harper and Brothers, 1860.

"Col. M. C. Taylor's Diary in Lopez Cardenas Expedition, 1850." *Register of the Kentucky State Historical Society* 19, no. 57 (September 1921): 79–89.

Craven, Avery, ed. *"To Markie": The Letters of Robert E. Lee to Martha Custis Williams.* Cambridge: Harvard University Press, 1933.

Crist, Lynda Lasswell, ed. *The Papers of Jefferson Davis.* Vol. 5, *1853–55*. Baton Rouge: Louisiana State University Press, 1985.

———. *The Papers of Jefferson Davis,* Vol. 7, *1861.* Baton Rouge: Louisiana State University Press, 1992.

C. V. [Cirilo Villaverde]. *General Lopez: The Cuban Patriot.* N.p., 1849.

Cuba Commission Report. *A Hidden History of the Chinese in Cuba.* Baltimore: Johns Hopkins University Press, 1993.

"The Cuban Expedition." *Brownson's Quarterly Review* (Oct. 1850): 490–517.

Davis, Jefferson. *The Rise and Fall of the Confederate Government.* 2 vols. 1881. Reprint, New York: Da Capo Press, 1990.

Davis, Varina. *Jefferson Davis: A Memoir.* New York: Belford Company [ca. 1890].

D'Hamel, Enrique B. *The Adventures of a Tenderfoot.* N.p., 1914.

Dickinson, John R. *Speeches, Correspondence, etc. of the Late Daniel S. Dickinson of New York.* 2 vols. New York: G. P. Putman and Son, 1867.

Dix, Morgan. *Memoirs of John Adams Dix.* New York: Harper and Brothers, 1883.

Emilio, Luis F. *History of the Fifty-Fourth Regiment of Massachusetts Volunteer Infantry, 1863–65.* Boston: Boston Book Company, 1894.

"Expedición del General Narciso López." *Boletín de los Archivos de la República de Cuba* 3, no. 1 (January–February 1904), 13–19.

Freret, William. *Correspondence between the Treasury Department, &c., in Relation to the Cuban Expedition, and William Freret, Late Collector.* New Orleans: Alex, Levy, and Company, 1851.

"General Lopez. The Cuban Patriot," *United States Magazine and Democratic Review.* February 1850, 97–112.

Gonzales, Ambrose E. *The Black Border: Gullah Stories of the Carolina Coast.* Columbia, S.C.: The State Company, 1922.

———. *In Darkest Cuba.* Columbia, S.C.: The State Company, 1922.

Gonzales, Ambrosio José. "The Cuban Crusade: A Full History of the Georgian and Lopez Expeditions." *New Orleans Times Democrat,* 6 April 1884, 9.

Bibliography / 479

———. *Heaven Revealed: A Series of Authentic Spirit Messages, from a Wife to Her Husband, Proving the Sublime Nature of True Spiritualism.* Washington, D.C.: McQueen and Wallace, 1889.

———. *Manifesto on Cuban Affairs Addressed to the People of the United States.* New Orleans: Daily Delta, 1853.

———. "On to Cuba." *New Orleans Times Democrat,* 30 March 1884, 9.

González, Diego. *Historia Documentada de los Movimientos Revolucionarios por la Independencia de Cuba de 1852 a 1867.* 2 vols. Havana: Imprenta "El Siglo XX," 1939.

Hagood, Johnson. *Memoirs of the War of Secession.* Columbia, S.C.: The State Company, 1910.

Hardy, Richardson. *The History and Adventures of the Cuban Expedition.* Cincinnati: Lorenzo Stratton, 1850.

Home for Incurables. *Twenty-Sixth Annual Report.* New York: Eugene D. Croker, 1892.

Howe, Julia W. *A Trip to Cuba.* Boston: Ticknor and Fields, 1860.

Jameson, J. Franklin, ed. "Correspondence of John C. Calhoun." *Annual Report of the American Historical Association for the Year 1899.* 2 vols. Washington, D.C.: 1900.

Johnson, John. *The Defense of Charleston Harbor.* Charleston: Walker, Evans, and Cogswell, 1890.

Jones, Samuel. *The Siege of Charleston.* New York: Neale, 1911.

Jones, C. C., Jr. "The Battle of Honey Hill." *Southern Historical Society Papers* 13 (January–December 1885): 355–67.

"The Late Cuba State Trials." *United States Magazine and Democratic Review* 30 (April 1852): 307–19.

Lee, Robert E. *Recollections and Letters of General Robert E. Lee.* Garden City, N.Y.: Doubleday, Page, 1924.

Long, A. L. *Memoirs of Robert E. Lee: Collected and Edited with the Assistance of Marcus J. Wright.* New York: J. M. Stoddart, 1886.

Manning, William R. *Diplomatic Correspondence of the United States: Inter-American Affairs, 1831–1860.* Vols. 7 and 11. Washington, D.C.: Carnegie Endowment for International Peace, 1939.

Martí, José. *Política de Nuestra América.* Mexico: Siglo 21 Editores, 1977.

*The Maynard Rifle.* Washington, D.C.: G. S. Gideon, 1860.

Meriwether, Colyer. *History of Higher Education in South Carolina with a Sketch of the Free School System.* Washington, D.C.: Government Printing Office, 1889.

O'Connor, Mary Doline. *The Life and Letters of M. P. O'Connor.* New York: Dempsey and Carroll, 1893.

O.D.D.O. [J. C. Davis]. *The History of the Late Expedition to Cuba.* New Orleans: Daily Delta, 1850.

*Organization of the Lee Monument Association and the Association of the Army of Northern Virginia, Richmond, Va., Nov. 3d and 4th, 1870.* Richmond: J. W. Randolph and English, 1871.

Porcher, Francis Peyre, M.D. "President's Address: Before South Carolina Medical Association." *Yellow Fever in Charleston, 1871, with Remarks upon Its Treatment.* Charleston, S.C.: Walker, Evans, and Cogswell, 1872.

Porter, A. Toomer. *The History of a Work of Faith and Love in Charleston, South Carolina.* New York: D. Appleton, 1882.

*Preliminary Report of the Commission Appointed by the University of Pennsylvania to Investigate Modern Spiritualism in Accordance with the Request of the Late Henry Seybert.* Philadelphia: J. B. Lippincott, 1887.

Prentiss, George L., ed. *A Memoir of S. S. Prentiss.* 2 vols. New York: Charles Scribner, 1855.

Prince, Walter Franklin. "Supplementary Report on the Keeler-Lee Photographs." *Proceedings of the American Society for Psychical Research* (March 1920): 529–87.

Reagan, John H. *Memoirs, with Special Reference to Secession and the Civil War.* New York: Neale, 1906.

Roman, Alfred. *The Military Operations of General Beauregard in the War between the States, 1861 to 1865.* 2 vols. New York: Harper and Brothers, 1884.

Rowland, Dunbar, ed. *Jefferson Davis Constitutionalist: His Letters, Papers, and Speeches.* Vol. 5. Jackson: Mississippi Department of Archives and History, 1923.

Schlesinger, Louis. "Personal Narrative of Louis Schlesinger, of Adventures in Cuba and Ceuta," *United States Magazine and Democratic Review* (September–December 1852): 210–24.

Shewmaker, Kenneth E., and Kenneth R. Stevens, eds. *The Papers of Daniel Webster: Diplomatic Papers.* Vol. 2, *1850–1852.* Hanover, N.H.: University Press of New England, 1987.

St. Barnabas Hospital. *1991 Annual Report.* New York: Wolk Press, 1992.

"A Trip to Cuba." *Russell's Magazine,* October 1857, 59–63; November 1857, 116–23; December 1857, 235–39; January 1858, 322–27; February 1858, 439–45; April 1858, 60–69.

Tulane University of Louisiana: Catalogue of Academical Department, 1884–85. New Orleans: L. Graham and Son, 1885.

United Daughters of the Confederacy, South Carolina Division. *Recollection and Reminiscences, 1861–1865 through World War I.* N.p.: South Carolina Division, United Daughters of the Confederacy, 1990–.

Velázquez, Loreta Janeta. *The Woman in Battle: A Narrative of the Exploits, Adventures, and Travels of Madame Loreta Janeta Velazquez, Otherwise Known as Lieutenant Harry T. Buford, Confederate States Army.* Richmond: Dustin, Gilman, and Company, 1876.

Wallace, Alfred Russel. *My Life: A Record of Events and Opinions*. 2 vols. London: Chapman and Hall, 1905.

Wiggins, Sarah Woolfolk, ed. *The Journals of Josiah Gorgas, 1857–1878*. Tuscaloosa: University of Alabama Press, 1995.

Wilson, Thomas W. *An Authentic Narrative of the Piratical Descents upon Cuba Made by Hordes from the United States, Headed by Narciso Lopez, a Native of South America*. Havana: n.p., 1851.

Woodward, C. Vann, ed. *Mary Chesnut's Civil War*. New Haven: Yale University Press, 1981.

Ximeno y Cruz, Dolores María. *Aquellos Tiempos . . . Memorias de Lola María*. Vol 1. Havana: El Universo, 1928.

## Secondary Sources

*Books, Pamphlets, and Articles*

Abbott, Martin. *The Freedmen's Bureau in South Carolina, 1865–1872*. Chapel Hill: University of North Carolina Press, 1967.

Ahumada y Centurión, José. *Memoria Histórico Política de la Isla de Cuba*. Havana: A. Pego, 1874.

Allardice, Bruce S. *More Generals in Gray*. Baton Rouge: Louisiana State University Press, 1995.

Alvarez Pedroso, Antonio. *Miguel de Aldama*. Havana: Imprenta "El Siglo XX," 1948.

Amann, William Frayne. *Personnel of the Civil War*. Vol. 1, *The Confederate Armies*. New York: Thomas Yoseloff, 1968.

Andrist, Ralph K., ed. *The American Heritage History of the Confident Years, 1865–1916*. New York: Crown, 1987.

Arthur, Stanley Clisby. *Old Families of Louisiana*. New Orleans: Harmanson, 1931.

Baez, Vicente, ed. *La Enciclopedia de Cuba*. Madrid: Playor, 1974.

Ballard, Michael B. *Pemberton: A Biography*. Jackson: University Press of Mississippi, 1991.

Barbour, George M. *Florida for Tourists, Invalids, and Settlers*. 1882. Reprint, Gainesville: University of Florida Press, 1964.

Basso, Hamilton. *Beauregard: The Great Creole*. New York: Charles Scribner's Sons, 1933.

Bauer, K. Jack. *The Mexican War, 1846–1848*. New York: Macmillan, 1974.

Beirne, Francis F. *St. Paul's Parish, Baltimore: A Chronicle of the Mother Church*. Baltimore: Horn-Shafer, 1967.

Beringer, Richard E., et al. *Why the South Lost the Civil War*. Athens: University of Georgia Press, 1986.

Bradford, Richard H. *The Virginius Affair*. Boulder: Colorado Associated University Press, 1980.

Bradley, Mark L. *Last Stand in the Carolinas: The Battle of Bentonville.* Campbell, Calif.: Savas Woodbury, 1996.

Brandon, Ruth. *The Spiritualists: The Passion for the Occult in the Nineteenth and Twentieth Centuries.* New York: Alfred A. Knopf, 1983.

Bresee, Clyde. *How Grand a Flame: A Chronicle of a Plantation Family, 1813–1947.* Chapel Hill: Algonquin, 1992.

———. *Sea Island Yankee.* Columbia: University of South Carolina Press, 1995.

Brown, Charles H. *Agents of Manifest Destiny: The Lives and Times of the Filibusters.* Chapel Hill: University of North Carolina Press, 1980.

Browne, Jefferson B. *Key West: The Old and the New.* 1912. Reprint, Gainesville: University of Florida Press, 1973.

Burton, E. Milby. *The Siege of Charleston, 1861–1865.* Columbia: University of South Carolina Press, 1990.

Butler, Pierce. *Judah P. Benjamin.* Philadelphia: George W. Jacobs, 1906.

Caldwell, Robert Granville. *The Lopez Expeditions to Cuba, 1848–1851.* Princeton: Princeton University Press, 1915.

Calhoun, Jeanne A., and Martha A. Zierden. *Charleston's Commercial Landscape, 1803–1860.* Charleston: Charleston Museum Archaeological Contributions 7, Sept. 1984.

Callahan, James Morton. *The Diplomatic History of the Southern Confederacy.* Baltimore: Johns Hopkins Press, 1901.

Camden County Historical Commission. *Camden County, Georgia.* Camden, Ga.: n.p., 1992.

Candler, Allen D., and Clement A. Evans, eds. *Georgia.* Atlanta: State Historical Association, 1906.

Carse, Robert. *Department of the South: Hilton Head Island in the Civil War.* Columbia, S.C.: State Printing, 1961.

Casso, Evans J. *Louisiana Legacy: A History of the State National Guard.* Gretna, La.: Pelican, 1976.

Cernuda, Ramon, ed. *La Gran Enciclopedia Martiana.* Miami: Editorial Martiana, 1978.

Chaffin, Tom. *Fatal Glory: Narciso López and the First Clandestine U.S. War against Cuba.* Charlottesville: University Press of Virginia, 1996.

Chitwood, Oliver Perry. *John Tyler: Champion of the Old South.* New York: Russell and Russell, 1964.

Chorley, E. Clowes. *The Centennial History of St. Bartholomew's Church in the City of New York, 1835–1935.* New York: private publisher, 1935.

Clark, E. Culpepper. *Francis Warrington Dawson and the Politics of Restoration: South Carolina, 1874–1889.* University: University of Alabama Press, 1980.

Clement, William Edwards. *Plantation Life on the Mississippi.* 2d ed. New Orleans: Pelican, 1961.

Coggins, Jack. *Arms and Equipment of the Civil War.* Wilmington, N.C.: Broadfoot, 1990.

Cohen, Stan. *Historic Springs of the Virginias: A Pictorial History.* Charleston, W.Va.: Pictorial Histories, 1994.

Coleman, Kenneth, and Charles Stephen Gurr. *Dictionary of Georgia Biography.* Athens: University of Georgia Press, 1983.

Collins, Richard H. *History of Kentucky.* Frankfort: Kentucky Historical Society, 1966.

Cooper, William J., Jr. *The Conservative Regime: South Carolina, 1877–1890.* Baltimore: Johns Hopkins Press, 1968.

———. *Jefferson Davis, American.* New York: Alfred A. Knopf, 2000.

Corbitt, Duvon Clough. *The Chinese in Cuba, 1847–1947.* Wilmore, Ky.: Asbury College, 1971.

Córdova, Federico. *Gaspar Betancourt Cisneros: El Lugareño.* Havana: Editorial Trópico, 1938.

Coulter, E. Merton. *Thomas Spalding of Sapelo.* Baton Rouge: Louisiana State University Press, 1940.

Crocker, Anne Ward. *St. Timothy's Protestant Episcopal Church: A History of the Original Church Building, 1881 to 1969.* Herndon, Va.: Herndon Historical Society, 1997.

Daddysman, James W. *The Matamoros Trade: Confederate Commerce, Diplomacy, and Intrigue.* Newark: University of Delaware Press, 1984.

Daniel, Larry J., and Riley W. Gunter. *Confederate Cannon Foundries.* Union City, Tenn.: Pioneer Press, 1977.

Davidson, Chalmers Gaston. *The Last Foray; The South Carolina Planters of 1860: A Sociological Study.* Columbia: University of South Carolina Press, 1971.

Davis, William C. *Jefferson Davis: The Man and His Hour.* New York: Harper Collins, 1991.

de la Cova, Antonio Rafael. "Cuban Filibustering in Jacksonville in 1851." *Northeast Florida History Journal* 3 (1996): 17–34.

———. "The Taylor Administration Versus Mississippi Sovereignty: The Round Island Expedition of 1849." *Journal of Mississippi History* 4 (winter 2000): 1–33.

Dickey, Dallas C. *Seargent S. Prentiss: Whig Orator of the Old South.* Baton Rouge: Louisiana State University Press, 1945.

Dollero, Adolfo. *Cultura Cubana (La Provincia de Matanzas y su evolución).* Havana: Imp. Seoane y Fernández, 1919.

Donald, David, ed. *Why the North Won the Civil War.* Baton Rouge: University of Louisiana Press, 1960.

Dorsey, John. *Mount Vernon Place.* Baltimore: Maclay and Associates, 1983.

Drago, Edmund L. *Hurrah for Hampton! Black Red Shirts in South Carolina during Reconstruction.* Fayetteville: University of Arkansas Press, 1998.

Du Bois, W. E. B. *Black Reconstruction in America.* 1935. Reprint, New York: Atheneum, 1992.

Dufour, Charles L. *Gentle Tiger: The Gallant Life of Roberdeau Wheat*. Baton Rouge: Louisiana State University Press, 1957.

Durkin, Joseph T. *Confederate Navy Chief: Stephen R. Mallory*. Columbia: University of South Carolina Press, 1987.

Eaton, Clement. *Jefferson Davis*. New York: Free Press, 1977.

*Encyclopedia of Occultism and Parapsychology*. Detroit: Gale Research, 1978–82.

Ervin, Eliza Cowan, and Horace Fraser Rudisill. *Darlingtoniana: A History of People, Places, and Events in Darlington County, South Carolina*. Columbia, S.C.: R. L. Bryan, 1964.

Ettinger, Amos A. *The Mission to Spain of Pierre Soulé, 1853–1855: A Study in the Cuban Diplomacy of the United States*. New Haven: Yale University Press, 1932.

Evans, Clement A., ed. *Confederate Military History*. Vol. 13, *Louisiana*. 1899. Reprint, Wilmington, N.C.: Broadfoot, 1988.

———. *Confederate Military History*. Vol. 6, *South Carolina*. 1899. Reprint, Wilmington, N.C.: Broadfoot, 1988.

Evans, Eli N. *Judah P. Benjamin: The Jewish Confederate*. New York: Free Press, 1988.

Fauquier County Bicentennial Committee. *Fauquier County, Virginia, 1759–1959*. Warrenton, Va.: Virginia Publishing, 1959.

*Fauquier White Sulphur Springs*. Warrenton, Va.: Warrenton House, n.d.

Feipel, Louis N. "The Navy and Filibustering in the Fifties." *United States Naval Institute Proceedings* (April and May 1918): 767–80, 1009–29.

Floyd, J. W. *Historical Roster and Itinerary of South Carolina Volunteer Troops Who Served in the Late War between the United States and Spain, 1898*. Columbia, S.C.: R. L. Bryan, 1901.

Foner, Eric. *Reconstruction: America's Unfinished Revolution, 1863–1877*. New York: Harper and Row, 1988.

Franck, Frederick William. "The Virginia Legislature at the Fauquier Springs in 1849." *Virginia Magazine of History and Biography* (January 1950): 66–83.

Fraser, Walter J., Jr. *Charleston! Charleston!: The History of a Southern City*. Columbia: University of South Carolina Press, 1989.

Freeman, Douglas Southall. *R. E. Lee: A Biography*. New York: Charles Scribner's Sons, 1934.

Fuess, Claude M. *The Life of Caleb Cushing*. 2 vols. New York: Harcourt Brace, 1923.

Garesché, Louis. *Biography of Lieut. Col. Julius P. Garesché, Assistant Adjutant-General, U.S. Army*. Philadelphia: J. B. Lippincott, 1887.

Gay-Calbo, Enrique. *El Centenario de la Bandera Cubana, 1849–1949*. Havana: n.p., 1949.

Goldfarb, Russell M., and Clare R. Goldfarb. *Spiritualism and Nineteenth-Century Letters*. Cranbury, N.J.: Associated University Presses, 1978.

Govan, Gilbert, and James Livingwood. *General Joseph E. Johnston, C.S.A.: A Different Valor*. 1956. Reprint, New York: Konecky and Konecky, 1993.

Graydon, Nell S. *Tales of Edisto*. Columbia, S.C.: R. L. Bryan, 1955.

Greene, Karen. *Porter-Gaud School: The Next Step*. Easley, S.C.: Southern Historical Press, 1982.

Guerra y Sánchez, Ramiro. *Manual de historia de Cuba: Desde su descubrimiento hasta, 1868*. Madrid: Ediciones R, 1975.

―――― et al. *Historia de la Nación Cubana*. 10 vols. Havana: Editorial de la Nación Cubana, S.A., 1952.

Guiteras, Pedro Jose. *Historia de la Isla de Cuba*. 2d ed. Havana: Cultural, S.A., 1927.

Hall, Richard. *Patriots in Disguise: Women Warriors of the Civil War*. New York: Marlowe, 1994.

Heinl, Robert Debs, Jr., and Nancy Gordon Heinl. *Written in Blood: The Story of the Haitian People, 1492–1971*. Boston: Houghton Mifflin, 1978.

Hellberg, Carlos. *Historia Estadística de Cárdenas*. 1893. Reprint, Cárdenas: Comité Pro-Calles de Cárdenas, 1957.

*Historia Gráfica de Cuba, 1492–1925*. Miami: Trade Litho, 1976.

*Historic Old Cedar Key*. Cedar Key, Fla.: Cedar Key Historic Society, 1994.

*A History of Cedar Key*. Cedar Key, Fla.: Cedar Key Beacon, 1994.

Holgren, Virginia C. *Hilton Head: A Sea Island Chronicle*. Hilton Head Island, S.C.: Hilton Head Island Publishing, 1959.

Holt, Thomas. *Black over White: Negro Political Leadership in South Carolina during Reconstruction*. Urbana: University of Illinois Press, 1979.

Holzman, Robert S. *Adapt or Perish: The Life of General Roger A. Pryor, C.S.A*. 1976. Reprint, Biloxi, Miss.: Beauvoir Press, 1992.

Hudson, Leonne M. "A Confederate Victory at Grahamville: Fighting at Honey Hill." *South Carolina Historical Magazine* (January 1993): 19–33.

――――. *The Odyssey of a Southerner: The Life and Times of Gustavus Woodson Smith*. Macon, Ga.: Mercer University Press, 1998.

Hudson, Linda S. *Mistress of Manifest Destiny: A Biography of Jane McManus Storm Cazneau, 1807–1878*. Austin: Texas State Historical Association, 2001.

Hughes, Nathaniel Cheairs Jr. *General William J. Hardee: Old Reliable*. 1965. Reprint, Wilmington, N.C.: Broadfoot, 1987.

――――, and Thomas C. Ware. *Theodore O'Hara: Poet-Soldier of the Old South*. Knoxville: University of Tennessee Press, 1998.

Hume, Edgar Erskine. "Colonel Theodore O'Hara and Cuban Independence." *Bulletin of the Pan American Union*, May 1937, 363–67.

Huxford, Folks. *Pioneers of Wiregrass Georgia*. Homerville, Ga.: n.p., 1851.

Inman, Samuel Guy. *Inter-American Conferences, 1826–1954: History and Problems*. Washington, D.C.: University Press of Washington, D.C., 1965.

Johnson, L. F. *The History of Franklin County, Ky*. Frankfort, Ky.: Roberts Printing, 1912.

Johnson, Robert Underwood, and Clarence Clough Buel, eds. *Battles and Leaders of the Civil War.* New York: Century, 1887.

Johnston, J. Stoddard. "Sketch of Theodore O'Hara," *Register of Kentucky State Historical Society,* Sept. 1913, 67–72.

Jones, C. C., Jr. "The Battle of Honey Hill." *Southern Historical Society Papers* (January–December 1885): 355–67.

Jones, Lewis Pinckney. "Ambrosio Jose Gonzales, A Cuban Patriot in Carolina." *South Carolina Historical Magazine* (April 1955): 67–76.

———. *Stormy Petrel: N. G. Gonzales and His State.* Columbia: University of South Carolina Press, 1973.

———. "William Elliott: South Carolina Nonconformist." *Journal of Southern History* 18 (August 1951): 361–81.

Jones, Robert H. *Disrupted Decades: The Civil War and Reconstruction Years.* New York: Charles Scribner's Sons, 1973.

Joslyn, Mauriel. "Well-born Lt. Col. Paul Francois de Gournay Was the South's Adopted 'Marquis in Gray.'" *America's Civil War* (September 1995): 8, 85–88.

Juárez Cano, Jorge. *Hombres del 51.* Havana: Imprenta "El Siglo XX," 1930.

Kerr, Howard. *Mediums, and Spirit-Rappers, and Roaring Radicals.* Urbana: University of Illinois Press, 1972.

Kirkland, Randolph W., Jr. *Broken Fortunes.* Charleston, S.C.: South Carolina Historical Society, 1995.

Klunder, William Carl. *Lewis Cass and the Politics of Moderation.* Kent, Ohio: Kent State University Press, 1996.

Knight, Lucian Lamar. *Georgia's Landmarks, Memorials, and Legends.* Atlanta: Byrd, 1913.

———. *A Standard History of Georgia and Georgians.* Chicago: Lewis, 1917.

Lamers, William M. *The Edge of Glory: A Biography of General William S. Rosecrans, U.S.A.* New York: Harcourt Brace and World, 1961.

Lander, Ernest McPherson, Jr. *A History of South Carolina, 1865–1960.* Columbia: University of South Carolina Press, 1970.

Latimer, S. L., Jr. *The Story of* The State, *1891–1969, and the Gonzales Brothers.* Columbia, S.C.: State Printing, 1970.

Linder, Suzanne C. *Historical Atlas of the Rice Plantations of the Ace River Basin, 1860.* Columbia: South Carolina Department of Archives and History, 1995.

Lonn, Ella. *Foreigners in the Confederacy.* Baton Rouge: Louisiana State University Press, 1940.

———. *Foreigners in the Union Army and Navy.* Baton Rouge: Louisiana State University Press, 1951.

Marbán, Jorge. *José Agustín Quintero: Un enigma histórico en el exilio cubano del ochocientos.* Miami: Ediciones Universal, 2001.

Marrero, Levi. *Cuba: Economía y Sociedad*. 15 vols. Madrid: Editorial Playor, 1971–92.
Marszalek, John F. *Sherman: A Soldier's Passion for Order*. New York: Free Press, 1993.
Martin, Richard A. "River and Forest: Jacksonville's Antebellum Lumber Industry." *Northeast Florida History* 1 (1992): 19–33.
Marvin, William. "Autobiography of William Marvin." *Florida Historical Quarterly* 36, no. 3 (January 1958): 180–202.
Mauncy, Albert. *Artillery through the Ages: A Short Illustrated History of Cannon, Emphasizing Types Used in America*. Washington, D.C.: Government Printing Office, 1990.
May, Robert E. *John A. Quitman: Old South Crusader*. Baton Rouge: Louisiana State University Press, 1985.
———. "Lobbyists for Commercial Empire: Jane Cazneau, William Cazneau, and U.S. Caribbean Policy, 1846–1878." *Pacific Historical Review* (August 1979): 383–412.
———. "'Plenipotentiary in Petticoats': Jane M. Cazneau and American Foreign Policy in the Mid-Nineteenth Century." In *Women and American Foreign Policy: Lobbyists, Critics, and Insiders*, edited by Edward P. Crapol, 19–44. New York: Greenwood Press, 1987.
McAulay, John D. *Civil War Breech Loading Rifles*. Lincoln, R.I.: Andrew Mobray, 1987.
McCaslin, Richard B. *Portraits of Conflict: A Photographic History of South Carolina in the Civil War*. Fayetteville: University of Arkansas Press, 1994.
McDonough, James Lee. *Stones River: Bloody Winter in Tennessee*. Knoxville: University of Tennessee Press, 1980.
McFeely, William F. *Grant: A Biography*. New York: W. W. Norton, 1981.
*Memoirs of Georgia*. Atlanta: Southern Historical Association, 1895.
Miller, Edward A., Jr. *Gullah Statesman: Robert Smalls from Slavery to Congress, 1839–1915*. Columbia: University of South Carolina Press, 1995.
Moore, R. Laurence. *In Search of White Crows: Spiritualism, Parapsychology, and American Culture*. New York: Oxford University Press, 1977.
Morales y Morales, Vidal. *Iniciadores y primeros mártires de la revolución cubana*. 3 vols. Havana: Cultural, S.A., 1931.
Moreno Fraginals, Manuel. *El Ingenio*. Vol. 3. Havana: Editorial de Ciencias Sociales, 1978.
Morton, Jennie C. "Theodore O'Hara." *Register of Kentucky State Historical Society*, September 1903, 49–56.
Muzzey, David Saville. *James G. Blaine: A Political Idol of Other Days*. Port Washington, N.Y.: Kennikat Press, 1963.
Nepveux, Ethel S. *George Alfred Trenholm and the Company That Went to War, 1861–1865*. Anderson, S.C.: Electric City, 1994.
Nevins, Allan. *The Emergence of Lincoln: Douglas, Buchanan, and Party Chaos, 1857–1859*. New York: Charles Scribner's Sons, 1950.

———. *Grover Cleveland: A Study in Courage.* New York: Dodd, Mead, 1966.

———. *Ordeal of the Union: A House Dividing, 1852–1857.* New York: Charles Scribner's Sons, 1947.

Nichols, Roy Franklin. *Franklin Pierce.* Philadelphia: University of Pennsylvania Press, 1931.

Opatrny, Josef. *U.S. Expansionism and Cuban Annexationism in the 1850s.* Lewiston, N.Y.: Edwin Mellen Press, 1993.

Owsley, Frank Lawrence. *King Cotton Diplomacy: Foreign Relations of the Confederate States of America.* Chicago: University of Chicago Press, 1959.

Pearsall, Ronald. *The Table-Rappers.* New York: St. Martin's, 1972.

Pemberton, John C. *Pemberton: Defender of Vicksburg.* Chapel Hill: University of North Carolina Press, 1942.

Perez, L. M. "Lopez's Expeditions to Cuba, 1850–51." *Publication of the Southern History Association* (November 1906): 345–62.

Perry, Milton F. *Infernal Machines: The Story of Confederate Submarine and Mine Warfare.* Baton Rouge: Louisiana State University Press, 1965.

Pfanz, Harry W. *Gettysburg: The Second Day.* Chapel Hill: University of North Carolina Press, 1987.

Ponte y Domínguez, Francisco J. *Matanzas: Biografía de una provincia.* Havana: Imprenta "El Siglo XX," 1959.

Portell Vilá, Herminio. *Historia de Cárdenas.* Havana: Talleres Gráficos Cuba Intelectual, 1928.

———. *Historia de Cuba en sus Relaciones con los Estados Unidos y España.* 4 vols. 1938. Reprint, Miami: Mnemosyne Publishing, 1969.

———. *Narciso López y su Época.* 3 vols. Havana: Compañía Editora de Libros y Folletos, 1930–58.

———. *Vidas de la Unidad Americana: Veinte y Cinco Biografías de Americanos Ilustres.* Havana: Editorial Minerva, 1944.

Potter, David. "Jefferson Davis and the Political Factors in Confederate Defeat." In *Why the North Won the Civil War,* edited by David Donald. Baton Rouge: University of Louisiana Press, 1960.

Powers, Bernard E., Jr. *Black Charlestonians: A Social History, 1822–1885.* Fayetteville: University of Arkansas Press, 1994.

Puckette, Clara Childs. *Edisto: A Sea Island Principality.* Cleveland: Seaforth Publications, 1978.

Quesada, Ernesto. *Primera Conferencia Panamericana.* Buenos Aires: Imprenta Schenone, 1919.

Quisenberry, Anderson C. *Lopez's Expeditions to Cuba, 1850–1851.* Louisville: John P. Morton, 1906.

Rauch, Basil. *American Interest in Cuba, 1848–1855.* New York: Columbia University Press, 1948.
Reddick, Marguerite. *Camden's Challenge: A History of Camden County, Georgia.* Jacksonville: Paramount Press, 1976.
"Register of St. John-in-the-Wilderness, Flat Rock." *South Carolina Historical Magazine* (April 1962): 109.
Reinecke, J. A. Jr. "The Diplomatic Career of Pierre Soulé." *Louisiana Historical Quarterly* (April 1932): 283–329.
Remos y Rubio, Juan J. *Historia de la literatura cubana.* 3 vols. Havana: Cárdenas y Compañía, 1945.
Rerick, Rowland H. *Memoirs of Florida.* Vol. 2. Atlanta: Southern Historical Association, 1902.
Ripley, Warren. *Artillery and Ammunition of the Civil War.* Charleston, S.C.: Battery Press, 1984.
———. *The Battle of Chapman's Fort.* Green Pond, S.C.: Ashepoo Plantation, 1978.
———, ed. *Siege Train: The Journal of a Confederate Artilleryman in the Defense of Charleston.* Columbia: University of South Carolina Press, 1986.
"Robb Papers Discovered." *Historic New Orleans Collection Newsletter* (winter 1986): 1–4.
Rodríguez, José Ignacio. *Estudio Histórico Sobre el Origen, Desenvolvimiento y Manifestaciones Prácticas de la Idea de la Anexión de la Isla de Cuba á los Estados Unidos de América.* Havana: Imprenta La Propaganda Literaria, 1900.
Roig de Leuchsenring, Emilio. *Los Primeros Movimientos Revolucionarios del General Narciso López, 1848–1849.* Havana: Municipio de La Habana, 1950.
Rose, Willie Lee. *Rehearsal for Reconstruction: The Port Royal Experiment.* New York: Oxford University Press, 1964.
Rosen, Robert N. *Confederate Charleston: An Illustrated History of the City and the People during the Civil War.* Columbia: University of South Carolina Press, 1994.
———. *The Jewish Confederates.* Columbia: University of South Carolina Press, 2000.
Ross, Ishbel. *Rebel Rose: The Life of Rose O'Neal Greenhow, Confederate Spy.* Marietta, Ga.: Mockingbird, 1992.
Rowland, Dunbar. *Mississippi.* Atlanta: Southern Historical Publishing Association, 1907.
Sanborn, Margaret. *Robert E. Lee: A Portrait.* Philadelphia: J. B. Lippincott, 1966.
Santovenia, Emeterio. *Huellas de Gloria.* Havana: Editorial Trópico, 1944.
Saville, Julie. *The Work of Reconstruction: From Slave to Wage Laborer in South Carolina, 1860–1870.* Cambridge: Cambridge University Press, 1994.
Scaife, William R. "Battle of Honey Hill." *Blue and Gray Magazine* (December 1989): 30–32.
Sheppard, William Arthur. *Red Shirts Remembered: Southern Brigadiers of the Reconstruction Period.* Atlanta: Ruralist Press, 1940.

Shields, Joseph D. *The Life and Times of Seargent Smith Prentiss*. Philadelphia: J. B. Lippincott, 1884.

Sifakis, Stewart. *Compendium of the Confederate Armies: South Carolina and Georgia*. New York: Facts on File, 1995.

Simkins, Francis Butler, and Robert Hilliard Woody. *South Carolina during Reconstruction*. Chapel Hill: University of North Carolina Press, 1932.

Skates, John Ray, Jr. *A History of the Mississippi Supreme Court, 1817–1948*. Jackson: n.p., 1973.

Smith, Elbert B. *The Presidency of James Buchanan*. Lawrence: University Press of Kansas, 1988.

Smith, Matthew Hale. *Sunshine and Shadow in New York*. Hartford: J. B. Burr, 1868.

Snowden, Yates, ed. *History of South Carolina*. Vol. 2, Chicago: Lewis, 1920.

South Carolina Synod of the Lutheran Church in America. *A History of the Lutheran Church in South Carolina*. Columbia, S.C.: R. L. Bryan, 1971.

Spencer, Ivor Debenham. *The Victor and the Spoils: A Life of William L. Marcy*. Providence: Brown University Press, 1959.

State Board of Agriculture of South Carolina. *South Carolina*. Charleston, S.C.: Walker, Evans, and Cogwell, 1883.

Stone, Elaine Murray. *Brevard County: From Cape of the Canes to Space Coast*. Northridge, Calif.: Windsor, 1988.

Strode, Hudson. *Jefferson Davis: American Patriot, 1808–1861*. New York: Harcourt Brace and Company, 1955.

Symonds, Craig L. *Joseph E. Johnston: A Civil War Biography*. New York: W. W. Norton, 1992.

Taylor, Alrutheus Ambush. *The Negro in South Carolina during the Reconstruction*. New York: 1924. Reprint, AMS Press, 1971.

Thomas, Dean S. *Cannons: An Introduction to Civil War Artillery*. Gettysburg, Pa.: Thomas Publications, 1985.

Thomas, Emory M. *Robert E. Lee: A Biography*. New York: W. W. Norton, 1995.

Thomas, Hugh. *Cuba: The Pursuit of Freedom*. New York: Harper and Row, 1971.

Tindall, George Brown. *South Carolina Negroes, 1877–1900*. Columbia: University of South Carolina Press, 1952.

Trelles y Govin, Carlos M. *Bibliografía Cubana del Siglo 19*. Matanzas: Imprenta de Quirós y Estrada, 1912.

———. *Matanzas en la Independencia de Cuba*. Havana: Imprenta "Avisador Comercial," 1928.

Trotter, William R. *Silk Flags and Cold Steel: The Civil War in North Carolina: The Piedmont*. Winston-Salem, N.C.: John F. Blair, 1988.

Urban, C. Stanley. "The Africanization of Cuba Scare, 1853–1855." *Hispanic American Historical Review* (February 1957): 29–45.

———. "New Orleans and the Cuban Question during the Lopez Expeditions of 1849–1851: A Local Study in 'Manifest Destiny.'" *Louisiana Historical Quarterly* (October 1939): 1095–167.

Walker, Laurence C. *The Southern Forest: A Chronicle*. Austin: University of Texas Press, 1991.

Wallace, Edward S. *Destiny and Glory*. New York: Coward-McCann, 1957.

———. *General William Jenkins Worth: Monterrey's Forgotten Hero*. Dallas: Southern Methodist University Press, 1953.

Warner, Ezra J. *Generals in Blue: Lives of the Union Commanders*. Baton Rouge: Louisiana State University Press, 1991.

———. *Generals in Gray: Lives of the Confederate Commanders*. Baton Rouge: Louisiana State University Press, 1959.

Weinert, Richard P. "The 'Hard Fortune' of Theodore O'Hara," *Alabama Historical Quarterly* (spring–summer 1966): 33–43.

Welch, Richard E., Jr. *The Presidencies of Grover Cleveland*. Lawrence: University Press of Kansas, 1988.

Williams, George W. *A History of Negro Troops in the War of the Rebellion, 1861–1865*. New York: Harper and Brothers, 1888.

Williams, T. Harry. *P. G. T. Beauregard: Napoleon in Gray*. Baton Rouge: Louisiana State University Press, 1954.

Williamson, Joel. *After Slavery: The Negro in South Carolina during Reconstruction, 1861–1877*. Chapel Hill: University of North Carolina Press, 1965.

Wilson, James Grant. *The Memorial History of the City of New-York: From Its First Settlement to the Year 1892*. Vol. 3. New York: New-York History Company, 1893.

Wise, John S. *The End of an Era*. Boston: Houghton Mifflin, 1902.

Wise, Stephen R. *Gate of Hell: Campaign for Charleston Harbor, 1863*. Columbia: University of South Carolina Press, 1994.

———. *Lifeline of the Confederacy: Blockade Running during the Civil War*. Columbia: University of South Carolina Press, 1988.

Woodworth, Steven E. *Jefferson Davis and His Generals: The Failure of Confederate Command in the West*. Lawrence: University Press of Kansas, 1990.

Young, Ida, Julius Gholson, and Clara Nell Hargrove. *History of Macon, Georgia*. Macon, Ga.: Lyon, Marshall, and Brooks, 1950.

Young, Leonard. *A Short History of St. Bartholomew's Church in the City of New York, 1835–1960*. New York: private publisher, 1960.

Zaragoza, Justo. *Las Insurrecciones en Cuba*. Madrid: Imprenta de Manuel G. Hernandez, 1872.

*Unpublished Material*

Glenn, Virginia Louise. "James Hamilton, Jr., of South Carolina: A Biography." Ph.D. diss., University of North Carolina, 1964.

Guinn, Gilbert Sumter. "Coastal Defense of the Confederate Atlantic Seaboard States, 1861–1862: A Study in Political and Military Mobilization." Ph.D. diss., University of South Carolina, 1973.

Harris, Sheldon Howard. "The Public Career of John Louis O'Sullivan." Ph.D. diss., Columbia University, 1958.

Jones, Lewis Pinckney. "Carolinians and Cubans: The Elliotts and Gonzales, Their Work and Their Writings." Ph.D. diss., University of North Carolina, 1952.

Scafidel, Beverly. "The Letters of William Elliott." Ph.D. diss., University of South Carolina, 1978.

Zuczek, Richard Mark. "State of Rebellion: People's War in Reconstruction South Carolina, 1865–1877." Ph.D. diss., Ohio State University, 1993.

# INDEX

Abbeville, S.C., 218, 254
Accabee plantation, Charleston County, S.C., 219
Acosta, Ricardo, 320
Adam Tunno House, Charleston, S.C., 255
Adams, James Hopkins, 97
Adams, John Quincy, 135
Adams & Damon, 273
Adams Express Company, 138, 244, 345
Adams Run, S.C., 103, 123–24, 126, 129–30, 134, 142, 162, 165, 177–78, 212, 215, 221, 226, 256, 259, 272, 282, 297–301, 312, 332, 428–29, 434
Adeno, Bernardo, 375
Adger & Co., J. E., 273
African Americans, 186, 188, 196–97, 200, 210, 231, 239, 242, 252–53, 256–59, 261, 264, 270–72, 274–75, 277, 279, 282, 285, 306, 318–20, 330–31, 334, 340–41, 345, 358, 447
African American servants: Jacob, 252, 256, 258, 264; Chloe (see Wilson, Chloe); Dick, 252, 256; Elias, 196; Frederick, 181, 219; Heyward, 358–60; Jacob, 252; John, 252, 254, 266
Africanization of Cuba scare, 116–17
Agramonte, Gaspar, 393

Aguado, Sec. Lt., 50, 55, 61
Aguedo Valdés, José, 375
Agüero Sánchez, Miguel de, 393
Agüero Sánchez, Pedro de, 7, 27, 35, 75, 393
Agüero y Agüero, Joaquín de, 92, 100
*Aid* schooner, 264
Aidelar, Pvt., 375
Aiken, Hugh K., 434
Aiken, S.C., 209
Aiken, William, 128
Airy Hall plantation, Colleton County, S.C. xiv–xv
Akin, Warren R., 84
Alabama, 82, 84, 86, 122, 216
*Alabama,* CSS, 249
Alaska, 14
Albergottie, Mrs., 155
Albing, James, 375
Aldama Alfonso, Miguel, 4–5, 129, 131, 265, 287, 302, 304, 308–9, 320
Alexander II, 330
Alexandria, Va., 36
Alfonso, Antonio Luciano, 375
Alfonso García de Medina, José Luis, 4–5, 129, 137
Algeria, 232
*Alice* blockade runner, 223
Allardice, Bruce S., xxv
Allen, A. R., 344

Allen, Bernard, 375
Allen, Gaston, 91
Allen, John, 50, 64–65, 369, 415
Allen, Pvt. John, 375
*Alliance* steamer, 286
Allston, Benjamin, 147
Allston Association, 307
Almerillo, Pedro, 401
Alsace-Lorraine Society of New York, 306
Alston, Charles, Jr., 182, 188, 197, 432
Altman Station, S.C., 317
*American Register* of London, 336
American Revolution, 106
American Society for Psychical Research, 350
Ames, William, 233
Amory, William, 146, 266
Ancejo, Miguel, 399, 401
Anchorage House, Beaufort, S.C., xv, 103, 161, 254
Anderson, John W., 76
Anderson, Joseph Reid, 154
Anderson, Richard H., 147
Anderson, Robert, 139, 142
Andrews, James, 369
Angulo, Laureano, 336
Angulo Guridi, Alejandro, 93, 409
Annandale plantation, Georgetown, S.C., 167
Anti-Masonic Party, 15
Appleton, William, 146
Aragón, Manuel, 375
Aragón, Ramón, 375
Argentina, 37, 131
Aristegui Vélez, Rafael de (Count of Mirasol), 47
Arlandes, Carlos, 399, 401
Armenteros, Juan, 12
Armenteros Muñoz, Isidoro, 9, 92
Armero Peñaranda, Francisco, 47–48, 66–68

Armstrong, Robert, 41
Armstrong, William H., 355–56
Arnao Alfonso, Juan, 358, 462–63
Arnao Alfonso, Ramón Ignacio, 316, 375, 463
Arnold, Benedict, 340
Arnold, Richard Dennis, 75–76, 130
Arnold, T. W., 375
Arrieta, Victoriano de, 75, 393
Arteaga, Carlos de, 393
Artús, Mr., 303
Ashe, Lt., 431–32
Ashepoo plantation, Colleton County, S.C., 267, 304
Ashepoo River, S.C., xiv, 102–3, 162, 167, 189, 195, 215, 220–21, 241, 261
Ashley River, S.C., 142, 162, 179, 181, 215, 219, 241, 246
Associated Press, 415
Astor House, N.Y.C., 13, 75
Atarés fortress, Cuba, 96, 129
Atlanta, Ga., 12, 25, 85, 211, 229, 232, 235, 415
Atlanta arsenal, 186–87, 190
Atlantic & Gulf Railroad, 324
Audubon, John James, 253
Augusta, Ga., 12, 184, 186, 231, 233, 235, 241, 243, 252
Augusta Arsenal, 187, 206
*Augusta Chronicle,* 346
Augustin, Jean Baptiste Donatien, 38, 72, 77
Augustine River, Ga., 87–88
Australia, 342
Averasboro, N.C., 248
Ayala, Esteban de, 393

Bachilder, John, 375
Bachman, John B., 253, 443
Bachman, William K., 196, 443
Bacon, A. B., 74

Bacon, Edwin H., Jr., 440
Badneih, Emerich, 375
Bahía Honda, Cuba, 36, 95
Bailey, David, 84, 293, 407
Bailey, David Jackson, 104
Bailey, John, 84
Bailey, Louise, xvi
Baker, J.D., 375
Baldasano y Ros, Miguel, 399
Ball, Mrs., 362
Ball, William H., 375
Baller, James D., 375
Ballou, Maturin, 134
Ballowville, S.C., 269
Balls plantation, Beaufort County, S.C., 102, 320
Balser, James Henry, 369
Balser, William, 369
*Baltic* steamship, 123
Baltimore, Md., 16, 18, 20, 24, 30, 124, 260–61, 263–64, 273, 281–83, 291, 294, 300–301, 303–4, 306–8, 310–15, 321, 323–25, 329, 334–36, 338, 388, 421
Baltimore Steam Boiler Works, 282
Bank of Louisiana, New Orleans, 183
Baracoa, Cuba, 368
Barckley, Flora, 290
Barnum's Hotel, N.Y.C., 75
Barnwell, Henry, 297
Barnwell, John G., 146, 173, 179, 431–32, 437
Barnwell, Robert, 164, 169
Barrera, Manuel, 401
Barstow, Elias Butts, 87–88, 91, 98, 408
Barton, William H., 369
Basinger, William S., 437
Bates, John C., 56, 369
Batista, Fulgencio, 368
Batres, Antonio, 357
Battery No. 1, 434

Battery No. 2, 213, 228
Battery No. 4, 213
Battery No. 5, 213–14, 434
Battery Beauregard, 191, 206, 224
Battery Bee, 191, 211, 224, 231
Battery Cheves, 200–202, 212–14, 433, 436
Battery Glover, 214, 436
Battery Gregg, 179, 191, 205, 210, 224, 231
Battery Harleston, 435
Battery Haskell, 200–204, 206–7, 212–14, 216–17, 228, 433, 436
Battery Island, S.C., 145, 166, 172, 214
Battery Kinloch, 224
Battery Marion, 224
Battery Marshall, 224, 231
Battery Pringle, 210–11, 214, 228, 436
Battery Ramsay, 207
Battery Reid, 213
Battery Ryan, 200, 203–4, 207, 212–14, 217, 433, 436
Battery Simkins, 199, 201–2, 212, 214, 228, 433, 436
Battery Tatom, 200, 203, 212–14, 217, 433, 436
Battery Tynes, 228, 436
Battery Wagner, xxii–xxiii, 178, 191–92, 198–201, 203–4, 207–8, 210–11, 224, 433, 435
Battery Wampler, 214, 435, 436
Battle of Averasboro, N.C., 248
Battle of Belén Gate, Mexico, 32
Battle of Brandy Station, Va., 457
Battle of Buena Vista, Mexico, 28, 35, 82
Battle of Chancellorsville, Va., 449
Battle of Chapman's Fort, S.C., 261
Battle of Churubusco, Mexico, 7
Battle of Cold Harbor, Va., 435
Battle of First Manassas, Va., 150–51, 164, 225, 306

Battle of Franklin, Tenn., 347
Battle of Gettysburg, Pa., 164, 422, 449
Battle of Honey Hill, S.C., 236–240, 249, 258, 440
Battle of Molino del Rey, Mexico, 7
Battle of Monterrey, Mexico, 7, 20, 32, 82, 311
Battle of Olustee, Fla., 223
Battle of Port Royal, S.C., 102
Battle of Secessionville, S.C., 174, 191
Battle of Stones River, Tenn., 422
Bawder, Louis, 375
Bay Point, S.C., 148–49, 155, 160
Bay Point plantation, Phillips Island, S.C., 103, 127, 130
Bayamo, Cuba, 38
Bayard, Thomas Francis, 333, 343
Bayne, Lt., 369
Beach, Ransom, 375, 410
*Beacon* newspaper, 116
Bean, Horace, 404
Beaufort, S.C., xv, 102–3, 139, 142, 146, 149, 155, 164–65, 172, 243, 247, 257, 278, 301, 307, 309, 313, 317–18, 325, 440
*Beaumont* estate, Flat Rock, N.C., 123, 227
Beauregard, A. N. Toutant, 201
Beauregard, Pierre Gustave Toutant, xxii, 3, 101, 130, 137, 149, 171, 177, 187–88, 190, 192–98, 199–200, 201–5, 207–13, 215–21, 223–24, 229, 250, 290, 339, 429, 438–39; back in Charleston 240–43, 246–47, 249–50; in Charleston 141–47; commands Military Division of the West, 231; in New Orleans 282–83; returns to Charleston 178–80, 182–84; returns to Virginia 225; in Virginia 147
Bechtold, Conrad, 375
Bee, Barnard Elliott, 150, 175

Bee, William C., 160, 223, 309, 313, 315, 317, 320–21, 436
Bee Hive plantation, Colleton County, S.C., 103, 418
Bee's Creek, S.C., 163, 234–37
Belk, Gary, xvi
Belk, Sherry, xvi
Bell, D., 375
Bell, Edwin Q., 375
Bell, George, 59
Bell, William H., 20, 53, 55, 58, 60–61, 72, 74, 77, 94, 108, 119, 369
Bellini, Vicenzo, 22
Belmont, August, 114
Benham, Henry W., 174
Benjamin, Judah P., xxi, 77–81, 163, 336
Bennett, James Gordon, 106
Benson, Rufus, 38, 48, 75
Bent, M. P., 375
Bergeron, Arthur, xvi
Bermuda, 196
Bermúdez, Anacleto, 6
Berry, George S., 375
Bestor, W. C., 136–38
Betancourt, Alonso, 17, 393
Betancourt Cisneros, Gaspar, 6, 9, 13–14, 27, 35, 75, 91, 110, 113–14, 122, 384, 393
Bethel plantation, Pocotaligo, S.C., 161, 165, 223, 252, 256, 263
Bettner, Annie C., 359–60
Biro, Michael, 375
Biscoe, Walter F., 21
Bissell, Henry E., 269, 325
Bissell Bros., 273
Black, John L., 165, 428
Black Code of 1865, 261
Black Island, S.C., 200, 203, 206, 224
Blaine, James Gillespie, 351–52
Blanco, Roque, 402
Blanding, James Douglass, 146, 366

Blanding, Ormsby, 228, 437
*Blasco de Garay* war steamer, 47
Bloxham, William, 338
Blue, Victor, 366
Blue Ridge Mountains, 22
Bluff plantation, Colleton County, S.C., 103, 254, 256, 258, 260, 266, 275, 318, 325, 330–31
Bluffton, S.C., 91
Blumenthal, Pvt., 375
Boatwright, John L., 164–65, 173, 180, 431–32
Boca Chica Key, Fla., 72
Boer war, 153
Boggess, Francis Calvin Morgan, 52–53, 55, 57, 66, 69, 369
Bohlen, Henry, 12–13, 387
Bolan's Church, Grahamville, S.C., 234–37, 239
Bolivar, Simón, 38, 118
Bombalier, Santiago, 116, 415
Bonham, Milledge Luke, 143–45, 183, 198, 200, 208, 216, 246
Bonnel, Thomas, 273
Bontila, George, 375
Borland, Solon, 116
Boswell, John, 375
Bostick, Agnes, 439
Boston, Mass., 20, 24, 266
*Boston* troop transport, 261
*Bothnia* steamship, 336
Bournazal, Pierre Charles, 375
Boyce, William Waters, 130
Boyd, Franklin P., 375
Boyd, Linn, 120
Boyd, Samuel S., 32
Boyd's Landing, S.C., 232–33, 235
Boyd's Neck, S.C., 240
Boyers, Breedlove & Co., 39
Boykin, E. M., 331
Boykin, Mary Whitaker, 151

Braddock's Point, S.C., 148, 161
Bradford, Lt., 369
Brady, James, 375
Bragg, Braxton, 222, 225–26, 249
Brandt, James, 375
Brannan, John M., 180
Brantford, Richard M., 261
Brazil, 252
Breckenridge, John, 415
Breckenridge, Newton C., 44, 369
Breckenridge, Robert H., 375, 410
Breedlove, J., 375
Breedlove, James W., 24, 26, 38, 41
Breton Bay, Md., 356
Brigham, James C., 375
Brisbane, William H., 253
Bristol, R.I., 11, 115
*British Empire* steamship, 331
Broad River, S.C., 146–47, 163, 232–33, 235
Broadway Central Hotel, N.Y.C., 362
Broadway Hotel, Cincinnati, 30
Brooke Gun Battery, 214
Brooks, Ulysses R., xxiv, 153
Brown, John, 395
Brown, J. Wesley, 437
Brown, Sgt., 89
Brown, Thomas D., 375
Brown's plantation, Charleston County, S.C., 173
Browse, David, 375
Brunswick, Ga., 444
Brux, James A., 432
Bryan, George S., 283
Bryan, Thomas, 375
Bryant, William Cullen, 460
Bryce, J. O., 375
Buchanan, James, 12, 127, 130–32, 134–35, 138
Buckman, Thomas E., 431–32
Buena Vista, N.C., 356

Bueno Blanco, José, 336
Buffalo Marsh, Va., 313–14
Buford, Harry T., 145
Buist, George Lamb, 160, 182–83, 228, 339, 342, 437
Bulet, James, 375
Bull's Bay, S.C., 147, 150, 164
Bulman, J., 375
Bulwer, Henry Lytton, 85
Bunch, M. J., 35–36, 45, 52–53, 59–63, 68, 72, 369, 398
Bureau of American Republics, 351, 354
Bureau of Refugees, Freedmen and Abandoned Lands, 254, 258
Burke, V. J., 369
Burnt Fort, Ga., 76, 86, 408
Burnet, Ebet, 284, 296–97
Burr, Aaron, 340, 384
Burtnett, Daniel Henry, 83–84, 86–87, 407
Bush, John G., 375
Butler, Ciro, 74
Butler, M. C., 366
Butler, Matthew Calbraith, 249, 322, 330–31, 333–34, 341, 366, 457
Butler, William, 437
Byrd, Theodore P., 55, 369

Caballero, Ramón, 402
Cabañas, Cuba, 36
Cabrera, Joaquín, 375
Caha, John F., 263
Cahaba, Ala., 82
Cajerman, James, 375
Calderón de la Barca, Angel, 23, 29–30, 41, 72, 74, 80, 85–86, 107–8, 118,
Caldwell, James H., 72
Caldwell, Robert, 375
Calhoun, John C., xxi, 17, 25, 97, 108, 111, 339
*Calhoun* steamer, 102

California, 31, 35, 97
Call, George W., 353
Call, Wilkinson, 353–54
Callawassie Island, S.C., 89, 91
Camagüey, Cuba, 91, 283
Cambray plantation, Camden County, Ga., 84, 407
Camden County, Ga., 84, 86, 89–90
Cameron & Barkley, 273
Cameron, William, 375
Camp Elliott, S.C., 166
Camp Jackson, Jacksonboro, 220
Camp Sturgeon, S.C., 166
Campbell, Charles C., 21, 390
Campbell, John A., 119
Campbell, Robert Blair, 10–11, 58, 64, 397, 400
Campbell, William H., 437
Canada, 7
Canadian annexation, 14–16
Canary Islands, 1
Cancio, Leopoldo, 399
Candelaria coffee plantation, 96
Canimar, Cuba, 292
Canky, Robert, 375
Cánovas del Castillo, Antonio, 365
Cape Hatteras, N.C., 151, 154
Capers, W. C., 369, 415
Capers, Henry D., 437
Capitular House, Cárdenas, 50–36
Cárcamo, Nicolás, 50
Cárdenas, xxi, 44, 47–55, 80, 129, 292, 339, 365, 398
Cárdenas Bay, 48
Cárdenas expedition: casualties of, 55–56, 64; expense of, 41; failure of, 69, 93; Filibuster uniform, 45; Kentucky Regiment, 39, 42, 45, 50, 52, 55–56, 60–61, 87, 94, 97; Kentucky Regiment, Company D, 53; Kentucky Regiment, Company H, 61; Louisiana Regiment,

41–42, 45, 51–56, 59–61, 69; Louisiana Regiment, Company A, 55; Louisiana Regiment, Company B; Louisiana Regiment, Company G, 44; Mississippi Regiment, 36, 43, 45, 51–52, 59–60; Mississippi Regiment, Company A, 51; Mississippi Regiment, Company B, 52; Mississippi Regiment, Company C, 52; Mississippi Regiment, Company D, 52
Caribbean basin, 343
Carlist War, 8
Carnahan, John, 351
Carolina House, Charleston, S.C., 265
*Carolina Sports,* 131, 267, 269
Carr, Lewis, 21, 26
Carrasco, Feliciano, 62, 64, 402
Carrasco, Rafael, 305
Carter, John, 375
Carter, Samuel Powhatan, 323
Cartera, Dr., 289
Caruthers, Charles F., 404
Cary, Lena L., 321–22
Casanovas, J., 375
Cass, Lewis, 131, 134
Cass County, Ga., 85
Cassadega, N.Y., 356
Cassville, Ga., 84
Castaño, Ambrosio, 401
Castellón, Pedro Angel, 6
Castillo, Francisco, 393
Castle Garden, N.Y.C., 331–32
Castle Pinckney, Charleston, S.C., 139, 188
Caston plantation, Jasper County, S.C., 181
Castro, Fidel, 368, 386
Castro, Francisco de, 11
Castro, Lucas de, 11
Castro, Rafael de, 7
Cat Island, Miss., 21, 23
Catalonia Volunteers, 288

*Catholic Mirror* newspaper, 421
*Catskill* monitor, 191
Causton's Bluff, Ga., 188
Cay, Eugene, 375
Cazneau, William, 131
Ceballos, J. M., 320
*Cecile* steamer, 147
Cedar Keys, Fla., 286
*Celestial City,* 350
Ceruti, Florencio, 50, 52–55, 58, 60, 66
Céspedes, Carlos Manuel de, 283
Ceuta prison, Spain, 99
Chaferman, James, 375
Chagres, Panama, 38, 41–42, 79
Chaplin, J. S., 163
Chaplin, Thomas B., 195, 209
Chapman, Lt., 89
Chapman, William H., 432
Chapman's Bluff, S.C., 261, 283
Chapman's Fort, S.C., 167
Charleston, S.C., 12, 16, 75, 81, 85, 89, 96–99, 102, 104, 106, 123, 129, 133, 136, 138–43, 146–47, 149–52, 154–55, 159, 162, 164–67, 172–73, 176–80, 185, 187–88, 190, 192, 194, 197–200, 204–8, 210–12, 216–19, 221–23, 225–27, 229, 231–35, 239–47, 249, 251–54, 257–60, 263–64, 267–68, 271–73, 275–85, 287, 289, 291, 293–94, 296–97, 301–2, 304, 307–9, 311, 315, 317, 325–26, 328, 331, 336–41, 354, 359, 433–34, 439
Charleston and Savannah Railroad, 142, 158, 161–62, 173, 180, 220, 232–33, 236, 240, 321, 436
Charleston Arsenal, 183–84, 186, 195, 201, 203, 212–13, 216, 221
Charleston City Railway Company, 260
Charleston Gas Light Company, 436
*Charleston* gunboat, 246

Charleston Harbor, S.C., 138–39, 142, 146, 148, 188, 194–95, 204–5, 207, 210–11, 222, 226, 229, 231–32, 244, 334; Union ironclad attack on 190–92
Charleston Home School, 296
Charleston Hotel, Charleston, S. C., 12, 96, 102, 104, 139, 141, 325, 338, 417
*Charleston Mercury,* 97, 144, 149–51, 154, 159, 162–63, 168, 179, 218, 223, 227, 231
*Charleston News and Courier,* 328–30, 332–33, 336, 339–42, 345–46, 355–56
Charlotte, N.C., 246–47
Charlotte Harbor, Fla., 69
Charlton, Robert Milledge, 71, 101, 403
Charmes, William, 376
Chase, Salmon P., 351
Chassagne, Julio, 376
Chechesee River, S.C., 233
Chehaw plantation, Colleton County, S.C., 124, 218
Chehaw River, S.C., 102–3, 221, 260, 265, 267, 269, 291
Cheraw, S.C., 244–47, 251, 260
Cheraw and Coalfields Rail Road Company, 245
*Cherokee* mail steamer, 92, 99
Chesnut, James, Jr., xxi, 142, 151, 166–67, 169, 235, 240
Chesnut, Mary Boykin, 141, 151, 160, 169, 386
Chester, S.C., 251
Cheves, Langdon, 433
Chew, Robert S., 142
Chicago Fire of 1871, 340
Chicopee Falls, Mass., 136–37
*Chicora* gunboat, 188, 246, 430
Child, Lydia Maria, 460
Childes, John, 376
Childs, Frederick L., 184

Chinese coolies, 256–57, 259, 263, 267, 296; Isidore, 259; Luis, 265, 269, 271, 274, 276, 280–81, 287; Wong Foo, 351
Chisolm, Alexander Robert, 142, 146, 268, 318
Chisolm, Anna, 282
Chisolm, Rebecca, 282
Christensen, Ulf, xvi
Christi, Thomas, 376
Christian, John H., xvi
Christy, George, 348
Church Flats Battery, 180, 215
Church of the Unity, Boston, 185
Ciceri, Joseph, 376
Cichler, Conrad, 376
Cienfuegos, Cuba, 9–11, 422
Cimarrones, Cuba, 384
Cincinnati, Ohio, 21–22, 30, 37, 59, 82, 93, 100, 115, 345
*Cincinnati Nonpareil,* 37
Cisneros, Hilario, 316, 320
Citadel, The, Charleston, 160, 337
City Hotel, New Orleans, 108, 110, 122
City Hotel, Savannah, 71
*City of Columbia* steamship, 331, 338
*City Point* steamer, 293
Claiborne, John F. H., 109, 113
Clarendon District, S.C., 225
Clay, Clement Claiborne, 130
Clay, Henry, 108
Clayton, John Middleton, 18, 23–24, 26, 28–29, 58, 71–72, 77, 80, 392
Clendenin, W. S., 82
*Cleopatra* filibuster steamer, xxi, 83–84, 86–88, 93, 99
Cleveland, Grover, 340–45, 347
Cline, John, 376
Clingman, Thomas Lanier, 236, 431–32
Cobb, Howell, 108, 431
Cohen, L. L., 209
Cohen, Solomon, 75

Colchett, Alexander M., 376
Colcock, Charles Jones, 181, 233–38, 439
Coleman, Patrick, 376
Cole's Island, S.C., 145, 166–68, 172, 190
Colin, Capt., 369
College of Charleston, 336
College of the City of New York, 320
Colleton, S.C., 102–3
Colleton River, S.C., 89
Collins, E. F., 376
Collins, J., 60, 369
Collins, Napoleon, 376
Collyer, B. H., 376
Colombia, 118
Colquitt, Alfred H., 214–15, 433–34
Colt, A. H., 139
Columbia, S.C., 97, 139, 163, 166, 168, 186, 195, 198, 240, 242–43, 246, 249, 329–30, 341–42, 357–60, 362–63, 366–67, 441
*Columbia Record,* 346
Columbus arsenal, 187
Columbus, Ga., 82, 85, 89
Columbus, Ohio, 100
Colvin, Andrew V., 98
Combahee Ferry, S.C., 189
Combahee Neck, S.C., 195
Combahee River, S.C., xiv, 172, 195, 215, 220–21, 226, 232, 241–42, 331
Commission for the Adjustment of Mexican Claims, 18–19
Compromise of 1850, 114
Compromise Tariff of 1833, 139
Confederate troops:
—Army of Tennessee, 187, 243, 245–47
—Beaufort Volunteer Artillery, 143, 149, 163, 165, 181, 232, 238, 428, 440
—Calhoun Artillery, 148
—Calhoun Guards, 173
—Charleston Light Dragoons, 160, 163, 166, 181, 219, 433
—Colleton Rifles, 148–49
—Cuban Rifles Company, 145
—Earle's Battery, Furman Light Artillery, 238
—Georgia 1st Infantry Regiment, 437
—Georgia 2d Infantry Regiment, State Line, 440
—Georgia 5th Infantry Regiment, 240
—Georgia 12th Heavy Artillery Battalion, 437
—Georgia 17th Infantry Battalion, 440
—Georgia 18th Infantry Battalion, 437
—Georgia 22d Siege Artillery Battalion, 437
—Georgia 32d Infantry Regiment, 235, 239, 440
—Georgia 46th Infantry Regiment, 188
—Georgia 47th Infantry Regiment, 239–40
—Georgia 54th Infantry Regiment, 217
—Georgia Chatham Artillery Battery, 198
—Georgia Militia, Athens Battalion, 235, 440
—Georgia Militia, Augusta Battalion, 235, 440
—Georgia Militia, First Brigade, 235, 440
—Georgia Militia, Second Brigade, 238
—Georgia Militia, First Division, 235, 439
—Georgia State Line Brigade, 235
—German Artillery, Hampton's Legion, 443
—Hampton Legion, 457
—Holcombe Legion, 225
—Horry Artillery, 183
—Jackson Rifle Battalion, Louisiana Legion, 145
—Jasper Greens, 278
—Kanapaux's Artillery. *See* Confederate troops: South Carolina Lafayette Artillery Battery

Confederate troops (*continued*)
—Louisiana 12th Heavy Artillery Battalion, 421
—McQueen Light Artillery, 182
—Nelson Light Artillery of Virginia Volunteers, 196
—North Carolina 8th Regiment Volunteers, 163
—Orleans Independent Artillery, 421
—Palmetto Guard, 151, 160, 163, 165–66, 172, 214, 226
—Palmetto Guard Artillery, 182
—Rutledge Mounted Riflemen, 160, 166, 172, 181, 228, 428–29
—17th South Carolina Volunteers, 253
—South Carolina 1st Artillery Regiment, 173, 228, 270, 437
—South Carolina 1st Cavalry Battalion, 165
—South Carolina 1st Cavalry Regiment, 428
—South Carolina 1st Infantry Regiment, 177, 437
—South Carolina 2d Artillery Regiment, 437
—South Carolina 2d Cavalry Regiment, Company E, 234
—South Carolina 2d Infantry Battalion Sharpshooters, 176, 428–29
—South Carolina 2d Infantry Regiment, 226
—South Carolina 3d Cavalry Regiment, 181, 233, 237, 239, 440
—South Carolina 5th Cavalry Regiment, 195, 428
—South Carolina 11th Infantry Regiment, 428–29
—South Carolina 17th Cavalry Battalion, 428
—South Carolina 17th Infantry Regiment, 430
—South Carolina 18th Heavy Artillery Battalion, 182
—South Carolina 25th Infantry Regiment, 198
—South Carolina 27th Infantry Regiment, 217
—South Carolina Lafayette Artillery Battery, 235, 237–38, 278, 440
—South Carolina Lucas' Heavy Artillery Battalion, 167, 437
—South Carolina Marion Light Artillery Battery, 221, 255
—South Carolina Palmetto Light Artillery Battalion, 203, 221, 224–25, 437, 440
—South Carolina Washington Artillery Battery, 220–21, 232
—St. Paul's Rifles, 148
—Washington Light Infantry, 294
Confiscation Act of 1862, 252
Congressional Cemetery, Washington, D.C., 415
Connecticut, 340
Conner, James, 319
Conolly, Edward, 376
Conrad, Charles M., 107
Conrad, Peter, 376
Consans, William, 376
Constantine, W. S., 376
Contoy Island, Mexico, 44–45, 47–48
Contoy Island prisoners, 48, 66, 75, 77
Cook, Cornelius, 376
Cook, Gilman A., 376
Cook, Maj., 440
Cooke, Samuel, 303
Cooper, James Fenimore, 460
Cooper, John, 376
Cooper, Samuel, 169, 171, 187, 228, 245
Coosaw River, S.C., 162, 165
Coosawhatchie, S.C., 161–63, 180, 189, 235, 240–41

Coosawhatchie River, S.C., 162–63, 181, 189, 195, 215, 240
Corcoran, William Wilson, 312–13, 315
Corinthia expedition, 368
*Correo de Ambos Mundos* newspaper, 116
Costar, G. H., 302
Coste, C. A., 196
Coste, N. L., 147–48
Costera, Andreas, 376
Council of Superior Government, 26
Count of Alcoy. *See* Roncali, Federico
Count of Mirasol. *See* Aristegui Vélez, Rafael de
Count of Villanueva. *See* Martínez de Pinillos, Bernabé
Courtney, William A., 339
Cousans, William, 376
Covington, Ky., 31
Coya, Manuel, 401
Craft, William H., 376
Crane, Dr., 336
Crapo, John R., 356
Crawford, John, 339
Crawford, Martin J., 130
*Creole* filibuster steamer, 38–39, 41, 43–45, 48–49, 54, 56, 59–61, 64–69, 72, 77–78, 84, 93
Crespo, José, 402
Cressey, Edgard, 376
Creswell, Charles M., 431–32
Crisler, L., 369
Crist, Lynda L., xvi
Crittenden, William Logan, 93–96, 99, 129, 376, 410
Crockett, Mary Gray, 251
Crovatt, William, 259, 279, 282, 444
Crowther, J. Roy, xvi
Cruse, Henry, 369
Cruz, Fernando, 351
Cuba: purchase of, 111, 114, 121, 131, 135, 140; slaves and slavery in, 5–7, 9–10, 24, 33, 37, 42, 49, 56–57, 59–61, 65, 68, 105, 111, 114–17, 129, 132, 134, 252, 283, 290, 316, 384
Cuba Mining Company, 5
*Cuba* steamer, 264
Cuban annexation to the U.S., xx, 5–7, 9–11, 14–18, 25, 36–37, 43, 92, 97, 103, 107–11, 113–14, 117, 130, 132–35, 140, 154, 271, 316–17, 339, 344, 352, 356, 364, 384
Cuban Bonds, 38, 41, 79, 82–83, 91, 120
Cuban Cigar Store, Baltimore, 308
Cuban Council, 6, 12, 26, 93
Cuban Council of Organization and Government, 27, 29, 35, 75, 393
Cuban flag, creation of, 20, 358, 361
Cuban Junta of New York, 113–15, 118–22, 320
Cuban Revolutionary Association, 368
Cuban Revolutionary Party, 365, 463
Cuban War of Independence, 283
Cucalón, José Ramón, 310
Cullough, James M., 177
Cully, Supe L., 376
Cumming, Hattie, 286
Cumming, Wallace, 286, 301
Cummings Point, S.C., 142–43, 145, 178–79
Cunningham, A. F., 431–32
Cunningham, John, 144
Curtis, William E., 355
Cushing, Caleb, xxiv, 15, 18–19, 110–13, 115–16, 118, 316, 387, 413
Custom House, Cárdenas, 56
Cuyler, George A., 77, 106, 108
Cuyler, Richard R., 101, 106

Da Costa and C. P. Madan firm, 264
Dahlgren, John A., 430
Daily, Charles J., 376
Daily, Thomas, 376

Dalcour, Agustín J., 292, 300, 307
Dalcour, Teodoro, 4, 300–301, 303–5
Dallas, George, 5, 15
Dantzler, Olin M., 184, 430
Danville, N.C., 249
Dargan, John, 248
Darling, Edward Irving, 347
Darling, Flora Adams, 346–47
Darlington, S.C., 247, 251, 253, 362
*Darlington* steamer, 136
Darwin, Charles, 342, 348
Daufuskie Island, S.C., 76, 89, 148–49
Daughters of the American Revolution, 347
*David* torpedo steamer, 215
Davidson, R. C., 376
Davies, Paul, xvi
Davis, Ezekiel C., 82
Davis, J. C., 42, 45, 54–55, 57–58, 108, 369
Davis, Jefferson, xvii, xxi–xxii, 17–19, 30, 108–9, 111, 115, 120–21, 125–26, 130, 144–45, 147, 153–56, 158–59, 164–65, 167–69, 171, 177–78, 182, 184, 199–200, 205–6, 216–17, 222, 229, 231, 241, 245–46, 250, 281, 283, 285, 340–41, 346–47, 363–64, 420, 428, 439
Davis, John E., 75–76
Davis, John N., 376
Davis, Reuben, 395
Davis, Varina, 126, 158, 222, 347, 388, 426
Dawson, Francis Warrington, 340, 342
Dawson's Bluff, S.C., 189
Day, J. L., 38
de Aragón, Ramón T., 145
de Gournay, Paul Francis, 92, 145, 421
de la Concha, José, 122, 129, 415
de la Cova, Carlina, xvi
de la Cruz Rivero, Francisco Javier, 9, 37–38, 43, 50, 370
de la Huerta, Santiago, 9

de la Luz Caballero, José, 4
de la Vega, Manuel, 289
de Loño, Angel, 353, 359, 462
De Pass, William Lambert, 203, 225, 440
De Treville, Richard, 253–54, 257, 263–64
De Treville, Robert, 437
De Wolf, Daniel E., 376
Dear, Joseph C., 61–63, 370
Decourt, G., 306
del Rijo, Tomás D., 393
del Rosario de Torres, Josefa María, 1
DeLeon, Harmon Hendricks, 339
Delgado, José G., 355–56
Delgado del Valle, Joaquín, 320
Delmonte, José, 289, 292
Delta plantation, Sagua la Grande, Cuba, 278
Denient, J. W., 119
Denmark, 311
Dennett, J. H., 370
Denton, John, 376
Denton, Thomas, 376
Department of Public Works, N.Y.C., 312
DeSaussure, Louis, 236
DeSaussure, Wilmot Gibbes, 144, 247, 339, 433–34
*Detroit Daily Free Press*, 134
Deville, Pedro, 376
D'Hamel, Enrique B., 145, 421
*Diario de la Marina*, 24, 355
*Diario de Matanzas*, 4
Díaz, Manuel, 376
Díaz de Villegas, José G., 9, 11
Dibble, Samuel, 334
Dickens, Charles, 291, 342
Dickinson, Daniel S., xxiv, 15, 29, 108
*Dictator* steamer, 293
Dillon, Patrick, 376
Dill's plantation, Charleston County, S.C., 174, 202, 210, 228

Dilworth, John Hardee, 90–91, 409
Dix, John Adams, 114
Dixon, Maj., 62, 370
Dobbin, James C., 129
Dodge, Charles Cleveland, 319, 364
Donelson, Daniel Smith, 165
Donnelley, Gaylord, xv
Donnelly, James D., 376
Doon, John G., 71, 89
Dorent, Pedro, 376
Dorr, James W., 339
Dorr, John M., 376
Douglas, Stephen A., xxiv
Douglass, Samuel James, 68
Douvren, Charles Augustus, 376
Downer, Charles Augustus, 376
Downman, Robert L., 82, 86, 95, 376
Doyle, Edmund T., 58
Doyle, John, 376
Drayton, Percival, 160
Drayton, Thomas Fenwick, 156, 158, 160–62, 166, 169, 425
Driller, Henry, 376
Drury, Allen, xvi, 408
Du Barry, Franklin B., 183, 430–32
Du Pont, Samuel Francis, 150, 159, 164, 185, 190, 192
Dubouchet, Blas, 56
Duffy, Cornelius J., 376
Dulce, Domingo, 286–87
Dumm, Pvt., 370
Dunbar, A. F., 404
Duncan, James J., 14, 20, 387
Dunn, Pvt., 376
Dunn's Lake, Fla., 99
Dunovant, John, 165, 177–78, 182, 428
Dupart, Victor, 376
Dupeau, Capt., 59
DuPont, Ga., 326
Durham, N.C., 249
Duryea, Robert S., 148

Duval, P., 37, 43
Dwin, James G., 376

Eames, Charles, 119
East Lines, James Island, 173, 180, 202, 211, 213, 217
Eaton, Clement, 158
Echeverría, José Antonio, 6, 10, 13, 289, 359
Echols, William H., 174, 192
Eclectic Institute, Baltimore, Md., 306–7
Ecuador, 113
Eddings Bay, S.C., 148
Eddings Island, S.C., 149
Eddingsville, S.C., 133, 418
Edgeworth School for Young Ladies, Baltimore, 82, 308, 321, 335
Edisto Island, S.C., 147–48, 151, 162, 165, 257
Edisto River, S.C., 103, 147, 220, 241
Edmonds, John Worth, 460
Edwards, David W., 432
Edwards, J. D., 148
Edwards, Lt. Col., 440
Egerton, George, 376
*El Cubano* newspaper, 116, 384
*El Filibustero* newspaper, 116
El León de Oro, Cárdenas, 59
El Morro, Havana, 129
*El Mulato* newspaper, 116, 415
*El Noticioso* journal, 415, 420
El Príncipe prison, Havana, 11, 129
El Recreo, Cuba, 62
El Salvador, 347
Eligio de la Puente, Antonio M., 386
Ellerbe, William H., 366
Elliott, Mrs. Ann, 102, 120, 123, 127, 130, 135, 185, 219, 225, 231, 243–44, 247, 253–54, 256–59, 263, 265–66, 272, 275–77, 280, 290–91, 293–96, 299, 303, 305, 309–10, 312–13, 315, 318, 321–22

Elliott, Anne, 122, 124–25, 127–28, 130, 135, 213, 247, 251, 253, 258, 265–66, 268, 272, 293, 296–97, 299, 318, 324, 329, 344, 354, 356–57
Elliott, Apsley, 277
Elliott, Arthur H., 317–18
Elliott, Belle, 277
Elliott, Caroline Phoebe, 123, 127–29,185, 252, 321, 329
Elliott, Emily, xxiii, 122–25, 127–30, 135, 137, 169, 189, 212–13, 218, 220, 244, 247–48, 251, 258–60, 265, 272, 276, 291, 293–301, 303–4, 309–14, 317, 320–21, 323–30, 332–38, 342, 344, 346, 351, 354, 364, 416–17
Elliott, George P., 146
Elliott, H. H., 313. *See also* Gonzales, Ambrose Elliott
Elliott, Harriett Rutledge. *See* Gonzales, Harriett Rutledge Elliott
Elliott, Mary Barnwell. *See* Johnstone, Mary Barnwell
Elliott, Phoebe, 416
Elliott, Ralph E. (1834–1864), 143, 226
Elliott, Ralph Emms (1797–1853), 101
Elliott, Ralph Emms (1835–1902), 103, 124, 126, 130, 134–36, 146, 149, 154–55, 163, 165, 175–76, 178, 185, 197, 219, 221, 223, 225, 227, 245, 251–54, 256–57, 260, 263–64, 266–67, 274–75, 282–83, 293, 295–97, 299–300, 306, 308, 314–15, 317–19, 324, 327, 329–33, 342, 344, 355, 357
Elliott, Stephen (1804–1866), 252
Elliott, Stephen, Jr. (1830–1866), 103, 149, 160–61, 172, 181, 190, 199, 204, 212, 220, 225–26, 238, 246, 428, 431, 437, 440
Elliott, Thomas R. E. (1843–1862), 179
Elliott, Thomas Rhett Smith (1819–1876), 103, 130, 143, 149, 155, 160–61, 165–66, 179, 182, 189, 223, 252, 256, 263–64, 266–67, 274–75, 277, 280, 282, 293, 296–97, 299–300, 306, 314, 320
Elliott, William (1761–1808), 102
Elliott, William (1788–1863), xiii, xv, 101–3, 122–23, 125–34, 137, 146, 155, 160–61, 185, 231
Elliott, William, Jr., (1832–1867), 104, 124, 128–29, 150, 155, 161, 163, 166, 176, 180–81, 189, 218, 220, 225, 251, 254, 257–58, 263, 266, 351, 418
Elliott, William (b. 1840), 130–31, 143, 151, 164, 223, 228, 256, 266–67
Elliott, William Waight (1831–1884), 181, 238, 431–32
Elliott, William, (1838–1907), 322, 458
Ellis, James, 376
Ellis plantation, Port Royal Island, S.C., 103, 161, 254, 354
Ellis, Robert H., 370, 376
Elwegg, Charles, Jr., 432
Emerson, Ralph Waldo, 420
Empire Mills, Fla., 92
England, 48, 106–7, 269, 339, 348
English Lutheran Church, Charleston, S. C., 253
Enna, Manuel, 96
Enterprise, Fla., 136
Episcopal Christ Church, Savannah, Ga., 101, 112
Escoto, Antonio, 9
Esnard, Peter J., 255–56, 263
Espada Cemetery, Havana, 98
Espartero, Baldomero, 8,
*Esperanza* war frigate, 47
Espinosa, Irene, 334
Essex, Preston, 376
Esteves, José, 401
Euhaw Creek, S.C., 237
Evans, John G., 366
Evans, Josiah J., 130
Evans, Nathan George "Shanks," 162, 165, 175, 188, 215, 433–34

Evansville, Ind., 59
*Evening Star,* 355
Exporting and Importing Company of South Carolina, 223

Fagan, James Burton, 376
Falcón, Antonio, 376
Falligant, John G., 101
Falligant, Sarah A., 324
Fallon, Christopher, 135
*Fanny* filibuster steamer, 21, 26
*Farniente* estate, Flat Rock, N.C., 103, 123, 155, 254, 318
Faunel, Charles, 336
Fauquier White Sulphur Springs, Va., 21–22, 24, 91, 124, 126
Faust, Jacob, 376
Fayetteville, N.C., 248
Fayssoux, Callender Irvine, 21, 49, 65, 94, 99, 188, 370, 376, 390, 415, 430–31
Fellows, Cornelius, 404
Fenwick Island, S.C., 162
Ferguson, George, 370
Fernández, J. A., 340
Fernández Cavada, Adolfo, 145, 422–23
Fernández Cavada, Emilio, 422
Fernández Cavada, Federico, 145, 422–23
Fernández de Cossío, Tomás, 56
Fernandina, Fla., 286, 338
Fiddes, James, 376
Field's Point, S.C., 232
Fike, Matthew D., xvi
Filibuster, meaning of, 24, 391
Filibuster trial in New Orleans, 77–82
Fillmore, Millard, xxi, xxiv, 15, 84–86, 89, 99, 107, 127
Finch, John H., 77
Finlay, Kirkman, xiii
Finney, Michael, 89
*Firefly* steamer, 148–49
First Euhaw Baptist Church, Grahamville, S.C., 234

First Presbyterian Church, Columbia, S. C., 346
First Regiment of Cuban Patriots, 94
First Regiment of Mississippi Rifles (Mexican War), 20
Fischer, Thomas, 376
Fish, Hamilton, 84
Fisher, Ellwood, 22, 391
Fisher, Lila Vance. *See* Lefebvre, Mrs. Hubert P.
Fisher, Newton H., 376
Fisher, Thomas L., 44, 370
Fisk, James, 460
Flat Rock, N.C., xvi, 103, 123–24, 130–31, 135, 137, 155, 197, 216, 254, 260, 291, 318, 324, 329, 332
Fleen, A. M., 376
Fleury, Manuel, 376
Flor de Cuba sugar plantation, 5
Florence, S.C., 243
Florida, 84, 90, 92, 110, 113, 133, 136–37, 187, 223, 304, 338, 353, 368
Florida Militia, 99, 395–96
*Florida* steamship, 99, 108
Flot, Nicolas, 376
Floyd, Thomas Bourke, 408
Floyd County, Ga., 85
Fludd, Margaret, 267–68, 271, 278, 280
Foley, James, 370
Folly Island, S.C., 190
Fonts, Jacob, 376
Fonts Palma, Hilaria, 129, 131, 304–5
Foote, Henry S., 22, 30, 110
Ford, Martin J., 437
Forstall, Louis E., 404
Forsyth, John, 82, 130, 407
Fort Beauregard, Bay Point, S.C., 148–49, 155, 160–61
Fort Elliott, Beaufort, S.C., 146
Fort Heyward, Ladies Island, S.C., 162
Fort Johnson, S.C., xv, 181, 183, 192, 200–202, 205–6, 212, 214, 228, 436

Fort Lamar, S.C., xv, 175, 202–3, 205, 211
Fort McAllister, Ga., 240
Fort Monroe, Va., 345, 422
Fort Moultrie, S.C., 138–39, 166, 175, 183, 187, 191–92, 199, 202, 224
Fort Palmetto, S.C., 145, 166–67
Fort Pemberton, S.C., xv, 168, 174, 179, 183, 198, 203, 205, 210–11, 213, 216, 221, 231, 436
Fort Pickens, Battery Island, S.C., 145, 166, 172
Fort Pulaski, Savannah, Ga., 149, 162, 441
Fort Schnierle, Beaufort, S.C., 146
Fort Sumter, Charleston, S.C., 139, 142–45, 166, 171, 175, 179, 181, 186–87, 191–92, 195, 199, 201, 202, 204–8, 211–12, 215, 217, 220, 225–26, 228, 231–32, 270, 345, 433–34, 437, 438
Fort Walker, Hilton Head, S.C., 160–61
Fortress Monroe, Va., 357
Foster, George W., 376
Foster, Henry C., 53, 370
Foster, John G., 226, 229, 232–34, 239–40
Fourniquet, Hery A., 376
Fragua del Calvo, León. *See* Mádan, Cristobal
Frampton plantation, Jasper County, S.C., 181
France, 10, 75, 106–7, 114–15, 123, 140, 151, 269, 339, 421
Franco-Prussian War, 306
Fraser, E. W., 233–34
Frederick, A. D., 437
Free Soil Party, 114
Freeborn, Isaac, 376
Freedmen's Bureau, 254, 258
Freemasonry, xx, xxiv, 6, 8–9, 13–15, 18, 20, 28, 32, 36, 39, 68, 74, 76, 80, 87, 89–90, 92, 101, 103, 130, 141, 263, 283, 296, 307, 317, 340, 353, 359, 394;
—Centre Lodge, No. 108, Baltimore, Md., 307
—Clarke Lodge, No. 51, Louisville, Ky., 44
—Clinton Lodge, No. 54, Savannah, Ga., 87, 90
—Dade Lodge, No. 14, Key West, Fla., 72
—Grand Lodge of Free and Accepted Masons of Florida, xvi
—Grand Lodge of Maryland, Baltimore, Md., 307
—Grand Lodge of Mississippi, 394
—Grand Lodge of New York, 394
—Grand Lodge of South Carolina, 394
—Halo Lodge, No. 5, Cahaba, Ala., 82
—Harmony Lodge, No. 22, Beaufort, S.C., 103
—Masonic Hall, N.Y., 316
—Orange Lodge, No. 14, Charleston, S.C., 103
—Perfect Union Lodge, No. 1, New Orleans, La., 38
—Quitman Lodge No. 76, New Orleans, 394
—Quitman U.D. Lodge, Veracruz, Mexico, 394
—Solomon's Lodge, No. 1, Savannah, Ga., 74–76, 101
—Solomon's Lodge, No. 20, Jacksonville, Fla., 93
—St. Marys Lodge, No. 126, St. Marys, Ga., 90
—Táyaba Masonic Lodge, Trinidad, Cuba, 9
—Union Kilwinning Lodge, No. 4, Charleston, S.C., 14
—Washington Lodge No. 3, Mississippi, 32
—Washington Lodge, No. 5, Charleston, S.C., 163

—Yazoo Lodge, No. 42, Yazoo City, Miss., 44
—Zerubbabel Lodge No. 15, Savannah, Ga., 71, 76, 108
Frémont, John C., 127
French Broad River, N.C., 123
French Institute, N.Y.C., 3–4
Frías Jacott, Dolores de, 8
Frías Jacott, Francisco de (Count of Pozos Dulces), 6, 9, 96
Friedling, Bernard, 370
Frobisher's College, N.Y.C., 336
Fry, Birkett Davenport, 281, 449
Fry, Joseph, 316
Fuller, Robert B., 309

*Gaceta de La Habana* newspaper, 24
Gadsden Purchase, 113, 115
Gainsville, Miss., 74
Gaite, Jacinto, 401
Gaither, E. B., 21
Gale, Levi, 404
Galliano, Maria, xvi
Galliano, Ralph, xvi
Gallup, David G., 370
García, Juan, 50
García-Menocal Martín, Aniceto, 312, 345, 357
García Ortega, Antonio, 64
Gardiner, George A., xxiii, 16, 18–19, 21, 26, 29, 37
Gardiner, John Carlos, 16, 19, 28
Gardner, John H., 432
Gardner, John L., 138
Garesché du Rocher, Julius Peter, 145, 387, 422
Garfield, James A., 332, 351
Garlington, E. A., 366
Garnett, James J., 64, 370
Garnett, John J., 225
Garniss, J. P., 363

Garrison, William Lloyd, 460
Garth, Patrick Abec, 376
Gauffreau Berault, Emilia, 4
Gaulden, William B., 71
Gauneaurd, Felipe, 59, 66
Geblin, Charles, 376
Geddes, John, 38
Geddes, Robert, 38, 77
Geiger, Michael, 376
Gener Buigas, Tomás, 2–3
Gener, José Tomás, 287
Gener Junco, Benigno, 17, 131, 134, 259, 264, 287, 445
*General Clinch* steamer, 146, 148
Genoa, Italy, 2, 354
George, Quash, 270
George Page and Company, 261, 278, 283–84
Georgetown College, 422
Georgetown, S.C., 143, 147, 150, 162, 164, 167, 172, 184, 218, 220, 434
Georgia, 80, 84, 110, 136–37, 179, 198, 220, 223, 225, 229, 240
Georgia Central Railroad, 86
Georgia Militia, 76, 82–84, 89–90, 94, 236, 240
*Georgiana* filibuster barque, 38–39, 44–45, 48, 75, 77, 79
Gerardeau, Thomas, 128
German Schutzenfest Festival, Charleston, 296
Gibara, Cuba, 368
Gibbes, Mrs., 324
Gibbon, Charles, 376
*Gibraltar* blockade runner, 205, 207
Gidding, Joshua, 460
Gillmore, Quincy A., 166, 198, 200–201, 204, 206, 210, 214, 222, 226
Gilmore, Benjamin, 376
Gilmer, Jeremy F., 207, 208, 211–13, 217–18, 225, 245

Gino, David, 376
Girardeau, I. W., 197
Gist, States Rights, 179, 194, 431–32
Gist, William H., 137–38, 143
Gladden, Adley Hogan, 97
Glen Cove, N.Y., 23
Glenn Springs, S.C., 356
Glover, Mr., 273
Godbold, F. M., 432
Goicouría, Domingo de, 5, 108, 110, 122, 415, 462–63
Goldsboro, N.C., 248
Gómez Pastrana, María, 384
Gómez, Máximo, 367
Gonzales, Alfonso Beauregard, 137, 266–67, 272, 274, 276, 282, 286–88, 290–92, 294–95, 299–301, 303–5, 308, 312, 317, 329, 331, 337–38, 344, 357
Gonzales, Ambrose Elliott, xv, 129, 134, 164, 178, 268, 274, 277, 282, 287–88, 291–301, 304–16, 318, 321–30, 335, 338, 344–45, 346, 355–60, 362–63, 367; employed by the *Charleston News and Courier* 343; facial paralysis 336–37; founds *The State* 354–55; learns telegraphy 317; in New Orleans 336; reconciles with father 333; and Red Shirts 319–20; Western Union telegraphist 331–32, writes Gullah stories 345;
Gonzales, Ambrosio José
—adjutant general to Narciso López, xx, 44
—and alcohol, 124, 126, 277, 297
—American citizenship, 18
—annexationist, 5–6, 14–15, 22, 33, 103, 107, 109–11, 113, 116–18, 132–35, 140, 317, 339, 343, 352, 364, 413
—arrival in the U.S., 3, 12
—articles and letters in newspapers, 97, 104, 112–13, 116–21, 133–35, 149, 151, 159, 168, 193–94, 339–40, 344
—attitude on secession, 138, 341
—attitude on slavery, 6, 103, 116, 132
—in Baltimore, 18, 30, 124, 261, 306–12
—billiard player, 103, 127, 346
—and births of children, 129, 133, 137, 231, 259, 289
—character of, 33, 44, 103, 109, 112–14, 118, 120, 128, 130, 136, 153, 155, 158, 258, 265, 268, 277–78, 284–85, 294, 296–97, 308, 315, 323, 325, 328, 335, 342, 347
—cigar smoker, 297, 308
—Civil War: acting inspector general on Morris Island, 144–46; aide-de-camp to Beauregard, 142–44; aide-de-camp to Gov. Pickens, 147, 154; aide-de-camp to Gen. Ripley, 159; chief of artillery on James Island under Taliaferro, 202–4, 208; chief of artillery under Beauregard, 178; chief of artillery under Hardee, 231; chief of artillery under Johnston, 248; chief of artillery under Pemberton, 171; chief of ordnance under Beauregard, 183; chief of ordnance under Pemberton, 175; and coastal defense of South Carolina, 121, 146–53, 159, 167, 171; commanding artillery at the battle of Honey Hill, 236–240; commanding artillery siege train, 159–60, 162–66, 171, 190, 194, 197; conceptualized artillery siege train, 152–53; court-martial of Dunovant, at, 177–78, 428; feud with Ripley, 182, 189–90, 194, 202, 208; invested in Confederate bonds, 213; pallbearer of Gen. Bee, 150–51; pallbearer of Lt. Col. Wagner, 175; paroled at Greensboro, 249; plan to strengthen Fort Sumter casemates, 192–93, 204, 207–8, 434; promoted to Confederate colonel, 176–77;

promoted to Confederate lieutenant colonel, 171; recommended for promotion to brigadier general, 145, 147, 164, 167–69, 188, 205, 222, 229, 245; recommended land mines and shotguns for Battery Wagner, 199, 201, 208, 433; reorganized Charleston artillery defense lines, 179–81, 185; takes presidential amnesty oath, 253
—commanded the retreating *Creole* expedition, 65
—courtship and marriage of, 124–26
—and Cuban bonds, 38, 41, 79, 82, 91
—and Davis, Jefferson, 17–18, 30, 108, 111, 120, 125–26, 130, 140, 145, 147, 153–56, 158–59, 165, 168–69, 171, 216–17, 222, 281, 285, 340, 347
—death sentence in Cuba, 75, 79, 122, 129, 140
—death sentence rescinded, 127, 254
—described by others, 14–15, 22, 44–45, 47–48, 54, 70, 74, 84, 104, 107, 112, 122, 134, 141, 144, 151, 154–55, 159, 175, 294, 307–8, 319, 347, 352, 361, 416
—early life and childhood, 1–4
—education, 3–4, 103, 109
—employed as translator, 120, 338, 351, 354–55
—employed in Patent Office, 124–25
—family feud, 293–98, 305, 308–12, 315, 320–29, 334–38, 360, 364, 416
—Federal fugitive, 86–89, 99
—filibuster organizer, 74–76, 81–87, 100, 110
—financial problems and bankruptcy, 101, 120, 122, 136, 264–65, 270–80, 283–85, 295–98, 300, 304–5, 309–15, 318–19, 322–23, 327–28
—and Freemasonry, 6, 32, 37, 68, 74, 76–77, 80, 90, 100–101, 103, 296, 307, 359

—Havana Club member, 5, 11
—illnesses and injuries of, 91, 103, 185, 269, 276–77, 294, 322, 335–36, 344, 355–63 inheritance of, 334–35, 337, 340, 344–45, 364
—indicted for violating Neutrality Act, 72, 89
—interpreter for Narciso López, 17, 45, 48, 69, 74, 78
—in Key West, 71–72, 359
—landmark admiralty case, 264
—last years and death, 358–63
—in London, 336
—in Macon, Ga., 74, 82–83, 86
—*Manifesto on Cuban Affairs,* xxv, 104–6
—marriage dowry bond of, 126, 256, 263, 283, 298, 344, 417
—and Martí, José, xvi, xix, xxiv, 352–53, 357, 359
—Maynard Arms agent, 136–40, 149
—in Mobile, 74
—multi-lingual, 4, 19, 103, 109
—musical talents, 103, 251, 302, 323
—in New Orleans, 12, 31, 35–37, 72, 74, 108–9, 339–40
—New York City, 13–14, 27, 261, 301–6, 312–22, 325, 331–37, 359
—oratory of, 22, 44, 103
—in Paris, 335–36
—patronage pursuer, 18–19, 108–15, 121, 127, 130–32, 134–35, 317, 319, 323, 340–47, 364
—physical appearance, 44, 101, 141, 194, 223
—and Pierce, Franklin, 119, 125, 127–28, 140
—as planter and saw mill operator, 135, 263–80
—political affiliation of, 103–4, 107–8, 112–14, 133, 140
—pseudonyms of, 6, 12, 33, 83

Gonzales, Ambrosio José (*continued*)
—and Quitman, John, 31–33, 35–36, 74, 108–10, 130
—religious beliefs, 101, 112, 127, 269, 284, 303, 323, 364
—in Richmond, Va., 108, 151–54, 228–29
—Round Island expedition, 21
—in Savannah, Ga., 74–76, 84, 87, 99–102, 104, 108, 136, 233, 293–301
—sharpshooter, 103, 135–36
—social life of, 101, 103, 122, 288, 307, 311
—as songster and choir member, 22, 101, 151, 303, 306–9, 311, 319, 322–23, 347
—in Spain, 5
—and spiritualism, 347–51, 354–56, 364
—at summer resorts, 22, 24, 104, 114, 120, 124, 126–27, 130, 137, 322–23, 331–32, 337, 343–46, 357–58
—surrenders to federal marshal, 72, 99
—as teacher, 4–5, 288–91, 297–98, 302, 306, 312, 319–20, 330, 334
—testifies at filibuster trial, 77–80
—at White House, 28, 78, 125–27, 140, 351
—wounded at Cárdenas, 54
Gonzales, Ambrosio José, Jr. *See* Gonzales, Ambrose Elliott
Gonzales, Ana Rosa, xiv, 289, 292, 301–2, 308, 310, 312, 314, 325, 329, 335, 338, 346, 357, 367–68. *See also* Gonzales, Harriett R. E.
Gonzales, Benigno Gener. *See* Gonzales, William Elliott
Gonzales, Esteban, 376
Gonzales, Gertrude Ruffini, 231, 258–59, 267–68, 274, 281–84, 286, 288, 291–92, 295, 306, 308, 312–13, 321–27, 329–30, 334, 338, 343–44, 346, 357–58, 362

Gonzales, Harriett Rutledge Elliott, 123–34, 160, 166, 177, 185, 220, 226, 231, 243, 247, 251, 254, 256, 258–60, 263–65, 267–84, 287–94, 303, 309–10, 315, 323, 342, 344, 348–50, 354–56, 364, 416
Gonzales, Narciso Gener, xiv, xxiv, 133, 178, 268–69, 282, 294–95, 298–301, 303–5, 308–8, 310, 311–12, 315–16, 322, 324–25, 327–28, 331, 356–57, 360; *Charleston News and Courier* reporter 329–30; cigar smoker 334; edits student newspaper 317; fight with Robert Smalls 334; founds *The State* 354–55; *Greenville News* reporter 329; at Grover Cleveland inauguration 341; learns telegraphy 318; murdered 367; reconciles with father 326–28; and Red Shirts 319–20; reporter in Charleston 336; reporter in Columbia 342, 346; survives yellow fever 292; temper of 309; Washington correspondent 332–33; writes bilingual letter to brother 314
Gonzales, Robert Elliott, 356
Gonzales, Sarah Cecil, 346, 356
Gonzales, William Elliott, 259–60, 267–68, 274, 276, 278–79, 282, 291–92, 301, 310, 325, 328–31, 356–57, 367, 416; cadet at the Citadel 337–38; employed by *Charleston News and Courier* 342; marriage of 346; reconciles with father 337; treatment for stammering 336
Gonzales, Woodward & Co., 255, 260–61, 263
González, Anacleto, 57, 66
González, Andreas, 376
González, Brígida de los Dolores, 2, 351
González, Gertrudis Dominga, 2
González, Isabel María, 2
González, José Antonio, 1

González, Juana, 337, 340, 344
González, Lola, 265, 275, 287–88, 290, 292, 313, 334–35, 345
González, Plutarco, 393
González-Abreu, Margot, xvi
González Echeverría, Manuel, 359
González Gauffreau, Emilio, 4
González Gauffreau, Ignacio, 4, 327, 337, 340, 344–45
González Gauffreau, Próspero, 4, 327
González Govantes, Pedro, 376
González Perdomo, Ambrosio José, 1–5
González Vigil, Antonio, 145
Good Hope Plantation, Jasper County, S.C., 237
Goose Creek Church, S.C., 444
Gorgas, Josiah, 179, 184, 186–87, 196, 208, 219–20, 225, 228–29, 231, 245
Gotay, Felipe N., 57, 60, 74, 95, 377
Gouldin, Dr., 74
Gourdin, Henry, 106
Gourdin, J. G. R., 193
Govín, Félix, 9
Graffan, Joseph A., 38, 48, 75
Graham, Burton, 370
Graham, E. H., 238
Graham, George A., 377
Graham's Neck, S.C., 232
Grahamville, S.C., 233–36, 240, 318, 324–25
Grahamville Democratic Rifle Club, 319, 325
Grammar School No. 25, N.Y.C., 320
*Granada* schooner, 390, 430
Grand Central Hotel, Columbia, S.C., 357
Grand Opera House, New Orleans, 339
Granma expedition, 368
Grant, Reuben H., 377
Grant, Ulysses S., 226, 232, 240, 243, 248, 316, 319
Grau, Francisco, 399, 401
Gray, M. M., 183

Gray, Pvt., 377
Greeley, Horace, xxiv, 460
Green, Moses, 377
Green, Thomas, 128
Green Pond, S.C., xiv, 221, 232, 267, 271, 280
Greenhow, Robert, 25
Greenhow, Rose O'Neal, 25
Greenlee, Lt., 370
Greensboro, N.C., 248–49, 341
Greenville, S.C., 189, 216, 251, 254, 281, 329, 457
*Greenville News,* 329
Greenwood, S.C., 434
Gregg, Maxcy, 97, 143–45
Gregg, Thomas E., 432
Gregorie, John W., 146
Gregory, James, 240
Gregory's Point, S.C., 240
Grider, Robert M., 377
Griffin, Ga., 12, 85
Grimball, Thomas, 187
Grimball plantation, James Island, S.C., 172, 187, 214
Grinevald, A., 195
Griswoldville, Ga., 235
Grove plantation, Port Royal Island, S.C., 103, 161, 254, 354
Guamacaro, Cuba, 53, 62
Guanajay, Cuba, 36
Guatemala, 347, 351, 354, 357
Guayabo, Cuba, 92
Güell Renté, José, 336
Guerin, Mantoue, & Co., 244
Guerra, Miguel, 377
Guiteras, Julián, 458
Guiteras Font, Antonio, 287, 289, 339
Guiteras Font, Eusebio, 289
Guiteras Font, Pedro José, 3–4, 123–25, 128–29, 345, 417
Guiteras Gener, Daniel, 339, 458
Gulfport, Miss., 21

Gunst, Joseph B., 377
Guy, John R., 370

*Habanero* war frigate, 48
Habersham, F. A., 449
Habersham, Isabel R., 76
Habersham, Leila Elliott, 286, 293, 346, 449
Habersham, William, 182
Hackley, William R, 69, 398
Haeckel, Louis, 377
Hagan, Louis, 377
Hagar, Frederick, 377
Hagood, Johnson, 144–46, 177, 202, 214–17, 322, 330, 431
Haiti, xxi, 7, 41, 43, 117, 132
Hale, Capt., 52, 370
Hale, Thomas G., 41, 48, 75
Hall, Capt., 370
Hall, Henry C., 292, 364
Hall, J. Prescott, 26, 86
Hall & Co., Charleston, 255
Halleck, Henry W., 240
Halpin, James, 277
Hamerik, Asger, 307, 311
Hamilton, Catherine A., 309–10
Hamilton, James, 42, 89, 91, 102
Hamilton, Samuel Prioleau, 89
Hamilton plantation, Beaufort County, S.C., 89
Hammond, James Henry, xxi, 130, 132
Hampstead Mall, S.C., 194
Hampton, Alfred, 366
Hampton, Ambrose Gonzales, xiii–xiv, xvi, 22, 367, 446
Hampton, Ann Fripp, xiii, 368, 446
Hampton, Joanne, xvi
Hampton, Mary, xiii
Hampton, Wade, 320, 322, 340–41, 366
Hampton Roads, Va., 159
Hanleiter, Cornelius R., 241

Hannd, Benjamin F., 377
Hanvey, George M., 437
Happoldt, J. H., 223
Harbele, Jacob, 377
Hardee, William J., xxii, 231–36, 240–43, 245–49
Hardeeville, S.C., 162–63, 235, 241
Harden, W. D., 431–32
Hardy, Richardson, xxiv, 39, 42, 44, 48, 55–57, 63–64, 69, 82, 370
Hardy, William, 31, 37, 52, 58, 65, 82, 100, 115, 370
Harkins, J., 370
Harllee, William Wallace, 167, 322
Harnley, William, 370
Harris, Pvt., 370
Harris, David B., 181, 192, 196, 202–4, 211, 213, 217, 220–21, 223, 429
Harris, David Golightly, 447
Harris, L. J., 404
Harris, R. A., 370
Harrisburg, Pa., 24
Harrison, Benjamin, 351
Harrison, Charles, 377
Harrison, George, 377
Hart, Henry B., 377
Hart, John L., 248
Hartford, Conn., 353
Hartnett, Thomas, 377
Hartwell, Alfred S., 233, 239
Hartwick Seminary, N. Y., 31
Harvard College, 102, 104, 300
Hasell, Thomas Morritt, 216
Haskell, Alexander C., 355
Haskell, Charles T., Jr., 433
*Hatteras* warship, 249
Hatch, John P., 232–35, 237, 239, 247
Havana, Cuba, 7, 10–12, 16, 47, 50, 64, 66, 83, 85, 94–96, 115, 122, 129, 260, 264, 281–82, 284, 286–88, 292, 302,

312, 326–28, 330–31, 334, 340, 353, 366–67, 387, 421, 422
Havana Chamber of Commerce, 367
Havana Club, 5, 7, 10–11, 13, 20, 27, 129, 137, 359
Havana Railway Company, 5, 265
Hawkins, Capt., 52, 370
Hawkins, Thomas Theodore, 28, 30, 39, 52–53, 55, 60–64, 72, 77, 93–94, 371
Hawley, Joseph R., 340–41
Hayden, George B., 53, 371
Hayden, J. A., 72
Hayne, Issac W., 431–32
Haynes, William Scott, 110, 377, 411
Hearsey, James H., 377
Hearsey, Thomas H., 377
Hebert, Louis, 249
Hefrou, Michael L., 377
*Helen* schooner, 148
Henderson, John, xxi, 28, 31–33, 35, 37–39, 41–43, 72, 77–82, 90, 108, 121, 130
Henderson, John, Jr., 31
Henning, H. E., 371
Henry, Timothy K., 377
Hepworth, George Hughes, 185
Herb, William K., 377
Herbert, John, 371
Hernández, Antonio, 377
Hernández, Bernardino, 57, 66
Hernández, Diego, 377
Hernández, José Elías, 17, 113
Hernández Canalejos, José Manuel, 37–38, 43, 91, 371, 415
Hernández Cano, Juan José, 38
Herndon, Va., 315, 317
Herre's Educational Bureau, Miss, N.Y.C., 320
Herren, Julio, 377
Heth, Henry, 164–65, 169
Hewlett's Hotel, New Orleans, 37

Heyward, Nathaniel, 256, 270, 276
Heyward, William C., 148–49, 428–29
Heyward, William Henry, 172
Hicacos, Cuba, 48
Hiern, R. A., 38
Higgins, John, 371
Higginson, Thomas Wentworth, 460
Hillsborough, N.C., 248–49
Hilton, Robert B., 339
Hilton, Thomas, 377
Hilton Head, S.C., 128, 146–49, 155, 160–61, 166, 181, 233, 254, 265, 318
Hinkel, Miss, 252
Hinsey, Nathaniel, 371
Hodge, Charles J., 377
Hodge, William L., 42
Hodges, John Sebastian Bach, 307
Hogan, William J., 377
Holdship, George, 377
Holland, Henry D., 106
Holmes, Isaac Edward, 14
Holmes plantation, James Island, 173
Holy Communion Church Institute, Charleston, 328
Holy Communion Episcopal Church, Charleston, 294
Home for Incurables, Bronx, N.Y., 360–63
Honduras, 347
Honey Hill, S.C., xv, xxiii, 236–38, 241, 247
Hood, John Bell, 243, 245, 438
Hope plantation, Colleton County, S.C., 103
Hopkins, Charles H., 76
Hopkins, Henry, 404
Hopkins, John L., 98
Hopkins, Mrs., 323
Hoppock, George H., 266, 289, 291, 300, 303, 317
Hoppock, Howell, 317

Horner, William H., 377
Horton, C. O., 371
Horwell, Charles H., 377
Hot Springs, Va., 358
Hotel Mathis, Coventry, England, 336
Hotel Windsor, Washington, D.C., 347
Hough, Fenton B., 377
Houston, Sam, 29
Howain, Pvt., 377
Howard, J. C., 371
Howard, Oliver O., 243
Howard, Sgt., 89
Howder, Pvt., 377
Hoy, Thomas P., 59, 371
Hudson, N.Y., 13
Hudson House, Hudson, N.Y., 13
Huger, Cleland Kinloch, 172–73, 180, 231, 273, 339, 437
Hughes, Joel D., 377
Hughes, Nathaniel C., xvi
Huguenin, David, 163
Huguenin, Thomas Abram, 165, 436–37, 438
Humbert, Richard H., 324
Humphreys, Andrew A., 422
Hundnall, Thomas, 377
Hungary, 29, 37, 44
Hunt, Randell, 339–40
Hunter, Beverly E., 377
Hunter, David, 168, 174, 190, 198
Hunter, John, 101
Hunton, Logan, 72, 77, 80–81, 404
Hunton, T. G., 371
*Huntress* steamer, 163
Hurd, Lt., 371
Hurd, William K., 377
Hurlbut, George P., 233
Huston, Felix, 122
Huston, Felix, Jr., 95, 377
Huston, Pvt., 371
Hutchinson, "Large Trees," 280

Hutchinson, Thomas, 102
Hutson, T. W., 166
Hygeia Hotel, Old Point Comfort, Va., 345

Ibáñez, Ginés, 402
Iglesias, Francisco, 377, 401
Iglesias, José, 377
*Ilustración Americana,* 415
*Illustrated Newspaper,* 459–60
Immaculate Conception Church, Cárdenas, 51
*In Darkest Cuba,* 367
*Independiente* newspaper, 116
Inglis, Frances "Fanny" Erskine, 23
Ingraham, Duncan N., 176, 179, 188, 414
Ingraham, H. Laurens, 180, 431–32
Inslee, William W., 377
Instituto de Matanzas, 289
Ireland, 21
Irish, J., 371
Iron Battery, Morris Island, 143
Irving, Washington, 460
Irving Hotel, Washington, D.C., 15–16, 20, 27–29, 110,
Irving House, N.Y.C., 25, 107
*Isaac P. Smith* gunboat, 187–88, 207
*Isabel* mail steamship, 69, 71, 81, 106, 129
*Isabella* steamship, 256
Island Home School, Beaufort, S.C., 309–10
Italy, 37
Iverson, Alfred, 130
Ives, Thomas J., 74
Izbert, Ibrahim, 377

J. & R. Geddes, 38
Jackson, Andrew, 5
Jackson, John, 371
Jackson, Maj., 440
Jackson, Miss., 31, 79, 82
*Jackson* revenue cutter, 147

Jacksonboro, S.C., 220
Jacksonville, Ala., 231
Jacksonville, Fla., 76, 84, 86–87, 91, 93–94, 96, 99–100, 106, 326, 338, 359, 366
Jacksonville Hotel, Jacksonville, Fla., 87, 92
Jacksonville, Pensacola and Mobile Railroad, 338
Jacques, Richard, 204, 206, 209, 215, 226
Jalapa, Mexico, 7
Jamaica, 7, 132
James Island, S.C., xv, 145, 165–66, 172–74, 180–81, 183, 187, 197–200, 202–7, 211–19, 224, 226–28, 433, 436
James Island Creek, S.C., 173, 198, 204, 206, 209, 211
James, Thomas B., 377
Jardines, Porfirio, 393
Jasper, Henry, 377
*Jasper* steamer, 98
Jefferson, Thomas, 5
Jefford, Robert J., 428
Jenkins, John, 234–36, 366, 440
Jenkins, Micah, 366
Jervey, Theodore Dehon, 160, 173, 328, 339
Jesse Mount House, Savannah, Ga., 324
Jessert, Jacob, 377
John Fraser & Company, 196, 272–73, 447
John's Island, S.C., 187, 213–14, 225, 228, 231
John's Island Ferry, S.C., 180
Johnson, Albert W., 39, 50, 61–64, 371
Johnson, Andrew, 253
Johnson, Benjamin Jenkins, 150
Johnson, David, 181–82
Johnson, David Sanford L., 315
Johnson, J. Z., 260
Johnson, John, 37

Johnston, John Carl, 371
Johnston, Joseph E., xxii, 247–49
Johnstone, Andrew, 137, 167, 227
Johnstone, Annie, 123, 291
Johnstone, Elliott, 125, 227, 291, 315, 323
Johnstone, Emma Elliott, 293, 306, 308, 311
Johnstone, Fannie, 306
Johnstone, Mary Barnwell, 123–24, 143, 171, 231, 251–52, 256–57, 263–64, 281–83, 294, 303–4, 308–9, 311–12, 314–15, 321–22, 323, 325–26, 336, 344
Jones, Ann Fripp. *See* Hampton, Ann Fripp
Jones, Charles Colcock, Jr., 220, 225, 320, 431, 437
Jones, E. L., 54, 371
Jones, L. C., 377
Jones, Lewis Pinckney, xiv, xxv
Jones, Samuel, xxii, 225–29, 233–35, 240–43
Jones, William Carey, 19
Jons Hotel, Philadelphia, Pa., 19
Jordan, Thomas, 179, 184, 189–90, 193–94, 196, 208, 220, 283, 290
Jorge, Eleuterio, 377
Josephs, A. A., 56, 371
Joster, George W., 377
*Juno* steamer, 203
Junta for Promoting the Political Interests of Cuba, 27–28, 37, 78

Kanapaux, J. T., 237–38
Kansas, 132
Keating, Capt., 52, 371
Keeler, Pierre L. O. A., 347–51, 356
Keeler, William M., 350
Keenan, M. J., 377
Keith, Willis "Skipper," xv–xvi
Keitt, Lawrence Massillon, 130, 210, 435

Keller boarding house, Valdosta, Ga., 324
Kelly, James A., 95–96, 119, 377, 411
Kelly, William, 59, 371
Kemper, Delaware, 202, 223, 225, 337, 431, 437
Kendall, J. H., 223
Kennedy, Thomas J., 56, 371
Kentucky, 110
*Keokuk* monitor, 191–92
Kerekes, Bela, 377
Kerr, Victor, 94, 377
Kershaw, Joseph B., 144–45
Kewen, Achilles L., 51, 371, 415
Kewen, Thomas, 371
Key West, Fla., 65–69, 72, 94, 96, 112, 125, 286, 339, 353, 358–59, 364, 462
Kiddle, Henry, 460
King, Col., 371
King, J. Gadsden, 255
King, Mitchell, 177
King, W. S. E., 119
King's Mountain Military School, Yorkville, S.C., 330
Kinsey, Robert, 275
Klinck & Wickenberg, 273
Knight, Capt., 61, 371
Knott, Clark, 371
Know-Nothing Party, 134
Koockogey, Samuel J., 89
Koss, Alfred, 377
Kossuth, Louis, 29, 86
Koszta, Martin, 414
Kummer, Sarah Agnes, 308

La Cabaña fortress, Havana, 64, 129, 287
La Croi, José, 377
*La Crónica* newspaper, 24
La Empresa school, 289, 292
La Fontaine, Jean de, 126
La Punta fortress, Havana, 129

La Rosa sugar mill, Cimarrones, Cuba, 384
La Sere, Emil, 109
*La Verdad* newspaper, 7, 15, 25, 120, 156, 320, 384
Laberinto sugar mill, Cuba, 9
Laborde Rueda, Juan Ygnacio, 41, 72
*Labrador* steamship, 335
Labuzan, Pvt., 377
Lacoste, Peter, 377
Ladies Island, S.C., 162
*Lady Davis* steamer, 163, 426
Lafitte, Edward H., 265, 272–73, 280, 283, 301, 303
Lafitte, John B., 305, 447
Lagunillas, Cuba, 57, 59, 62
Lainé, Francisco Alejandro, 377, 415
Lainfiesta, Francisco, 347, 351
Lallande, Joseph, 404
Lama, John, 75, 77, 85–86, 88–89, 92, 113, 405
Lamar, Charles Augustus Lafayette, 82
Lamar, Mirabeau Buonaparte, xxi, xxiv, 80, 82, 85, 108, 131, 420
Lamar, Thomas Gresham, 173–75, 427
Lamont, Daniel Scott, 344
Landa, Manuel, 377
Landier, Antoine, 43
Landinghan, Pvt., 377
Lane, E. D., 371
Lane, Gregory, xv, 439
Lane, Harriet, 131
Lang, Catherine, 407
Lang, Isabella, 84, 407
Lang, Kevin, 407
Langan, Nicholas, 371
Lanier, Sidney, 83
Lapeyre, J. M., 404
Las Pozas, Cuba, 95–96
Latimer, S. L., Jr., xxv
Laughlin, William, 404

Laurel Grove Cemetery, Savannah, 408
Laurel Hill Island, S.C., 167
Laurel Isle cotton plantation, Georgia, 90
Laurent, Emilio, 368
Laury & Alexander, 273
Lawton, Mrs., 336
Lawton plantation, James Island, S.C., 179, 206
Lawton, Thomas, 372
Lay, George W., 179
Lay, Jennie, 356
Leathers, James, 372
Le Count, H. M. & W., 301
Lecompton Constitution, 132
Lee, Fitzhugh, 341, 365–66
Lee, Francis D., 147, 179
Lee, Mary Custis, 311
Lee, Robert E., xx, 18, 25–26, 161–65, 167, 169, 182, 195–96, 225, 247–48, 296, 311, 346, 365, 388
Lee, Thomas H., 377
Lefebvre, Mrs. Hubert P. 308, 311
Lefrow, Michael L., 377
Legare, Joseph J., 189, 431–32
Legare, W. W., 432
Legare's Island, S.C., 228
Legare's plantation, James Island, S.C., 436
Legare's Point, S.C., 201, 213, 216
Legare's Point Place, S.C., 187–88
Le Havre, France, 335
León, Pedro de, 16
León Infantry Regiment, 49–50, 58, 61–62, 401–2
Leonardtown, Md., 356
Leslie, Frank, 415, 459–60
Levis, Clara, 20
Levy, J. St., 377
Levy, Mr., 244
Lewis, Armstrong Irvine, 43, 62, 65, 67, 72, 77, 84, 87, 93–94, 99–100, 104, 372, 377, 412

Lewis, Capt., 372
Lexington, Va., 126
Liberty County, Ga., 71
*Liberty* steamer, 264, 287, 303–4
Light House Inlet, S.C., 196–97, 198, 203, 210, 226
Limehouse, H. L. "Buck," xiv–xv
Lincoln, Abraham, 138–39, 142, 248, 350–51
Lincoln, Benjamin, 246
Lincoln, Mary Todd, 460
*Littell's Living Age,* 127
Little, Thomas, 377
Little, William B., 377
Little Folly Island, S.C., 198
Liverpool, England, 123, 272, 336
Livry, James H., 377
Llanes, Félix, 57, 66
Logan, John A., 61, 64, 66, 372
Logan, Thomas L., 249
Logan, W. T., 197
Lombardy, 44
London, England, 131, 336
Long, Armistead L., 169
Long Island, S.C., 226, 231
Longfellow, Henry Wadsworth, 420, 460
López, Francisco, 402
López, Juan, 401
López, Miguel, 377, 401
López, Narciso, xx, 7–13, 16, 19–21, 24–33, 35–39, 41–44, 47–48, 71–72, 74–79, 81–83, 86–98, 100, 109, 114, 116, 118, 129, 133–34, 138, 140, 283, 302, 357–58, 386, 388, 421, 463; and annexation of Cuba, 9–11, 27, 30, 133; captured and executed, 98; at Cárdenas 49–64, 65–69; character of, 8, 10–11, 27, 52–53, 100; and Freemasonry, 9, 32, 74, 76; indicted for violating Neutrality Act, 72; physical description, 54, 72; pseudonym of, 30, 89; in Savannah,

López, Narciso, (*continued*) 71, 74–76, 87–88, 90; sentenced to death, 19, 78
López, Pedro Manuel, 43, 377
López de Santa Anna, Antonio, 10, 19
López, Sotero, 377
López Moreno, Víctor, 378
Losner, William, 378
Loudon Park Cemetery, Baltimore, 421
Louisiana, 84, 113, 133, 249
Louisiana Legion, 38
Louisville, Ky., 28, 30, 44, 93
Louisville Hotel, Louisville, Ky., 30
Lovell, Mansfield, xxiv
Lowers, John A., 378
Luaces, Antonio, 145
Lucas, James Jonathan, 215, 437
Lucas Muro sugar warehouse, Cárdenas, 49, 51
Luce, Alonzo B., 75–76, 99, 113, 293
Ludwing, Ansell R., 378
Lumpkin, John H., 130
Lyons, Michael, 378

Macfarlan, Allan, 245, 251, 441
MacGath, Patrick, 378
Machado, Gerardo, 368
Macías Sardina, Juan Manuel, 16, 27–28, 30, 37, 43–44, 50, 55, 59, 62–63, 75, 81, 92–94, 97, 99, 108, 316, 356, 372
Mackays Point, S.C., 146, 180, 189, 232, 235
Macon, Ga., 74, 82–83, 85–86, 89, 196, 235
Macon arsenal, Ga., 187
Macon & Western Railroad Company, 83, 85
Mádan, C. P., 264
Mádan Gutiérrez, Joaquín, 384
Mádan Mádan, Cristóbal F., 6, 26–30, 35, 43, 75, 384, 393

Magnolia Cemetery, Charleston, S.C., 185, 266, 303, 321, 444
*Magnolia* steamer, 88
Magrath, Andrew Gordon, 246, 441
Mahan, Francis C., 378
Maine, 32, 39
Mallory, Angela, 339
Mallory, Stephen Russell, xxi, 68, 71–72, 107, 111–12, 115–16, 125, 130, 132, 137, 180, 200, 286, 339
Maltby House, N.Y.C., 302–3, 312–13, 318–19
Managua, Cuba, 1
Manicaragua, Cuba, 9
Manifest Destiny, 7, 384
Manigault, Arthur Middleton, 147, 162, 197
Manigault, Edward, xiv, 197–98, 201, 204, 212–13, 437
Manigault, Mary, 293
Manigault, Peter, xiv
Mann, Robert, 372
Manning, John Laurence, 142, 166
Manning, S.C., 225
Manresa, Agustín, 93
Mantoue, Benjamin, 244
Mantua, Cuba, 36, 65, 69
Manville, James L., 378
Marble, Daniel, 372
March, Thomas, 372
Marcy, William Learned, 7, 111–12, 115, 125, 414
Mariel, Cuba, 36
Marriott, James C., 25, 372
Marsh, A. W., 372
Marshall, Mrs., 299
Marshall House, Savannah, Ga., 293, 295, 298, 301
Martello Tower Battery, 214, 435
Martí, Carlos, 320
Martí, José, xvi, xix, xxiv, 352–53, 357, 359, 363–65, 368, 463

Martínez, Agustín, 372
Martínez, Antonio, 402
Martínez de Pinillos, Bernabé (Count of Villanueva), 49
Martínez, G. J., 28
Martínez Fortún, León, 62
Martínez, Manuel, 378
Mason, John Y., 14
Massachusetts, 87, 311, 356
Massachusetts Bay Colony, 102
Matanzas, Cuba, 1–4, 7–9, 11, 37, 47, 50, 59–60, 66, 129, 256, 260, 264, 281–82, 288–90, 292, 295, 298, 301, 305, 312, 329–30, 333, 384, 417, 462–63
Matanzas City Council, 2
Matanzas General Cemetery, 5, 66
Mathews, Beverly, 44, 372
Mathews, John Raven, 261
Mathisson & Co., W., 273
Matthew's Bluff, S.C., 233
Maxwell, J. A., 241
May, A. H., 339
May, Robert E., xvi
Mayfield, Pvt., 372
Maynard Arms Company, 135–38
Maynard rifle, 135–37, 142, 149, 244, 297
Mayo, George Upshur, 173, 216, 223–24, 437
McAlly, Alexander, 378
McBeth, Charles, 197
McCaleb, Theodore, 72, 77, 80
McCann, John McFarland, 37, 62, 372
McClellan, George Brinton, 166, 421
McClelland, Thomas, 378
McCleskey, George A., 87, 302
McCleskey, Thomas J., 87
McCord, D. J., 97
McCormeck, J. J., 372
McCoy, Pvt., 209
McCrady, John, 196
McDerman, John, 372
McDonald, Edmund H., 378

McDonald, Sgt. Maj., 63, 372
McEwen, William A., 82, 100, 115
McFarlan, William W., 26
McFarland, W. A., 372
McFarland, William P., 136, 138
McGreggor, Pvt., 59, 372
McGuffin, Lt., 372
McHenry, Jonathan D. Rush, 60, 372, 378
McHenry, William Henry, 378
McIntosh County, Ga., 76, 84
McIver, Harriet, 248
McKensey, William H., 378
McKenzie, Mark, 407
McKenzie, Maryann, 407
McKinniop, John, 378
McKivigan, John, xvi
McLawrin, L., 372
McLaws, Lafayette, 225, 242, 246
McLeabe, Bernard, 378
McLean, James L., 404
McLean, William B., 89
McLeash, Archibald, 270
McLeod, William Wallace, 198, 209, 434–35
McLeod's plantation, James Island, xv, 198, 209, 433
McManus, Thomas, 353, 364
McManus, William Telemachus, 384
McMaster, Cecilia, xiv
McMaster, Col., 332
McMaster, George H., 366
McMaster, Richard, 366
McMullen, Mark J., 437
McMullin, Peter D., 378
McMurray, C. A., 378
McNeill, Thomas L., 378
McOunegie, Pvt., 372
McPherson, James M., 158
Mc Pherson House, Washington, D.C., 344–46
McPhersonville, S.C., 434
McQueen, John, 130

Means, William B., 256
Medical College of South Carolina, 148, 353
Melesion, Martin, 378
Mellichamp, Stiles, 173, 427
Mellichamp's plantation, James Island, 173, 203, 436
Memphis, Tenn., 35, 328
Méndez, Vicente, 368
Mendive Daumy, Rafael María de, 6
Menllen, Martin, 378
*Mercedita* steamer, 188
Mercer, Hugh Weedon, 181, 183, 431–32
Merino, Felipe, 401
Meriwether, R. L., 416
Mestre, José Manuel, 320
Metcalfe, George E., 378
Metcalfe, Henry B., 378
Mexican War, 5, 7, 12, 14–15, 19, 21, 26, 28, 31–32, 35–36, 39, 53, 59, 64, 81–83, 87, 97, 118, 137, 143, 178, 197, 366, 387, 390, 422, 439, 449
Mexico, 113, 115, 132, 368, 415
Mexico City, 7, 19, 32
Micawber, Wilkins, 342
Middle Place plantation, Colleton County, S.C., 103, 254
Middleton, J. Izard, 172
Mikell, J. M., 207
Milanés, José Jacinto, 3
Miles, William Porcher, 142, 200, 339, 342–43
Miles Planting and Manufacturing Company, La., 342–43
Milford, Pa., 355
Miller, William, 378
Mills, Samuel, 378
Mills, William H. C., 99
Mills House, Charleston, S.C., 139, 145, 185, 266, 283
Miranda, Angelina, 359, 363
Mississippi, 84, 120, 208

Mississippi High Court of Errors and Appeals, 32, 41
Mississippi militia, 31
Missouri, 164
Mitchel, John C., 228, 438
Mitchel, Ormsby M., 180
Mitchell, Lt., 372
Mizell, A., 36, 53–54, 59, 372
Mobile, Ala., 12, 20, 23, 72, 74, 90, 104, 222, 412
*Mobile Register* newspaper, 130, 407
Mongin, William Henry, 76, 89, 101–2
*Monmouth* steamer, 98
Monroe Doctrine, 117
Monroe, Thomas R., 378
*Montauk* monitor, 191
Montefiore Hospital, N.Y.C., 360, 362
Monterrey, Mexico, 7
Montgomery, Ala., 12, 110, 146, 241
Montgomery, Cora. *See* Storm, Jane McManus
Monticello plantation, Cuba, 292
Montojo, Patricio, 50, 399
Montoro, Agustín, 378
Moore, Archibald B., 48, 77, 80
Moore, Capt., 372
Moore, Carol, xiii
Moore, J. Miller, 366
Moore, John, 372
Moore, Maj., 119
Mora, Ignacio, 393
Morales, José N., 62
Morales, Sebastian, 19–20
Morales, Vidal, 9
Mordecai & Company, 264, 301, 304
Moreno, Manuel R., 352–54, 356, 358–59
Morgan, M. J., 372
Morgan, Matthew Somerville, 345, 459
Morrillo, Cuba, 94–97
Morris, Ambrose Gonzales, xiii, 368
Morris, Col., 123

Morris, Lt., 373
Morris, Page, xiii
Morris Island, S.C., 142–46, 171, 178–79, 191–93, Union attack on 198–207, 209, 212, 215, 217, 219, 221, 225–27, 232–33, 340, 420, 433–34
Morse, A. E., 200
*Moultrie* dredge boat, 188
Mount Holly, S.C., 275, 280
Mujeres Island, Mexico, 44, 79
Muñoz Castro, Manuel, 6
Murphy, John, 378
Murray, W. M., 146, 148
Murry, Capt., 44, 60, 373
Murtigh, John, 378
Myers, James, 378
Myers, Joseph, 378
Myrtle Bank Plantation, Hilton Head, S.C., 103, 124, 128–29, 161, 254, 318

Nadal, Pedro, 56
Nagle, Louis, 378
*Nahant* monitor, 191–92
Nance, W. F., 199
*Nantucket* monitor, 191–92
Napoleon III, 120
Nápoles Infantry Regiment, 49, 62
Nashville, Tenn., 32
Nassau, Bahamas, 179, 244, 287
Nassau River, Fla., 98
Nast, Thomas, 459
Natchez, Miss., 31, 33, 113
National City Bank, New York, 305
National Express and Transportation Company, 263
Navarro, José M., 57
Nelson, Patrick H., 144
Nelson, Richard, 378
Neutrality Act, xxi, xxiii, 32, 69, 71–72, 74, 77, 80–81, 88–89, 100, 119, 353, 368
Nevins, Allan, 158

New Bridge at Charleston, 181, 183, 188, 232
*New Ironsides* ironclad frigate, 185, 191, 204, 215, 430
New Lines, James Island, 202, 204, 211–14, 217, 228, 434
New Mexico, 422
New Orleans, La., 11–12, 20, 23, 25–26, 29, 31, 33, 35, 37–38, 43, 47, 69, 72, 76, 79, 83–84, 86–87, 89–91, 93–94, 99–100, 107–8, 114–16, 119, 122, 185, 188, 224, 264, 281–82, 336, 397, 404, 420–21, 431, 439
New Orleans Customhouse, 82, 93–94
*New Orleans Delta*, 31
*New Orleans* filibuster steamer, 26
New Orleans Gaslight and Banking Company, 397
Newport, Ky., 28, 93
Newport, R.I., 114
New River, S.C., 166, 241
New Smyrna, Fla., 136
Newtown Cut, S.C., 174
New York, 114, 252, 340
New York City, 3–4, 11, 13, 20–21, 23–26, 29, 75–76, 83–84, 86–88, 91, 93, 99–100, 107–8, 110, 115–16, 122–23, 137, 260–61, 266, 272, 283, 287, 301–3, 307, 312–16, 318–23, 325, 330–33, 335–38, 343, 350–51, 359–63, 368, 384, 415, 420–21
New York City Board of Education, 319
*New York Evening Post,* 352
New York Hotel, N.Y.C., 305, 359
New York Life Insurance Company, 359
New York Stock Exchange, 315, 336
*Niagara* frigate, 146
*Niagara* steamship, 335
Nicaragua, 116, 390, 395–96, 409, 415, 430, 449, 463
Nichols, George S., 101
Nicholson, Alfred, 130

Nicoll, John C., 71
Niesman, William, 378
Niskos, Janos, 378
Nixon, James O., 108
Noble Brothers & Co., Rome, Ga., 179, 429
Nolasco de Zayas, Pedro, 378
Norfolk Navy Yard, Va., 149
Norriss, John, 378
North Carolina, 94
North Edisto River, S.C., 146, 164–65, 168, 190, 227, 232, 244
Northeastern Rail Road, 245, 275
Northrop, Lucius, 169
Norway, 311
Norwich, Conn., 384
Nott, G. W., 183, 339
Null, Charles, 378
Núñez del Castillo, Carlos, 5

Oak Grove Cemetery, St. Marys, Ga., 409
Oak Lawn plantation, Charleston County, S.C., xiii, 102–4, 123–26, 128, 132, 135, 137–38, 142, 146, 160, 164–66, 169, 177–78, 185, 189, 195, 197, 212–13, 216, 218–21, 223, 225, 243–44, 252, 254, 256–60, 265–69, 272, 275, 278, 280, 282, 289, 291–97, 300, 303–5, 308–10, 320–21, 324–25, 327–30, 332, 335, 341, 357–58, 367–68, 412, 416; auctioned for taxes 317; destroyed by Union troops 247; redeemed 318
Oberto Urdaneta, Ildefonso, 94–95, 378
O'Brien, John Thomas, 378
Ocean House resort, Newport, R.I., 13, 25
Ogeechee River, Ga., 102
Ogier, Thomas L., 185, 231, 277
Oglevie, James, 378
Oilnette, Frank, 119
Ojeda, Antonio, 51

Old Lines, James Island, 212
Old Point Comfort, Va., 345
Old Town Creek, S.C., 173
Oleaga, Juan, 378
Oliva, Rafael, 289
Oliveira, William E., 89
Ollis, Elira J., 378
Olmstead, Charles H., 437
Opelika, Ala., 12
Orbegoso, Juan de, 120
Ordaz, Pedro, 57, 66
Oregon annexation, 16
Organization of American States, 351
O'Reilly, Pvt., 378
O'Rourke, Patrick, 378
Orphan Home and School, Charleston, 294, 311
Orr, James L., 134
Otis, Elias J., 378
Otter Island, S.C., 162
Our House and Restaurant, Savannah, 76
Our Lady of Guadalupe School, Havana, 4
Owen, Allen Ferdinand, 85
Owen, J. L., 317
Owen, James G., 378
Owen, Robert Dale, 460
Owens, Pvt., 378
O'Connor, Michael Patrick, 278, 283, 322
O'Donnell Jorris, Carlos, 8
O'Donnell Jorris, Leopoldo, 8
O'Hara, Theodore, 21, 28, 30–31, 42, 50–51, wounded 52, 58, 60–61, 65, 69, 72, 77, 87, 92, 97, 373
O'Neall, John B., 126
Order of the Lone Star, 386
Oriente, Cuba, 283, 290
Orkney Springs, Va., 346–47
O'Sullivan, John Louis, xxi, 20–21, 24–25, 28–30, 72, 76–77, 83–84, 87–88, 93, 99–100, 104, 114, 116, 384

O'Sullivan, Mary, 384
Otranto plantation, S.C., 444

Padrines, Dr., 56
Páez, José Antonio, 113
Paicuriche, Antonio, 378
Paine, Mrs., 127
Palatka, Fla., 99, 136
Palma Romay, Ramón de, 6, 289
*Palmetto State* gunboat, 188, 246, 430
*Pampero* filibuster steamer, 93–94, 96–101, 106, 114, 122, 353
Pan-American Congress, 351–53
Pan-American Union, 351
Panama, 36
Panama Canal, 120
Panic of 1873, 315
Panknin, Charles F., 273
Paratolt, Conrad, 378
Paris, France, 123, 129, 335–36
Paris Exhibition, 123
Paris, Ky., 37
Parish, Lt., 373
Parker, Francis L., 337
Parker, Sgt., 373
Parr, George, 378
Parrish, Robert A., Jr., 26
Pascagoula, Miss., 20
Pass Christian, Miss., 31, 38
*Passaic* monitor, 191
*Patapsco* monitor, 191
*Patria* newspaper, 359–63, 368
Patton, Bryson, 366
Paulding, Hiram, 129
*Pawnee* gunboat, 214
Peabody Academy of Music, Baltimore, 307
Peabody, H., 373
Peabody Institute, Baltimore, 306
Pedro, Pvt., 378
Pee Dee River, S.C., 167
Peeples, William B., 237

Pemberton, John Clifford, xxii, 162, 164, 167, 169, 171–79, 192, 203, 205, 208, 211, 229, 245, 312
Pendleton, Simeon, 41, 44, 48, 398
Penn, Alexander Gordon, 107, 112
Pennal, R. A., 273, 277
Pensacola, Fla., 23, 38, 339, 421
Peoli, Laura, 339
Peraza Sarausa, Fermín, xxv
Perdomo, Rosalia, 1
Pérez, Vicente, 402
Perkins & Company, 282
Perkins, J. C., 373
*Perrit* expedition, 290
Perry, M. C., 11
Perth Amboy, N.J., 86
Peteri, John, 378
Petersburg, Va., 12, 225, 430
Petigru, James Louis, 125, 169
Petipiers, Pvt., 378
Peuquet, Hyacinth, 3
Peuquet, Louis, 3
Pezuela, Juan Manuel de la, 117
Phelps Dodge & Co., 319
Philadelphia, Pa., 19–20, 23–24, 125–26, 164, 313, 384, 386–87, 422
*Philadelphia Manufacturer*, 352
Philippines, 20, 365
Phillips, Asher J., 378
Phillips, J. B. W., 197
Phillips, Michael, 378
Phillips Island, S.C., 103, 127, 146
Pickens, Francis Wilkinson, xxi, 139, 141–42, 144–49, 153–54, 158, 167, 171, 177, 182, 208, 428
Pickett, John T., 21, 28, 30, 37, 39, 50, 55, 58, 60–61, 64, 72, 77, 93–94, 107, 110, 114, 116, 119, 373, 415, 457
Piedra Key, Cárdenas, 48, 66
Pierce, Franklin, xx, 104, 107–8, 110, 113, 115, 118–19, 125–28, 339, 343

Pierce, Mrs. Jane Means Appleton, 125–26, 128
Pierra, Fidel G., 351
*Pilot Boy* steamer, 301
Pinar del Río, Cuba, 36, 60
Pinckard, George M., 404
Pinckney, Charles, 431–32
Pinckney, Charles Cotesworth, 125, 304, 310, 328–29
Pinckney, Mary, 416
Pinckney's Agency, Mr., N.Y.C., 320
Pineberry, S.C., 221, 226
Pinillos Plaza, Cárdenas, 61, 64
Piñeyro Barri, Enrique, 336
Pittfield, Mr., 84
Pittsburgh, Pa., 14, 30
*Pizarro* war steamer, 47–48, 66–68
Planos, Angel, 378
*Planter, The,* steamer, 150, 334
Plymouth, England, 336
*Pocahontas* armed steamer, 160
Pocotaligo, S.C., 161, 163, 165–67, 172, 181, 189, 223, 234–35, 240–43, 256
Pocotaligo River, S.C., 147, 161, 181, 189
Polk, James Knox, 5, 14–15, 18, 32, 387
Polk, Leonidas, 153
Pomeroy, Augustus A., 378
Pon Pon River, S.C., 136, 226
Ponce, Antonio, 75–77
Pope, A. F., 431–32
Pope, J. J., 176, 183, 431–32, 437
Port, N., 378
Port Hudson, La., 353, 421
Port Royal, S.C., 151, 154–56, 160–61, 164, 168, 171, 181, 194, 252, 254, 366, 425
Port Royal and Augusta Railway Company, 331
Port Royal Ferry, S.C., 163, 165
Port Royal Island, S.C., 103, 161, 354
Port Royal Railroad, 317–18

Port Royal Sound, S.C., 146, 149–50, 155, 159–60, 162
Portell Vilá, Herminio xiii, xxv, 9, 60, 386
Porter, Anthony Toomer, 294, 307, 328–29
Porter, J. G., 378
Porter-Gaud School, Charleston, 294, 328
Porto Zarazate, Miguel F., 81
Portocarrero, José A., 28
Portugal, 8, 114
Posse, José R., 393
Potter, David M., 158
Potter, Edward E., 233–34, 237
Poujaud, Carlos, 305
Poujaud, Cecilia Michel, 291–92, 295, 303
Pragay, Janos, 29, 86, 94–95
Preble, George Henry, 233–34
Prentiss, Seargent Smith, 72
Preston, William Ballard, 23
Pringle, James R., 205
Pringle, Motte A., 189
Prioleau, Charles K., 447
Pritchard, William R., 437
Proctor, Stephen, 201
Pruitt, John T., 378
Prussia, 306
Pryatt, J. B., 318
Pryor, Roger Atkinson, 142, 320
Puebla, Mexico, 7
Pueblo Nuevo, Cuba, 288–90
Puerto Príncipe, Cuba, 7, 92, 95
Puerto Rico, 7, 9, 47, 57, 75, 118, 365
Pulaski House, Savannah, 74, 84, 108
Purcell, Joseph, 283
Purnell, Stephen Howard, 378
Purysburgh, S.C., 166

*Quaker City* steamship, 261, 266
Quash, Joseph, 231
Queen María Cristina, 8
Queen Victoria, 460

Queipo, Joaquín, 51–52
Quesada, Cayetano V. de, 393
Quesada, Gonzalo de, xxiv, 351–52, 357–59, 361, 363
Quick, George, 378
Quin, John, 373
Quincy, Fla., 323
Quintayros Plaza, Cárdenas, 49, 51, 53, 57, 60–61
Quintero, José Agustín, 6, 145, 420
Quitman, Frederick Henry, 29–30, 108
Quitman, John Anthony, xxi, xxiv, 21, 28, 31–33, 35–36, 39, 41–42, 47, 65, 69, 72, 74, 77, 81, 108–10, 113–14, 118–20, 122, 130, 156, 394, 439
Quitman Rifles, 420

R. G. Dun & Co., 31, 76, 255, 293
Radcliffe, James D., 163
Radnitz, Pvt., 378
Rains, George W., 184, 206
Raleigh, N.C., 240, 248–49
Ralston, Richard, 89
Ramsay, William M., 174, 204, 434
Randall, Edwin M., 339
Randolph, George W., 176–77, 180
Rantoles Station, S.C., 215
Raoul & Lynah, 273, 283
Rawlings, C. H., 373
Raysor, H. C., 234
Reading, John B., 373
Reagan, John H., 159
Real, L., 336
Reconstruction Act of 1867, 267
Red Bluff, S.C., 166, 195, 215, 221
Red Shirts, 319–320, 325, 330, 456
Red Top House, Lawton plantation, 206, 209
Redding, William, 50, 77, 373
Redoubt No. 1, James Island, 173, 207, 212, 213–14, 436
Redoubt No. 5, James Island, 173

Reed, James, 373
Reed, John, 77, 373
Reed, Samuel, 378
Reeves, Wilson L., 378
Regiment of Noble Neighbors militia, Havana, 47
Reid, Capt., 287, 303
Renean, J., 85
Reni, Guido, 361
Rensselaer Polytechnic Institute, Troy, N.Y., 312
Reunión Deseada plantation, Cuba, 300
Revolutionary War, 163, 176, 246, 406
Rhett, Alfred Moore, 144, 190, 205, 270, 273, 433–34, 437
Rhett, Andrew Burnet, 223, 431–32, 436–37
Rhett, Edmund, 155
Rhett, Haskell S., 267, 270, 318
Rhett, Marie, 270
Rhett, Robert Barnwell, Jr., 144, 270
Rhett, Robert Barnwell, Sr., 437
Rhett, Thomas S., 164
Rice, James Henry, Jr., 412
Richardson, George W., 378
Richardson, John P., 346, 356
Richland Volunteer Rifle Company, 346
Richmond, Va., 12, 22, 108, 151, 167–68, 174, 179, 184–87, 192, 196, 200, 205, 207, 211, 217, 220, 224–29, 231, 245
Richmond Arsenal, 190, 205
Richmond State Guards, 341
Rico, Benigno, 320
Ridgeway, S.C., 246
Rion, James Henry, 144–45
Riordan, Bartholomew Rochefort, 328
Ripley, Roswell Sabine, 150, 159–60, 162–64, 166, 169, 171, 173, 182–84, 188–90, 193–94, 197, 199, 201–4, 207–8, 218, 431, 433
Ritchie, Thomas, 21
Rives, Ramón, 378

Roach, Roseanna, 267, 271, 277–78
Robb, James, 42, 397
Roberts, Hiram, 86, 88–90, 408
Robertson, Beverly H., 212, 219, 239–40, 247
Robertson, William H., 115
Robertsville, S.C., 235
Robinson, Charles A., 378
Robinson, Henry H., 37, 53, 61, 82, 93, 100, 115, 373
Robinson, John, 378
Rockbridge Alum, Va., 126
Rockwell, William S., 437
Rodgers, John, 142
Rodríguez, José Ignacio, 344–45, 351–52, 355
Rodríguez, Juan, 401
Rodríguez-Capote y de la Cruz, Alejandro, 51, 55, 399
Rodríguez-Capote, Regla, 55
Rodríguez Mena, Manuel, 6
Rogers, Samuel St. George, 89
Rogers, William T., 84, 87
Roig de Leuchsenring, Emilio, 8
Roman, Alfred, 339
Román, Felipe, 401
Rome, Ga., 179, 429
Romero, Antonio, 378
Roncali. Federico (Count of Alcoy), 26, 30, 47, 50
Ronquillo, José, 401
Roosevelt Hospital, N.Y.C., 360
Rosales, Juan Antonio, 378
Rosario Pineda, José, 57, 66
Rosecrans, William S., 422
Roseland plantation, Jasper County, S.C., 163
Rosis, Tomas M., 75–76, 81
Ross, N., 379
Rossini, Gioacchino, 10
Rough Riders, 366

Round Island expedition, 23–28, 37–38, 388
Round Island, Miss., 23, 25
Rousseau, David Q., 379
Roy, Thomas B., 246
Royall, Crosky, 173
Royall's House, James Island, xv, 173, 204, 206, 210, 213, 216–17
Rubio, José, 379
Rues, Tim, xvi
Ruffini, Paul, 2
Rufín, Bernardino, 1
Rufín de Torres, Cresencia Josefa Gertrudis, 1
Ruiz, Cándido, 264–65
Rusk, Thomas Jefferson, xxi, xxiv, 17–18, 130, 388
Russell, J. R., 431–32
*Russell's Magazine,* 129, 132
Russia, 330
Rutledge, Benjamin Huger, 160, 339
Rutledge, F., 278
Ruvira, J. B., 379
Ryan, William H., 433

Sabanilla Rail Road Company, 264–65, 445
Sagua la Grande, Cuba, 278
Saint Timothy's Episcopal Church, Herndon, Va., 315
Saint Timothy's Home School for Boys, Herndon, Va., 315, 317
Sainz, Felipe, 401
Sainz, Francisco, 401
Salas, F. P., 264, 291
Salisbury, N.C., 249
Salkehatchie, S.C., 241, 271
Salkehatchie defense line, S.C., 244
Salmerón, Eduardo, 379
Salmon, James, 379
Salt Sulphur Springs, W.V., 124
San Carlos Cathedral, Matanzas, Cuba, 1
San Carlos theater, Key West, Fla., 353

San Cristóbal de Carraguao High School, Havana, 4–5
San Fernando de Camarones, Cuba, 9
San Juan Hill, Cuba, 366
San Juan River, Cuba, 288
San Luis Potosí, Mexico, 19
San Luis sugar mill, Cuba, 9
*San Marcos* steamer, 359
Sánchez, Edward, 281
Sánchez, Pedro Gabriel, 10–11
Sánchez White, Calixto, 368
Sánchez Yznaga, José María, 9–11, 16–17, 20, 37–38, 41, 43–44, 69, 71, 74–76, 83–84, 87, 93–94, 99, 108, 373, 386, 398
Sánchez Yznaga, Pedro M., 393
Sánchez Yznaga, Saturnino, 9, 11
Sand Key lighthouse, Fla., 67
Sanderra, Juan, 401
Sanders, James, 379
Sandy Hook, N.J., 86
Sanford, J., 379
Sank, John G., 379
Santa Bárbara Sugar Mill, Cienfuegos, Cuba, 9
Santa Cruz, Agustín, 379
Santee, S.C., 128, 167
Santee River, S.C., 147, 162, 246–47
Santiago de Cuba, 7, 316, 366–67, 386
Santo Domingo, 35, 41
Santovenia, Emeterio S., xxiv
Sapelo Island, Ga., 76, 160
Saratoga Springs, N.Y., 134, 137–38, 331–33
Sartain, George F., 373
Sartorio, Félix, 363
Sartorio, Ricardo, 363
Satilla River, Ga., 84, 86, 89
Sauls, Benjamin, 317–18
Saunders, A. L., 119
Saunders, Romulus M., 116
Saunders, Walter, 373
Saunders, William J., 179, 431–32

Savannah, Ga., 69, 71, 74–76, 81–82, 84–94, 96–97, 99–102, 104, 106, 108, 112, 122, 136, 147, 162, 167, 179, 181, 183, 188, 193, 195, 198–99, 218, 225, 231–36, 240–41, 243–44, 247, 284, 286, 291, 293–98, 301–2, 310, 315, 323–24, 346, 359
*Savannah* ironclad ram, 241
*Savannah Republican,* 239
Savannah River, Ga., 88–89, 162, 166, 181, 199, 233, 241
*Savannah* steamer, 293
Saxton, Rufus, 229, 254, 443
Sayer, William M., 160
Sayle, Henry, 379
Sayre, Burwell B., 25, 64, 373
Scharlock, Amanda, 282, 284, 294
Scharlock, John, 282
Scheiprt, Zyrtack, 379
Schimmelfennig, Alexander, 186
Schlesinger, Louis, 84, 87–88, 93–95, 100, 379, 415
Schluht, Harbo, 379
Schmidt, George, 379
Schmidt, Henry, 379
Schuets, Robert, 379
Schulz, Frederick C., 221
Sckneck, C., 379
Scott, Isaac, 83, 85
Scott, Malbon K., 379
Scott, R., 373
Scott, Robert K., 258, 443
Scott, Samuel S., 373
Scott, Winfield, 12
Screven, James P., 101, 130
Screven's Ferry, S.C., 88
Scully, Luke, 379
*Sea Gull* filibuster steamer, 21, 26
Sea Island Circular, 252
Seabrook Island, S.C., 190
Seabrook plantation, Colleton County, S.C., 273, 276

Seaside Plantation, James Island, 434
Seay, Dandridge, 379
Sebastian, William K., 130
Sebring, Cornelius, 379
Secessionville, S.C., 172–74, 202–3, 211, 213–14, 217, 226, 231, 436
Seddon, James, 197, 205–6, 224, 229, 245
Sedgwick, Theodore, 108
Segura, Andrés, 50, 55, 61
Seilert, John, 379
Seminole War, 7, 20, 143
Semmes, Raphael, 249
Semple, James A., 306
Semple, Letitia Tyler, 306, 312
Serrano, J. Vicente, 351
Sestar, Andrés, 379, 401
Seward, William H., 15
Sewanee, Tenn., 328
Sewer, Frederick S., 379
Sexias, Lt., 54, 66, 373
Seybert Commission, 348, 350
Seymour, Truman, 200
Sharkey, William L., 32
Sharratt, F. T., 164
Shaw, Robert, 200–201
Sheen, Mrs., 271
Shell Point, S.C., 199
Shell Point plantation, Port Royal Island, S.C., 103, 161, 254, 354
Sherman, E. H., 280
Sherman, Hugh M., 39, 373
Sherman, Thomas W., 159, 162, 166, 168
Sherman, William Tecumseh, 229, 232, 234–35, 240–44, 246–49, 252, 340–41
Shilling, William, 379
Shiver, Sarah Cecil. *See* Gonzales, Sarah Cecil
Shoup, Francis A., 245
Shulz, Frederick C., 195
Shute's Folly Island, S.C., 139

Siege Train of Georgia, 183, 198–99, 214, 228, 437
Siege Train of South Carolina, 182, 187, 190, 193–98, 201, 207, 212, 216–17, 219, 221, 232, 432, 437
 Company A, 182, 187, 197–98, 217, 432
 Company B, 182, 187, 197–98, 432
 Company C, 182, 197–98, 432
Siegling, Rudolph, 339
Sigur, Laurent J., 31, 37, 39, 41–42, 72, 77, 80–81, 83, 86, 96–99, 121
Sila, Manuel, 401
Simkins, John C., 433
Simmons, "Old," 271–72
Simmons, William, 273
Simons, James, 143–44, 322
Simonton, Charles H., 198, 202, 322
Simpson, J. P., 379
Simpson, William D., 341
Singletary, Robert L., 232
Skirving, Ann Rebecca, 102–3,
Skirving, James, 102
slaves and slavery, 5, 15–17, 25, 28, 31–32, 41, 75–76, 88, 103, 140, 155, 161, 163, 176, 178, 181–82, 192, 197, 212, 214, 217–19, 234–35, 242, 247–48, 252–53, 257–58, 271–72, 329, 334, 434
Slidell, John, xxi, 112, 130, 134–35
Smalls, Robert, 334, 458
Smart, William, 67
Smith, Ann Hutchinson. *See* Elliott, Mrs. Ann
Smith, Benjamin B., 428–29
Smith, B. Burgh, 148
Smith, Charles, 373
Smith, Charles N., 379
Smith, Charles F., 13
Smith, Cotesworth Pinckney, 32, 35, 41, 43, 72, 77, 395
Smith, Duncan P. *See* Daniel Henry Burtnett

Smith, Edward, 373
Smith, G. N. L., 44, 373
Smith, Gustavus Woodson, xxiii–xxiv, 13, 122, 235–38, 240–42, 246, 249, 439–40
Smith, James, 379
Smith, John T., 379
Smith, Joseph, 373
Smith, L. Jaquelin, 183, 431–32
Smith, Persifor F., 390
Smith, Peter, 35–36, 59, 61–62, 74, 373
Smith, Robert A., 83
Smith, S. P., 432
Smith, Thomas Rhett, 102
Smith, William Duncan, 176
Smith Benjamin B., Jr., 428
Smithfield, N.C., 248–49
Social Hall plantation, Colleton County, S.C., xiv–xv, 103, 135, 254, 256, 260–61, 263–65, 267–83, 296, 304–6, 310–12, 322, 330–31, 342, 344; auctioned for taxes 318, 326–27; redeemed 335
Solomons, S. S., 245
Somers, Henry, 379, 411
Soulé, Pierre, xxi, xxiv, 111–12, 114, 116, 134
South Atlantic Blockading Squadron, 159, 164, 192
South Carolina, 110, 136, 179, 233, 236, 240, 242, 246, 249, 267, 360, 362
South Carolina Executive Council, 166–67, 171
South Carolina Land Commission, 318
South Carolina Military Academy. *See* Citadel
South Carolina Military Commission, 167
South Carolina Military Districts: First, 182, 184, 194, 198–99, 202, 204, 217, 224, 433–34, 438; Second, 175, 196, 219, 221, 226–27, 434, 437–38; Third, 178, 195, 198, 215–16, 219, 221, 233, 434, 438; Fourth, 223, 232, 434, 438; Fifth, 217, 224; Sixth, 215, 217, 221, 224, 226, 437–38; Seventh, 215, 217, 224, 437–38
South Carolina Militia: First Brigade of South Carolina Volunteers, 153; First Regiment South Carolina Volunteers, 146; Second Regiment South Carolina Volunteers, 146; Ninth Regiment South Carolina Volunteers, 148; Fifteenth Regiment South Carolina Volunteers, 160; 13th Regiment, 136; Hamilton Guards, 149
South Carolina Ordinance of Secession, 139
South Carolina Railroad, 243
South Carolina Railroad workshops, 186
South Carolina State Convention, 139
South Carolina Volunteers, 143–44
South Edisto River, S.C., 221
South Island, S.C., 167
Southern Express Company, 244, 255, 276
Sowers, John A., 379
Spain, Albertus Chambers, 247–48
Spalding, Randolph, 76, 160, 425
Spalding, Thomas, 76
Spanish-American War, 366–67
Spanish espionage in U.S., 23, 29–30, 42, 83, 108, 122
Spanish Gaslight Company, 397
Spartanburg, S.C., 447
Spaulding, Joseph W., 22
Spiritualism, 347–51
Sprague, John T., 7, 13
Springer, Joseph A., 344
Springville, S.C., 247–48, 250–54, 258
St. Andrews, S.C., 142, 162–63, 173, 194, 211–12, 216–19, 433–34
St. Augustine, Fla., 136, 168, 197
St. Barnabas Hospital, Bronx, N.Y., 361

532 / Index

St. Bartholomew's Episcopal Church, N.Y.C., 303, 319
St. Charles Hotel, New Orleans, 12, 35, 37–39, 41, 77, 80, 110
St. Davids Cemetery, Cheraw, S.C., 441
St. Helena, S.C., 103, 150, 162, 164, 195, 254
St. James Episcopal Church, Charleston, 427
St. John in the Wilderness Episcopal Church, Flat Rock, N.C., 123, 131, 227, 291
St. Johns River, Fla., 76, 87, 92, 99, 136
St. Joseph's Church, Charleston, 224
St. Louis, Mo., 21, 404
St. Louis Hotel, New Orleans, 86, 90, 108
St. Marys, Ga., 87, 90, 409
St. Marys River, Ga., 90
*St. Mary's* steamer, 136
St. Michael's Church, Charleston, 175–76
St. Nicholas Hotel, N.Y.C., 138
St. Paul's Episcopal Church, Baltimore, 306–7
St. Paul's Episcopal Church, Charleston, 151, 228
St. Paul's Episcopal Church, Richmond, 108
St. Paul's Parish, S.C., 103, 317
St. Stephen, S.C., 246
Stanley, James M., 373
Stanley, M., 431
Stanly, M. B., 182, 432
Stanmire, Henry, 379
Stanniford, Clement, 379
Staunton, James, 379
Starke, William Edwin, 165
*The State* newspaper, 354, 357, 367–68, 412
State of Georgia Floating Battery, 199
Staten Island, N.Y., 26
Stearns & Curtis, 361
Steed, Abner C., 55, 61, 373
Steenbergen, Bessie B., 108
Stender, F. W., 209
Stevans, Joseph, 379
Stevens, Thaddeus, 15
Stevens, Theodore, 379
Stevenson, Thomas, 59, 66, 373
Stewart, Daniel H., 90
Stewart, William H., 379
Stezinger, Pvt., 379
Stockwell, Mr., 314
Stoneman, George, 249
Stoney Creek Cemetery, Yemassee, S.C., 320
Stoney Creek Church Yard, Pocotaligo, S.C., 266
*Stono* gunboat, 188
Stono Inlet, S.C., 145, 164, 168, 172
Stono River, S.C., 145, 162, 166, 168, 172–74, 180, 183, 187, 202–3, 205, 210–11, 215, 228
Storm, Jane McManus, 7, 15, 28, 131, 384–85
Stoughton, Francis, 84
Stoval, Sgt., 374
Stowe, Harriet Beecher, 460
Stratton, Mrs., 355–56
Stuart, Charlie, 325
Stuart, Henry Middleton, 149, 232, 238–39, 439, 440
Stubbs, John, 379
Stubbs, W. W., 275–76, 279
Stull, Surg., 374
Sullivan's Island, S.C., 138, 160, 165, 191, 206–7, 218, 224, 229, 231, 366, 433
Sumner, Charles, 15
Summerville, S.C., 190, 429
Super, John, xvi
*Susan Loud* filibuster brigantine, 41–45, 48, 75, 77–79
Swamp Angel Battery, 206, 224

Sweet Springs, Va., 124, 323
Syracuse, N.Y., 362

Tacón Theater, Havana, 129
Talbot, John, 379
Taliaferro, William Booth, 202–4, 206, 208, 212, 215, 217, 246, 249, 433–34
Tallahassee, Fla., 323, 338
*Tallahassee Floridian,* 339
Tallmadge, Nathaniel, 460
Tammany Hall, N.Y.C., 107
Tampa, Fla., 359
Tampico, Mexico, 19, 390
Tatnall, Josiah, 160, 163
Tatom, William T., 433
Taylor, Alrutheus Ambush, 270
Taylor, Conrad, 379
Taylor, George T., 374
Taylor, Marion C., 55, 64, 69, 374
Taylor, Richard, 438
Taylor, Zachary, xxiv, 14–15, 18, 20, 23–24, 26, 29, 42, 70, 80, 82
Tejedor, Galo, 402
Tennessee, 85, 110
Terrell, L. F., 437
Teurbe Tolón, Emilia, 384
Teurbe Tolón, Miguel, 6, 19, 20, 27, 114, 384, 393
Texas, 135, 415, 420, 422
Texas annexation, 5, 16, 89, 133
Texas Republic, 21, 89
Texas statehood, 14
Thixton, Lt., 374
Thomas, L. C., 60, 374
Thomason, H. J., 379
Thompson, Hugh Smith, 341, 359
Thompson, James R., 374
Thrasher, John Sidney, 56, 58, 119, 122, 415
Thunderbolt, Ga., 87

Tiffany-Fisher House, Baltimore, 308
Tift, Asa F., 72
Tillman, Benjamin Ryan, 355, 367
Tillman, James H., 367
*Times-Democrat, The,* New Orleans, 339, 344
Tirry Lacy, Juan, 3
Tirry Loynaz, Guillermo, 16
Titus, Henry Theodore, 37, 39, 56, 63, 65, 76, 84, 87, 91–92, 97–99, 374, 395–96, 415
Titusville, Fla., 395–96
Togno, Rosalie Acelie, 123
Toombs, Robert, xxi, xxiv, 130
Torres Hernández, Anselmo, 379
Tosca, Basilio N., 59, 66
Toucey, Isaac, 131
Toumey, John, 273
Tower Battery, Secessionville, S.C., 172, 174–75, 191
Towns, George Washington, xxi, 80–82, 100, 406
Trapier, James Heyward, 169, 431–32, 434
Tredegar Foundry, Richmond, Va., 151, 154
Trenholm, George A., 447
Trenholm, William L., 160, 228, 428–29
Trezevant, James Davis, 212
Trinidad, Cuba, 7–9, 12, 93, 117, 422–23
Triplett, R. S., 28, 374
Trist, Nicholas Philip, 5, 14
Troy, N.Y., 312
Tryon, N.C., 324–25
Tucker, John R., 179
Tufts, William, 404
Tulane University, 339
Tullifinny Creek, S.C., 240
Tullifinny River, S.C., 147, 162
Tupelo, Miss., 243
Turks Island, 21
Turla, Leopoldo, 92

Turner, Henry, 110
Turner, John W., 204
Turner, Richard T., 77
Turnpike Road, S.C., 173
Tyler, John, 306

U.S. Department of State, 120
U.S. Naval Academy, 121, 131
U.S. Patent Office, 124–25
*USS Celtic*, 366
*USS Maine*, 365
Ujházy, László, 29
Unger & Keen, 302
Union troops:
—Army of the Cumberland, 422
—3d New York Artillery, 233, 238
—3d Rhode Island Artillery, 233
—4th Massachusetts Cavalry Regiment, 233–34
—8th Connecticut Volunteers, 292
—23d Pennsylvania Infantry, 422
—25th Ohio Regiment, 233, 237
—26th U.S. Colored Troops Regiment, 233
—32d U.S. Colored Troops Regiment, 233–34
—35th U.S. Colored Troops Regiment, 233, 258
—50th Regiment of Pennsylvania Volunteers, 172
—54th Massachusetts Infantry Regiment, xxii, 200, 233, 237
—55th Massachusetts Infantry Regiment, 233, 239
—56th New York Regiment, 233
—75th Pennsylvania Infantry, 387
—102d U.S. Colored Troops Regiment, 233
—114th Pennsylvania Infantry, 422
—127th New York Regiment, 233, 237
—144th New York Regiment, 233, 237
—157th New York Regiment, 233, 237, 247
United States Arsenal at Charleston, S.C., 186
United States Hotel, Washington, D.C., 28
*United States Magazine and Democratic Review,* 30, 384–85
United States Military Academy, West Point, 13, 21, 25–26, 36, 142, 153, 155, 164, 169, 207, 225, 253, 422, 429, 439
University of Georgia, 90
University of Havana, 4, 384
University of Louisiana, 339
University of Modena, 2
University of Nashville, 36
University of New York, 353
University of Pennsylvania, 348
University of the South, 328

Valdés Sierra, Gerónimo, 8
Valdespino, Antonio, 401
Valdosta, Ga., 324, 326, 328–29
Valenzuela, Francisco, 402
Valiente, Porfirio, 113, 122
Van Vechten, Philip S., 379, 411
Vanderbilt, Cornelius, 460
Varadero, Cuba, 48
Varnville, S.C., 318
Vaughan, William H., 379
Vázquez, Antonio, 59
Veazy, T. B., 379
Velasco, José Antonio, 4
Velazco, Pedro, 11, 43, 63, 374, 379
Velázquez, Loreta Janeta, 145
Venable, Abraham Watkins, 17
Venezuela, 7, 110, 112–13, 115, 119, 140
Veracruz, Mexico, 114, 116, 390
Veranda Hotel, New Orleans, 37, 108
Vermont, 357
Vernon, E., 374

Vesey, J., 374
Vicksburg, Miss., 31, 179, 205, 211, 229, 245, 421
Vicksburg and Jackson Railroad, 31
Vienne, H. T., 379
Villa Clara, Cuba, 93
Villalón, José R., 351
Villarino, Luis, 379, 401
Villaverde, Cirilo, 6, 9, 20, 25, 27, 28, 30, 36, 75, 90–93, 99, 289, 357, 386, 393
Vingut, Francisco T., 393
Vining, Mrs., 338
Viñas, Luis, 401
Virag, Janos, 379
Virginia, 145, 195, 225–26, 231, 240, 248, 296, 314, 323, 428
Virginia General Assembly, 22
Virginia Hotel, Shenandoah Valley, Va., 358
Virginia House, Fernandina, Fla., 286
*Virginius* affair, 316
Von Schlicht, H., 379
Vuelta Abajo, Cuba, 9, 47, 60, 65, 69, 94

Waccamaw River, S.C., 167, 172
Waddy, John R., 195, 219, 223, 437
Wade, Benjamin, 460
Wagner, John A., 160
Wagner, Theodore, 447
Wagner, Thomas M., 174–76, 178
Walker, David, 339
Walker, Leroy Pope, 142, 144
Walker, Robert J., 14, 82
Walker, William, 115–16, 283, 390, 395–96, 415, 430, 449, 463
Walker, William Stephen, 195, 199, 215–16, 431, 434
Wall, William H., 72
Wallace, Alfred Russel, 347–48
Walter, George H., 220
Walterboro, S.C., 271, 277, 304, 330
Wampler, J. Morris, 435

Wappoo Creek, S.C., 173, 180, 198
Wappoo Cut, S.C., 215
War of 1812, 7, 89
Ward, Mrs. N. L., 338
Warley, Frederick F., 437
Warner, George, 59, 374
Warren, John D., 267, 304–5, 318, 326, 335
Warren Green Hotel, Warrenton, Va., 343
Warrenton, Va., 21, 91, 312, 343, 346
Warrenton Springs, Va., 21, 104, 107, 120, 128, 130
Washington College, 296
Washington, D.C., 13–14, 16, 20, 22, 24–25, 27–29, 85, 97, 104, 107, 110, 114–15, 120–22, 124, 126–32, 134–35, 137, 139, 214, 315, 322–23, 332–34, 340–42, 344–45, 349–51, 353, 356, 422
*Washington Post,* 355
Waterloo, 3, 62
Watervliet, N.Y., 13
Waties, Thomas Davis, 183
Waverly plantation, Ga., 89
Wayne, James Moore, 120, 126
Wayne, Mary W., 120, 126, 133
Wayne, Richard, 75, 101, 113
Webb, Alexander Stewart, 320
Webb, Benjamin C., 432
Webster, Daniel, xxiv, 15, 74, 85, 99, 108
*Weehawken* monitor, 191
Weiss, Edward, 379
Welles, Gideon, 150, 164, 185
Welsman, James, 447
West, Henry, 379
West & Jones, 273
West Chester, Pa., 14
West Lines, James Island, 173–74, 198, 202–3, 211
West Siege Lines. *See* New Lines
West Point. *See* United States Military Academy

Westbrook, Theodoric R., 343
Western Union Telegraph Company, 331–33, 336
Weyler, Valeriano, 365
Weymouth, J. B., 379
Wheat, Chatham Roberdeau, 36–37, 41–42, 60, 65, 72, 77–78, 93–94, 108, 119, 122, 374, 415; wounded 52–53, 58
Wheat, John Thomas, 36
Wheaton, John F., 198
Wheeler, Joseph, 374
Wheeler, Gen. Joseph, 242
Wheeling, W.V., 308
Wheeling, Robert, 374
Whippy Swamp, S.C., 279
Whitcomb, T., 379
White, Mrs. Ashton, 127
White, Edward B., 437
White, George William, 21, 23–24, 26, 28, 390
White, Miss, 124–25, 128
White, William H., 38
White, William P., 101
White, William W., 337
White Hall, S.C., 218
White Hall Ferry, S.C., 161
White House, Washington, D.C. 28, 78, 125–27, 131, 140, 351, 460
White Point, S.C., 183, 227, 232, 244
White Point Battery, Charleston, 173
Whiting, William H. C., 143, 145
Whitmarsh Island, Ga., 231
Whittier, James Greenleaf, 460
Wichman, Mr., 318
Wier, Edward, 26
Wigfall, Louis Trezevant, xxi, 142
Wilcox, William, 48
Wildt, Edward A., 238
Wildwood plantation, Camden Co., Ga., 408

Wilkinson, William L., 379
Wilkinson, Willis, 173, 179, 432
Willard's Hotel, Washington, D.C., 16, 28, 30, 110, 112, 115, 122, 124, 126–27, 130–33, 334
William C. Bee & Company, 160, 328, 338
*William Roe* filibuster sloop, 86
Williams, Harney, 379
Williams, Henry, 71, 99
Williams, John Stuart "Cerro Gordo," 26, 29, 393
Williams, Martha Custis, 311, 321–23
Williams, W., 374
Williamsburg, Va., 306
Williamson, Capt., 98
Williamson, Joel, 259
Williamson, Lt., 374
Willis, Col., 440
Willis & Chisolm, 272, 283
Willtown Bluff, S.C., 221
Wilmington, N.C., 12, 97, 154, 205, 222, 244
Wilmington Island, Ga., 87–89, 91, 97–98
Wilmot Proviso, 5, 17, 25, 406
Wilson, Chloe, 161, 252, 258, 301, 329
Wilson, A. H., 374
Wilson, Clytus, 329
Wilson, F. W., 432
Wilson, Fielding C., 61, 374
Wilson, George, 379
Wilson, James, 440
Wilson, James M., 379
Wilson, John, 374
Wilson, Phineas O., 84
Wilson, Samuel J., 197
Wilson, W., 379
Wilson, William, 379
Wilson, Woodrow, 367
Winburn, David, 379
Winchester, Va., 312–13
Winston, Thomas M., 374

Winthrop, Joseph, 432
Wise, Edward, 379
Wise, Henry Alexander, 215, 221
Wise, Stephen R., xv
Woer, Amand R., 379
Wolfe, Thomas R., 37
Wolff, Abraham, 38
Wood, L. D. C., 404
Woodbine Plantation, Camden County, Ga., 84, 86, 90–91, 407
Woodbury, Levi, 53
Woodbury, Wergener, 53
Woodhull, Victoria, 460
Woodlawn Cemetery, Bronx, N.Y., xvi, 363
Woodruff, William E., 374
Woodward, W. T. J. O., 138, 244, 255, 263
Woodworth, Steven E., 159
Woolfolk, Lt., 374
Work, Thomas, 374
World's Fair, 354
Worth, William Jenkins, xxiv, 7, 10–14, 20, 387, 439
Wragg, B. W., 379
Wragg, Thomas, 374
Wyman, Mary Armistead Byrd, 311
Wyman, Samuel Gerrish, 311

Ximeno, Blas, 9
Ximeno, Dolores "Lola" María, 288
Ximeno, José Manuel, 4, 288
Ximeno, María Dolores, 288–89

Yañez, Tomás, 401
Yates, Joseph A., 212, 437
Yazoo City, Miss., 44
Yemassee, S.C., xv, 331, 439
Yorkville, S.C., 330
Young America, 111, 114
Young, H. E., 278
Young, Robert, 89
Young, William, 379
Young's Agency, Miss, N.Y.C., 320
Yucatán, Mexico, 17, 21, 44, 80
Yue, A. H., 257
Yulee, David Levy, 5, 286, 338
Yumurí River, Cuba, 288
Yumurí Valley, Cuba, 129
Yznaga, José Andrés, 393
Yznaga Borrell, José Aniceto, 6, 35, 393
Yznaga del Valle, Antonio, 12, 29, 393
Yznaga Hernández, Alejo, 11
Yznaga Hernández, Pedro José, 11, 25, 29, 393

Zaragoza, Justo, 64
Zayas, José Antonio, 379
Zealy, C. J., 237–38
Zornes, A. J., 374

www.ingramcontent.com/pod-product-compliance
Lightning Source LLC
Chambersburg PA
CBHW030558230426
43661CB00053B/1762